Third Marine Division's

Two Score And Ten

History

U. S. MARINE CORPS

"Two Score And Ten" *is dedicated to* **Colonel Austin P. Gattis, USMC (Ret.).** *A Marine of indomitable spirit, innovative ideas, and unflagging determination, who (despite ongoing and debilitating medical problems) never lost sight of the ultimate goal, thereby inspiring all who have contributed to the fruition of this project. Col. Gattis developed the concept for this book of anecdotes and personal stories about Third Division marines... and led the task force who carried through to the book's completion.*

The materials were compiled and produced using available information; Turner Publishing Company and Mark A. Thompson regret they cannot assume liability for errors or omissions.

Co-produced by Mark A. Thompson,
Independent Publishing Consultant
for Turner Publishing Company

Book Author: Third Marine Division Association, Inc.
Book Design: J. Robert Cudworth
　　　　　　　Elizabeth Dennis
ISBN: 978-1-68162-185-2
Library of Congress Catalog
Card No. : 92-61187

Limited Edition

Contents of Two Score and Ten

The Drama Of
Two Score And Ten

This is the story of a division formed of necessity in the early days of World War II—triumphant and gallant at Bougainville, Guam and Iwo Jima.

... deactivated after World War II was over.

... reactivated during Korean hostilities,, demonstrating courage and camaraderie in a difficult Vietnam War, saddled with "slow downs" and a mostly non-supportive citizenry back home.

... showing continued readiness and being a formidable protective force in the Pacific during the 1970s and 1980s.

... providing several units to the Persian Gulf in 1990 and 1991, as well as being a ready back-up should more trained Marines be needed.

The year 1992 marks the 50th anniversary of the 3rd Marine Division. This book is highlighted with anecdotes from the people who made up the "Fighting Third" and who carried its colors so gallantly. There are stories of many, many brave men which we unfortunately do not have. We would tell them if we had them.

But enjoy the book. Laugh with us. Cry with us. Reminisce with us. Marines are great story-tellers ... and their memories help to record some of our country's great periods of history. The 3rd Marine Division Association has worked hard to keep alive the spirit and camaraderie of the "Fighting Third."

We Thank You

We appreciate all of the personal anecdotes, "sea stories," ideas, and photos that were sent to us for this book.

It was necessary to edit and rewrite some of these pieces to conform to our book style that was established. For example, some of the salty language was de-salted a bit—not because we didn't want realism, but in deference to our children, grandchildren, friends and associates who we hope will be reading this book ... and might be offended. After all, most of us don't talk now as maybe we did then anyway. Nor, could we use all of the photos that were sent, but we do appreciate the thoughtfulness in sending them.

Why A Two Score And Ten?

I have often said that a man who has not had some military service is missing an important element of life. The Marine Corps saw to it that I had and shared that experience way back in World War II and I have been grateful for it ever since.

Boot camp at Parris Island, SC provided our passage to manhood. We were all stripped of any vestige of baby fat, whether physical, mental or spiritual, sweated down to basic bone and sinew, and rebuilt to more nearly resemble the image of ourselves we had previously imagined.

The experience was mind-boggling for the youngster from the Bronx (who had never even been on a train until his 24 hour trip by rail to Yemasee), and to those of us who had been nurtured in environments best described as hot houses. The whole process began with a bang when a salty Marine sergeant, Sir, first greeted us as we lined up beside the train that had transported us to this other world.

Everyone who has been privileged to have worn the proud uniform of the Corps will remember his experiences with a relish. That is markedly true of the old-timers—serving under venerated Generals Hal Turnage and "Blood and Guts" Erskine—who survived the World War II campaigns of Bougainville, Guam, and Iwo. Nothing could be more heartwarming to us than to get together with the fellows who shared our experiences ... and retell them—sometimes with beer-injected embellishment. It was equally stimulating to listen to younger generations of Marines ... of their lives on Okinawa, their struggles and close shaves in Vietnam, their lost buddies, NCO's and officers, and often their almond-eyed companions—whose recalled beauty grew in intensity each time in the telling.

Four Years Ago ...

We had just finished a fine annual reunion of the 3rd Marine Division Association in Tampa, FL in 1988 when Colonel Jerry Brown (then our association president) and I were enjoying a beer together. We were lamenting the increasing number of deaths in our ranks and the fact that way too many fine sea stories were being irretrievably lost to eternity, when Jerry said, "Right! You select a committee and figure out how we can best preserve our past."

For a beginner, our efforts could contribute to the recognition of the 50th anniversary of the division which would be coming in 1992. We had a pool of some 3,000 members of the association from whom anecdotal contributions could be garnered.

It was an easy job to recruit my task force, but they were far from lolling around with nothing to do. Vince Robinson, a retired colonel, administrative assistant to the association president, was elected to be co-chairman. Cy O'Brien, to be editor, had been a longtime professional writer, a war correspondent from World War II, and an editor with a fertile imagination. Ben Byrer, a "grunt" from the original 9th Marines on Bougainville, one of the most respected fine artists in the States, assumed the graphics responsibilities. Bob Cudworth, busy with his writing, editing, and publishing commitments, would contribute his expertise. And, Jim Kyser, a retired Master Gunnery Sergeant from the new breed, who could furnish the needed public relations and administration, even though he was "up to here" in his job as Manager of the Association, agreed to join in.

Our task force agreed we would publish an anecdotal history of the 3rd Marine Division covering the full span of 50 years since being commissioned in 1942 and accept a target date of 1992 for publication. We would solicit contributions from membership of the division and any Marine or corpsman, doctor, chaplain, dog handler, UDT member, Seabee, shore party, boat coxswain, flyboy, or any other who had served in support of the division.

In the meantime our task force has been greatly enhanced by the addition of contributing editor/writers Bob Van Atta, Carroll Garnett, and Dr. Harry A. Gailey, Professor of History, San Jose University, CA; Tom Bartlett, Associate Editor, *Leatherneck* magazine; Nicholas Kominus, 3rd Division Information Officer in Japan; and writer and psychologist, Paul F. Colaizzi, Pennsylvania state licensed psychologist.

The task force has donated thousands of hours to this project. No chairman has ever formed a more capable or dedicated group. Also of great importance, is our fine publisher, Dave Turner, president of the Turner Publishing Company, who has published some of the finest military books of our times.

The format of our history is unique. Bare bones of history appear on some pages and that "skeleton" is fleshed out by vignettes, the experiences of our veterans. The anecdotes cover a gauntlet from humor through tragedy, pathos, human relations, bravery, and just plain nostalgia.

It is with great satisfaction and pride that we place in your hands *Two Score and Ten*, an anecdotal history of the 3rd Marine Division in celebration of the division's 50th anniversary.

We hope you will share our enthusiasm.

Austin P. Gattis, chairman
Colonel, USMCR (Retired)

The Task Force For Two Score And Ten

Austin P. Gattis

Austin went through Parris Island in 1942 and training at New River and quartermaster training at Quantico, was selected for OCS and was graduated as a second lieutenant.

He was assigned to the 12th Marines on Guadalcanal, and participated in the Bougainville and Guam campaigns—acting as S-4 on Guam. He was released from active duty in early 1946.

Gattis was a member of several VTU's and was executive officer and commanding officer of a VTU that published *Toward The Marine Corps University, 1975*—a study performed at the request of Marine Corps Commandant General Wallace Green.

He chaired the By-laws Rewriting Committee of the 3rd Marine Division Association and then served two consecutive terms as president of the Association 1969-1970 and 1970-1971.

After 32 years of active and reserve service, he was retired in 1976 as full colonel.

He has been the leading force in initiating *Two Score and Ten,* the anecdotal 50 year history of the 3rd Marine Division, and has chaired the writing, art, and production group.

(Photo of Austin P. Gattis was taken of a painting produced by artist Ben Byer at the completion of Colonel Gattis' second term as president of the 3rd Marine Division Association.)

Vincent J. Robinson

Vince Robinson, co-chairman of the 50th Anniversary Book Committee, has had the job of making initial agreements with the publishing company and providing liaison to the board of directors of the association.

A native of New England, he joined the Marine Corps in mid-1942. After Parris Island and Quantico he served with the 14th Defense Battalion in New Caledonia, then joined the 3rd Marine Division on Guadalcanal immediately following the Bougainville Campaign. He served with the 12th Marines throughout the Guam and Iwo Jima campaigns.

He returned to the States in December 1945, and subsequently served with the 10th Marines at Camp Lejeune ... with a mobile training team in Tokyo in 1950 ... and later in South Korea on a reconnaissance mission. Robinson again became a member of the 12th Marines when the entire 3rd Marine Division was re-established at Camp Pendleton in 1952.

In a cross-service tour he was a supporting arms instructor at Fort Sill, OK and was later artillery advisor to the Chinese Nationalist Marine Corps, headquartered on Taiwan.

His final active duty assignment, prior to retirement, was commanding the 14th Marines, a unit that was expected to be activated as part of the 4th Marine Division during Vietnam hostilities. Civilian opportunities have kept Colonel Robinson involved in various DOD contracts and activities.

Cyril J. O'Brien

Cy O'Brien has served as editor for *Two Score and Ten*, writing and rewriting a multitude of anecdotes and "sea stories" that have come in from veterans of the 3rd Marine Division.

An early veteran of the 3rd Marine Division, Cy served in a line company of the 3rd Marines on Bougainville, and as a Marine Corps Combat Correspondent on Guam and Iwo Jima.

After discharge he was a Washington correspondent for newspapers in New Jersey, and was on the editorial staff of newspapers in New York and Pennsylvania. He recently retired as media representative of the Johns Hopkins University Applied Physics Laboratory in Laurel, MD.

A frequent traveler overseas, he has made jaunts to Bougainville and Guam. Cy has written and edited several of the reunion journals of the 3rd Marine Division Association in past years.

He lives in Silver Spring, MD.

Benjamin J. Byrer

Ben Byrer, a native of Canton, OH, served with "A" Company, 9th Marines as a BAR man during the Bougainville and Guam campaigns. After discharge he quit the steel mill where he was working and entered Art School in Pittsburgh because art had always been a paramount interest.

Ben later left Art School to work for a commercial art agency and has been completely involved with art ever since. He has been Buhl Science Center's staff artist for more than 30 years, creating moons, planets, and other worlds in panoramas for the science center's sky shows.

A versatile artist and careful researcher, Ben's paintings and art work of many thousand pieces have ranged widely: from railroading and industrial to shore scenes, horses, colonists and Indians, food and fashion. His realistic drawing of combat scenes and other aspects of military life in *Two Score and Ten* will be nostalgic for all Marines.

His artwork is on display at the Island Art Gallery in Manteo, NC and in many private collections in the United States, Europe, Canada, and Australia. Those people who have attended reunions of the 3rd Marine Division Association will have seen his gigantic mural about the 3rd Division on display at the annual banquets.

Ben and wife Vi live in Valencia, PA near Pittsburgh. Their daughter Lili operates Lili's Artisan's Showplace in Pittsburgh.

Gattis

Robinson

O'Brien

Byrer

| Cudworth | Kyser | Van Atta | Garnett |

J. Robert Cudworth

A native Central New Yorker, Bob Cudworth handled page layouts for *Two Score and Ten.*

After graduation from Syracuse University, he went to "boot camp" at Parris Island, then OCS at Quantico. He joined A Company, 9th Marine Regiment at Guadalcanal in January 1944, and served with A-1-9 in the Guam and Iwo Jima campaigns as a rifle platoon leader, company exec, and company commander.

Following active duty he worked with newspapers in Upstate New York as a writer and editor, and a magazine editor and writer for companies and advertising agencies. He has conducted a writing and editing service in Camillus, NY for about 20 years.

James G. Kyser III

Retired Master Gunnery Sergeant James G. Kyser III is 3rd Marine Division Association Manager and has been editor of the *CALTRAP* and *Reunion Journal* since January 1988. Members who have attended reunions of the association remember him for the multitude of photos that he takes for *CALTRAP.*

He enlisted in the Marine Corps in 1952 and completed recruit training at MCRD San Diego. During his 21 year career, he served with the 1st Marine Brigade, 1st Marine Division, and 3rd Marine Division. His duty has also taken him to MCAS, El Toro; Marine Barracks, Pearl Harbor; Marine Barracks, 8th and I, Washington, DC; Marine Corps Schools, Quantico; two tours at Headquarters Marine Corps, and with the Military Assistance Command, Vietnam (MCAV).

While in Vietnam he was awarded the Campaign Medal with three stars. As a Rifle Expert he has received 11 awards and seven awards as Pistol Expert.

Following his retirement in December 1973, he spent four years as public relations staff advisor with the American Petroleum Institute and was director of advertising for the Association of American Railroads in Washington for ten years.

A native of Napa, CA, he is married to the former Virginia Detweiler of Easton, PA. They have two children and reside in Dumfries, VA.

Robert B. Van Atta

Robert B. Van Atta served in the Marine Corps during both World War II (two years in Pacific) and on Reserve recall in the Korean War. He was a founder of the Parris Island newspaper, *Boot,* in 1943. For several months in 1945, he was acting regimental sergeant major of the 12th Marines.

A native of Pittsburgh, and graduate of the University of Pittsburgh, he is currently history editor of the Greensburg (PA) *Tribune-Review* after retiring in 1986 from a public relations management position with an electric utility.

He is an honorary life member of the Marine Corps Combat Correspondents Association, a charter and sustaining member of the Marine Corps Historical Foundations, and has served as a trustee of the 3rd Marine Division Association Memorial Scholarship Fund since its founding in 1968.

Active in many aspects of civic, community, and educational life, he was

recently awarded the Silver Good Citizenship Medal by the Sons of the American Revolution for outstanding military and community service, and inducted into the Westmoreland County (PA) Sports Hall of Fame. He is also a retired scholastic and collegiate sports official.

He has authored a number of books on regional and sports history. He and his wife, Beatrice, reside at Greensburg, PA.

Carroll M. Garnett

A Virginia native, Carroll Garnett served as a medical corpsman (PhM 1/c) for about three years with the 1st Battalion, 21st Regiment of 3rd Marine Division during the Bougainville and Guam campaigns. He was awarded the Bronze Star Medal with combat distinguished device, a permanent citation from the Secretary of the Navy, and the Purple Heart Medal.

In civilian life he retired as a special agent of the FBI after 26 years, having served in bureau offices in Buffalo, Boston, Washington, Omaha, and Richmond, VA.

Garnett is a long-time special feature correspondent for the *Rappahannock Times*, Tappahannock, VA and the current chairman of the Essex County Tricentennial Commission.

He is married to the former Cornelia Morris. They have four children, eight grandchildren and reside in Chester, VA.

Carroll has provided anecdotal stories for the World War II period as well as researching and compiling story data for the Okinawa Period in *Two Score and Ten.*

Origin Of The Caltrap
... Designed By A Pharmacist's Mate

The 3rd Marine Division Caltrap insignia—tri-shaped and three colors—has been in existence since 1943.

When the 3rd Division was stationed in New Zealand a contest was held to select a suitable shoulder insignia (shoulder patch) for the "hell for leather" division to carry into combat.

The late Major General C.D. Barrett, division commander, then Brigadier General Hal Turnage, and chief of staff, Colonel A.H. Noble selected the tri-shaped design from among more than 50 submitted by members of the division and affiliated units.

It had been designed by Pharmacist Mate Robert Harding Miller Jr. U.S. Navy "A" Company, 3rd Medical Battalion.

Even though thousands of Marines wore Caltrap shoulder patches and the scarlet, gold, and black insignia was seen in numerous locations throughout the world, a few years later no one knew who had submitted the original design.

T.O. Kelly, longtime secretary and former president of the 3rd Marine Division Association, set out to learn who had designed the insignia. After seven years of persistent searching and checking, he found in 1967 that it was former Pharmacist Mate Robert Harding Miller Jr. who had submitted the Caltrap design.

Miller, from Haleyville, AL had joined the Navy in February 1942. He trained at Camp Elliott, CA and served with the 3rd Marine Division until he returned to the United States later in 1944. He died in 1960.

The words, "Fidelity, Honor, and Valor" were added to the insignia later; and the association banner (standard), complete with the design and wording was presented by the Pittsburgh Chapter of the Association.

Probably few wearers of the Caltrap insignia were aware that the caltrap (or caltrop) was an ancient military weapon designed to slow down foot soldiers or horses. Made of two metal spikes, twisted to form four spiked points, it was designed so that when three spikes were touching the ground another would be pointing straight up.

Now, nearly 50 years later, the Caltrap (caltrop) insignia is one of the best known escutcheons in American military history.

Proud Trail Of The Caltrap

(Carroll Garnett summarizes this article from the "Okinawa Marine" September 16, 1983 issue which tells about the history of the Caltrap insignia).

The logo of the 3rd Marine Division is easily recognized by nearly every American resident of Okinawa. Few, however, understand the symbolism of the division insignia.

The three-pronged figure featured in the insignia is a "Caltrop" ... now known as Caltrap. The device was an ancient military weapon consisting of an iron ball and four metal spikes. It was designed so that when three spikes were touching the ground another would be pointing straight up.

The Caltrop was used primarily to impede enemy cavalry and easily adapts to the 3rd Marine Division motto, "Don't Step on Me." On the division logo, the caltrop is viewed from ground level. Its three points and the triangular shape of the logo allude to the number three.

The scarlet and gold colors of the logo are the official colors of the United States Marine Corps. The black is for contrast.

The division logo became official on August 25, 1943. Less than three months later, Marines wearing that insignia received their baptism of fire on the Northern Solomons Island of Bougainville. Following Bougainville, the 3rd Marine Division logo was branded into the history books at the battles for Guam and Iwo Jima.

World War II ended while the division was preparing to invade mainland Japan. In December 1945, the 3rd Marine Division was deactivated on Guam. Since first leaving the country in 1942 to this date, the 3rd Marine Division has not returned to the United States.

Due to the escalating events in Korea seven years later, the 3rd Marine Division was reactivated and sent, with its logo, to Japan to support the 1st Marine Division in the defense of the Far East.

Okinawa has been "home" for the 3rd Marine Division since the initial division units left the Korean area of operations and began arriving here in mid-1955.

The division logo was bloodied again as members of the 3rd Marine Division became the first American combat troops committed to the Republic of Vietnam in 1965. Forward elements of the division made the first amphibious landing at Da Nang, Vietnam on March 8, of that year.

During the conflict in Southeast Asia, Marines bearing the 3rd Marine Division insignia conducted more than 120 combat operations and the division earned the distinction of having the longest tour of continuous combat service of any similar unit in the history of the U.S. Armed Forces.

Today the 3rd Marine Division logo is known throughout the Far East as the division undergoes continuous training in the Western Pacific as part of America's "Force in Readiness."

Wherever members of the 3rd Marine Division serve, their logo becomes easily recognizable. With a knowledge of what the insignia represents, the presence of 3rd Marine Division Marines also becomes more understandable. **Okinawa Marine, September 16, 1983**

"Don't Step On Me"

A Message From Commandant Mundy

As we begin commemorating the 50th anniversary of the battles and campaigns of World War II, it is only fitting that the illustrious history of the 3d Marine Division be re-told. The story of the Division's service at Bougainville... Guam...Iwo Jima... and in Vietnam... and of its unit's service in Beirut and Kuwait is a distinguished one.

The 3d Marine Division holds a special place in the hearts of all Marines. It represents a crossroads of sorts...where through the years, as so often happens, the "old Corps" and the "new Corps" come together to write new chapters in our proud history. It began when the Division was first activated in September of 1942 and its infantry regiments were built around cadres of officers and NCOs from the 1st and 2d Marine Divisions. That special and unique relationship continues today as units rotate from the United States to Okinawa for service with the Division.

This monograph adds to the Corps rich and treasured heritage of service to our Nation. There is no finer tribute than to be remembered by one's comrades. Marines of the past, present, and future are well-served by this history.

C.E. Mundy Jr.

C.E. Mundy, Jr.
General, U.S. Marine Corps
30th Commandant

A Message From Commandant Gray

Marines everywhere join me in extending warm greetings to all those who have had the privilege of serving our Corps and Country as a member of the 3rd Marine Division. From its activation on September 16, 1942, the 3rd Marine Division's commitment to excellence has ensured that our Corps has remained America's premier force-in-readiness.

The Marine Corps has maintained its reputation as the world's finest for more than 215 years by meeting adversity on familiar and foreign ground with relentless mettle; perpetuating discipline, loyalty and professionalism.

Today the 3rd Marine Division can proudly add Operation Desert Storm to a legacy of valor that has included such historic chapters in Marine Corps history as Bougainville, Guam, Iwo Jima, the Vietnam War, and the rescue of the crew of the SS *Mayaguez*.

Along with coalition forces, more than 90,000 Marines serving with I Marine Expeditionary Force, including units of the 3rd Marine Division, pounded and defeated the Iraqi army of Saddam Hussein and liberated the country of Kuwait. As always, Marines fought on the land, at sea and in the air.

Where future hostilities will occur is never certain. What is certain is that the 3rd Marine Division remains ever ready to carry out its mission upon which the very security of the American people depends. We can rest easy knowing they continue to maintain the spirit of the triad-valor, honor and fidelity.

To each of you I extend my sincere appreciation for your continued loyal support. Many things change in this world, but the affection and respect of Marines for their Corps is constant. May it always be so.

A.M. Gray
General, U.S. Marine Corps
29th Commandant
During whose watch this history was begun.

"Heroes Of Our Times"
... says Bob Hope

A group could get no finer tribute from a public figure than one from Bob Hope, because he was right where the action was, time and time again. This letter, which he sent to Austin Gattis, tells how he loved and respected the Marines.

Following his letter are some of his appearances where he and his USO troupe entertained troops, with 3rd Marine Division personnel likely to have seen and enjoyed, his shows.

Dear Colonel Gattis,

It's nice to hear from you and also to read about the 3rd Marine Division. They've been in the Far East, World War II, Korea and Vietnam, also in the Gulf War.

Well, there's no way you could have missed me because I did shows in all those wars. I have stories that are something else.

For instance, in 1944 I was on a little island called Benetka in the South Pacific. A Marine came up to me and said, "I'm from the First Marine Division and we're over here on an island called Pavuvu, 15,000 of us. We've been training for six months and never had a show. If you would come over and visit us, it would be a great asset."

So I said, "What are you training for" and he said, "We're going to invade Peleliu."

I said, "Well, how do we get there?"

He said, "We have no airport, so you have to land on the road in a Piper Cub." So, there were eight of us in the unit and we all got in a Piper Cub and flew over to Pavuvu. We looked down and there were 15,000 Marines and everytime we flew over the road they would cheer. We landed, and did a show for them just before they invaded Peleliu.

Not less than eight months later, I was asked to go up and dedicate an amphitheater at the Oak Knoll Hospital in Oakland, CA, which I did.

After the dedication, the doctor asked me if I could go in and visit in some of the wards, and I said, "Absolutely." The first ward I walked into a man stuck his hand out and said, "Pavuvu."

I said, "You're kidding. This whole ward?" And he said, "All of us." I shook hands with everybody, and I got to the back where there was a single room and they were fanning a Marine who had just come out of an operation, to get some air to him. He opened his eyes and said, "Bob, when did you get here?" It was all I could stand. I had to walk out of that room.

About 1956, I was in Pyongyang, the North Korean Capitol. Les Brown and the band, Carolyn Maxwell and I found out that the Marines were going to be in Wonsan, North Korea. So I called headquarters and asked if I could go over there. They said, yes, you can go over there at 1100 hours on a certain date. So we got on our plane, landed there at 1100, and there was no one around, except this one guy in a jeep who took us over to the hangar.

We stood there about 15 minutes, when General Walt, Admiral Strugel, and about 10 of these VIPs walked up and shook my hand, and asked when had we got there? I said, we'd been there about 20 minutes.

He said, "You're kidding, we just made the landing." As we were flying in, I saw all these boats and I thought, "Boy, they're coming in to see the show." We didn't know they were making a landing, or I never would have landed that plane ahead of them.

I was overseas since 1941, so you know I've played for a lot of Marines, and they're the greatest audience yet.

Anyway, I've received the Marines' highest honor, and it was one of the great moments of my life to be decorated by these marvelous, valiant ... I call them, heroes of our time.

Regards,

Bob Hope

Postscript From Gattis:

The "Bob Hope USO Tour Highlights" is a tremendously impressive list. From it I have excerpted those occasions where troops from the 3rd Marine Division were most probably present. This may help you in your recollecting.

Late 60s-Hope, along with Raquel Welch and Ann Margret, toured battle-scarred Vietnam. His USO stops included the USS *Ranger*, the USS *New Jersey*, and the USS *John F. Kennedy*.

1972-Hope's last overseas Christmas Tour (until 1983) included a side-splitting lampoon of the Navy. Hope and Redd Foxx played two seamen. The dialogue went like this: REDD: I got it all planned. I'm gonna cash in on everything they taught me in the Navy. HOPE: How? REDD: I'm gonna open a chain of washrooms.

December 1957-Honolulu, Okinawa, Korea, Tokyo, Kwajalein, Wake Island, Guam. His troupe included: Jayne Mansfield, Erin O'Brien, Carol Jarvis, Jerry Colonna, Hedda Hopper, Mickey Hargitay, Arthur Duncan, Les Brown.

1962-Japan, Korea, Okinawa, Taiwan, Philippines, Guam. Troupe included: Lana Turner, Janis Paige, Anita Bryant, Jerry Colonna, Amedee Chabot (Miss USA), Peter Leeds, Les Brown and band.

1964-Vietnam, Thailand, Philippines, Guam, Korea. Troupe included: Jill St. John, Janis Paige, Anita Bryant, Anna Marie Alberghetti, Jerry Colonna, John Bubbles, Anne Sidney (Miss World), Peter Leeds, Les Brown and band.

1965-Vietnam, Thailand, Guam. Troupe included: Les Brown, Diana Lynn Batts (Miss Virginia), Joey Heatherton, the Nicholls Bros., Peter Leeds, Anita Bryant, Jerry Colonna, Carroll Baker, Kaye Stevens.

1966-South Vietnam, Thailand, Guam. Troupe included: Phyllis Diller, Joey Heatherton, Vic Damone, Anita Bryant, Diana Shelton, The Korean Kittens, Les Brown, Reita Faria (Miss World).

1967-Vietnam, Thailand, USS Ranger and Coral Sea, Philippine Islands, Guam, Midway. Troupe included: Raquel Welch, Barbara McNair, Elaine Dunn, Madeleine Hartog-Bel, Phil Crosby, Earl Wilson, Les Brown.

1969-Berlin, Italy, Turkey, Vietnam, Thailand, Taiwan, Guam. Troupe included: Neil Armstrong, Connie Stevens, Romy Schneider, The Golddiggers, Teresa Graves, Suzanne Charny, Eve Reuber-Staier, Les Brown, Hector and Ted Pierro.

1970-England, Germany, Crete, Thailand, Vietnam, Korea, Alaska. Troupe included: Ursula Andress, Johnny Bench, Lola Falana, Gloria Loring, Jennifer Hosten, Bobbi Martin, The Golddiggers, The Dingalings.

1971-Hawaii, Wake, Okinawa, Thailand, Vietnam, Spain, Citmo. Troupe included: Jim Nabors, Sunday's Child, Vida Blue, Charley Pride, Jill St. John, Jan Daly, Suzanne Charny, Brucene Smith, The Hollywood Deb Stars, The Blue Streaks, Rear Admiral Alan B. Shepard, Don Ho, and Les Brown.

1972-The Aleutians, Japan, Thailand, The Island of Diego Garcia, South Vietnam, Guam. Troupe included: Dolores Hope, Redd Foxx, Lola Falana, Redy Carcenas, Belinda Green (Miss World), Fran Jeffries, 12 American Beauties, and Ingeborg Sorensen (Miss Norway).

"I Admire Them Very Much"

When the USO asked us to visit the troops in Saudi Arabia we immediately said yes. We had no idea where we would be going, and they wouldn't tell us, for security reasons, so we signed on for a mystery week in the desert. First stop on the mystery tour was a 15 minute meeting with General Schwarzkopf. He said the only reason he joined the Army was because of the soldiers, and their welfare was his primary concern, and that was why we were there. Some of them had been in the desert for 60 days and could do with a visit from someone other than a general. He was a man you had to believe.

We visited the Marines on the front lines. It was like Mad Max. The engineers had dug compartments for each truck out of the sand. The mess area was under netting, the cots were under netting, the headquarters tent with the field telephone was under netting, and there was a map of the desert squared off with string in the sand. We were so close to the Iraqi army we could have dropped in for tea. Apart from the flies, the scorpions, and the camel hours between eleven and three when even the camels lie down, what the Marines liked least were the sand vipers. They called them Mr. No-Shoulders. Going back to the helicopter I had a lump in my throat all the way. It was their faces. As the troops gathered around Steve, I talked to the men near me. One of them had left his bride in Hawaii on three hours notice. He asked me to call Ronda and tell her he loved her. Davis was a single parent and asked me to call his son Mike and tell him his dad was O.K. It always ended with each man stepping up to get his picture taken with his arm around my shoulders. And each one said, without fail, "My wife will kill me!"

When the troops thanked us for coming, we felt that we didn't deserve it. We were there to thank them. I abhor violence. I'm not happy they were there. But as men and women I admire them very much.

Red, White And Blue Achievement

Thanksgiving 1990 will always be special to me. Thanks to an invitation from President Bush, I was able to spend Thanksgiving Day with our Desert Storm troops in the Persian Gulf. No doubt about it, sharing turkey dinner on the sands of Saudi Arabia with the finest fighting men and women on earth is something I will never forget.

We spent time with every branch of the service and it was easy to see what they all had in common: professionalism, patriotism and pride. Despite the tough conditions and the prospects of combat, morale was sky high a tribute to their training and leadership.

Of course, the Marines were there—standing tall on the frontlines, as usual. Once the ground war started, Saddam Hussein and his over-matched army got a real taste of Marine fury. It was something the Iraqis will never forget.

And America will never forget the sacrifice of each and every one of its Desert Storm heroes and their courageous families back home.

I am proud to help salute the legendary 3rd Marine Division on its 50th anniversary. From the beaches of Iwo Jima, to the airfields of Da Nang, the 3rd Division has carried on the proud tradition of the United States Marines. Although not fully deployed to the Persian Gulf, many of the 3rd's units were augmenting to the 1st and 2nd Divisions.

No doubt about it, for the 3rd Marine Division, it is a 50 year record of red, white and blue achievement. Congratulations.

Senator Bob Dole

With Best Wishes on the 50th Anniversary of the Third Marine Division Association,

The 3rd Marine Division

by Edwin P. Hoyt

In all its exploits of which I am aware the 3rd Marine Division has conducted itself in the finest tradition of the U.S. Marine Corps, always under the most trying of circumstances.

I have written at some length about the 3rd Marine Division in books that deal with Bougainville and Guam. In the former, the fighting around Empress Augusta Bay was formidable, for they faced some of Japan's crack troops who had the advantage of knowing the terrain, and a leader, Colonel Hamanoue, who was determined to win a victory.

He did not, because of the bravery and fortitude of the men of the 3rd Marine Division.

In the slippery volcanic sands of Iwo Jima the men of the 3rd Marine Division again covered themselves and the Marine Corps with distinction.

They did not get into the Korean War as a division, but they were in Japan and ready.

And then came the Vietnam War and the 3rd Marine Division was the first American fighting unit to be employed.

Da Nang Air Base, Quang Nam province, and then Quang Tri, and Thua Tien. Even Americans who know virtually nothing about the Vietnam War know the name of Khe Sanh and some have heard of the Rockpile.

The 3rd Marines, the 4th Marines, the 9th Marines, the 12th Marines, and the 26th Marines all left their mark on the division's grand record. (The 26th, of course, was attached from the 5th Marine Division.)

Vietnam — 29 Medals of Honor for one division (and the 26th Marines). Need any more be said?

But more was said, the 3rd Marine Division, evacuated when the American administration decided to wind down the U.S. involvement in the Vietnam War, was brought back again to manage the evacuation of Pnom Penh and Da Nang and Saigon in those strange days of 1975, and later to rescue the stranded Mayaguez.

As Marine divisions go, the 3rd Marines is a baby outfit, but a look at the record shows them standing right up three with the finest of the corps.

Edwin P. Hoyt, among the most popular writers on the Pacific War and Japan, is the author of numerous books on the European war including the dramatic portrayal of the legendary battles between the HMS *Hood* and the *Bismarck*. This article was written especially for the history of the 3rd Marine Division.

Third Marine Division

Two Score and Ten

*This drawing by Ben Byrer, representing action on Iwo Jima, is one of several he has produced for **Two Score And Ten** to graphically portray the action and life during 50 years of 3rd Division Marines.*

Training at Camp Pendleton for Action in the Pacific

A Story Told To Me Some 50 Years Ago

A United States Naval Surface Air exercise is said to have taken place in 1930 or 1931. At that time there was considerable discussion as to the efficacy of air as a military component. Old line admirals thought that air could never effectively challenge sea supremacy.

Younger "air admirals" were almost willing to bet that surface fleets could be obliterated by weapons delivered by planes, and that a far greater part of the naval budget should be directed toward development of that great potential. The war games exercise was designed to determine the most plausible winner of the argument.

Naval Air was based in southern California. COMSOPAC was at Pearl Harbor and the United States Pacific fleet was based there too.

It was a given fact that any military exercise in the Pacific area would be closely monitored by Japanese agents, and it was assumed that these agents abounded in Hawaii, if not along the west coast of the States at the same time. Additionally, representatives from the other branches of United States military were "invited" to observe the exercise, i.e. the Army and the Marines.

It turns out that the NAVLEX was short-lived. Right after the whistle blew for the games to begin, Naval Air made a surprise attack on Pearl Harbor on a Sunday morning, catching the bulk of the Pacific fleet moored in the harbor and, on paper, rendered the fleet destroyed.

The critique made as much noise as did the "fighting." The more senior admirals cried "foul" and said, "It could never happen!" The air admirals, naturally, remained unconvinced but had to downplay their rubuttals to their seniors.

The Japanese agents reported the entire exercise to the headquarters of the Imperial Japanese Navy in Tokyo, and due notice was taken as to the outcome of the opposing American forces.

The United States Marine observers, lieutenant colonels, reported to their assigned billets in Research and Development Center, Marine Corps Base, Quantico, Virginia. To them, the paper defeat of the Navy surface fleet by air was plausible, and they took it on themselves to rationalize the possibility of a World War II starting in just that manner in the Pacific ... and what courses of action would be available to United States forces to overcome the catastrophe. The study scenario, centered around the most probable national response and the overall strategy, with particular emphasis on the role of the Marine Corps.

There emerged the concept of amphibious operations, from island to island, gradually closer to the Japanese homeland, with, finally, an assault on Japan itself to bring the conflict to a successful conclusion.

Almost no one outside of the Corps recognized the visibility of the study. The Corps numbered a scant 17,000 officers and men, and their budget could allow little beyond "paper studies." But, the Corps and the country could be thankful for that groundwork when fact overcame fiction on December 7, 1941. **Austin Gattis, Washington DC**

Who Remembers ...
The Mail Guards

Major Wilbert F. Morris brings to memory the long-ago Old Corps duty of mail guards which harkens back to President Warren G. Harding and the "protect the mail" orders of none other than Major General John A. Lejeune, then (1921) Commandant. Then Secretary of Postmaster General, he had made it clear: "When our men go in as guards over the mail, that mail must be delivered or there must be a Marine dead at the post of duty." Among those who served in mail guards, Major Alexander Vandegrift, Captain Clifton Cates and Lieutenant Evans Carlson. Major Morris, United States Marine Corps (Retired) of Watsonville, California, a member of the mail guards in 1926, served with the 12th Marines in World War II.

Nippon On The Move And The Need For A 3rd Marine Division

Japan had expanded in lightning thrusts across the Pacific and even threatened Australia. Her armies—experienced, able and superbly equipped—easily disposed of most opposition, even to undoing defenders sometimes thrice the number of Japan's assault forces. At the time, her pilots, and the Mitsubishi Zero made the deadliest team in the world, and amateurs of Western airmen.

The Rising Sun was over resource-rich Java and Sumatra and Japan's Hideki Tojo had only to turn the spigot to tap the bubbling reserves of oil and rubber up and down the fat Dutch East Indies. In Tokyo, there was euphoria. The Europeans were out of the Orient and their centuries-long colonialism would now give way to the Greater East Asia Co-Prosperity Sphere.

Worse, the intentions, resources and energies of the West were being husbanded for the eventual onslaught on Nazi fortress Europe. President Franklin Delano Roosevelt, Army Chief of Staff George C. Marshall, and the Chiefs of Staff viewed the defeat of Adolph Hitler the principal goal with rescue of the Pacific decidedly in abeyance .

Into this world of woe was born the 3rd Marine Division on September 16, 1942 at Camp Elliott, California. It was almost to the day when Field Marshal Eric Rommel (though virtually out of fuel) was holding back all drives of the British Eighth Army, on the same day of a big new German thrust on Stalingrad, and very near the first discussion of President Franklin D. Roosevelt and British War Minister Winston Churchill of the terrible new weapon that could be made by splitting atoms.

It was also a month when men who would otherwise return to school were enlisting. Enlistments in the United States Marine Corps doubled in the eight months from January through October, 1942. In Pittsburgh at West View Danceland, zoot suiters were stomping out the lively bars of Harry James, Glenn Miller, *One-*

(Continued on page 21)

Liberty Ship

Don't Always Listen To Old-Timers

I enlisted in the Marines before WWII, and after six months was stationed at the Naval prison in Philadelphia. I had enlisted from the Pittsburgh area, and there was another fellow from Pittsburgh, on base. He owned a car; can you imagine having a car on $21 a month? Of course, he was still a private after being in the Marines for quite a while. I should have known we'd get in trouble.

One bad winter night he asked me if I'd go into town with him. I didn't know what he was going for, but I went along on the trip. He finally said, "I want to get a pint of whiskey."

I said, "You can't drink on duty." But him being in for a couple of years, I thought he knew what he was doing. We stopped at the liquor store, just around the corner from the place he parked the car. A lady in the store said, "Are you fellows going back to the Navy Yard?"

When we said yes, she said "Would you take me back to the yard because the busses aren't running any more?"

We said, "Yes, we would." It turned out, she was the wife of a major on the base. When we reached the gate, the guard saluted us through without question. Evidently, he recognized the major's wife.

When we got back to our post, the private said, "Have a drink." Well, I don't drink much, and there was another fellow at our post ... we all had a little taste ... and as you know, that's a no-no on duty.

An Optimist ... And Generous

Captain Philip C. Ferguson, recalls Conrad Fowler, was one of those unusual patriots in WWII who resigned his seat in Congress from Oklahoma to take a commission as captain in the Marines.

A rancher and banker from Woodward, OK, he eschewed a desk job and went to Scout and Snipers School, joined the 9th Marines, and once borrowed several thousand dollars to provide emergency loans so unpaid Marines could have money for liberty.

Fergie was always an optimist and took everything in stride. Word came that he had been shot up on New Georgia while with a Raider battalion. He was up front telling a corporal in a big booming voice where to put his machine guns. The Japs opened up on him and he was seriously wounded.

He sought duty and obligation from the establishment, and active duty for himself on the front lines. We cannot but respect him for that ... **Conrad M. Fowler, Lannett, AL**

Somehow the major found out about the drinking on duty. Next thing we knew, the sergeant called us in. Someone told about our unauthorized trip to town, and we were in real trouble.

When you're on duty at the brig, you already have one foot in the brig and the other on a banana peel. The result was, we got three days with bread and water, with a little salt on the third day. This was all before WWII, and the private got a summary court martial.

In those days, anyone doing much time in the brig ended up in Philadelphia. Every day we marched 100 prisoners out into the yard. We even had men from two German U-boats ... they had come right up to the main gate and turned themselves in. We put them in rooms with high windows so they couldn't see out, then in a week or so they were transferred out.

When we marched the prisoners to the mess hall, it was three floors down. One day a sailor broke away and jumped over the rail, and fell to his death with a broken neck. Couldn't take prison, I suppose. We also had a prison break that year of 1940-41. We found the escapees quickly. They were hiding in lumber piles, which were stacked at the yard because of all the building that was going on in the area. Guard duty was easy duty, compared to the war.

When we were stationed in New Caledonia during WWII, we had a lot of lamb to eat, three times a day. I'd have the duty of taking the platoon to chow. On passing the other platoon returning from chow, we'd say "How was chow, boys?" They would answer, "It wasn't baaaaaaaaaad." **John Habay, Boca Raton, FL**

Train From New River

We boarded a long train, the front occupied by black troops, perhaps slated for stevedore duty somewhere. The back half, all white, was designated a "replacement battalion."

In the middle of the train there were two dining cars and on the tail a bedroom/roomette car for the officers. Other than that one car for us, all the cars were "day coaches." This was in 1943. The Marine Corps, except for the stevedore units and maybe a few other special duty assignments, was lily white. There was simply no thought of integration.

... Our company had only lieutenant officers and I know you must have heard the old adage that "seniority among lieutenants is like virginity among whores." Well, it wasn't that bad for us as one was a first lieutenant. He was named company commander and the rest of us bowed to his superior rank. Must admit that what decisions we faced were decided most democratically. We'd all find out shortly that that's not exactly the way the corps was run, though.

... Sometimes it seemed like we were traveling north, sometimes south, but generally it must have been west because eight days later we pulled into San Diego station, disembarked, formed up and marched through a tile-lined passenger tunnel singing the Marines' hymn at the top of our lungs ... Our morale definitely received a shot in the butt. We were through with that train ordeal.

We shared relief at saying good riddance to that train. The odor of dirty bodies and filthier clothes so permeated all the coaches I doubt any amount of scrubbing or Lysol could ever obliterate.

... How we survived those eight days and nights? The diners threw most of the meals at us ... I mean threw. The front half, blacks, being fed in the first diner and the back half in the second. Those boys had to wolf down their rations, too, to make room for the next seating.

Often the train stopped for long periods ... and when we knew it would be long enough ... we broke out the troops along the siding for some organized grab-ass the Marine term for physical training ... We (officers) got regular sack time in roomettes and could manage to shave and wash to a degree. But, heaven help the boys. They seldom saw water, let alone had enough to more than wet their faces. They tried desperately to get comfortable and get some sleep, assuming grotesque positions on the seats, on the deck and the aisles and even in the overhead luggage racks ... We wished for a clothespin for the nose as we passed through that train.

San Diego ... home of the brave ...

Officers had foot lockers they could lay uniforms out flat. For the men, things were different. They each had a seabag, period. Our boys had to don their greens if they were to get any liberty and pass the MPs at the front gate.

Austin Gattis explains how rolling them properly in the seabag didn't work ... didn't hold the creases, not enough to pass gate inspection. The post tailor said he could press 56 greens in a month. Gattis bundled up the greens took them to an off base tailor and the greens got proper creases.

Now there is an absolute regulation in the corps that an officer can have no financial dealings with an enlisted man ... "what so ever!" Infringement: punishable by firing squad ... But screw the regulations ... maybe nobody would catch me.

Gattis was able to draw three months pay ... delve it out to the troops for at least one good liberty now that they had pressed greens ...

"Someday, I want the money back! Understood?" He got back every cent.

Sharpshooters ... On Land Or Sea

Somewhere in the Pacific ... with 21st Replacement Battalion ... en route to Guadalcanal and New Caledonia.

Somewhere out there in the Pacific we heard the Navy was going to have practice gunfire. We got the word that our major told the ship's captain that his Marines could do better with their rifles than the ship's gunnery crew—and they made a bet.

When the day came for the challenge, there were 13 of us (I was one of them) who were expert riflemen of Scouts and Snipers School. The major selected us plus some sharpshooters. We were split in four groups. A black balloon was released and when it reached a certain height we were to fire.

Of the four balloons sent up we (the Marines) shot down three of them. I don't remember how many the Navy gunners brought down but it was less than three and our major won the bet ...

He may have won the bet but we, the Marines, lost. From then on until we landed in New Caledonia, we were cut down from three to two meals a day.

How's that for sportsmanship ... you swabbies? **George Walden**

O'Clock-Jump or In The Mood. They'd be doing the same things at "Tops" out on Route #101, San Diego, all in unknowing preparation for the drill cadences that would command their steps at Parris Island or San Diego recruit centers.

Many enfolded into the 3rd Division which would in little more than a year be emerging from a jungle battleground, older than their fathers with heads full of strange sounding names like Tassaforanga, Koromokina or Torokina.

Ready For Combat

The 3rd was a division composed in a hurry. Regiments, units, contingents consigned to the 3rd Division were oceans apart. They would be put together as a physical and numerical whole, nine months after inception, far far away on foreign soil in Auckland, New Zealand, near Wacky-Racky (Waiaraka) Park, and Whangarei.

It was there in June, 1943, that Commanding Officer Major General Charles Dodson Barrett, declared the feisty 3rd Marine Division ready for combat. Soft-spoken, unassuming, General Barrett had heard his first fire in the Meuse-Argonne offensive.

With its clean bill of health, the new division spent July (1943) and part of August as bits and pieces snuggled down in the plush comforts of such passenger liners as *George Clymer, American Legion, Hunter Liggett*—bound for a remote coconut grove near Tetere Beach on Guadalcanal.

It was not far from the beached Japanese troop ship *Kinugawa Maru* and right on the cruise route of *Washing Machine Charlie*, the Japanese airplane that did cause concern but did little harm.

Submerged in everybody's war problems from Stalin to Haile Selassie, Admiral Ernest J. King, Commander in Chief, U.S. Fleet, always had his heart with the Navy in the Pacific. Well he could sympathize with the often-times vexatious General Douglas MacArthur, Allied Supreme Commander in the Southwest Pacific, who told how Filipinos watched smoke pillars from their burning villages hold up the sky, while United States interests were directed first and foremost to Britain and Russia.

The policy and plan of the United States was to remain defensive in the Pacific, although Admiral King strained with anxiety to get the Pacific offensive

(Continued on page 23)

They Saw Samoa of Samoa

The 3rd Regiment sailed aboard the SS *Lurline* from San Diego on September 1, 1942 and arrived September 14 at Tutuila, American Samoa.

A few months after we had been there, we awoke in the middle of the night because a bugle was blowing like crazy. None of us fairly new to the Marine Corps knew what the bugle was trying to tell us. All most of us knew were Taps, Reveille, Chow Call, and Pay Call.

Finally an older sergeant in our hut said, "I think that's fire call."

We all asked, "What do you do at fire call?" The sergeant said we should all get our buckets (we had just been issued a canvas collapsible bucket that held water quite well) and run to where the bugler was, because the fire would be there.

Our hut passed this word to the other huts and our whole headquarters company ran in our "skivvies" with our buckets down to the bugler.

I knew we were in trouble when we all fell into ranks in front of the bugler because standing beside him was our Colonel G.O. VanOrden. Both he and the bugler were dressed and had their steel helmets along with weapons and cartridge belts.

"What are you people doing with those buckets!"

Our sergeant stepped forward, I think very bravely, and said, "We're here for fire call, Sir!"

With that the colonel exploded. "That wasn't fire call, that was CALL TO ARMS. Go back, get dressed, and get your weapons and report back here on the double. We have a report that a Jap submarine has surfaced off the harbor at Pago Pago, and we don't know if it means an enemy landing on Samoa or what!"

As I recall, the Jap sub might have fired a few shells from its deck gun, and then left. We never heard anymore about it. But, at least, we knew what was not fire call.

Colonel VanOrden was our battalion commander the nine months we trained in Samoa. His code or nickname in combat was "The Beast," and he was promoted up to regiment or division just prior to our battalion landing at Torokina Point, Bougainville. **Edward N. Bieri, Alliance, Ohio**

Hemorrhoids Or Fear

During our 3rd Division organization days, in a newly formed platoon, we were "gifted" with a beau-brumellian, aloof type of old reservist. He slowly rose in the ranks of NCO's attaining the designation of acting platoon sergeant.

We often looked with cynicism and sarcasm upon his ineptitude, and as we shipped out to California (Las Pulgas-San Onofre canyons, Camps Elliott and Pendleton) it grew to disdain.

We really commenced to reject him during a training exercise in the boondocks. We were hiked to a gradual steep cliff that ended in a sheer drop, and shown how to reppel down this cliff to the bottom. He was the only one to refuse and his fright was obvious. His own diagnosis was hemorrhoids.

Our arrival and assignments to bivouac areas used by New Zealand troops in Kaipara Flats, New Zealand signalled a campaign to have our sergeant replaced since combat was imminent.

Our wooden huts were for four men,

while NCO's had two people per hut. On the right occasion I ushered a farmer's Irish wolfhound into the sergeant's hut with a generous supply of Spam (our fillet mignon, at the time) and locked the door from the outside. This dog was of King Kong proportions and could easily have passed for a Clydesdale horse in the night.

On our return, he opened the door, the dog galloped out, and the shock that froze his features was undescribable ... followed by cries of the saltiest, seagoing language. We retreated to the confines of our huts and laughed the discomforts of the two day maneuvers away.

The other sergeant that shared the hut with platoon sergeant "Hemorrhoids" immediately pictured the scenario and moved into another hut. The incident quieted down soon so as not to alert the command of his situation. We did not believe that they knew, and that a change was inevitable.

Later a three-day exercise was scheduled; and being in the 3rd Squad, 3rd Platoon, I happened to be the last to leave the hut area. At his hut I opened the hasp, removed the pin that kept the hut

door closed, threw in bread and chicken feed, and placed containers of water on the floor. Our rear echelon lads, knowing the plans, had captured a dozen hens from a farm. We tossed the chickens into the hut, closed the hasp, and inserted the pin.

When we returned from the three-day maneuvers, I concluded that we had discovered a new term called, "chicken shit." As a youngster on the farm during the summer I cleaned and whitewashed some chicken coops. The chickens we released into the hut that day must have been constipated when they arrived; but the bread, feed, water, and confines of the area did cause them to release and overcome that condition.

In short time, our Sergeant was transferred to Graves Registration where he performed his duties efficiently and admirably. To our relief, we entered combat with a highly-regarded platoon sergeant from Texas. **John J. Wlach, Commack, New York**

The Goat

While some of us were based at New River (Jacksonville), North Carolina in 1942 we frequently went on liberty in the town of Kinston. We had to travel to and from there by bus. One character in our battery by the name of Cracker somehow procured a goat and was convinced it should be the battery mascot. He wanted to take it back to camp on the bus.

At first the bus driver was reluctant to accommodate the goat. But since there were only Marines on the bus, and a good number of them had had a little too much to drink, the driver decided the best policy was to get back to Jacksonville as fast as he could.

Cracker kept the goat until we shipped out of New River for the West Coast. **George G. Green, Webster Groves, MO**

Crocker and his goat.

Direct To The Colonel

When T.O. Kelly reported to New River Training Center from Quantico Marine Base in late summer of 1942, his reputation as "one of the toughest sergeants - major in the Marine Corps" accompanied him.

Kelly reported to New River as sergeant major of the newly formed 21st Marines. Colonel Campbell was commanding officer, Captain Bob Kriendler was adjutant.

When the 3rd Marine Division was formed in September, the rear echelon headquarters was housed in the officers country of the 21st Marines; personnel were carried on the headquarters roles of the 21st Marines. General Allen H. Turnage was commander; Colonel "Tex" Butler was the three (operations officer); Colonel Carvel Hall was the four. Clerical personnel occupied several huts in officers country.

On an occasion, General Turnage directed Corporal Brent I. Hancock to Colonel Campbell. One of Kelly's favorites was "Go back and start over."

To young Hancock he barked: "That's the trouble with young corporals, they don't know what they're doing."

It took one Irishman to handle another Irishman. Next time "Sergeant Major Kelly, Corporal Hancock requests the Sergeant Major's permission to talk to Captain Kriendler, Sir?" "Granted" grunted Kelly.

"Captain Kriendler, Corporal Hancock requests the captain's permission to talk to Colonel Campbell, please, Sir?" Captain Kriendler served with Sergeant Major Kelly in Quantico. "What are you trying to do, Hancock, impress Kelly?" asked an amused adjutant. "Anytime the general sends you with a message for Colonel Campbell, you report directly to the colonel."

Which "The Hershey Kid" did. **Brent I. Hancock, Hershey, PA**

Weapon Try-Out At Camp Elliott

We of D Battery, 3rd Special Weapons Battalion, had been issued along with other weapons, a number of Riesing sub-machine guns, manufactured by the Daisy BB-Gun Company.

One Sunday afternoon my buddy, (Raymond Brinkley from Tennessee) and I decided to give them a tryout in the nearby hills just outside of Camp Elliott. We overlooked the necessity of asking permission to fire live ammo ... knowing full well that no such permission would be granted anyway.

With the guns loaded, and pockets bulging with ammo clips, we walked a little distance from camp and found the ravines we were looking for. Firing the two Riesings simultaneously at full automatic into imaginary enemy troops on the opposite embankment, we made the most of that Sunday afternoon ... the only Sunday afternoon I can remember with real fondness at Camp Elliott.

Ammunition expended, we decided that we'd better take a precautionary, circuitous route back to camp. It's just as well that we did.

We had been back at the barracks maybe 10 minutes when we were told that the MPs were searching the entire countryside high and low for "someone out there shooting anything and everything in sight."

I still don't know why the Marine Corps returned the Riesing guns to the Daisy factory, never to be seen again. I thought they worked pretty damn good.

Confession is good for the soul, especially when one knows that the law of limitations has long since run out. **Maury T. William Jr., Dayton, OH**

George Green of Webster Groves, Missouri and a Riesing sub-machine gun during practice on the range.

---

underway. But then came the Battle of the Coral Sea (May 1942). A stand-off, it still blunted Japan's ambitions on Australia and created uncertainties where only confidence and arrogance had stood before. With the Japanese debacle at Midway (June 1942), and loss of four carriers, circumstances opened the way for increased offensives in the Pacific. The Joint Chiefs of Staff agreed to grind through the Solomons, New Guinea, New Britain and the Admiralties.

Such a thrust would open the way for General MacArthur, with, of course, naval ground and airpower to strike through New Guinea and to his beloved Philippines. As Admiral King saw it, the ambitious amphibious war would isolate the great Japanese bastion of Rabaul and leave the Gilberts, Marshalls and the Marianas up for grabs. With a foothold in the Marianas, American airpower would be posed for the Japanese jugular, the home islands. Any military man or woman in Japan knew that.

Amphibious ... Is The Key

In fact, the extended amphibious plan was already well in motion when in that cool California September the 3rd Marine Division was initiated. By then, the First Marine Division, certainly not having all its way, was on Guadalcanal well over a month and with improved air support was prepared to stay there.

Gathered into the 3rd was the salty old 12th Marines, spit'n polished in the old 3rd Brigade by the redoubtable Philadelphian Smedley D. Butler. Included was the 3rd Regiment with ribbons from the Dominican Republic where it chased bandits. It was already at work under instructions of Admiral King to garrison Samoa and discourage any Japanese intentions there.

The Ninth, also with vintage, was first activated in World War I and sent to the Caribbean to protect American interests, particularly the critical wartime sugar sources. Its first CO on reactivation was Lemuel C. Shepherd who would be USMC Commandant.

The 21st was a war baby, born, activated July 14, 1942, filled with boots from PI, buttressed by a handful of lean long salts from the Sixth Marines with traditions back to Belleau Wood.

The Nineteenth was different: engineers, seabees, bankers, welders, piledrivers, surveyors, some of the old 25th Naval Construction Battalion from Port Hueneme, California. This outfit came with rubber bridges, dark rooms,

(Continued on page 25)

The Mail Must Get Through

When troops of the 3rd Marine Brigade (later to join the 3rd Marine Division) were on Samoa in late '42 and early '43, this was their thatched roof post office. (Official U.S. Marine Corps Photo)

These mascots were getting a famous "bath in a helmet." And even though their "tub" was about the right size, they didn't seem to care for all the attention as their Marine owners got them ready for their next port of call. (Official U. S. Marine Corps Photo)

Samoans who served in the U.S. Armed Forces on Samoa wore garb like this... with hat, white T-shirt, cartridge belt, wraparound, and bare feet. This Seaman mounts guard over the office of the Military Governor, American Samoa. Fita-Fita Marines on Samoa wore the traditional fore-and-aft caps. (Defense Dept. Photo-Marine Corps)

"To him who is in fear, everything rustles..."
Sophocles

needles, cranes, bulldozers, dynamite, refrigeration and places to stow beer. Without them no island would be worth the keeping.

As for readiness and amphibious war, the Pacific offensive could not have come at a more opportune time (if war is ever opportune) for the United States Marine Corps.

Since 1933 when the Marines withdrew from Nicaragua, leaders in the Corps had been refining the art and tactics of amphibious warfare. Charles Dodson Barrett, to command the 3rd Division, was one such pioneer.

Jeter A. Isely and Philip A. Crowl in the Quantico officer candidate text (the United States Marines in Amphibious Warfare, Princeton University Press) make clear the Marines nor any other fighting unit was fully prepared for landing operations on hostile shores ... but ..."a detailed doctrine of amphibious warfare had indeed evolved, tested, improved, and found to be sound in its main principles. The remaining test was the ultimate battle."

"Amphibious warfare represented the turning point in the history of the United States Marine Corps," explains Major General Carey A. Randall (Ret.), Jackson, Mississippi, decorated in all three 3rd Marine Division campaigns, and executive officer to Secretary of Defense (1951)" ... and when Admiral William F. Halsey, Commander, South Pacific Area, and General Douglas MacArthur, Commander in Chief, South West Pacific, agreed to storm the beaches, there was no one more prepared than the United States Marines."

Major General Allen H. Turnage took command of the 3rd Marine Division on the death of Major General Charles D. Barrett, just prior to the Bougainville campaighn. (Defense Dept. Photo-Marine Corps)

Alouette, Revisited!

Training was rigorous in New Zealand and the officers who set the pace knew these young boots would need it, and soon, into the not-too-far jungles of the Solomons. Having our troops there was great for the New Zealand girls, but some farmers complained when their gardens were trampled by boondockers.

Carroll M. Garnett, now of Chester, Virginia, was there as a PhM 1/C in Baker Company, 1st Battalion, 21st Marines, and after retiring from the FBI returned to his old training ground, 10,000 miles away.

Travelling several miles from the old camp, I suddenly found myself on a narrow, dirt road threading through numerous beautiful, rolling slopes. Standing at the crest of these undulating hills, there was a flashback ... this was the route we, in Baker Company, had frequently travelled on our many forced marches. Over 200 men, maintaining columns on each side of the road and well spaced, our lines extended about one-fourth mile.

Upon urging, "Frenchy" from New Orleans, who naturally spoke fluent 'Cajun' would sing out the words to that popular French-Canadian folk song, *Alouette*, pausing after half-phrases to be repeated as a chorus by all the troops. The powerful strains of *Alouette* reverberated over the rolling hills.

Only "Frenchy" knew the meaning of this spirited song, but its rhythm and meter was especially suited for marching by the troops. We adopted *Alouette* for our very own.

Memories of this company now become poignant ... recalling in July 26, 1944 on the island of Guam ... some seven battalions of Japanese enemy conducted all-out wild banzai attack against our regiment and other selected units of the division.

This desperate charge was obviously designed to throw us back into the ocean. Baker Company, defending in the middle of the 1st Battalion, 21st Regiment, received the main thrust of this fanatical charge. Since its ranks were already severely depleted, a breakthrough by the enemy occurred. The company was decimated ... its manpower reduced to 18.

To counter this drastic situation, the battalion commander called upon every weapon at his disposal. Ordering the remaining troops of the 1st Battalion to withdraw, the vacated area was then blanketed with a devastating barrage of fire from rocket, rifle, artillery, tank, machine gun, mortar, and offshore naval vessels (especially battleships with their powerful 16-inch guns). Numerous flares lit up the area, so the enemy's onslaught was clearly visible.

The Japs had committed themselves to what for them was to be an avenue of death; now, there was no escape. The next morning revealed some 1,800 of the enemy lay dead in the battalion front, especially in the sector defended by Baker Company. One member of the company, "Poncho" of Mexican extraction, was found in a deep coma in his foxhole; exhausted from five sleepless nights. He had slept through the entire battle, being presumed dead by the enemy when the area was overrun.

The Japanese charge was clearly broken, but not until one platoon had infiltrated into the command post of the 1st Battalion, 21st Regiment. At the base of an almost perpendicular 75-foot cliff, the CP area was a maze of closely knit foxholes. Here the enemy was able to cause considerable problems with knife, bayonet, saber, and light Nambu machine gun. Hand-to-hand struggles were the order of the night ... but there was occasional machine gunfire, grenades bursting, and voices of desperation. It was sheer madness, but at the edge of dawn the remaining Jap elements attempted to escape. For them, too, there was no escape..

Standing here in New Zealand and thinking of the terrible sacrifice by gallant Baker Company and others, memories become bittersweet, again recalling the mighty roll of young voices across these hills in their favorite marching song, *Alouette*. Of all the places in the entire world; at this time, at this moment, this is where I would most like to be: "Alouette, gentile Allouette; Allouette, Je te plumerai!"

He Followed Me Home

My mother had put her foot down after I had been out with so many Americans that she said, "no more ... that's enough."

So I made up my mind I'd do as she said ... which I usually did. I tried to be a very good daughter. But, he wanted to take me home, so he did, and when I went in the house my mother looked at me very, very cross. So I turned around and said, "Mum, I couldn't help it. He followed me home. There was nothing I could do ..." **Beryl Colvin**

Her mother later recalled ... Chuck was a chap you never could have spoilt ... he was a delightful soul. Then they decided to get married. I said, "If that's what you want ... but you know what you're letting yourself in for ... you are going to go a long, long way from home." ... But they were very much in love ...

For Love Of A Soldier
(Annete Potts-Lucinda Strauss)
ABC Enterprises, Australia
Courtesy of Miss Colleen Greer

"Down Under" ...Sometimes Like Home

Here are stories and memories of old time (our time) New Zealand era, bulging with the same nostalgia as our stateside recollections: girls, romances, big bands, and farewells.

Some almost diary-like recollections are in Harry Bioletti's *The Yanks Are Coming* who tells about the 12th Marines (and more) around Whangarei. And there are poignant memories of war brides in *For Love of a Soldier* (Annette Potts-Lucinda Strauss) given us by Colleen Greer of Arlington, Virginia. (Her mother was an Australian war bride).

It was a day of stricter and more simple mores, and even sheer naivete (and, what's venereal disease?) ... of American heroes-all (while New Zealand's sons were away protecting the Crown) ... and of bonds and friendships still firm a half century later.

Marines Come Ashore

Ambassador to United States Denis Bazeley Gordon McLean (1991) recalls, "As a boy of 11 in Wellington I saw the Marines coming ashore ... it was a moving occasion and made an overwhelming impression on a young boy ... Perhaps there was even more emotion in the arrival of the Americans, because just two years before these same New Zealanders watched their own sons leave for other foreign fronts. Now these Ameri-cans were coming when New Zealand was defenseless."

Today Bioletti writes to Austin Gattis, "How old timers remember the Marines at Gum Town, at the dance hall in Kamo, or at their favorite drinking hangouts—the Settlers Hotel, the Grand, or the Whangarei Hotel."

Shirley of Maungatapere remembers the Americans ... "I was 13 and at school and the camp was right beside the school. The Marines used to hang over the fence and talk with us; that is, until the teacher hunted us out of it.

"We lived near the camp. One time we were playing football, and I called out to Dad ... 'I think the Yanks are pinching eggs out of the fowl house.' Dad blew his whistle. The Yanks thought it was the MPs, and got the hell out of it.

"Did I go out with Americans? Hell, no! My father was too strict for that."

Leo (a friend) remembers the Marines, too. "They were so well-behaved. No brawls that I can recall. Farmers hired horses to them. When they returned the horses, the dollars were always tucked in the saddle bags.

Always Paid Their Way

"They always paid for the milk too, when they came down to the milking shed. All doors were open to them locally. Sometimes they would route march to Dargaville 20 miles away, and camp out for the night. The kids loved the Americans, and the Americans were always handing out the stick gum and candy. Sometimes they would say to a family, 'We'll come and cook for you tonight.'"

Wilfred recalls, "I was at school at Whangarei Boys High. The army hospital was in our quadrangle. Some of our classrooms were taken over, and we were farmed out to other schools. The ambulances were in and out with the sick, both Americans and New Zealand troops. The latter were on duty in the Far North.

"There were special displays and marches in town. I was always aware of the different sounds between the New Zealand military band and the Americans with their sousaphones and saxophones. Of course the Americans came through as being meticulously dressed. Our own boys looked as if their uniforms were made out of chaff-sacks.

"Whangarei was very much a frontier town in those days. Mother had an exercise book with all the names and addresses of the American boys who came to our home. She wrote to their mothers in the States.

"The American presence was very obvious, but there was always friendly activity on both sides. The kids of 8, 9, or 10 were not shy in confronting the Yanks, asking them if they had any Marine badges or candy. The troops of course loved the attention they attracted.

"Then all of a sudden, just as they had suddenly arrived, they were suddenly gone. We wondered where to, and what happened to them. Looking back it was all so unusual it seems to be unreal."

The Marine Aircraft

The only United States Marine Corps aircraft to visit New Zealand were the 27 Douglas SBD-3 or SBD-4 Dauntless Dive Bombers VMSB-141, which were stationed at Seagrove Aerodrome (near the Manukau Harbour) from May to July 1943. The Marines handed these aircraft over to the RNZAF for the use of the newly formed No. 25 (Dive Bomber) Squadron.

"V" indicated 'heavier than air' as opposed to airships. "M" indicated Marines. "S" indicated Scout, and "B" was Bomber. The RNZAF flew these planes extensively until January 1944, and for a time they retained their original United States markings. So, viewed from the ground, observers would have concluded the planes were being flown by United States personnel.

Douglas Dauntless

Code Talkers At Play

The Townhouse Brothers ham it up in off-time entertainment for their fellow Navajo Indian code-talkers. The act is from a traditional ceremony commemorating the importance of turkeys in the food-chain of ancient Navajo life. The brothers took along in their seabags the feathers and other props for the ceremony.

Ducks And Romance ... And Where Is Harold Finney?

In the summer of 1987 a West Virginia grandmother took her grandson to Auckland, New Zealand (as a high school graduation present) to visit her son and family. As an added bonus and history lesson for him, they visited Pearl Harbor.

But I am getting way ahead of my story ...

Bill Linger, my distant cousin from Weston, West Virginia, joined the Marines in August 1942. One of Bill's friends overseas was the camp cook, a hash-slinging Marine named Harold Finney.

I joined the Marines in December 1942. Bill and I were to cross paths on Guadalcanal in 1943, when he arrived there via New Zealand and I came by way of the 15th Replacement Battalion on Samoa. After Bougainville we went our separate ways, and I did not see him again until after the war when we stood in line to sign up for the old 52-20 Club. If memory serves me right, the VFW with their slot machines and booze got it all.

But we shook off the effects of this binge and found work. Now retired, we get together to swap tales, poke fun at wet pant legs, corn cob pipes, and politicians. One day Bill said he wanted to locate his old friend, Finney, and we decided to try. Let Bill tell you a story.

As Bill Remembers ...

We had gone out as usual Friday morning (Harold Finney and myself, Bill Linger) for another day of field maneuvers among the sheep-dotted hills of New Zealand. There seemed to be 10,000 for every human. Needless to say we were fed mutton and corn willy to the retching point, and the damn mutton followed us all over the South Pacific after we left there.

There was a small stream and farmhouse about a quarter of a mile from camp. With nothing to do, I rigged up a line and pole, hoping to augment the wooly diet somewhat. Finney gave me some scraps of mutton for bait and I headed for the river. Two hours later, without a bite, I found the damn fish didn't like mutton any

better than myself. I threw my bait to seven ducks floating around close by.

Ducks! a fowl idea began to form.

Back in camp I told Finney and "Motta" from New Jersey, about my plan to go duck hunting. Finney agreed to cook. Motta said he would pluck. We slipped out of camp with my M-1 and clip of ammo. When we arrived the big white birds floated around like they didn't have a care in the world. I flipped down on the bank and at about 25 yards I had duck heads flying all over the river; except for number seven, a big gander who only got creased and was swimming around in circles faster than a PT boat at full throttle with a broken rudder.

I emptied the clip, missed, and went plowing into the river after him. He wasn't too hard to catch but got me black and blue before I was able to wring his neck.

Getting back inside camp with a load of fowl took a bit of doing, but we made it. Finney took over and I headed for the shower, thinking of nothing but roast duck for dinner.

When the bugler blew chow formation, I came around the side of the barracks, pulling on my jacket. I saw my captain standing with a New Zealand couple in front of my squad. I knew then it was going to be mutton for dinner, perhaps bread and water.

As I approached, I heard the captain say, "Would you recognize the Marine who shot your ducks?"

The lady pointed a finger at me and yelled, "That's the bloody bastard, I'd know him anywhere." The nice gentleman also got his "bloody bastard" in. Needless to say I felt like a mongrel dog caught with a mouthful of chicken feathers.

"Did you shoot Mr. and Mrs. Street's ducks?" asked the captain.

The feathers had my mouth so dry I cold only mumble, "Yes sir."

"And where are they now?" asked Captain Haufpower.

"Finney is cooking them, sir."

Turning to Mr. Street, the Captain asked, "How much do you want for the ducks, sir?" "One pound six pence should cover them," he replied.

"Well, I hardly think that is enough for the trouble Private Linger has caused you. He gets paid in one week and will give you $50 for his expert rifle demonstration." After the Streets departed, the captain said to me, "Private Linger, you report to the mess hall. You, Finney and Motta, will be serving duck to officers mess tonight. If you do a good job, we will save you the piece that went over the fence last."

Needless to say the officers had a great "quack quack" dinner that night, and they left damn little for us to sample. We cleaned up the mess and hit the sack around midnight.

Pay day rolled around and with 50 fresh duck bills in my pocket, the captain and I paid a social call on Mr. and Mrs. Street. Protesting that 50 dollars was too much, they finally accepted after the captain told them what such a wonderful dinner would cost in the States. No 90-day wonder, my captain. "Old Corps," all the way.

News From Stateside ...

After a spot of tea, cookies, and some small talk, Mr. Street invited Finney, Motta and myself to visit any evening for stateside news on his short-wave receiver at 6 p.m. So every evening possible, we went over for news and music from home. We became good friends ... the ducks became laughing matter, and so did I for some time. Some smart-ass even drew one on my target at the rifle range.

One evening after the news Mr. Street said, "Boys, this weekend we are going to have some friends in for a little party. I will bring a keg of beer, and three of my nieces from Auckland will be coming up for the weekend. I hope you can make it."

We sure as hell made it. The nieces were attractive girls by the names of Gaye, Faye, and Raye. Also arriving was a guy who played the concertina ... kinda one man band. We had a ball and were such good boys we got invited back for the next week, and every one thereafter until our departure for Guadalcanal. Faye and Raye were always there.

Gaye and Finney were seeing quite a bit of each other ... and his cooking was no longer up to par back at camp. The midnight snacks he used to bring back from the galley were a thing of the past. But we shipped out for Guadalcanal a short time later.

We were at Coconut grove, getting ready for the Bougainville Invasion. Finney had a big bulldog tattoo on his arm, which Gaye had got his promise to have removed at the first opportunity. It was done by a Navy surgeon within a week of our arrival on Guadalcanal ... they grafted skin back and the bulldog barked no more. Finney told me when it was all over he was going back to New Zealand to live.

After Bougainville, Finney was transferred out and I headed for Guam, never to see him again.

We Look For Gaye ...

Some months after Bill told me this story, I learned of Grandmother Potter's trip to New Zealand with her grandson. Would it be possible I wondered for Bill to get a line on Finney, after all these years?

I got a picture that Bill had of himself and Finney in New Zealand and went to visit Mrs. Potter. She was a voluble and active lady, still working, with a vivid memory of World War II. We chatted for some time and she got a big kick when I related the story of the "Mighty Duck Hunters." I asked if her daughter-in-law in New Zealand would mind checking and try to find Finney. She said, "Sure, Dee is just the one to dig into this. It's down her alley?"

She wrote ... and Dee agreed to carry on the search around Auckland. Things looked bleak at first, as some time went by. Then by strange coincidence,

"The Mighty Duck Hunters" Harold Finney, left, and J.W. (Bill) Linger visited zoo in Auckland, New Zealand in 1943

Dee ran into a member of the Street family, there from England doing a genealogical search. From her, Dee learned the location of Lorraine, 13 year old daughter of the Streets in 1943. Mr. Street was still living on the farm, and Lorraine was able to provide Dee with the location of Gaye, her cousin. As it turned out, Gaye (now Gaye Sewell) was widowed and lived at Dee's father's old family address in Auckland ... a coincidence.

Contacting Gaye was a touchy matter, as it was not known how she would react to this reminder of the past. However, a picture of Bill and Finney was mailed to Gaye, and Dee soon received a call. According to Gaye, who was a teenager in 1943, she never heard from Harold Finney again after he departed ... nor anything about him. She eventually married.

I had hoped for a different end to this tale. Instead, this story had led us over half the globe and maybe is ending too soon. Finney and Linger were in the 3rd Medical Battalion, 21st Marines from January to March 1943 at Warkworth, New Zealand. He would like to end this story by finding Finney. So, Harold if you are out there, help us out. Bill Linger said he would treat you with a roast duck dinner. His telephone number is 304-296-1066, and address is 242 South Main Street, Weston, West Virginia 26452. **William D. Carothers.**

Spirit Of ... Sabatinelli

In May of 1943 the division had been through the squad-platoon-company-battalion-regimental training phase cycle; and I understand the division commander felt we were in pretty good shape as far as training was concerned. But ... he wanted to be sure we were all well-conditioned with the physical stamina to withstand combat. So, in the month of May we engaged in hiking with weapons, combat packs, etc.

Usually we would do 20 miles ... go out 10 miles and then back. Spend a day in camp, then the following day go out 20 miles ... camp overnight ... then come back the 20 miles the next day. Spend a day in camp, then go out 20 miles for three consecutive days, and return to camp.

Many complications flowed from this activity.

Good men who had bad ankles or knees—often from high school athletics—often suffered, and we lost good men as a consequence. We were using Japanese—type rations part of the time, as I recall, raisins and rice; traveling light and trying to see how the Japanese did it, perhaps.

A young fellow in A Company, named Sabatinelli, was a real short fellow, probably minimum height to get in the Marine Corps. But he was a good man, a jolly man, and did his best on these hikes. We would hike 50 minutes and take a 10 minute break. Sabatinelli would fall behind and catch up with us while we were taking a break ... and get on us being lazy, having to take a break, and this sort of thing. He would pass on through the company, and perhaps the battalion. Then we would get on down the road and pass Sabatinelli ... but he had a great spirit.

Sabatinelli was a casualty on Guam. The first night of combat was his last, as a Marine and on this earth. He was a fine young man. **Conrad Fowler, Lanett, AL**

Down Under
(As An American Sees It)

This land they call Down Under,
In the South Pacific sea,
And what on earth can be its worth
Sure beats the likes of me.

People call the bar-rooms pubs,
They close at six each night.
Ask a person here a question
And he calmly says 'Too right."

A sidewalk is a footpath,
A buggy is a pram,
And above all things a street car
They call a blinking tram.

They call you 'Yank,' or 'joker,'
Frankly I don't get the jest,
And if you act American
You're a 'queer bloke' at the best.

If you're tired you're 'all knocked up,'
And when you're sick you're crook.
There's no defining words like these
In any blinking book.

The movies here are called 'the flicks,'
Old wine is ten days brewed,
It costs a chap a quid a quart
(Paid only when you're screwed).

One thing about this country though
It rains but once a year,
And to prove that what I say is true
It hasn't stopped since I've been here.

Lest you want to be a pauper, boys
Take heed to what I say,
Don't try to save your money
For New Zealand's rainy day.

(An anonymous serviceman's recorded reaction to New Zealand.)

In God We Trust Exclusively

Maureen Kimber around Pukekohe, New Zealand recalls the avid Marine with some homesickness always came around to help her dad with the horses. So Dad invited him and his buddies for dinner.

Frank was his name (no recall on first) and he was quick to say thanks and add, "Love my buddies, and I'd die for them, but I wouldn't trust any of them with your daughters ..." **Harry Bioletti, Warkworth, New Zealand**

.. Deja Vu For New Zealand War Bride

Sheila Roberson returned to Wellington, New Zealand not long ago. She had left there in 1946 as a war bride. Her recollections here, stirred by the music of returning Marine veterans, quite succinctly relate the typical feelings of a young girl who greeted these wartime leathernecks as heroes and saviors—with their own boys away on other fronts.

My leisurely mid-morning coffee in a cafe on Lambton Quay was interrupted by the stirring sounds from fifes, pipes and drums coming from an unknown source on the wide thoroughfare outside. No, not from an unknown source, for the nostalgic sounds evoked memories of parades in and around Washington, DC, my home ever since I had left these New Zealand shores as a war bride in 1946.

To me, music from such instruments could only mean an American marching band. I was not disappointed, for there along Lambton Quay, Wellington, I saw the Stars and Stripes fluttering in the breeze as it was held proudly aloft by an aging Marine veteran, who headed the little parade of veterans who were visiting the land they held such fond memories of. A gathering crowd watched as these men marched to a long ago drum, for suddenly it was the early 40s all over again.

Back Then——The quiet wartime thoroughfare became electrified that afternoon by the emotions of the crowd watching the recently disembarked Marines marching eight abreast down Lambton Quay. New Zealand was now safe from the threat of Japanese invasion. Our fighting men overseas—long gone and far away in North Africa, in the skies over Great Britain, and on the Atlantic and Indian Oceans—could rest a little easier during the coveted moments of rest they had. Their loved ones back home were protected and secure. The Marines had landed.

In the 40s, curiosity replaced surprise and gave way to gestures of welcome. New Zealand hearts and homes opened wide to these American servicemen. Very soon, most homes in and around the city had entertained one or more of these "wonders" in their smartly pressed and creased green uniforms, replete with rows of shiny sharpshooter medals.

They came offering candy, gum, and goodwill. It seemed that every unattached girl in Wellington was dating an American Marine or a Navy man.

Many a perplexed mother wondered what to feed these young men when her daughter asked permission to bring home a Johnny, Joe, or Mac serviceman. My own mother thought that apple pie would be the appropriate dish to serve as part of the meal, for didn't all Americans eat apple pie? Wasn't it almost a national dish?

An apple pie to be cooked suitable in a big deep pan, and covered with fluffy (French) pastry was duly baked.

To make the young man feel even more at home, a large fire was lit in the drawing room of our home. Much was being rationed at the time, but the coal was heaped on the fire to demonstrate the warmth of the welcome extended.

By now, members of the American Red Cross had arrived in town and "set up shop." Among other things, they offered advice to the lovelorn and those contemplating marriage. Many hurdles had to be overcome, and much American red tape had to be addressed, before the marriage ring and the minister could be approached.

Questions which seemed too degrading to most New Zealand girls had to be answered on lengthy questionnaires; questions regarding venereal diseases and criminal activity. Heck, most of us didn't even know what venereal disease was, let alone have a criminal record.

Gossip was rife in the little capital city, and rumor was rampant. Scandals broke out and had to be hushed up. Office girls filled milk bars at lunchtime and imparted titillating news, ... the heir to a hat factory and fortune, an officer with the division, was marrying one of the hometown girls ... an aide de camp, attached to the general staff, headquartered in a hotel on Willis Street was a Hollywood star.

To reinforce their numerical superiority and advantage, Marines made it known around town that any girl who made the mistake of dating a "swabbie" would be cold shouldered by the corps from then on. It is doubtful, though, that such a threat would be taken seriously by fair-minded, independent thinking girls such as the New Zealanders.

(Continued on page 32)

Battle Training "Down Under"

Much of what the salts and boots of the new 3rd Division were to learn of amphibious warfare doctrine and getting there "fustest with the mostest" came in the absolute school of hard knocks at North Island, New Zealand. There was enough spiny, unpleasant brush around for realism. Gardens were trampled and some sheep became hostage to violence and good taste.

But there were pleasantries, mostly girls. Much like the sweethearts they left at home, they were. And their brothers and boyfriends were away fighting Rommel or keeping safe the Empire and the Queen while Marines and a handful of Doggies were the only safeguard between these lovely civilians and those awful Japanese. The girls were pretty, danced to big band sounds, and not a few of them became Yankee wives.

Exercises with live rounds and cold steel took a serious turn at the camps around Auckland, Pukekohe, and tongue-twisting native-name towns, streets, and boonies. We were soon training again on Guadalcanal with sharp black mountains, butterflies as big as little birds, and people with yellow hair and teeth.

Jungle on the "Canal" was dense but as skirmishers you'd later appreciate that the bush on Guadalcanal was an English garden compared to the matted rain forest on nearby Bougainville, in the same Solomon Islands chain.

You knew you were heading for combat but nobody told you where. Maybe General Barrett did know, but he didn't say to you, the enlisted, at least. Before our baptism of fire he was picked to head the First Marine Amphibious Corps. He was waiting at home in Noumea, New Caledonia when he died in a home accident.

The command of the 3rd Marine Division then went to wise, professional, and kindly Major General Allen Hal Turnage, North Carolina (Pitt County). University of North Carolina, an old China hand. Colonel Oscar R. Cauldwell was named assistant division commander.

.. Deja Vu For New Zealand War Bride
(Continued from Previous Page)

General Vandegrift was in town for a time and was given God-like qualities by impressionable young women.

To entertain the troops, existing service clubs enlarged their membership, and new ones sprang up. One of the most popular enlisted men's clubs was the ANA (Army, Navy, and Air Force Club) which held Friday night dances at its quarters on lower Cuba Street.

Marines on leave, came into Wellington by train from Paekakariki, where they were based, and many attended these popular functions.

Sheila Roberson now lives in Rockville, Maryland. Sheila Roberson (nee Litchfield-Green) in 1944 married Donald W. Roberson (now deceased) who served at Pearl Harbor December 7 on the USS *Raleigh*. She came to the United States in 1946 aboard the SS *Lurline* well known to Marines. A writer, she has been president of the social Southern Cross Club, and lives in Bethesda, Maryland.

> *"The opinion of the strongest is always the best..."*
>
> Jean de La Fontaine

No Rose Garden On The Liner Mormacport

Read this memorable piece that recalls all the bitchin', the better shipboard food of the officers, the sticky salt water showers, the late night ceremony of dumping garbage at sea, and the long steel unsightly trench for a head.

This sea story concerns the voyage upon the SS *Moore McCormack* Lines, Dutch motor ship *Mormacport*, sister to the *Bloemfontein*. These two cannot be compared to the fine luxury liner SS *Lurline* that ferried most of the 3rd Marine Division to Auckland, New Zealand.

The *Lurline* surely was the love boat of 1942-1946 with its staterooms, swimming pools (used only to stow gear), the Coca Cola, fine food, and pogey bait—plus a fast ship with only 10 days to New Zealand.

Those of us who had to endure the grueling trip on the *Dutch Treat Twins* had 16 days non-stop, no escort. We had been the so called advanced echelon of the 21st Regiment and others.

The rear echelon on the *Lurline*, and their 10 day crossing, were waiting for us on the docks at Auckland and shouting, "Where the hell you guys been; did you get ambushed? What the hell you got on that ship?" We asked the merchant sailors. The answer, hi-explosive ammo and gasoline. No wonder they sent us alone.

Lieutenant Colonel Fry was in charge of our troops which included an advanced echelon unit from the 21st Regiment, 25th SeaBees, B Company, 3rd Tankers, and Special Troops. The ship could cruise at 15 knots, and at 19 full speed could outrun the subs. Had a Dutch captain and Navy gunners in the crew.

Someone really screwed the troops on food rations. The crew and officers mess ate top shelf; but for the troops, it was a combo of knockwurst, sauerkraut, boiled spuds, a bean or two, corn cake, and some kind of cereal. The East Indian cook was fat, greasy, with a large, black beard, and wore a black and white apron. He was always a mite sweaty. The head mess Jock carried a 38 sidearm to make sure order was maintained.

We lined up for morning chow at 0800 hours and got to eat cold (whatever it was) at about lunch time or 11:30—so you had brunch, not breakfast. Went down below three holds, and the canvas hatch covers were always wet and slippery from slop that had been spilled from a G.I. can of garbage to be dumped over the fantail at night (because of the subs).

We stood at five foot high plywood tables—no seats. With your mess gear full, you might get to eat half of it by the time you slid over the hatch to a table. Then as the ship rolled and pitched, you might be eating from your next buddy's mess gear ... as you held on to the table with one hand and a spoon in the other. But, at least it was better than a lot of guys in the jungle were getting.

We never got enough food. I saw a platoon sergeant begging one of his men on mess duty to get him an orange. A piece of fruit was like gold.

After 10 days the men were getting mean, and fights broke out. It was pinochle games all day, as you sat on your life jacket.

One of the cooks in the ship's galley and officer's mess used to place food near a port hole, then make a gesture to any Marine standing by. One day as I was sitting on the hatch cover near the bridge, he motioned me to come over. I looked about to see if anyone was looking ... no, they were all playing pinochle. So, I stuck my long arm down to the pan of muffins that he placed on a table; figured I'd grab at least three muffins. Two more long arms suddenly came over my right shoulder and the grabbing started. I ended up with a handful of precious muffin crumbs, but I really enjoyed them.

We had "Shellback" initiation as we crossed the Equator. I'm sorry to say I lost my small yellow Shellback card, signed by Lieutenant Colonel Fry when someone clipped my wallet years later.

Shellback initiation was some ceremony. A little rough, one guy fell and broke his collarbone. We all had our hair shaved half off or partly chopped away.

Later when we got to New Zealand, the company barber, Sergeant Frank Kemp (Kempinski) gave us all "baldies" right down to the skin. When we went on liberty to Auckland shortly later we went to the Peter Pan Dance Hall where the sailors told the gals to stay clear of bald headed Marines ... that they all had VD. You should have seen the looks as we removed our caps at the dance hall.

Who can forget the salt water showers and heads on the weather deck, fore and aft? ... where if you had to crap, the only safe place was in the middle of the trough when the ship was rolling in rough waters. Except, of course, when some guy lit a newspaper and tossed it in the trough at the end. As the ship moved from side to side, it made for a real hot seat. I always figured someone was trying to make the troops lean and mean, and that did just that. A lot of guys were seasick in the holds.

One has to look back and be grateful to those master sailors of the Merchant Marine who got us to our destinations safely. How many trips did those men have to make?—a hundred perhaps, and some not smooth and quiet. Like the old man said, "Give me a company of bitching Marines and I'll show you a happy, good outfit." **Thomas F. Murphy.**

A Trusty Shellback

Packed with over a thousand United States Marines, representing the 1st Battalion, 21st Regiment of the 3rd Marine Division, Fleet Marine Force, the SS *Mormacport* gently sailed out of the harbor at San Diego during the afternoon of February 20, 1943 and set a course southwestward for the high seas. Our destination was 7,000 miles away at Auckland, New Zealand where we would receive additional combat training in the hinterlands before being committed to battle against the Japanese in the British Solomon Islands.

We sailed alone, and with no fleet escort, this was risky business as enemy submarines were on the prowl in the south and central Pacific. As it developed, the threat of submarine attack soon was reduced to secondary concern as scuttlebutt ran rampant throughout the ship that we were approaching the equator and the "lowly pollywogs" aboard would have to undergo a rugged initiation provided by the "loyal shellbacks." Questions were quickly asked, "Who is a pollywog?" and "What is a shellback?" The answers came back with a degree of shock to over 99 percent of us aboard for we were "pollywogs," having never crossed the equator, and the "shellbacks" were those who had already undergone this initiation and had been accepted into membership of the "Solemn Mysteries Of The Ancient Order Of The Deep."

Well, as the story goes, Davey Jones (the spirit of the sea) was pretty much irritated a load of pollywogs, such as we, had the audacity to venture, without permission, into the kingdom of Neptunis Rex, ruler of the raging main, and therefore his trusty shellbacks would have to punish us.

In a preliminary move, the shellbacks required the officers to climb the mast and periodically call out, "Davey Jones, where are you? Come aboard, please!"

We schemed, feeling if we so outnumbered the shellbacks, why shouldn't we overpower and control them? However, we had, for a moment, forgotten we were still in the military and under strict discipline. Still, we wondered how 30 shellbacks could control over a thousand hardened Marines. We had, indeed, underestimated their organizational abilities for they utilized several principles, reduce our numbers to small pockets, and require us to remain below deck until commanded to appear topside; and allow the newly initiated to join them in the ceremony.

It was not long before our unit was called to topside, ordered to strip, and then we were blasted with a wall of salt water from pressure hoses; here, we desperately tried to keep our balance and, at the same time, with our arms and hands, tried to break the force of these powerful streams upon our bodies.

We were then ordered to run through a "hell line," composed of numerous eager shellbacks armed with large paddles and eagerly waiting to strike the posterior ends of the swift pollywogs who sought the end of the line and comparative safety.

We had to then present ourselves to the "royal electrician" who required us to grip iron pipes which had an electrical current attached, the resulting jolt was accentuated by our wet bodies. We then had to visit the "royal barber" who proceeded to give a shampoo of flour and water followed by an outrageous, slicing haircut. This was followed by a visit to the "royal doctor" who administered medicine which turned our mouths inside-out.

The "royal baby," represented by a strapping Marine with an oversized stomach, required all pollywogs to "kiss the baby's belly." We were then escorted to a ducking pool which had been erected on the fo'c's'le. One had to sit in an elevated chair with his back to the pool, the chair would then be triggered to flip backwards, casting the occupant into the pool where he was roughly grabbed by several shellbacks who repeatedly ducked him.

Our freedom was at the end of another "hell line" which had grown considerably as each newly initiated shellback joined this line in order to pass on some of the abuse he had just received.

Our graduation present was a certificate, signed by "King Neptunis Rex" himself, showing each pollywog had been duly initiated and then the King ordered, "All my subjects to show due honour and respect to him wherever he may be."

Therefore, all Marines of this 1943 voyage of the 1st Battalion, 21st Regiment are certified as "trusty shellbacks" and we speak as one voice, "Bring on the lowly pollywogs." **Carroll M. Garnett**

A Harley ... Not A Kawasaki

Third Division Marines who visited New Zealand in the early 40s will enjoy these comments on Rotorua and Auckland made by Carroll Garnett, who returned to New Zealand and published several articles in the Rappahannock (VA) Times.

America's influence in New Zealand may be seen or heard everywhere: radio, television, movies, the press, music and shops. Their mother tongue is English spoken with a definite British slant, but occasionally even an Americanized expression may be heard. An example follows:

Travelling on the highway one early morning, a stop was made at an all-night restaurant where I was served by an engaging young waitress who was about to go off duty. Upon leaving, she came up with a delightful parting phrase, "Have a nice holiday, Cheer up! Cheer up!"

The expression stuck, but after leaving, I wasn't quite sure what the ending meant. So, on the return trip, a stop was again made at the same restaurant. The original waitress was not present so an inquiry was made of the attending attractive Maori (native) waitress. She said "Cheer up" is sometimes used in parting and means something like "smile, have fun, keep a happy face."

When asked if she used this expression, she replied, "Oh no, I say bye-bye!" She then said, "Are you going back to the States soon?"

"Yes," I answered. "Well," she continued, would you do me a big favor?" "Surely." She quickly added, with a grin, "Send me a new Harley-Davidson!"

While seeking direction at a "petrol station," a jalopy driven by a teen-ager entered the lot and sounded his horn. To my astonishment, I was hearing the tune "Dixie" some 10,000 miles from the southland of the States.

The young driver noted my more than passing interest and identified the musical arrangement as "Dixieland of America." When asked if he knew the significance of this spirited tune, he said he did not but would like to know. Accordingly, he was given a history lesson on the spot and afterwards said, with assurance, he would in the future sound the horn frequently.

Visit any New Zealand home, day or night, and one is almost certain to hit upon one of their six meal times; breakfast, morning tea, lunch, afternoon tea, dinner, and supper. Besides their three regular ones, in-between meals are served mid-morning, mid-afternoon, and at about nine in the evening. These extra meals consist of tea and/or coffee, thin sandwiches, cookies and cakes. One exception is that Sunday evening tea takes the place of dinner and is made up of an assortment of various left-over foods.

New Zealanders are most hospitable and nothing pleases them more than to have visitors ... so guests will surely not escape joining the household at one of their mealtimes.

Soccer is a principal athletic activity in schools, and I was flabbergasted to observe a rural school soccer match, under teacher direction, between about 14 boys, all playing barefooted. Even more startling, at the conclusion of the game, there were no injuries—especially to the feet. They train their youths to be rugged in New Zealand.

There is no mistake about arriving in the vicinity of Rotorua as the pungent odor of sulphur permeates the air. Jets of steam, in the form of water vapor, shoot from pipes sunk into the ground, and geysers rush skywards—in the town itself, along the streets, and in public parks.

The whole area takes on an aura of enchantment, mystery and wonderment ... and Rotorua is geared for tourist trade as its streets are laced with numerous hotels, motels, restaurants and boutiques.

The area of Rotorua is considered the center of a large geographical thermal zone which reaches hundreds of miles.

On one chilly day several native (Maori) children were warming their hands over a street thermal pipe which emitted steam. One of the young boys was wearing a cap with a patch, which I instantly recognized. When asked what the badge stood for he said he did not know. He too received a history lesson, he was wearing an emblem of the battle flag of the Confederacy.

The original natives of New Zealand, of Polynesian origin, are the Maoris, and settled in the country about 1100 A.D. The first permanent English settlement occurred in 1840. The influence of the Maoris is present everywhere as such native names as Onehunga, Waikowhai, Otahuhu, Papatoetoe, and Owairaka are seen frequently.

The first permanent English settlement occurred in 1840. The population of the entire country now is only about 3,500,000 with Maoris numbering about 300,000. **Carroll Garnett**

A New Zealand Social Hall

Entertainment in this social hall in New Zealand was provided by the Navajo Indians (the code talkers) in a ceremonial dance. CWO Thomas O. Kelly is fourth from right of the men, up against the wall. Colonel I. Robert Kriendler (with hand to face) is between two women center. Ladies are Red Cross Nurses.

The Driver's Test... And A Gentleman Of The Highway

Orders had been received from 3rd Marine Division headquarters that everyone driving a Marine Corps motor vehicle must be issued a special USMC Driver's License, following a familiarization and testing on New Zealand's traffic regulations and customs.

The responsibility for carrying out these orders had then filtered down to the 12th Marine's Motor Transport Dispatcher, Staff Sergeant William A. Poncavage.

Poncavage was of Russian descent and spoke with a bit of an accent. He was an experienced Marine, having served in China prior to World War II with the Shangai Legation Guard and the Yangtze Patrol. In 1939, he was light heavyweight boxing champ of the Pacific Fleet. While he was a big, jovial fellow, well-liked by all who knew him, Poncavage was a dedicated career Marine, and he carried out his duties to the letter.

On this particular morning, Dick Hannon and I were waiting beside our assigned radio jeep when Sgt. Poncavage approached with a clipboard under his arm.

"Good morning, 'Poncie,'" said Hannon. "How ya doing?" "Well, if it isn't PFC Hannon and PFC Kerins. Dis morning, gentlemen, I am Staff Sergeant Poncavage, and I am here to check you out on de operation of dis Marine Corps vehicle and de traffic rules of New Zealand. Kerins, get your ass in de back seat. Hannon, get behind de wheel and do a right turn at de front gate," he ordered, climbing in on the passenger's side.

As per instructions, Hannon drove to the main entrance, and after checking out of camp with the gate guard, he turned out onto the road. Driving along in the left-hand lane, as is customary in New Zealand, he answered various questions as the sergeant checked them off on his clipboard.

The road was bordered by stone hedges as it meandered over the rolling countryside. An elderly New Zealand farmer waved to us as he tended his sheep, and Hannon returned the gesture.

"Gotdammit, Hannon," roared Poncavage, "keep your friggin' hands on de wheel and your eyes on de road. Don't you realize dat you're drivin' a vehicle belonging to de Marine Corps and purchased by de taxpayers of de United States? And don't you know dat you're drivin' dat vehicle on foreign soil and dat your gotdam sky-larkin' could cause a friggin' accident and create an international incident?"

Poncavage's ranting continued, suggesting, among other things, that such an incompetent bonehead should be assigned to stable duty with the Horse Marines, and never allowed to drive again. Farther down the road, he ordered Dick to pull over and to get into the back seat where he belonged.

"Kerins, get your ass up here behind de wheel. Turn dis vehicle around and take us back to camp."

As he had done with Hannon, the sergeant queried me on New Zealand's traffic regulations and on the operation of the jeep. Everything was going fine until we reached the site where Dick had committed his unpardonable sin. The same farmer was tending the same sheep, and once again, he waved to us.

"No way I am going to get chewed out," I thought to myself. I kept my hands glued to the steering wheel and my eyes on the road ahead. When we had passed the farmer, I slipped a quick glance at Poncavage to see if he approved of the way I was concentrating on my driving. It was difficult to tell, for he was staring intently at me with a frown on his face.

"Gotdammit, Kerins," he bellowed, "Didn't you see dat man wave to you? Don't you realize dat you're a member of de United States Marine Corps and a representative of your country on for-

eign soil? You're supposed to be an ambassador of goodwill, and you should also be a gentleman of de friggin' highway."

Once again, Poncavage carried on a lengthy dissertation, referring to the fact that it was beyond his understanding how two such irresponsible idiots as Hannon and Kerins could have been accepted in the Marine Corps in the first place.

When we arrived at camp, we parked the jeep and stood beside it, waiting for Sergeant Poncavage's final comments.

"De friggin' moron dat recommended you two for de rank of PFC ought to be court-martialed," he said. "You are now dismissed, and if you should happen to have liberty dis evening, you can buy me a beer down at de local slopchute."

The following day, Hannon and I were summoned to the first Sergeant's office and handed our Marine Corps driver's licenses. They were signed by our regimental commander, Colonel John Bushrod Wilson, and countersigned by Staff Sergeant William A. Poncavage, Regimental Motor Transport Dispatcher. **Jack Kerins, Terre Haute, IN**

This isn't the road that PFC Hannon and PFC Kerins drove for their road test, but these are 3rd Marine Division personnel on Maneuvers in New Zealand in terrain that's almost mean enough to be the real thing.

How The "Doc" Sees It ... From Pendleton to Bougainville

Horace L. Wolf Sr., Captain (Medical Corps) USNR (Ret.) was battalion surgeon with the 21st Marines ... from Camp Pendleton until after the Bougainville Campaign. Here's the other side of atabrine, sulfa drugs, and short arms.

It is difficult for me to separate incidents with the 6th Marines at Camp Pendleton (after Iceland) from those later with the 21st Marines. However, one recalls the bitterly cold winter nights there when we were camped out. Yet, by noon, even on winter days, it could be sweltering hot.

A flash flood, after a heavy rain, at Las Pulgas Canyon (Canyon of the Fleas) cost our battalion a lot of lost gear.

We sadly learned in obstacle course training that a nitro starch block (simulating a hand grenade explosion) could cause very serious injuries if it exploded next to a man ... two serious casualties on that maneuver.

Time for Shots

In giving preventive vaccines before overseas departure, we were very lucky that the yellow fever vaccine issued us contained no contaminating hepatitis virus. Army studies, continuing after WWII, proved that similar yellow fever vaccines given to some Army units were carrying the hepatitis virus.

On board the *Mormacport*, several thousand Marines endured 17 days zigzagging from San Diego to Auckland, New Zealand. Even isolated on the Pacific it is difficult to complete the necessary "shots" for each Marine in the battalion. Some hide out from the corpsmen. Meals were served continuously on the ship, two meals, morning and evening.

One Marine came down with meningitis, putting the whole ship at risk. Sulfa drug is the only antibiotic available. We treat him intensively, and put all the troops on sulfa prophylaxis. He recovers and there are no more meningitis cases. We inspect our medical supplies. Even though there were guards on the holds aboard the *Mormacport*, the tins of ethyl alcohol we brought from San Diego are now missing. Field mice have chewed up our beautiful white Navy blankets to make nests in the middle of the packs.

I drive our jeep ambulance to Auckland on trips to see our sick and injured that we evacuated to the large Navy Hospital there. Passing other cars from the left side of the road while seated in a left-sided driver's seat is dangerous ... and making 90-degree turns in downtown Auckland is awkward.

In Auckland I am continuously asked for cigarettes by the New Zealand citizens. It seems that tobacco is a wartime scarcity for them. Food is good in Auckland restaurants ... but meat cuts differ from ours, and there are dishes such as steak and eggs, steak and tomatoes.

The nearest town to our base camp is Warkworth. On one occasion a pickup musical band from our battalion (I play a flute brought along from the States) plays a concert from a bandstand there, for a local celebration.

I'm invited to have tea with a New Zealand farm family and enjoyed a rich pastry called "trifle" cake—a cake with fruit and whipped cream. I also used to bring back cakes to camp from Auckland, and when we sailed for Guadalcanal I weighed 180 pounds—the most ever in my life.

The battalion had some wonderful hikes on North Island. But my assistant battalion surgeon's flat feet could not hold up, and we had to transfer him out.

New Zealanders had a high incidence of dental problems. At intermission time in Auckland movie houses, the ads often were for dentures.

A "Doctored" Sample

My new assistant battalion surgeon turned out to have quite a teasing sense of humor. I learned I had been promoted to lieutenant commander from lieutenant while on the troop transport from Auckland to Guadalcanal. This required a physical exam. My new assistant "doctored" my urine sample to show four plus albumen, and announced to all that I had failed my physical.

On another occasion—a division training maneuver at Guadalcanal—General Turnage inspected our battalion aid station. My assistant battalion surgeon had to be persuaded by me not to "capture" the general because he was not wearing the temporary insignia of our side in the practice combat maneuver.

I recall peering over the edge of my foxhole at our base at Guadalcanal to watch in the distance the nightly bombings by Japanese of the shipping at Lunga Point. Searchlights ... anti-aircraft gunnery explosions ... fires aboard damaged ships.

Malaria prevention—doses of atabrine—made many of us look yellow. Netting over one's cot, freon bombs. We tried them all, yet we all became infected with the parasites ... and in the stress of combat later, many of us came down with chills and fever.

Regimental Adjutant Kriendler initiated Friday night services for those of us in the regiment who were Jewish.

Up In The Air

I learned that a bottle of whiskey could get one an airplane ride at Henderson Field. So on a Saturday afternoon, when off-duty—I enjoyed the first flight of my life (the cool air on high was a great relief from the sweltering heat on Guadalcanal) on a B-25.

When Eleanor Roosevelt visited Guadalcanal, her DC-3 (with three fighter planes on each side as escorts) flew in right over our 1st Bn. 21st Marines base camp.

Available water at our camp on Guadalcanal was brackish. Washing one's clothes in it left them stinking. A better way was to swim in the nearby ocean and tow the clothing along with individual items tied together. I was always a bit scared when ocean swimming at the "Canal;" memories of a banded sea snake (probably poisonous) seen by several of us from our transport when we first anchored offshore after the trip from New Zealand via New Caledonia.

Killing fish with hand grenades in a river near our base camp became a sort of "fishing" for some Marines. I was called to pronounce dead one man who had "frozen" a grenade in his hand.

Intro to Tooba

I think many of us were glad to leave Guadalcanal. It had looked like a tropical paradise on first arrival ... with Lever Brothers coconut plantations ... coconut milk to drink, and coconut meat to eat. But we soon tired of these sources of nourishment, although some clever guys learned how to ferment coconut tree sap into a more potent beverage (Tooba or Tuba).

Off to Bougainville on an APD (troop carrying destroyer). I enjoyed a salt water shower en route.

First night on the beach, we were bombed. Fear taught me that in my foxhole the metal part of my helmet made an emergency "potty."

We set up a first aid station on the trail leading to the perimeter (front lines). A booby trap was found nearby and explosive experts defused it. We treated a lot of "jungle rot" cases (fungus skin infections).

Along the trail one day came H.V. Kaltenborn (world famous radio newsman) and Gene Tunney (former champion boxer, in charge of Navy athletics programs).

Money was of little use on Bougainville, but one day I bought a loaf of fresh bread from a SeaBee for a $10 bill. It tasted like cake, compared with the rations we had been on for weeks.

I went out with a patrol to bring in a wounded Japanese. As was true of medical personnel on Bougainville, I wore no red cross arm band, and was armed with a carbine and ammunition. We found a small, unconscious Japanese. The squad wanted to kill him, then and there. I persuaded them that if our medical and intelligence people could take care of him, he might yield information that would save American lives. So with a makeshift litter, and a unit of plasma dripping into a vein, we brought him back to our unit.

I came down with fever and chills, plus severe diarrhea, and had to turn in to our division field hospital—but was back on duty after 24 hours chock-full of quinine and paregoric.

In the combat area of Hellzappoppin Ridge we replaced a Marine paratroop battalion. My medical section took over their battalion aid station. Several of my brave corpsmen were wounded in this action. The regimental band musicians made brave litter bearers.

I still remember the terrible odor of our dead in the tropical heat. The smell pinched one's nostrils and clung to clothing. During combat in the swamps, about all one could do to try and purify water to drink was to put two drops of iodine solution in a canteen.

Night was the worst, when we could not evacuate our sick or wounded. For the latter ... morphine, plasma, dressing changes, boosting of morale were all we could offer until daybreak came and we could use our jeep ambulance. It had room for two litters, two seated ill or wounded, and a driver.

If one could ride to the air strip on the jeep ambulance to put sick or wounded on evacuation planes, one could see a female (Navy or Army nurses) for the first time in many months. **Horace L. Wolf Sr., Amarillo, TX**

The bandstand in Warkworth, New Zealand was a familiar sight to 3rd Marine Division Marines and other servicemen stationed in the area. Complete with a memorial plaque on it to Queen Victoria and to King Edward VII, this is probably the same bandstand Dr. Horace L. Wolf Sr., attached to the 21st Marines, played his flute in a pickup band for a local celebration. (Photo Courtesy of George J. Green, CWO)

"Pants Pressed While You Wait"

Lloyd T. Betzler of Colorado Springs, Colorado—a veteran of A Company, 9th Marines during World War II—ran across a 1943 issue of *Leatherneck Magazine* recently in which Fred Beldkamp had written an article, *The Marines Down Under*.

It told how Marines had adapted to New Zealanders, and how New Zealanders learned to understand Marines. One anecdote provides a humorous example, in which Beldkamp wrote ...

"... Soon he (the Marine) is thoroughly converted, and isn't surprised to find tea being served at almost any time and place.

"While waiting for my trousers to be pressed, I recently shared a dressing-room in a tailor's shop with a newly-arrived Marine on a similar mission. He had just handed his trousers through the curtain and was standing uncertainly in his shirt, shorts, socks and shoes when a man's face appeared between the drapes, stared directly at the new customer, and demanded; "Cream and sugar?"

"The Marine finally mumbled "What?," and after the question had been repeated I explained that tea and cookies were about to be served (at no extra charge, by the way.)

"Once over his amazement, he settled down to enjoy the cookies and tea, which he had little trouble balancing on his bare knee. The man who did the pressing came in and joined us, doubtless out of a desire to be sociable with the visiting Americans, and chatted along for about 20 minutes on New Zealand's history, landmarks, and colorful birds. Finally, pleading pressing business, he took his leave.

"The signs in the tailor's windows "Pants Pressed While you Wait" are followed to the letter, with particular attention to the last word: but the cookies and tea have a remarkable soothing effect."
Lloyd T. Betzler, Colorado Springs, CO

The Indomitable T.O. Kelly Or Don't Fool With T.O

CWO T.O. Kelly, United States Marine Corps was no man to fool with, especially as Sergeant Major of the 3rd Marine Division. Officers and men from bird colonels to privates first class would hear in one way or another if they stepped out of line.

Here are a couple of Tom's recollections. He lives now in La Plata, Missouri.

"The headquarters was moving into a house which we were to use as my office. We were waiting for our gear. You may remember Bill Marsh (a full ancestral Iroquois Indian) one of my clerks.

"Bill was assigned his office space and was quietly sitting around awaiting his field desk etc. when in walked a busy-body officer. He said to Marsh, "Why aren't you at work?" Marsh had black eyes, black hair, a really good man.

"Suddenly, Marsh threw up his hands and said "How!" in his best Mohawk, got up out of his chair and gave a good imitation of an Indian chief doing a war dance.

"The busy-body clanged into my office to tell me that the Indian in my office had gone crazy ... major so and so, I said, you'd better get out of here before he takes his bow and arrow and holds target practice on your rear end. No more interference from that quarter."

Actually, not long after on Bougainville, the Indians of the 3rd were particularly valuable as scouts and as Indian talkers ... without any coded transmissions they spoke by radio in their own tongue to offer the Japanese a cryptic challenge they wouldn't have broken until this day.

In my office in Waikaraka, New Zealand, the adjutant and sergeant major rated a chair each.

One busy morning I had to see the commanding officer, Snuffy E.O. Ames. When I returned to my office I found all of Bob Kriendler's (Major Irving R.) buddies crowded around my field desk having a field day going through my papers. Malcolm Beyer was holding the group spellbound with a long lie. I walked to the desk and no one noticed me. With one hand I swept the desk clean. With the other hand I scattered the papers I had just brought in from the CO's office.

Paper all over the floor, I sat down, crossed my legs and started to assemble the mess so I could work on them.

It was then that Beyer arose from my chair to say ... "Oh, sergeant major, was I sitting in your chair?"

"No, major," I said. "I always work on the floor." I was almost crushed in the rush for the door. No one ever got close to chair or desk after that.

Liberty

Here it was liberty ... a nearby town ... and the nearest home away from home for 21st Marines, and a few other regiments in New Zealand. Warkworth is 60 miles from Auckland. Some of its young ladies are now and have long-been Yankee wives. (Photo Courtesy of James M. Galbraith)

LURLINE
... A Luxury Liner

One of the nation's busiest troopships, the SS Lurline carried much of the 3rd Marine Division to its first overseas post, Auckland, New Zealand. During 1942-1944 the Lurline made 24 trooper voyages carrying GI's to numerous ports in the Pacific. In June 1945 she picked up troops in Europe and deployed them to the Pacific. After World War II, the Lurline returned thousands of troops stateside.

A luxury ship, the Lurline even had a swimming pool, but it was used to stow gear when carrying troops. During wartime service the Lurline could carry up to 4,037 troop passengers under somewhat crowded conditions. As the peacetime Lurline, however, she accommodated 750 paying passengers.

Built at Quincy, Massachusetts by Bethlehem Shipbuilding Corp., the ship was operated by the Matson Navigation Company before, during, and after the war. In 1963 she was sold to a Greek shipping company, renamed SS Ellinis, and for several years carried emigrants from Greece to Australia.

'Twas In The Cards On The SS Lurline

She was the queen of the Matson Line, premier passenger service in the Pacific, recalls Austin Gattis. She must have catered to a highly-moneyed class of passengers. The first class dining room was tremendous, ten times the size of the second class dining salon and there were no other classes."

That magnificent first class salon in its conversion was stripped of tables, chairs, carpet, etc. and in place of the finery, the floor was tiled and stand-up tables were installed. That was the dining area for troops.

Officers ate in the second class salon which had been left pretty much intact. They enjoyed the luxury of sit-down tables for four, linen cloths and napkins, china, glassware, and flatware. We could hardly appreciate the fact that this would be the last such civilized dining we would enjoy for many months.

Our destination was Samoa, a distance of some 3,200 miles and, as the *Lurline* cruised at about 20 knots, it was considered safer for her to sail alone than in a 10-knot convoy ...

One day a Marine from my platoon asked permission to see me ... "At ease, son, what can I do for you?"

"Would you hold some money for me ... I'd hate to have it stolen ..."

"I didn't expect to see the huge roll

he produced from inside his dungarees. It could have choked a horse—$2,500—a lot of money in 1943.

And the young Marine added ... "There's no trick to shuffling or dealing that I don't know, there's no way I can lose at cards. Oh, sir, I won't play with our fellows, even though they beg me to. They believe they could catch me cheat-

ing. But, no sir, I'm not going to cheat our fellows."

The young anachronism from the gilt and glitter of Mississippi faro tables had plenty of room to keep his word. His platoon was not 1/100th of the 3,500 troops on the Lurline ... and most of them thought they could beat him ... Barnum or not. **Austin Gattis, Washington, DC**

Although mutton became a much-maligned item of food as we traversed the islands of the Pacific, sheep were a mainstay of the New Zealand economy ... for their meat and their fleece.

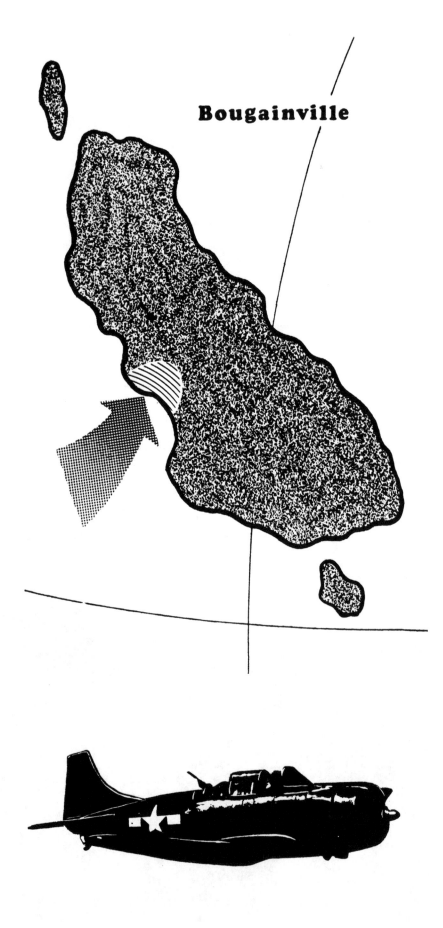

Bougainville

Bougainville Island Of Co-operation

Bougainville taught the Marines a lesson in survival, how to live in swamps, jungle, mud and rain with little more than K-rations, some aspirin and atabrine tablets. The omnipresent Jap was there, too, and he was proud of being able to fight in these miasmic surroundings. Here, dry socks and dry underwear were luxuries, "The best deal was to sleep with a buddy and share ponchos—one above and one below. If you could keep the water out, you had it made."

On rare occasions when the combat situation and the bivouac areas allowed, Marines got together over stews of C-rations, bouillon soup and tomato juice heated in helmet over a fire. In terms of weapons, Marines of the 3rd Division learned something of the tools of their trade, 60mm mortars could be registered within 25 yards of their positions; 81mm mortars and 75mm pack howitzers within 50 yards; and 105mm howitzers within 150 yards. But BAR and riflemen had to get within 10 yards, so well concealed and bunkered was the enemy.

The battle for Bougainville continued through to the hill mass between Piva and Torokina, when Marine Air contributed to the success of the mission by providing close support.

On December 10, the 2nd Battalion, 21st Marines, took up positions on Hill 600 to the right flank of what came to be known as Hellzapoppin Ridge, between Cibik's Ridge and the Torokina River. Here the Japanese elected to make their last stand until December 18, and it was a good one. Artillery and mortar were unable to dislodge the enemy, and it was not until Major A.C. Robertson of VTMB-34 led an attack of six torpedo-bombers to within "yards" of the front lines that the Japanese were beaten back, 48 100-pound bombs crashed into enemy positions within 75 yards of the 21st Marines. The second strike was the crusher, and it was at this juncture that the 1st Battalion, with bayonets and grenades, charged-and Hellzapoppin was American

Talks Of Torpedoes And Whistling Bombs

This succinct recollection of leaving Tetere Beach, coming ashore at Empress Augusta Bay, Jap air attacks, life around a foxhole and artillery on Bougainville, seems as personal and casual as a few pages from a diary. It was written by **George J. Green**, Fox Battery, 12th Marines, St. Louis, Missouri.

"At Tetere Beach, Guadalcanal, We had loaded our firing battery, guns, and equipment aboard LST 341 ... The three-day trip was uneventful and we all enjoyed the pleasure of sleeping on soft bunks, some fresh food and rest. The only discomfort was the heat below in the troop compartments.

"(November 17, 1943) ... Reveille and general quarters were sounded well before daylight with sufficient time for the Navy to serve us the last hot meal, steak and eggs. We then packed all our gear and made ready to go ashore ... All of the personnel were to remain below decks until the LST beached ... The only thing between the troop compartment and the sea was a piece of 3/8 inch steel. It didn't take me long to reason that if we were torpedoed or hit a mine, I would be better off topside in one of the vehicles with a machine gun. At the same time, general quarters was in effect and Jap aircraft were in the vicinity.

"Just as I stepped between the two blackout curtains of the exits to the main deck, all hell broke loose with the 50-caliber machine guns and Navy anti-aircraft firing. It scared the daylights out of me; and before I knew it, I quickly found a position on a truck as an assistant.

"All of the LST's and LPD's were heading into the beach along parallel lines and you could just make out the LST on our starboard and the USS *McKean* on the port side as the darkness was turning into dawn.

"You could make out the island mass off the bow along with the tropical aroma. We could hear Jap planes in the air, and all of a sudden from the starboard side and in front of the bow, a Jap torpedo bomber flying about 500 feet appeared.

"Tracers lit up the whole sky around the plane. He was close enough that you could see plainly the pilot and the rear crew member moving. Shortly after leaving the range of our guns, he dropped his torpedo and all of a sudden you could see a bright flash and a red spot in the sea. The torpedo had hit the USS *McKean* and eventually it sank. We had about 20 Fox Battery personnel aboard ... Somehow, all of our men who were below decks when the torpedo hit, were able to get off except Private First Class Rogers, who was last seen topside."

Green and people of Fox Battery spent the next day clearing jungle for positions and field of fire, and by nightfall he still had not had time to dig a foxhole. More accommodating circumstances allowed it hazardously after dark. Still he was better off than most of the 3rd Battalion. Twenty-one crowded into a swamp where they couldn't dig foxholes at all. For protection the 3rd Battalion employed the surplus logs the 12th discarded. Green recalls one break, the jungle had so many hooks you could hang up everything you owned.

"After dark sitting around the foxhole drinking coffee and listening to the jungle sounds, we tried to get a good night's sleep. I just closed my eyes when the sirens started wailing. Condition RED for air attack! I think this was as terrifying as anything else I had experienced so far.

"Searchlights picked up a group of three bombers in V formation ... All our (AA) shells were bursting behind the Japs. It wasn't long before we heard the bombs singing their way down ... and (we) got deep as we could into our foxholes. A series of explosions went through our positions, with one bomb landing just outside our No. 1 section. This caused a tremendous flash fire (igniting powder charges) ... No one was hit as everyone was well dug in."

Semper Fi ... All Around

About a week or so after landing on Bougainville, 2nd Lieutenant Tom Pottinger joined our battery (F-12th Marines), as a forward observer. It was almost dark by the time he arrived, too late for him to go forward.

Captain Clement Newbold, battery commander, was at the 21st Marines ... and we didn't expect him back that night. Captain Newbold was rather unusual and demanding and we prepared a nice foxhole for his use ... I told the new Lieutenant to take it which he readily accepted ... not expecting Captain Newbold to return. But the (demanding) Captain Newbold returned and found the lieutenant in his foxhole. He did not appreciate this. He was then rather hard on Lieutenant Pottinger, not knowing I was the one who told the lieutenant to use the captain's foxhole.

Bougainville ... Next Stop

Looking out over the waist-high chain rail of the American Legion ... across the ever-baby-blue waters at the boringly slow pace of the eight other transports ... you grasped ... despite the shipboard indolence ... that your first firefight was dead ahead.

All you had been told of bunching up, lunging a bayonet, huggin' the ground and "snoopin' and poopin'" was about to converge into reality. The scene of your first battle star was 130 miles long, and was stocked with 40,000 Japanese—experienced veterans most, some who had been in on the infamous "Rape of Nanking."

In the big picture, you were never more important. Chief of Naval Operations King saw you as critical. Pacific Fleet Commander Chester W. Nimitz and General Douglas MacArthur, Supreme Allied commander, probably never would agree more, ... Bougainville, or some of it, had to be wrested from the organized Japanese enemy. Bougainville protected Rabaul, 225 miles away, and it sat athwart the Australian-American lifeline. A secure and dangerous Rabaul hovered like an avenging angel over any idea, plan, quest or thought of any Western action in the South Pacific.

The Japanese knew it too.

Rear Admiral Injuin, commander of Destroyer Division Three at Rabaul, said openly, "If Bougainville falls, Japan will topple."

Between the day-long chow lines you "learned" unofficially, as well as in rare briefings, about the tough, seasoned veterans of Japan's 17th army who didn't like guests. You did not learn as your mother and sister were told at home that all Japs had buck teeth, glasses, and carried bloody daggers—but were durable dangerous enemies, beyond brave. At Torokina you learned the scuttlebutt was right on the money.

Lieutenant General Haruyoshi Hyakutake, who commanded Bougainville's defenses, knew as others there and in Japan that the axe would fall on Bougainville. He had no idea where ... certainly not in that hellish rainforest of Empress Augusta Bay.

(Continued on page 43)

"He's Not Heavy ... He's My Brother"

When I turned eighteen in September 1942, my brother informed my father that he and I wanted to join the Marines Corps ... We boarded a train in Manchester, New Hampshire and two days later we arrived in the unforgettable Parris Island, South Carolina ... Well, needless to say, I was scared to death and learned to keep my mouth shut.

...There were tough times also. We did all the things Marines do, like seeing if we could buy a beer; trying "Sugar Hill," a house of ill repute and at my tender age, I was so scared that I ran out of the place. Maybe not typical Marine, but I wasn't alone ...

A lot of us had jobs assigned aboard ship ... USS *Rixie*, to New Zealand ... I was in the laundry ... and put bleach instead of starch in with my khakis and they turned out a light shade of lavender ... I was relieved of that job ... Officers didn't have much sense of humor.

American Samoa was beautiful ... and we left for New Zealand ... I'll never forget the homemade bread at chow in the morning ... Sometimes we would bring hamburgers home from Queen Street, and heat them in a coke stove in our tent ...

Then on Bougainville (November 1), we were all apprehensive, all scared ... We had a few encounters ... every few days ... Once I saw three coming down a path, then I saw them after they were dead of artillery fire ... You could smell them because of their filthy foxholes ...

On November 24 the 3rd Battalion and L Company ran into hand grenades ... We were on the left flank and a Jap stood up and when I saw him I fired ... At that time a blast went off to my right and I fell.

As I looked around I saw my brother lying there and I turned him over, but he was gone. I cannot describe the feeling. I was led back to an area where the wounded were being treated and my brother was brought and laid beside me. I was lost ... We lost a lot of people that day ..." **Howard E. Long, Northport, FL**

Serviceable -Not Fancy- Footgear

A bunch of us were visiting on Bougainville in a little spot of sun. You remember how we would get congregating in a little patch of sun. Upon that Hill One Thousand there wasn't much sun to speak of.

This stranger came up, put his foot on one of the rocks where we were sitting, rested a few minutes. One of the guys said, "I'll sure be glad when I get back to where I can buy a decent pair of shoes. These old Marine Corps shoes aren't any good" ... (or something like that).

This man finally said, "Well, when I was a damn second lieutenant I thought the same thing, but as I've got older, I realize these Marine Corps shoes are the best you can get." He just took his foot off the rock, and went on about his way.

Somebody said, "Who the hell is that?" It turned out he was our new battalion commander (Lieutenant Colonel Carey Randall). We learned a lot about him after that, and it was all good. **Jay Strode, A Company, 9th Marines**

INTRODUCTION TO COMBAT

My name is Edward N. Bieri, Alliance, Ohio and on that morning November 1, 1943, I was in the same Higgins boat as our 1st Battalion commander, Major Leonard Mason, nickname "Spike," serving as his radio operator.

Shortly after running off the Higgins boat, Major Mason had been hit, I believe in both legs below the knees by machine gun fire. Although following close behind him, I somehow didn't get hit.

As I recall, the first few hours were pretty hectic due to rifle companies A and B landing further to the left than planned and C Company and headquarters company landing in their place.

I remember being close by Major Mason prior to his being evacuated and his stating how long he had trained himself and his battalion for combat and now he was leaving before the first day was over.

I next reported to Major Steve Brody (John P.) or Major Bailey (Charles J. Jr.), I forget which, was now in command of the 1st Battalion.

Just after this, I kept hearing someone hollering a language I didn't understand and finally decided it was Japanese. I asked one of my buddies how that Jap could still be so close without getting shot, and I was told the voice was (of) Captain Gordon Warner who could speak Japanese and was yelling goofy orders to confuse any Japanese troops in hearing range.

(It was later verified that the 'goofy' orders of Captain Warner did indeed confuse a few of the defenders).

I managed to survive the Japs and the jungle and was happy to leave Bougainville on Christmas Day (1943) when our battalion, 3rd Marine Regiment, returned to Guadalcanal.

Several years following discharge from United States Marine Corps at Parris Island as a MTSgt (clerical line),

my family and I visited General A.H. and "Miss Hannah" Turnage in Alexandria, VA.

Having served the general as chief clerk from New River Training Center to Guam (3rd Marine Division, reinforced), the general shared some fond reminiscences.

"Hancock," he related. "Remember the time Halsey (Admiral William "Bull" Halsey) flew up from New Caledonia (HQ, 1st PhibCorps) to Bougainville. He and I went into the blackout tent for a conference."

Continued the general, "I never told you. But I bet you men thought our conference in the blackout tent was on high-level strategy."

With a twinkle in those wonderful North Carolina (Fayetteville) eyes, the general continued, "Actually, Halsey figured my supply was running a little low. He had brought me two bottles of bourbon." **Brent I. Hancock, Hershey, PA**

Keep
it dry

Our War ... Bougainville

Our war was a little war. No one remembers the anonymous battles on the anonymous island except a very few old men whose ranks are rapidly declining. Even those who served there remember little about it; most saw no combat there. We were the unfortunate victims of chance. We were in the wrong place at the right time to die.

Our war was a little war. When we fought the nation was preparing for war in Europe. That was the primary objective of Roosevelt and Churchill, the defeat of the madman Hitler. The Japs would come later.

Our war was a little war so we were given obsolete arms, obsolete equipment and obsolete food. I will never forget the day we received Hardtack. Hardtack! The handful of Doughboys who still survive remember it well. Fellows, there was some you didn't use up. We got it in our war. We had not eaten in two days due to a particularly dirty patrol and the Raiders brought us Hardtack.

Our war was a little war and was not spectacular. There were no pictures of massive warships blasting enemy emplacements before we stormed the beaches. There was no vast armada of carriers and battleships in our war. We often traveled alone, with no protecting warships. A well placed Japanese sub could have leisurely blasted the whole regiment with a few torpedoes during our wanderings.

Our war was a little war so we often went hungry, and the food, when we could get it, was too often spoiled. I remember carrying a case of C rations to the company and after an ardous trip from the beach, I pried open the lid only to find the cans swollen and their contents stinking. We went hungry that day.

Our war was a little war. We did not have the glamour of being the first to fight, like the war on Guadalcanal. We did not have the terrible casualties like at Iwo and Tarawa and Pelileu and D-Day and the Battle of the Bulge. But men died just as finally in our war as they did elsewhere.

Our war was a little war and did not get the publicity of the big wars. There was never a spectacular picture taken in our war. No flag raising as at Iwo Jima. No picture of bodies floating obscenely in the surf as at Tarawa. No pictures of terrified civilians leaping to their deaths as at Okinawa. The picture I have in my mind is of a Marine, sitting on a tree stump, forlornly picking maggots from his poorly bandaged open leg wound.

Our war was a small war with a terrible loss of men to disease and infection. I saw men drink filthy water and once saw men trying to eat a bird they had killed. Raw. Of our company only five men did not die or get hospitalized for wounds or disease on the anonymous island.

But our war gave us each other. I am proud to have served with such men; some weak, some strong. My memories of our war are fading now. But the memories of my beloved comrades remain and grows stronger as I near my final muster. **Douglas G. Lyvere**

Douglas Lyvere served with Love Company, 3rd Regiment, was wounded and received the Bronze Star. He is a retired businessman in Hollister, California which gives him time for golf.

The Jap Warships

Lieutenant Colonel Victor H. Krulak's (2nd Parachute Battalion of the 1st Parachute Regiment) noisy diversionary raid at Choiseul (where the Japanese were shot up just to get their attention) and the occupancy of the Treasuries by the 8th New Zealand Brigade under Brigadier R.A. Row, gave the Japanese no clues.

So when American rockets emblazoned the black hills beyond Empress Augusta Bay, scene of the landing, and when United States dive bombers chewed up the dismal mud, the Japanese Admiralty was sure where and when the transports and Task Force 31 (under Rear Admiral Theodore S. Wilkinson) had arrived.

Immediately, the Japanese command at Rabaul dispatched, post-haste, a covey of cruisers and destroyers to mince the beachhead. Right behind the warships came six destroyer transports with more than enough manpower to recover any lost real estate.

It was a sunny November 1. We had been up before 0400, allowing time for chow and prayers, and more than enough time to stand and wait—leaning with full pack and M-1 on the rail to stare at the pointed hills. A feather of steam waffled over Bagana, the nearest volcano (Balbi is the other). Suddenly you were descending those infernal cargo net ladders with pack and piece while your legs groped for a landing craft that was either rushing up to crush you, or dropping away from your feet.

First troops were out of the transports President Jackson and President Adams. They boarded a narrow strand where the jungle came right to the water. It was one of the 12 assault beaches on a 7,500 yard front.

With 1/3 (Major Leonard M. Mason) thoroughly engaged on Torokina Point and an island by the same name, the 2nd Raiders (Lieutenant Colonel Joseph P. McCaffery) and 2/3 (Lieutenant Colonel Hector de Zayas) pretty equally involved, a toehold remained unsure. The 2/3 (Lieutenant Colonel Ralph M. King) was toward the center, close to the 9th Marines. They were up against the 1st Bn., 23rd Japanese Infantry (17th Division) under firebrand Lieutenant Harawa Ichikawa housed in 25 pillboxes and supporting trenches. The job was no easier for the 3rd Raiders (Lieutenant Colonel Fred D. Beans) against a stubborn platoon of the same 23rd on Puruata Island which held out until 1800 the next day.

(Continued on page 45)

Tears On Turkey Day

We considered ourselves a salty bunch—we in H Battery, 3/12, a Pack Howitzer Battalion—veterans of Samoa, New Zealand, and Guadalcanal ... all training areas. But the war really came home to us on November 24, 1963, Thanksgiving Day, at the battle of Piva Forks.

Our battery was in support of the 1st Battalion, 3rd Marines on the left flank ... that morning, on the 24th, our batteries had saturated the Nip lines with the heaviest fire of the campaign. The forward observers believed that the opposition had been obliterated.

The attack was made through dense jungle over 800 yards wide. At the time of the jump-off we were met by absolute silence which reinforced our thoughts that this would be a careful "cakewalk." However, within half an hour, the Nips had reorganized and hit the first and third battalions with heavy counterbattery fire.

At first we assumed that our own rounds were falling short. It was during this carnage that Private First Class Paul Parks, Private First Class John Peters and 1st Lieutenant Stanley Pollard from our battery (as a forward observer party) were mortally wounded.

When you are 18, you really believe that you are immortal; and probably because of this—when later in the day of the 24th, our battery was served real turkey dinner in honor of the holiday, there were very few dry eyes, nor much conversation in any of the sections—word had spread so fast.

I can still recall the sadness of the faces as we sat around eating that turkey. In particular, I recalled Paul Parks and how on our first night of arrival when we joined the battery on American Samoa, he allayed all our fears and told us what an excellent group of officers and men we had the great luck to be part of.

I was a member of other outfits that were to lose fine Marines ... but those three were the first, and we shall never forget them as long as any of us remain alive. **Frank W. Nunan, Lansdowne, PA**

"Was He Ticked!"
Anxious Rifleman Shoots Air Cover Pilot In Leg

I remember going back to the beach to fill the canteens late ... on the first day ... Only one ship was to be seen. It probably had run aground. The captain of our ship said we were to land in uncharted waters. All of the others had pulled out to sea (air raids). Higgins boats as far as we could see were beached in the high surf. What a sickening feeling!

There were big dogfights by planes on the 2nd and 3rd days. We were on an outpost about 200 yards in front of the lines between a swamp and the sea. A Jap Zero strafed the beach with an American plane on his tail. Someone on the beach shot at the Jap but instead hit the pilot of the American plane in the leg. He came back and landed in the ocean near us. We waded out and got him out of the plane. He was ticked off.

I was in D Company attached to B Company ... at this time. Later we were closer to the Piva trail. Closing the lines late one afternoon, we crossed a small river and walked into an ambush. Luckily, no one was killed but a couple were wounded. We pulled back across the river and spent a miserable night.

The next morning we moved up, but the Japs had moved out. We found ten dead Marines, a patrol that had been ambushed a few days earlier. We buried them. I often wondered if they were ever found.

After we got up on the mountain, we saw beautiful sunsets across the jungle almost every day. The smoke drifting down the sides of the volcano would change colors in the late afternoon sun.

We saw some planes coming in from a patrol over the lines. Later we learned it was Pappy Boyington's Black Sheep Squadron. We'll never forget being awakened at daybreak by an earthquake. **Charles Cork, Birmingham, AL, machine gun corporal A-1-9**

How To Get A Head ...

On Bougainville, January 1944 ... some of the comforts of home were always available if some genius were present to figure it out. To make this creation, cut out both ends of an oil barrel, half bury it in the ground and make the seat from ammunition crates.

"Hellzapoppin Ridge"

The battle for Hill 1000 "Hellzapoppin ridge" was one of the bitterest campaigns on Bougainville. Elements of the 9th and 21st Marines engaged a dug-in and determined enemy. Marine veterans, who relieved the 1st Parachute Regiment, remember crawling over the bodies of the newly dead.

Placid, quiet grandfathers today recall that all-too-little recognized encounter. Here's the memories of one of them. **John F. Pelletier, A-1-21 Marines, Warner, NH**

On or about December 10, 1943, we received word to pack up and move out. We were going to relieve a paratroop outfit that had suffered heavy casualties on Hill 1000.

As a lead scout, I was one of the first members of Able Company to arrive on Hill 1000. There were dead paratroopers all over the side of the hill. Dead Japanese were still hanging in the trees and it appeared that no Marines had been able to cross over the crest and live to tell about it.

My squad leader, Sergeant Oliver, moved us into position to take Hill 1000. He ordered me to move out, through the bamboo, and check the top of the hill. As I moved out, the only Japs I found were dead, and the ones in foxholes I shot to make sure. We secured the hill that afternoon, but did not go down the ridge to what is now known as Hellzapoppin Ridge.

The next morning Sergeant Oliver told me to advance down the ridge as we were going to secure the point. That point was to become our most costly battle of Bougainville. We moved down the center until we were within 20 feet of the point. We had been checking all the trees and ground area. The ridge was less than 100 feet wide and several hundred feet high, with trees all around.

The Japs hit us with machine gun, rifle, and mortar fire. They popped up out of spider holes and were dug in under trees. We were in a horseshoe-shaped ambush. We were firing as fast as we could when Sergeant Oliver pulled me in back of a tree. He gave the order to pull back up the ridge, but he didn't make it.

We counter-attacked within minutes but were unable to secure the point. Within a very short time, we had lost Gunnery Sergeant Puckett, Sergeant Oliver and Sergeant Rice. Our platoon leader, Lieutenant Averitte, was wounded. Lieutenant Averitte's head wound caused the loss of his eyesight and I was assigned to take him back down off Hill 1000 to the Battalion Aid Station.

The next day I went to the hospital and was not present when Hellzapoppin was secured. I can't help but feel that I was the first Marine to cross over Hill 1000, and the first to enter out onto Hellzapoppin Ridge to within 20 feet of the tip, and live to tell about it.

Nose To Nose

All across Torokina's front, the 3rd and the Raiders were nose-to-nose with the Imperial troops. On the beach it took Ka-Bar, throttle, fist and flamethrower to pry them out of bunkers and trenches. Wounded on the first day, Mason was exasperated ... "After all that training, I'm out the first day." Major John P. Brody took over.

Not far from Mason's battalion, a 75mm mountain gun on Torokina Point had sunk six landing craft and the men in them. Sergeant Robert A. Owens of 3rd Battalion, 3rd Marines stormed into the bunker, destroyed the gun and gunners, and was himself cut down. Sergeant Owens was from Spartanburg, South Carolina. He was awarded the Medal of Honor.

"It (Bougainville) was one of the first occasions in which amphibious troops encountered and occupied an organized beach practically untouched by preliminary bombardment," say authors of the "3rd Marine Division," published by *Infantry Journal* in 1948.

At sea, November 1 and 2, the threat worsened.

American snooper planes found the vengeful Japanese fleet out from Rabaul steaming at high dudgeon for the Yankee transports. In it were the heavy cruisers Myoko, Haguro, and Agano, the veteran workhorse light cruiser Sendai, and an entourage of destroyers.

A bare 20 miles from Empress Augusta Bay, Task Force 39, under Admiral Stanton "Tip" Merrill, intercepted the Japanese ships. Marines ashore mistook the flashes and gun thunder for heat lightning. In the United States greeting force were the Desron 23 "Little Beavers" destroyers of Arleigh Burke, later Chief of Naval Operations. It was in that engagement that Burke won the nickname "31-knot Burke" for the speed he managed to send the redball fleet bloodied and back to Rabaul—and the beloved Sendai to the bottom with 325 men.

With all that over, there was relative quiet (for a battlefront) on that miserable rain forest side of Bougainville until November 7-8. by then the six Japanese destroyer transports took station off the Laruma River (Atsinima Bay) with good intentions of reversing the fortunes of the Emperor around Empress Augusta Bay.

(Continued on page 47)

Landing boats move toward the beach at Bougainville. (Defense Dept. Photo-Marine Corps)

A Friend Indeed!!!

The Navy established an operations post at Yellow Beach #3 on November 1, 1943, the day 3rd Marine Division landed at dawn on Bougainville Island.

I was stationed there as one of the Beachmasters. Our assignment was to provide boat transportation up and down the coast for the Marines and to set up a boat repair unit. Boat Pool #11 was in operation on Puruata Island out in the bay not many days after the initial landing.

Early in November, I was requested to provide two LCVP's for a patrol which was to scout enemy activity in the vicinity of the bay not many days after the initial landing. With a crew of three on each boat, we transported the patrol to the left bank of the river and waited until the patrol was ready to return. The sailors were fishing, and I was reading a paperback.

Not much time had elapsed when a sailor spotted a patrol coming down the right bank of the river in single file. There was a ridge of sand dunes along the beach so only heads could be seen. As the patrol neared the beach, the column turned left proceeding along but still behind the sand dunes. The sailors returned to their fishing.

I continued to watch and wondered why it crossed the river.

When the dunes tapered off, the fatigues of the men could be seen from the knees up. I waved to them, but they did not acknowledge it. When the column reached the end of the dunes, it suddenly made a sharp left turn into the jungle. Then I noticed the men were wearing leggins ... It was a Jap patrol.

Then along came a PBY and they yelled down if I had seen a Jap patrol ... I said yes I did and had waved to them ... We never saw the Japs again ... but our people would never forget my camaraderie. Ask Frank, they'd say, if a question on Japs came up ... He is on friendly terms with them.

Not so ... I always said ... They never waved back. **Frank E. Ross, Framingham, MA**

First Hours Ashore

Everything was going in our favor until we closed on the point (Torokina) Gunny Sergeant Snead noticed that the jungle foliage had recently been cut. The Japanese use this foliage to camouflage their positions and had set up an ambush for "B" Company. As Gunny Snead raised his carbine to shoot...he was hit and wounded. This alerted "B" company that we were under enemy fire. Gunny was evacuated...and died from his wounds.

By his seeing the camouflage and acting as he did, he saved the company from many casualties.

We dug foxholes and strung barbed wire...on which we attached C-ration cans with stones in them to make a noise if anyone hit the wire. Hand grenades with trip wires were also laid in front of our perimeter.

During my watch, in the middle of the night, one of the men in our foxhole was sleepwalking and headed towards the barbed wire. I just happened to recognize him...and woke the other two men so we could try to call him back to safety. We did. When we told the sleepwalker what had happened, he didn't believe us. **Joseph J. Barishek Lewes, DE**

Chow Hound And Laundry Bag

Mike Riley was the platoon chowhound when we were part of C Company, 3rd Regiment. This was evident early on—even when we were in Samoa, before we were on short rations.

On Guadalcanal, when we were preparing to move out for the invasion of Bougainville, Mike managed to squeeze into his pack one extra pair of socks, or can of chow, but not Riley. A laundry bag, for him.

After a couple of weeks of small unit combat, including a platoon combat patrol engagement at the Nouma Nouma trail, we were short of ammo, water, and rations. A detail was sent back to the beach to replenish our supplies. Riley and I were among those on the detail.

At the beach we picked up water cans. Riley disappeared, re-emerging with the laundry bag, slung over his shoulder, filled with C-rations. When we rejoined the company, somehow this was overlooked by the officers.

Shortly after, we were assigned to capture a ridge. Riley and I were BAR men and along with Sal DiStasio and another BAR man, we went up first. The rest of the platoon fixed bayonets.

The enemy set down a withering fire from Nambu machine guns and rifles. It was tough going.

Suddenly, the fire was directed away from us. Riley had laboriously covered the last few feet to the top, still carrying his laundry bag of chow over his shoulder! He was a perfect target, and we were ignored. The remaining three of us were able to go over the top of the ridge and seek cover while the enemy concentrated on Riley. We succeeded in capturing the ridge.

There was an official name given to this minor battle, but to C Company, it will always be the Battle of Laundry Bag Ridge. (Riley now lives in Pittsburgh, PA). **Douglas Lyvere, Hollister, CA**

A Night Visitor

One night in 1943 on Bougainville, two of us were spending the night in a dugout next to our guns. We were always on the lookout for Japanese infiltration in our lines (we even caught one in our chow line at one time.)

This night all was quiet 'til we heard a noise that sounded like someone crawling on coral gravel. My gunmate and I tensed, alert for a Jap infiltrator. We cocked our rifles and waited for him to come in. Shortly—though it seemed like an eternity—a koala bear poked his head into the dugout. Needless to say, we were two relieved Marines. **Richard M. Coulter, Daingerfield, TX.**

Rain helped

Jungle And Mud

The Japanese (17th Division) came ashore in barges, some almost directly in front of the 9th Marines who reacted violently. Out of this landing resulted the battle of the Koromoka River, fought in waist-deep water with mud over your ankles. "Never," said General Turnage, "had Marines ever had to fight over such terrain ... where you couldn't hear, see, or smell a man five feet away."

The landing of the Japanese over a wide front, with little time or room to organize ashore, isolated Marine outposts and patrols. A platoon under Lieutenant Orville Freeman (he later became Secretary of Agriculture) was caught by the Japanese on the distant banks of the Laruma River. He was able to melt into the jungle and slow the Japanese by ambushes. It took the Freeman patrol 30 hours to get back to the K Company, 3rd Battalion, 9th Marines lines.

It was in the battle for the Koromokina that Sergeant Herbert J. Thomas of Columbus, Ohio, while pressing an attack to destroy two well-concealed machine gun positions, threw himself on a hand grenade which he had thrown but rebounded from jungle vines. He spared the lives of his comrades and assured the destruction of the enemy positions.

Actually it was K Company of the 9th that took the brunt of the Japanese invasion attempt. They fought longest and hardest, then got the relief of elements of the 3rd Marines who had almost as hectic a fight. Fire fights were at distances of 10 to 50 yards. The Japanese using their superb skills of camouflage set up, almost immediately, defenses you couldn't see.

It took two major barrages of the 12th Marines (Colonel John B. Wilson) and the 3rd Defense Battalion to destroy both Japanese personnel and intentions. When the 21st Marines came ashore, and Lieutenant Colonel Ernest W. Fry's 1st Battalion attacked into that no-man's land, there was no opposition. There were shreds of men in trees and the bodies that weren't dead were groggy, disoriented, and no longer soldiers.

In a day or so, about November 8, the battle of the Koromokina was over. The Japanese attempt to secure the beach by sea had failed. And now the Army's New Georgia battle-hardened 148th, the 145th, and 129th were aboard, or due soon to Bougainville, to serve with the 3rd Marine Division and the 1st Marine Amphibious Corps. They were under

(Continued on page 49)

THE CHOW LINE

... And Washing Mess Gear

Field Dishwasher

Since our initial landing at Empress Augusta Bay on November 1, 1943, our only food had been the individual, wax sealed packets of K-rations, a few small tins of C-rations, and an occasional D-ration chocolate bar. Needless to say, it was with considerable enthusiasm that we welcomed the arrival of our rear echelon when it moved up to join us from Guadalcanal. The cooks set up their field galley just behind our position, and prepared our first hot meal in three weeks.

On this particular day, Private First Class Dick Hannon, Private First Class George Christie and I had our TBX radio dug in near the Regimental Fire Direction Center. We were kept busy receiving fire commands from our air observers in the SBD Dive Bombers that flew up daily from the Munda airstrip on New Georgia Island. Since two of us could handle the radio's operation, we decided to rotate one man at a time, going back for his chow and bringing it to the foxhole. A flip of a coin determined that Christie would go first, I was second, and Hannon last.

At the head of the chow line, the cooks had set up a row of four burners. Over each of these fires was a square, aluminum tub. The first contained boiling, soapy water for washing our mess gear, and the next two held clear water for rinsing. In the fourth, our chief cook, Sergeant George Campbell, was brewing coffee. Next in line, was a row of tables holding the large pots and pans of food being served by the cooks.

"Be sure to give your mess gear and canteen cup a good dousin' before you start through the chow line," Christie told us when he returned with his food. "There's a corpsman checking on it, and if it doesn't pass his inspection, he'll send you back for a better wash job, and you'll have to start through again from the end of the line."

When I returned with my food, Hannon and Christie were busy sending and receiving the messages of an artillery fire mission. As soon as there was a lull in the action, Christie took over the radio, while I cranked the generator. This relieved Hannon so he could go get his chow.

"You better get a move on," I told him. "The line was getting short when I went through."

Although there was plenty of food left, by the time Dick made the trip down the jungle trail to the galley tent, there was no one in the chow line. Even the corpsman had secured his inspection detail and had gone off to enjoy his own food.

Dick walked up to the first burner he came to, and as he was dousing his mess gear up and down in the tub's steaming contents, he looked up to see Sergeant Campbell glaring down at him.

"Hannon, you stupid bastard, get your friggin' mess gear out of my coffee," he roared.

"Gee, Sarge, I'm sorry," said Dick. "I guess I started at the wrong end of line."

With a red face, Hannon walked along on one side of the serving tables, holding out his mess gear, while Campbell moved along the other. Each item of food the sergeant slapped on Dick's trays was accompanied by such irate ravings as, "You Goddamn idiot," "Friggin' imbecile," "Dim witted moron," and "You bird brained miscarriage," among others. When he finally sputtered out of descriptive adjectives, the mess sergeant waved a big serving spoon under Hannon's nose.

"Damnit, man, is my coffee so bad you couldn't tell the difference between it and the friggin' dish water?"

When word of the humorous incident passed through the outfit, it helped ease the tensions of living under combat conditions in one of the world's densest rain forests. **Jack Kerins, Terre Haute, IN**

The Gun That Failed

An embarrassing and possibly lethal incident occurred when our heavy machine gun platoon of D/9 under Lieutenant James S. Bowling was attached to Captain Frank K. Finneran's C/9 Rifle Company on Bougainville.

The 1st Battalion 9th commanded by Lieutenant Colonel Carey Randall (retired Major General) was ordered to attack up a steep hill on November 25 that was later to be known as the "Battle of Hand Grenade Hill."

As the lead elements ascended the hill, they encountered fierce small arms and hand grenade resistance. Our machine gun section with Sergeant Orrin R. Wade was ordered forward to assume a firing position for our first fire fight of the war.

We low mounted our 30 caliber gun on a huge long fallen tree. Simultaneously, the Japs were rolling grenades and Nambu firing machine guns which were the other side of our tree—and our gun failed to fire.

After 16 months of training on our gun and being able to field strip and assemble it blindfolded, it didn't fire because the barrel packing was improperly adjusted.

Prior to this time, we hadn't tested or fired our gun for fear of disclosing our position. You should have heard Sergeant Wade curse, but that didn't help.

Fortunately for us it was getting late and since the battalion advance had been repulsed, we were ordered back to the base of the hill to regroup and prepare for another attack the following morning.

With our gun in firing condition as we prepared to assault, word filtered down that scouts found the Japs had evacuated during the night.

They left 32 dead, while we had six killed and 42 wounded, including the writer from shrapnel from a mortar tree burst.

Ironically, that was the only opportunity we had to fire our gun during the entire 57 day campaign ... **Walt Wittman, Bradenton, FL**

High Octane Cover

Frank Ross ... his story of beaches, bombers, and bombs follows ... was in the Amphibious Force Boat Pool, and the Navy Boat Pool. He was an ensign ... USNR ... and was the one who waved to the Japanese patrol, and got ribbed for it.

The Japanese had an irritating habit of diving out of the bright early morning sun to bomb Puruata Island. The raids were frequent, and happened when the LSD's had landed supplies the day before. The mainland beaches were not suitable for landing cargo, so our little island became a supply depot until the bomber strips were completed on the mainland.

One morning I was walking along the beach ... probably checking on boat availability ... when the air raid siren went on. I dove into the jungle, seeking cover, just as the bombs were falling. I discovered that I might as well have stayed on the beach. I was comfortable nestled in a cache of high octane aviation gasoline for the fighter strip. **Frank Ross, Framingham, MA**

For Want Of A Nail

I was in the Third Division, after being transferred out of the Second, and we had made the landing on Bougainville. We were just removing our gear, when my haversack was stolen from my gear—and my only pair of shoes were taken by some Japanese. My only recourse was to steal a pair of shoes ... a pair of Japanese shoes as it happened.

About three weeks later I was called down by the battalion commanding officer, Lieutenant Colonel Ernest W. Fry, Jr. who told me my missing haversack was found on this Japanese soldier. He had everything—my shoes, my serial numbers and miscellaneous clothing.

It had been a long hard three weeks in Japanese shoes with the water still coming out of them ... and was a real pleasure getting my shoes back. **Robert J. Healy, Placentia, CA**

This Japanese flag captured on Hill 1000 during the fighting on Bougainville was displayed by, from left—unidentified corpsman, the Rev. Bal Schomer, chaplain and Lieutenant Commander Munroe J. Epting, Naval Medical Corps. Photo by S/Sgt. Angelo Dugo. Submitted by Lt.Col. Al Jenson

Maj. General Robert J. Beightler, Co., 37th Army Division.

Luckily, the most intense of the fighting would now shift to the right flank of the beachhead. There'd be a taste of fire in patrol action. The Army would have to temporarily wait for another day—but it would come with a vengeance.

We Needed Air Fields

Remember, there was no American attempt to conquer all of Bougainville. All we wanted was a perimeter for fighter and bomber strips, and enough elbow room (eight to 10 miles) for protection and movement. The Japanese would keep the rest ... although they couldn't enjoy it. Yet, we had to remain unfriendly to keep the Japanese from bothering us. Colonel Robert Blake, Chief of Staff, insured the probing and belligerence in managing the patrols.

A limited goal was something the Japanese never understood, neither did the Bougainvillians who thought the Americans "daft" for settling down in that God-awful rainforest. There were drier palm-fringed beaches, fields near Kieta on the other side of the island. On some higher ground you even had to wear a sweater.

Yet, seeing what we brought ashore, 16,000 troops and 6,500 tons of gear in two days, it certainly appeared we were coming to stay. The Japanese showed their resentment first with strong bursts of air power. Colonel Edward A. Craig's 9th Marines (now Lieutenant General United States Marine Corps, retired, El Cajon, California) were hard hit by strafing Zeroes.

On securing the beach, division assigned probes inland to find immediately a site for the first fighter airfields. Patrols found a likely spot for Seabees and Engineers to appraise. By November 9 earth was being turned for the first strip on Torokina.

Now with the perimeter out to 1,500 yards (November 11-12), Major General Roy S. Geiger, 1st Amphibious Corps Commander, ordered further extension of the area. The new ground would house the bombers or other strips.

The task of extending the perimeter up near the Piva and East-West Trails flung the 3rd Division into bitter engagements. (It was in these encounters at a roadblock that, sacrificing his own life on a grenade, Private First Class

(Continued on page 51)

A Visit By The Natives

Colonel Evans O. Ames, commanding officer of 21st Marines, poses with two of the Island of Bougainville natives. Only after the 21st had reached its objective, and settled down to await relief by Army American Division, did any of them show up in the area of our final command post which was called Evansville in honor of the regimental commander. (Submitted by Al Jenson)

Seabees Know All About Water

Lieutenant Colonel Albert L. Jenson spent his teen years on a farm in Kansas, enlisted in the Marine Corps in 1917, and found himself on an old coal burning battleship. He helped organize and train the 21st Marines to enter the 3rd Marine Division ... was regimental ammo officer for Bougainville and Guam operations ... wounded on Guam ... lived through banzai charges ... helped to normalize Guam after the war ... retired in December 1954 with 32 years in the Corps. Here are some of his recollections:

On Bougainville in December 1943, fierce fighting was going on in the heavy jungle. I was a hundred yards behind the front line with my ammunition section. Foot and amtrack traffic jammed the narrow trail, ankle to knee deep mud everywhere.

A very officious acting major came tramping toward the front, muddy only from the knees down, his gold leaves shining like "for inspection," dungarees clean and looking almost pressed above the knees, wearing both jacket and shirt—while most of us were stripped to the waist. He must have just landed as we were only about a half mile from the beach.

Overtaking two young, very muddy and tired looking Marines carrying perhaps 30 canteens strung on bamboo poles between them, he said "What have you there, son?" The lead Marine said, "Drinking water, sir."

The major said, "Are you sure it is potable?"

The lead Marine looked back at his buddy who was staring at the major, but quickly rose to the occasion and said, "It must be, sir, the SeaBees are taking showers in it."

A Board This Long

Everyone made do with what he had on hand. Common tools were scarce.

For instance, one day the company carpenter of H & S 21 was seen striding earnestly among the tents, his arms outstretched two and a half feet apart, looking right and left. Asked what he was about, he said "I am making a desk for the colonel and all I need to finish is a board this long ..." I doubt if he ever found it.

A Pole Charge Without The Pole

We were crossing a dry stream bed. The point had crossed. I was about 30 yards to the right rear of the point when a sergeant yelled, "Duck, Captain" and began firing his rifle from the hip.

I saw what looked like a flat piece of board come sailing through the air directly at me. All I could do was hit the ground, flat on my belly, with my head toward the oncoming object which landed and exploded about 15 feet in front of me.

Luckily the ground was sort of sandy and contained no pebbles or other hard objects. It jammed my helmet hard down on my head and covered me with dirt. I was almost knocked out, but was able to continue the mission.

The object thrown was a pole charge from which most of the pole had been cut, leaving a short handle, enabling the thrower to swing it like a hammer throw .. thus gaining more distance. These charges, designed to be carried on the end of a light bamboo pole when attacking against barbed wire entanglements.

The charge was thrust ahead under the wire and detonated by pulling an attached cord. In this case, some of the Japs carrying the charges were probably in the rear of the banzai group and crossed the wire by simply walking over the dead bodies of the leading chargers killed by intense machine fire ... and had no opportunity to use that pole charge ... so they threw it.

And The Target Was ...

We formed a skirmish line and approached the hut. We received intermittent fire until about 150 yards from the huts when a sudden increase of rifle shots covered our entire line, wounding three men, one very seriously. The fire could have come from the huts or the sharply rising brush beyond.

Seeing what I knew to be units of the 9th Marines in reserve on high ground to the right, I sent a sergeant to request a few 60mm mortar rounds on the hut. The first well aimed round dropped through the thatch roof and exploded on the ground ... triggering much movement in the lower part of the walls.

To our amazement, only a sow and nine or ten piglets burst out through the walls not taking time to find a door or any other opening.

I never heard the ribbing cease about that until I left the 21st some months later.

"L" Company On Hellzapoppin Ridge

As we landed at Empress Augusta Bay on D plus 17, Jap torpedo bombers attacked landing boats and troops and the APD McKeen (carrying nearby Company K) was sunk resulting in 52 casualties ... more than had occurred on D-Day.

We in L Company had our first casualties the next night in a reserve area when a Jap bomber dropped a bomb in the 1st Platoon area, killing six and wounding six more.

For the next three weeks we did extensive patrolling, but there was little enemy contact. But on December 11 we were ordered to clear a spur on Hill 1000, in front of B Company, of what was reported to be a squad. An artillery barrage was to precede, but several tree bursts fell among us, causing several casualties.

On the next day, again preceded by an artillery barrage, which could not be very effective because of terrain and jungle, L Company attacked along the length of the spur—only 40 yards wide with very steep slopes on each side and covered by dense jungle growth. The Paramarines had made the first contact with the enemy dug in on the spur (later named Hellzapoppin Ridge) and had left their dead there. After these few days they had become very unpleasant reminders of what faced us as we crawled forward ... in many instances, right next to them.

Progress was slow as one platoon moved straight ahead 50 yards and was pinned down. The other two platoons in enveloping action could make only slow progress because of the machine gun crew. When ordered to pull back to tie in, the platoons dug in and used some existing Jap foxholes. In one of the Jap foxholes, a dead Jap major was found. He had a map that showed a platoon—not a squad—on the ridge.

Further attacks were made on the Jap fortified ridge, but were unsuccessful because the Marines could not see the Japs ... but the dug in enemy machine gunners and snipers could see the Marines, and cause casualties.

A few days later the ground was taken by another battalion with the aid of artillery and dive bombing. It had taken 11 days and 158 casualties to take Hellzapoppin Ridge. **John W. Yager, 1st lieutenant, 3-L-21**

After the war, John W. Yager went to law school, then returned to Korea where he won the Bronze Star. He practiced law and became Mayor of Toledo, Ohio.

87 Pounds Soaking Wet

... Only the memories lived after our platoon leader, Lieutenant Leon Stanley of B Company Tanks was shot by a sniper from a spider hole alongside that trail. He was the first officer in the 3rd Tank Battalion to die in combat ... his tank was crippled by a mine ... two of the crew are still alive with us now ...

Yes, I shall always remember that little Jap lad who was captured that evening by two Raider scouts when he came down at sunset for water at the river. He was tied in that tree for one month with his rifle, rice, and canteen. He was covered with a horsehair cape so no one could detect him ... weighed in at 87 pounds at the Aid Station ... and was covered with jungle rot.

He was very scared as he feared some of those Marines would do him in. Well guarded, he was removed to the rear lines. He was the one who shot my buddy, Sergeant Barnish in the foot. **Thomas F. Murphy**

Henry Gurke merited the Medal of Honor.) The Japanese gave only inches in those vicious encounters in combat hand-to-hand and tree-to-tree.

Piva Forks

New action November 5 thru 9 was on the Piva Trail where the 9th Marines and 2nd Raiders met the first attempt of the point of the two 17th Army battalions sent down the Numa-Numa and East-West trails from Buin to wrest the beachhead on the right flank from the Americans. Actually the Japanese idea was to coordinate the landing at Laruma with an attack down the trails. But things didn't work out. The Japanese heroes of the day were now the elements of the 17th Army from Buin on the trails.

Obviously the fighting would be bitter here. Control of the trails meant control of the war. Worse, they were narrow footpaths mostly, only inches from the water. They were surrounded by swamp, often waist-deep water. Flanking was not an option. The fighting would be head-on, a collision course.

So to protect the beachhead, to overcome the Japanese, to build airfields, to bomb Rabaul, the trails had to be protected. The fierce task in the most miserable conditions, fell to the 3rd Marine Division, at first the 2nd Raiders and the 9th Marines.

It was early on into the second week that General Turnage ordered Colonel Edward A. Craig, 9th Marines to clear the trail ... and up to the Piva Village where the Japanese had settled in. Colonel Craig assigned the mission to Alan Shapley's 2nd Raiders who called in tanks and halftracks, of little avail in the swamps and muck. Concentrated artillery support and U.S. Marine Corps persistence and superior firepower did get us to the junction of the Piva and Numa-Numa Trails.

Much was made possible by the 12th Marines, the Army's 105 and 155mm Field Artillery Battalions, and the long guns of the 3rd Defense Bn.— so often a guarantor of victory. Their shells and shards literally left strips and pieces of foemen on the ground and in the trees.

Basically, the Japanese thought of everything we did, but were no match for our firepower and aggressiveness, that even their bushido could not overwhelm.

It was our aggressive patrolling (November 17-18) that found the enemy roadblocks on the Numa-Numa and East-

(Continued on page 53)

Finding The Jap Life Line

The Finest Of Overseas Architecture

Slit Trench

Any trail was critical, if not vital, on Bougainville, and few approached Empress Augusta Bay through the morass of swamp and jungle. The East-West Trail was a highway across most of the island and a life and death path for the Japanese. It was this trail that two junior enlisted men of the 21st Marines, Harry Noble and Jack Post, were sent to find.

With a squad on a reconnaissance patrol, east of the Torokina River, we came upon the East-West Trail ... the Japanese supply route. Because of dense tall trees and thick jungle, it was not visible from the air.

Our patrol was now one and one-half miles forward of the lines. The trail had never been mapped. As scout, it was my job to do ... my good buddy, Jack Post, volunteered to go with me.

We split from the rest of the patrol, blackened our faces and hands with mud from a nearby creek and headed west along the trail's edge. Jack on one side and I on the other. We took an azimuth at each bend, and kept a record of the distance covered. At one point we came upon a Japanese bivouac area which looked as though it had recently been occupied ... luckily it was now abandoned.

As we continued, we kept looking back and motioned various hand signals to an imaginary patrol ... hoping that if any of the enemy were watching, they would hold their fire and save it for the main body behind us. There was no one.

Hours later, at about 200 yards to our front, the trail ended by a sheer drop ... at the base was the Torokina River. We had to reach the end of the trail and take azimuth readings in the direction of hills, in order to locate the trail's end.

Suddenly there were flashes of artillery exploding on the other side of the river. (We found out later it was on Hellzapoppin Ridge.) The 12th Marines were pounding the ridges in preparation for an attack. Their enfilading fire started to creep toward ... it came across the ravine and soon we were pinned down by our artillery. When we saw it coming, we took cover in some nearby foxholes ... thanks to the Japs.

After the rolling barrage lifted, we took our compass readings and headed back east. It was getting dark and we took refuge in an abandoned Jap dugout which we had passed earlier. The dugout was lousy with lice, but at least we were safe.

The 12th was busy that night laying down harassing fire. Sleep was impossible, but we felt secure that there would be no Japanese returning during the display that our artillery was putting on.

At the break of dawn, Jack and I slipped into the jungle and found our way back to our squad's rendezvous point. To this day I still feel that those "Cannon-Cockers" of the 12th ... who never knew we were out there ... saved our lives that night. THANKS GUYS!

Footnote: Two days later a Marine combat patrol was in the same area we had reconnoitered. They ran into two Jap machine gun nests and lost several men before silencing those guns. **Harry K. Noble, Sarasota, FL**

> *"War is cruel and you cannot refine it."*
> William Tecumseh Sherman

Seabees Dedicate A Highway

When the Army's American Division relieved the 3rd Marine Division in January 1944 on Bougainville, the Marines were pleasantly surprised to find this completed highway only a few hundred yards behind their front lines. It provided passage back to the beach in minutes through heavy jungle growth that had taken two months of hard fighting to penetrate. Left to right are Corporal John L. Riley, Sergeant Peter Pavone, and Corporal O. Nelson—all of H&S, 21st Marines. (Photo by Sergeant Angelo Dugo, submitted by Lieutenant Colonel Al Jenson.)

TO OUR VERY GOOD FRIENDS THE FIGHTING MARINES WE DEDICATE This HI-WAY

MARINE DRIVE

BUILT BY 53RD N.C.B. 1ST M.A.C.

Bougainville Vignettes

Surviving ... Thanks

Eugene Rosplock is happy enough that his whole gang has survived the war, age and time. Of Bougainville he remembers most: the rain and the Banzai ... **Eugene J. Rosplock, Elmira, NY**

God And Country— From The First Gun

The Reverend H. Scot Thompson, CCGM, for a man of the cloth who could well have found safer havens, has a long list of service ... the envy of any salt of any period.

He was Marine Corps, 1939-1947 with 1st, 2nd, 3rd Marine Divisions, with time for the North Atlantic European Theater, not counting New Guinea, Cape Gloucester, Peleliu, and Okinawa. He's a live member of the DAV. Served in Korean conflict. Retiree of the USN Chaplain's Service, he's with Christ Church, Mount Vernon, New York. **The Reverend H. Scot Thompson, CCGM**

I Remember ...

A few incidents are still etched in my memory. The strafing by Japanese Zeroes ... bringing back wounded Marines to field hospital ... four of us crawling with a stretcher ... sniper firing ... slushing swamps, firefights, earthquakes ... Guadalcanal ... rest and reorganization. **Bert Daly, Williston Park, NY**

Cpl. William Coffron, USMC, firing at sniper from behind a tree, during barrage being laid down on Torokina Island on November 5, 1943. (U. S. Marine Corps Photo taken by Capt. P. O'Sheel)

West trails. We reduced both, clearing the area to move on. But, in so doing, we were also able to fix the positions of enemy strength. Our movement here also helped us occupy a commanding ridge, which gave us virtual-everywhere observation.

Problem was that the hill was ahead of our movement, and Lieutenant Steve Cibik's platoon of F Company, 2nd Battalion, 3rd Marines moved up to occupy it. Why the Japanese had left it vacant, even today surprises student strategists ... but the Japanese fought all night to try to get it back.

The Battle for Cibik's Ridge and the rest of the high ground was, says 3rd Division, Infantry Journal Press, 1948, a significant prelude to the Battle of Piva Forks.

That battle was preceded by an artillery preparation in an 800-square yard area. It included basically anything that would fire—machine guns, mortars, and the whole gamut of Marine and Army artillery.

Entering the Japanese zone, on attack, it was clear that devastation was complete. The 2nd and 3rd Battalions of the 3rd Marines, recalls Combat Correspondent Alvin Josephy in Uncommon Valor, advanced through a "weird, stinking, plowed-up jungle of shattered trees and butchered Japs ..."

"By the time Lieutenant Colonel King's 3/3 had advanced 500 yards," explains the original text history of the 3rd (Robert A. Arthur, Kenneth Cohlmia, and Robert T. Vance) "the enemy was ready and launched a desperate counterattack. There in the jungle, in the brutal hand-to-hand, tree-to-tree struggle, the Japanese flanking force was completely destroyed. The 2nd Battalion neared its objectives too ... closed with the enemy reinforcements ... Colonel de Zayas' men cut the enemy down to the last man ..."

On November 25 a strong pocket of Japanese was wiped out by the 1st Battalion, 9th Marines under Lieutenant Colonel Carey A. Randall.

Th Battle of Piva Forks had broken the back of organized resistance in the Empress Augusta Bay area.

Piva Forks was a critical area dominating the landscape. While in enemy hands, no beachhead or airfield was safe or could be called our own. The battle did determine the future ownership of the bomber strips to come—really the reason why we were on Bougainville at all.

Major (later Brigadier General) Donald M. Schmuck, who led F Com-

(Continued on page 55)

Running "The Slot" And Diving From Bombs

Brigadier General John Seymour Letcher was appointed a second lieutenant in the Marine Corps in 1927 and a member of the "Old Corps" who served in Nicaragua. He landed with H&S 12th Marines on Bougainville.

I was in command of the half of our regimental headquarters personnel to land with the second echelon. On the afternoon of the fourth (November 1943), we embarked from Tetere Beach. With other details, making in all about 200 hundred officers and men, we boarded a destroyer. With darkness we headed northwest and made a fast run up the "Slot" the channel between the islands of Guadalcanal and New Georgia on the south and Santa Isabel and Choiseul on the north was called.

The destroyer was terribly crowded and we had to sit up all night because there was no room to lie down. With all ports and doors closed, so that no lights would show, the interior of the ship was hot and close. The padded life jackets, which we were required to wear, did not add to our comfort.

At daybreak, we were at Empress Augusta Bay. As it grew light we moved in toward the shore where we could see the white sand beach and the coconut palms of Puruata Island and Cape Torokina. Behind them was the forest and several miles inland there were mountains, the highest of which was Mt. Bagana, an active volcano with a small cloud of white smoke hanging above it.

A mile or so from the shore the destroyer "lay to" while boats came alongside to take us to the beach. We landed just north of Cape Torokina ...

Colonel John Bushrod Wilson, regimental commander and operations officer, briefed me on everything that happened since Guadalcanal. The D-Day landing had been alright except that numerous boats had been wrecked by the waves (surf) ... there had been sharp fighting on Puruata Island and at Cape Torokina ... our infantry had been pushing inland to enlarge the beachhead. I visited the battery positions and battalion headquarters ... got into my jungle hammock to get a good night's rest.

I had been asleep no longer than an hour, because exactly at nine o'clock the air raid sounded. I pulled the zipper on the side of my hammock, jumped out and got down in my wet and muddy slit trench. The planes arrived a few minutes later and came over the beachhead while our anti-aircraft guns filled the air with ineffective shells. I heard the bombs whistling as they came down, but they fell near the beach a quarter of a mile away from where I was.

When the "All Clear" sounded I got back into my hammock, but there was no rest that night because there were a dozen raids. During the first ones I would get down into the trench as soon as I heard the siren, but I noticed that it was always 15 minutes after the siren, before the planes arrived. So after the fifth or sixth raid, I decided to wait until I heard the planes.

I must have gone back to sleep because I awoke just as I heard the planes overhead and the backfire of their engines, which occurred when their bombs were released. I had no time to find the zipper as I heard the whistle of the bombs. With one frantic sweep of my arm I tore out the whole side of the mosquito netting, and dived into the slit trench ... getting into it a fraction of second before the bombs landed.

They were large ones and came relatively close, landing on the division command post ... a short hundred yards away. There were some screams and groans from that direction.

For the rest of the night I was out of my hammock with the first note of the siren. In the morning we learned that several men at the 3rd Division command post had been killed in their hammocks.

Marines landing on Bougainville race from the landing boats across beach to the jungle. (Defense Dept. Photo-Marine Corps)

Marines and Sailors trying to control fire and other general coverage after a big Jap air raid made a direct hit on a gas and oil dump at Empress Augusta Bay. (Defense Dept. Photo - Marine Corps)

Vignette Of Valor
Remembering "Doc" Shepard

On November 24, 1943, the 3rd Battalion's I Company (3rd Regiment) was out in front of the Regimental Aid Station on Bougainville. A call went out—"Corpsman, Corpsman." We rushed out and found seven or eight wounded, amongst whom were Platoon Sergeant Campisi and Corpsman George Weiss. They had been hit by machine gun fire. We administered first aid and directed the stretcher bearers to take the wounded directly to the "C" Medical Hospital, not too far away. The Japs continued to pour in mortars, artillery, and machine gun fire on "I" Company and dangerously close to the hospital.

As we approached the hospital with the wounded, I noticed 15 or 20 wounded Marines in front of the hospital waiting their turns to be attended to. This is when I noticed Dr. Shepard ... Dr. Duncan Shepard. The flaps on the hospital tent were open and there was Dr. Shepard operating away ... so calm, so brave, so courageous as though he was back in the Mayo Clinic in the States where he had trained.

It was getting dark and he asked for volunteers to dig foxholes inside the hospital tent in which to place the wounded. He continued to operate all through the night. Tom Murphy of the 3rd Tankers was also a witness to all of this and we talk about it often. Dr. Shepard received a Silver Star for this outstanding performance of duty.

A member of the 3rd Marine Division Association, he passed away on October 30, 1989. I spoke with the doctor's widow after he died. She made mention of a Marine tie clasp that was given to him by some Marines in 1942. He wore this tie clasp till the day he died. I am sure that many of the Marines of the Bougainville and Guam campaigns will remember Dr. Duncan "The Knife" Shepard.
Andrew Barnard (Navy Corpsman)

Lament Of A Squad Leader

What hurt me most about combat was all of the young that came in ... didn't know as much as me, and got killed. And, I didn't know nothing. **Sergeant William Otho McDaniel, Paragould, AR**

pany of the 3rd Marines tells how the 3rd in that engagement met and destroyed a whole Japanese regiment ... "which fought, counter-attacked from commanding positions with artillery, mortars, machine guns, mines and booby traps ... The action, which determined the fate of the largest and most heavily defended of the Solomon Islands, deserves its place on the roll of Leatherneck victories." He lives in Wyoming.

If in the jargon of William Tecumseh Sherman's *War is Hell*, the brutal scrimmages at Piva reaffirmed it. Major Smoak recalled the young tow-headed Marine swinging a BAR, saying ... "Nips, 30 or 40 ... not counting the pieces."

Keep The High Ground

Cibik's move hampered the Japanese defense dearly, and even today students wonder why the Japanese ever left it unoccupied. They hit Cibik with everything ... from every approach ... in a desperate onslaught to get it all back. But ... too late. Cibik held it long enough to let the rest of the regiment catch up.

But the fight was not over. There were still unoccupied and unfriendly hills some 2,000 yards to the front, which dominated the area between Piva and the Torokina Rivers. Generals Geiger and Turnage concurred. This hill mass was a continual threat to the perimeter. It had to be ours. What's more, the hills provided an excellent defensive position ... blocking anyone or anything that came down the East-West Trail or the Torokina River ... virtually the only way in or out.

So the 21st (under Colonel Evans O. Ames) first sent a patrol and then strong outpost to sit and hold until the perimeter could be expanded. This was to be the last line of defense, the edge of the perimeter, and final "dare-you" mark in the sand where the Japanese would face the 3rd Marine Division.

It involved, for your memory, Hills 500, 600, 600-A, and 1000 to be secured by elements of the 3rd, 9th, 21st Marines and the 1st Marine Parachute Regiment (under Lieutenant Colonel Robert H. Williams. *History of the United States Marine Corps in World War II* (Volume II) tells about the specific Hellzapoppin Ridge, a high spur extending east from the heights of Hill 1000. Some veterans, however, apply the name to the general hectic and bitter mass of hills.

The strong Japanese position resisted all efforts by Marines for six days. The bitter fighting on the ridge, which

(Continued on page 57)

Now And Then
'Code Talkers,' U.S.M.C.*

The date-the place-the event: July 20, Banquet Room, Stouffer Concourse Hotel, Denver, CO., 37th Annual Reunion of the 3rd Marine Division Association.

Over a thousand veteran Marines and their wives arose as one and gave a rousing ovation which lasted several minutes to two Navaho veteran Marines, who earlier, were identified as "Code Talkers" and ordered to report to the podium for recognition.

Attired in native dress, but with World War II medals and the 3rd Marine Division patch prominently displayed on both cap and shoulder, they presented a different but intriguing view.

The assembled group had already been given a rich, emotional program, designed to renew, once again, the eternal bond established by Marines through combat on Bougainville, Guam, and Iwo Jima. A Medal of Honor winner, Colonel Barney Barnum, acted as master of ceremonies, the Marine band gave a "patriotic review," and the Commandant of the Marine Corps, General Al Gray Jr., gave the reunion address. By this time, the group was at an emotional high but the pitch would go even higher.

The two Navaho Marines, each spoke for several minutes and then one broke into singing, in Navaho, the Marines' Hymn: "From the Halls of Montezuma, to the shores of Tripoli." As noted, the entire room reverberated with applause, followed by a general buzzing sound of appreciation and good cheer. The epitome of emotional pitch had been reached.

During the early stages of World War II the Japanese were known to intercept Marine radio messages transmitted, of course, in English. So a unique tactic was created to offset this. The Navaho language has never been recorded; so a group of Navahos were taught how to take a military message written in English, translate it mentally into Navaho and send it in the clear over the radio. It worked! And one can imagine the difficulties this caused the Japanese.

This select group of ten Navaho Marines received, after basic training, special instruction for about eleven weeks at Camp Elliott, San Diego, California. But first they had to qualify: understand English and be able to blend it with their Navaho language, read, write, accurately spell, print with speed, and learn to send and receive English messages diligently. Oral communications were sent in Navaho and written in English at the other end, already decoded, "The reason," voiced a former 'codeman,' "it was so effective." This program was originated by an Anglo-Indian who, along with three Navaho instructors, trained the group of ten at Elliott. Their course even included vocabulary, alphabet, alertness, and becoming proficient in military jargon. At this base, the language codes were formed and developed by all Navahos. As a "codeman expressed it, "We were definitely trained as a walking, human code." At the completion of this unique program, the ten were transferred to different units within the 3rd Marine Division operating in the Pacific.

An ironic twist is present here for as a "codeman," whose formative years were spent on a reservation, expresses it: "I never felt in my life someday I will be using my own language in a war against an enemy. This same language when at school, if we talked we were punished!" And then in a spirit of pride, coupled with exaggeration, he said: "But this same language won the war for the Marines."

After many years, the "Code Talkers" received long deserved recognition by a representative group of Navahos being honored at the White House. Reportedly, their distinctive service to the Marine Corps and to the Nation, had not been publicly acknowledged earlier because their singular contribution had been classified.

That great World War II British leader, Winston Churchill, is especially noted for having "mobilized" the English language and sending it into battle. This being the case, there certainly is a strong corollary here because the Navaho "Code Talkers" literally sent their language into combat.

*Names of Navahos have been omitted by request since a writing commitment regarding "Code Talkers" has already been made. **Carroll M. Garnett**

The Navajo—Their arrows probably didn't add much, but their language certainly did. Navajo code-talkers on our primitive walkie-talkies confused the Japanese ... and the Marines. No cryptic could have broken that ancient code.

The Combination Was Miserable

"There were other campaigns where the fighting was harder and the casualties greater than on Bougainville, but the oppressive heat, the continuous rain, the knee-deep mud, the dark overgrown tangled forest with the nauseous smell of the black earth and the rotting vegetation—all combined to make this one of the most physically miserable operations that our troops were engaged in during the war." **Brigadier General John Seymour Letcher USMC (retired) in** *One Marine's Story*—McClure Press

Bougainville Today— Sadness

There are many World War II vintage Bougainvillians who remember well the American landing on Empress Augusta Bay and the loud "boom, boom" of the cannons on the big ships.

Up to a few years ago, Bougainville could boast a high standard of living with the Papua New Guinea kenya running at $1.28 to the American dollar, largely bolstered by the economy of Bougainville which provided over 50 percent of the whole PNG GNP.

Heart of the Bougainville economy was the Bougainville Copper Mine one of the largest in the world, surely the largest open pit mine, five miles around, employing some 3,000 and some of the most sophisticated technology in the world. It brought a whole new dimension of prosperity gleaned from the big high dark Crown Prince Range in the south central section of the island.

Rich already in cash crops of cocoa, copra, and palm oil, Bougainville that you left just after Christmas 1943, became on September 16, 1975 a province (Northern Province its new name) of Papua New Guinea with the capitol in Port Moresby.

Until 1988 all seemed well and people literally from the bush became computer operators, handled heavy equipment, went into public relations and away to foreign colleges and universities.

Meanwhile tourism began to boom— literally none from the United States—bringing visitors all the way from Munich, Germany to the plush and air conditioned island resort.

But all was not well in paradise. Landowners were awed at the pollution of the Jaba River where today not a fish can live. They who wanted to hunt, fish, and roam their sacred hills cringed at the great gashes made by the terrible blades of modern machine monsters and land movers.

So led by a former mine surveyor named Francis Ona, the locals, the natives, the landowners turned against the mine and eventually all outsiders, especially the big expatriate white landowners.

The Bougainville Revolutionary Army was formed with red bandanas as their insignia. Weapons, at first, were crude. Mine operators were shot by bows and arrows. Ancient shotguns were brought to bear. From PNG came the Army, helpless against the phantom guerrillas who melted into the jungle they knew like the palms of their hands.

Eventually the rebels blew up the electric pylons of the mine, attacked employees. The PNG Army was soon accused of killing innocent civilians. In May of 1989, the big mine was shutdown—and some say it may never open again.

Today Bougainville is still in revolt, governed not by PNG but the Bougainville Revolutionary Army. You can not get a telephone call, a letter, a fax or anything into Bougainville or out. Most of the white landowners escaped with their lives, some to Australia or England to begin life again.

Bougainville today still suffers hostilities perhaps more painful and woeful than the war we brought to them. The current unpleasantness has no end in sight.

Mrs. Noell Mason, now of Sydney, Australian widow of one of the much-decorated coast watchers, Paul Mason, who saved so many lives on Bougainville and in the entire Solomons by reporting on the Japanese, is heartbroken. Rebels even burned the bamboo, rattan and pandanus home she owned and the outbuildings.

"Poor little Bougainville," she said woefully." It is such a beautiful island with very handsome people. This island meant everything to Paul. I pray to God the troubles will work out." **C.J. O'Brien**

did not even show on the terrain maps provided by IMAC, earned the stronghold the name of Hellzapoppin Ridge.

Marines sometimes called the position "Fry's nose" for 1/21 Commander Ernest W. Fry or "Snuffy's Nose" for Colonel Evans Ames, 21st commander, the name "Hellzapoppin" was more indicative of the fierce fighting that marked attempts to capture the spur.

John W. Yager, 1st lieutenant (L/3/21) of Perrysburg, Ohio remembers how every inch to the top of any ridge hill in that mass was facing into the muzzles and faces of the Japanese.

The Hills ... Christmas ... Return to Canal

The battle for the hills went right into Christmas. Reinforced K Company, 21st Marines attempted to clear Hill 600-A. Even with a new concentration of fire, the 21st could not move. The Marines retired for the night. On Christmas Eve, the 21st attacked again, but nobody was up there. The Japanese had gone. What a Christmas present!

Even Ceasar and his lieutenants proclaimed (not just said) that no man knows war like the grunt on the front line. But Bougainville also typified just how much the point is dependent on the people behind ...

Had it not been for the amphibian tractors (their C.O. was Lieutenant Colonel Sylvester L. Stephan) says Marine Corps Historical ... "The supply problems would have been practically insurmountable. Amphib trails were broken in swamps, in jungle where before no man could (or did) pass. All roads were painfully squeezed up from the surrounding sloughs, even though communications and signal units reviled them for tearing up their precious wire."

The roads and evacuation routes for the dead and wounded were built— and almost immediately—by engineers and Seabees and Pioneers—much of the effort under command of Colonel Robert M. Montague.

It was much the same "miracle making" for the 2nd Medical Battalion under the command of Lieutenant Commander Delbert H. McNamara.

What we had come for happened on December 10. The Torokina Airstrip became operational... in time to support the attacks on Hellzapoppin Ridge and environs. The Piva Bomber field was a reality December 19. They were the work of the 36th Naval Construction Battalion ... and later the 77th and 53rd

(Continued on page 59)

Torokina bomber strip ... Won by United States Marines after bitter jungle fighting. The grass strip over World War II matting is in normal times used by over-mountain bush planes to connect remote villages. The once prosperous island, and key stepping stone to Japan in World War II, is now being gutted by a revolution and not even mail, telephone calls or fax messages can reach the island.

Raider Road Block

Lieutenant T.P. Hunter's 3rd Platoon of E Company, 9th Marines had been assigned a recon patrol on Bougainville to find any signs of enemy activity.

We left early morning down Numa Numa Trail to Piva Village area. There were many signs of the enemy—rice, empty cans, sake bottles, and extinguished small fires.

On our way back along the trail all hell broke loose. The Japanese had moved a sizable force between us and our lines ... we engaged enemy frontally on the trail ... fire fights produced casualties on both sides ... acts of heroism—especially that of Private Delaney—were common.

Lieutenant Hunter ordered our troops off the trail into almost impregnable jungle. A torrential rain came up to help cover our movement. Our force was split in two groups ... my part of the platoon moving in directions of our lines.

I was point man with machete, hacking away at the thick jungle growth. Sergeant Burton wanted us to move faster so he took over my position. He hadn't gone 50 feet, before disappearing with a scream. We never saw him again.

We continued to cut our way through the jungle growth for the next two hours ... and suddenly came upon a clearing littered with enemy bodies ...

Marine Raiders had established a road block. They stared at us in disbelief as we looked down the barrels of their machine guns and automatic weapons.

The memories still linger of the patrol action, and the Marines who took part. **Michael F. Ryan, 31 High Valley Drive, Chesterfield, MO**

Through the Bougainville mud and muck, a Marine artillery unit carries food to the forward gun positions. (Defense Dept. Photo-Marine Corps)

Gasoline Alley

... Loading Of An LST

The sunrise splashed brightly over Guadalcanal's Iron Bottom Bay, silhouetting Tulagi on the horizon and the dozen or more Navy ships lying silently at anchor in Sealark Channel. Languid waves washed quietly over the gray sand of Lunga Point. Glancing frequently at my GI-issue Elgin I waited for my LST with trepidation.

This LST would be the first of its kind I had ever laid eyes on; APA's, AKA's I had ridden, but no LST. And yet I was going to be the loading officer of this one for the invasion of Bougainville. All that I knew about the seagoing landing ships was from the diagrams and mockups in the G-4 tent deep in the coconut grove at division headquarters.

For weeks the loading officers of the various ships, lieutenants and warrant officers, none of whom had volunteered for the job, had pored over diagrams of cargo decks, fitting in jigsaw puzzles with little templates representing tanks, trucks, jeeps, artillery pieces and other rolling stock.

Then came the long, bouncing jeep rides over the dusty coastal road, from Tetere to Tassafaranga, conferring with unit supply officers who would have cargo to load on the LST, and locating staging dumps in the beach loading area.

Finally, the ship came into view, a low hulk creeping silently towards the beach, the clam-like doors of its bow already parted. It grounded itself 30 feet offshore. The loading ramp swung down on its cables revealing the dark emptiness of the cavernous tank deck. A petty officer inside motioned me to come aboard.

The longest day of my military career began agreeably with a cup of coffee in the wardroom. After that it was one snafu after another; Murphy's Law was being updated. Supplies that were supposed to load first (first on, last off in landing operations) showed up late, in some cases not at all. Truck drivers with loads of 155 millimeter projectiles went to the wrong beach; loading crews couldn't find beach dumps; vehicles broke down on the ship's ramp.

All the while the skipper, a lieutenant, kept a running reminder of his schedule. Precisely at the next high tide he would pull away from the beach, loaded or not. By late afternoon things finally began to go right and my hopes rose that everything would fall into place in spite of the snafu. Along the length of the ship, on the port and starboard sides next to the bulkheads were tons of bombs, artillery shells, rations and cases of explosives stacked over rows of 55-gallon drums of aviation gasoline. We were a floating disaster waiting for a Jap torpedo, but by golly we were ready, except for that damned barb wire. The beach crew had not been able to

locate the stash of barb wire belonging to an assault infantry battalion.

Despite my pleas for an extra hour to locate the wire, the skipper started the engines, reversed his screws, and pulled away to an anchorage in the channel. In a couple of hours we would join the invasion convoy. Wearily, I flopped into a bunk and crashed.

The next thing I remembered someone shaking me and shouting something about the barb wire, interspersed with threats of a court-martial. In the dim light of the cabin, I caught sight of the gold leaves of the major, the executive officer of the infantry battalion whose barb wire reposed in some remote supply dump.

We awakened the skipper, but he was not moved by my pleas to return for the wire, nor by the major's menacing presence.

If there was confusion in loading at Guadalcanal, D-Day at Empress Augusta Bay was a chaos by comparison. The narrow beach at Torokina Point was so cluttered with supplies that we were directed to off-load on the tiny island of Torokina some 500 years from the main island.

... with the theme getting the Marines to Japan "on the roads the Seabees built."

With consolidation of the ridge, and as additional Army troops continued to arrive, Admiral Halsey directed the Army's XIV Corps to take over, commanded by General Oscar W. Griswold.

Most of the troops did have turkey on Christmas Day, but two days later the whole 3rd Division was relieved by the Americal Division. The 9th was relieved by the 164th, the 3rd Marines by the 132nd, and the 21st by the 182nd Infantry—all seasoned veterans.

By January 16 all of the 3rd Marine Division had returned to the palm fringed lagoons of Guadalcanal. There again were such merry places for R&R as Tassaforonga, Kokumbona, Tulagi (if you could get a boat and find a girl).

And now we waited with "corn willie sandwiches" and Spam deluxe for a new adventure—at government expense—like Guam or Kavieng. **Cyril J.O'Brien**

Unloading of LSTs. Photo courtesy of Frank E. Ross, USNR of Framingham, MA.

A harried beach-master at Torokina put us on short notice to unload. A Japanese naval task force posed a threat to the beachhead and air raids were imminent. There was barely room on the cluttered little island for the vehicles. Bomb craters impeded movement. The rest of the cargo was wrestled off the LST by a sweating working party and piled helter skelter wherever room could be found.

All the while our nervous LST skipper was looking at his watch with one eye and at the sky with the other, growing more apprehensive by the minute. The struggling work crew was down to the barrels of aviation gasoline when his patience was exhausted. "Get your men off, we're pulling out," he shouted. The barrels, he said, would be pushed off over the bow by the ship's crew. They would float on the water and the tide would wash them ashore where they would be recovered.

Apparently the beach-master con-

curred with this hare-brained idea or he just didn't give a damn. Reluctantly, we waded ashore as the LST pulled away. Some 500 yards out the first barrels could be seen bobbing up and down on Empress Augusta Bay. But where would they land? Not on our Bougainville beachhead. Perhaps not even on any of the Solomon Islands. A swift current was taking the barrels out to sea, far out. We watched until the drums were mere specks on the water.

I still wonder where they eventually came ashore, New Guinea perhaps.

Before nightfall I joined my artillery unit and reported the details of the barb wire to my battalion commander. An understanding man, he shrugged and told me not to worry about it. We had a battle to fight.

The next day I learned that the irate major had been killed in the fighting along the Piva Trail. **Cliff Cormier, Gainesville, FL**

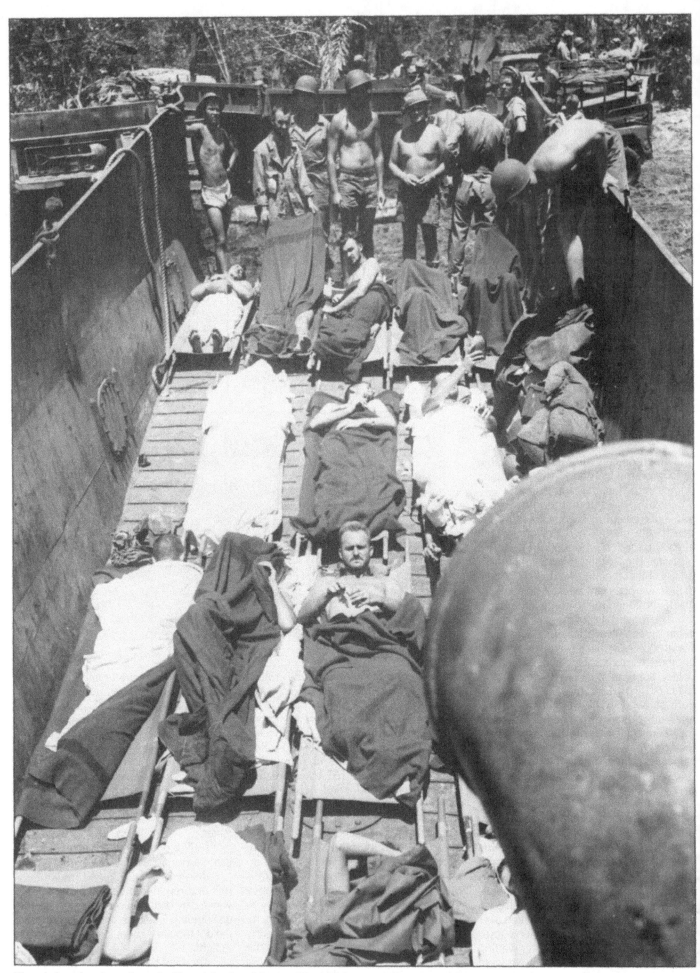

Wounded ready to leave Bougainville mainland are lying in a Higgins boat to go out to a hospital ship. Stretcher bearers have just carried them from the trucks and jeeps seen in the background. (Defense Department, Marine Corps photo)

Dealt A Bum Hand

I was sworn into the Marines with a fellow I met as we waited for the ceremony to take place at the recruiting office on Broadway in New York. He was a native New Yorker, born and raised on the island of Manhattan. He was older than most of those present, probably close to 30, while the rest of us were nearly ten years younger.

Lou was a short, swarthy man, very husky and tough as nails. He had a heavy beard and once on Parris Island, the D.I. made Lou shave every night with a dry razor, until he realized that Lou had a permanent 5 o'clock shadow.

We were shipped to Samoa where we joined L Company of the 3rd Marines. One day on Samoa I was watching a poker game when Lou came in to borrow some lighter fluid. He stood next to me watching the game. We were behind a newly arrived rookie who was losing steadily. After awhile the man got up shaking his head; he had been cleaned out.

Lou immediately took his seat. He proceeded to win until he had won the exact amount that the boot outside had lost and then quit the game. He took the boot outside and gave him the money he had won, and advised the man not to play poker again with the old-timers. He did not say he had been cheated, just that he was out of his class. He warned the man not to say anything to the rest of the men about his card playing skills, and the man complied.

A few days later I walked into Lou's tent. He was sitting on his cot with a deck of cards, cutting them using only his right hand. I watched fascinated until he looked up. When he saw the expression on my face, he laughed. I asked him about his skill with a deck, and he asked me not to say anything about it. He said he could do virtually anything he wanted with a deck of cards. He would not play cards with the other men, because he did not want to cheat them. He sat at a table and dealt out five Seven Card Stud hands. He turned over his hole cards and exposed three aces. When he saw the astonished look on my face, he smiled and scooped up all five hands, shuffled the deck, and dealt out five more hands. He then read off all the hole cards before turning them over. I promised never to reveal his skill to anyone, and I never did until I returned to the states. I never saw him play another hand of poker, but I know others saw him practising his dexterity.

The day of the Battle of Piva Forks, Lou was pinned down on the bank of the river with the rest of his squad. They managed to get across the river while losing one man to wounds. They were already short several men and were down to five men in the squad. The squad leader was hit and Lou took over the control of the remaining squad members.

The men followed Lou's orders as if he had always been the squad leader. He got one BAR man in position to fire at an enemy bunker. He ran to a spot behind a banyan tree from where he signaled the other BAR man and his assistant to his flank. The gunner and his assistant scrambled to the spot Lou indicated. The firing was intense, and Lou could not make himself heard, so with much arm-waving and pointing, he signaled the last man up.

A bullet caught Lou in the right hand. He raised his hand in disbelief and stared at it. Several more rounds struck him, two in the hand he had raised, and his hand disintegrated. Lou turned and looked toward the enemy. He picked up his M1 and slowly stood erect. Blood was coursing down his side from his wrist. He walked out from behind the banyan tree and placed the rifle to his shoulder.

He tried to pull the trigger with his missing hand. Lou is buried in the Punchbowl Cemetery on Hawaii ... Most of him. **Douglas Lyvere**

Douglas Lyvere of Hollister, California was awarded the Bronze Star and Purple Heart during the Bougainville Campaign. He is a retired businessman and amateur writer.

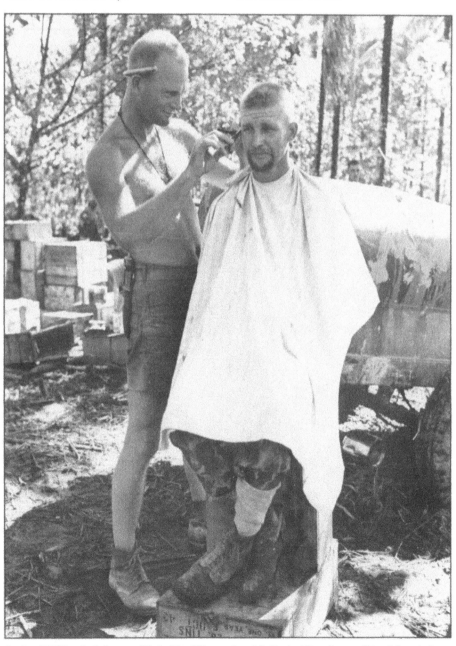

Barber PFC Leslie Johnson of Cain City, TX gave Sgt. J.R. Beck of Hawthorne, CA a deluxe haircut on Bougainville. (Official U.S. Marine Corps Photo)

300 Mile Hike ... 300 Miles?

I came from Columbiana, Alabama where my family resided, and Howell Heflin—now a United States Senator—was platoon leader, third platoon of A Company, 9th Marines. His father was a Methodist minister and, at one time, was in the adjoining county to where my parents lived.

I had a letter from my father saying he had received a letter from Howell Heflin's father, as well as a phone call from him. The call said, "Mr. Fowler, I have the most amazing news from Howell. He said in the month of May, the 3rd Marine Division went on a 300-mile hike." It was his impression that it was one continuous hike. As a matter of fact, we did a total of 300 miles, but it was in increments of 20 miles on a daily basis.

I don't suppose we ever had more than four or five miles of hiking in a combat situation. The longest I can recall is on Bougainville, having been relieved from the front to the rear ... and then having to go back up on the front lines.

We were in the act of enjoying a very late Thanksgiving Dinner, as I recall, turkey and trimmings—something we had been looking forward to. And it fell to my lot as company commander to tell the company lined up for chow to go back, get their packs and get ready ... we had to move out.

So the company did go back, get their packs, and pass through the chow line with their mess gear open. There we went down the trail eating our Thanksgiving Dinner. Total march from dinner to the front lines was probably not more than four miles. **Conrad M. Fowler, Lanett, AL**

Jungle Chow Line

Marines carry chow and ammunition up Hand Grenade Hill on Bougainville. This was the only means of transportation that could make its way through the jungle as the trails were too narrow for even a jeep. (Defense Dept. Photo - Marine Corps).

The Right Time For Goofy Orders

From 1991 to November 1, 1943 is a long time to look back but this is how I remember the landing on Torokina Point, Empress Augusta Bay ... My name is Edward N. Bieri, Alliance, Ohio, and on that morning ... I was in the same Higgins boat as our first battalion commander Major Leonard Mason, nickname "Spike," serving as a radio operator.

Shortly after running off the Higgins boat, Major Mason had been hit, I believe in both legs below the knees by machine gun fire. Although following close behind him I somehow didn't get hit.

As I recall the first few hours were pretty hectic due to rifle companies A and B landing further to the left than planned and C Company and headquarters landing in their place.

I remember being close to Major Mason prior to his being evacuated and him stating how long he had trained himself and his battalion for combat and how he was leaving before the first day was over ... I next reported to Major Steve Brody or Major Bailey.

Just after this I kept hearing someone hollering in a language I didn't understand and finally decided it was Japanese. I asked one of my buddies how that Jap could still be close without getting shot. I was told it was Captain Gordon Warner, who could speak Japanese, yelling goofy orders to confuse any Japanese troops in hearing range ... **Edward N. Bieri, Alliance, OH ...**

Relished By The Best Of Men

Each 1640 General A.H. Turnage, CG, 3rd Marine Division, reinforced, would take a break for a round of bourbon with his staff.

One day on Bougainville, the general, summoned me to the command center; quietly suggested that two bottles of bourbon should be stashed in my field desk. "Supply is getting little low" he commented.

Thereafter, when the field was clear (staff officers not present) the general would give me a wink.

From the battered field desk would come one of the prized bottles of bourbon.

The general, on occasion, would summon me following an evening movie (with Sergeant T.O. Kelly). Then the two of us, the two-star general from Fayetteville, North Carolina, and the master sergeant reservist from Hershey, PA, would chat while we enjoyed a beer.

We never shared a bourbon until several years following my discharge from the staff of "The Boot" at Parris Island. On that occasion my young family and I visited with the General and "Miss Hannah" in Alexander, Virginia. Then, we shared a few laughs, and a shot of bourbon! **Brent I. Hancock, Hershey, PA**

It Happened On Bougainville

Proud Of Cannoneers

E.E. May won't forget Bouginville nor his old 12th Marines. He was a little older than the rest and they called him "Pappy" ... though he hadn't turned 30. The gunners he knew turned the tide in many an encounter, and May's proud to tell anybody in Indianapolis about that.

Kept Us Free

Clifford L. Jones of McGehee, Arkansas is a charter member of the 3rd Marine Division and is proud of that. He came to the division early on, transferred out of the Second Division. His platoon leader in those days was a young lieutenant named Louis H. Wilson (later Commandant). Jones lost a lot of friends in the islands, but he feels "we are free people today because of them."

The Big Crawler

James G. Helt, Pittsburgh found on Bougainville that the Japanese were not the only enemies. How about the centipedes, three fingers wide? Yessirree!! One ran over Jimmy's neck, stuck him with a rear end fang ... and Helt became sick and vomited for a week.

Jimmy was quite a dancer, jitterbug that is ... He says his girl "Becky" back home was the best swing dancer in all of Pennsylvania.

50 Years

E.K. Brinkley of Rogersville, Alabama, as most of us, can't believe it's nearly 50 years ago ... "All of those young men I knew so many years ago ..."

Almost Unconscious

In 60mm mortars with F Company of 2/3, **Norman Tyndall**, later of H&S 2/3, will never forget when the friendly Naval officer lined the bunk of one of the poor enlisted with enough alcoholic delights to send a company to Nirvana. Norm went there too ... and it's not a thing he can forget, for he can't remember it either. Norman retired recently from Immigration and Naturalization Service. **Norman Tyndall, Coeur D'Alene, ID**

Glad You Made It, Phil ...

Thiac (Sergeant Phil) looked like death warmed over, and his stretcher bearers didn't think he would be alive very long.

Word came that L Company is almost wiped out; all the officers KIA or WIA, and they trudged on through the muddy jungle to the Aid Station. It was

Yamamoto's Aircraft A Jungle Curiosity

Admiral Yamamoto Plane wreckage Buin. Japanese Commander in Chief travelling in a Betty Bomber, shot down by US fighters April 1943. Photo: Warren Scott

Here are the remains of the Betty bomber which carried the Japanese beloved Admiral Isoruku Yamamoto, commander in chief, Combined Fleet. It lies where it fell nearly 50 years ago (April 18, 1943) in the Bougainville jungle.

The shooting down of Admiral Yamamoto by a strike force of 18 P-38 aircraft seriously damaged the morale of the Japanese military. The planner of the attack on Pearl Harbor, and actually described as a "friend" of the United States, Yamamoto opposed the war but when called upon he did his utmost for his country.

Still difficult to approach, the ruins have been visited by Japanese who are expected eventually to make a shrine of it.

later learned that L Company was in fair shape, perhaps a little disorganized as happens in battles such as Piva Forks.

Glad to know you made it, Thiac. See you at the reunion. **Charlie "Boondocker" Kennedy, platoon sergeant, F Company, 3rd Marines 1942-1943, captain, USMC (Retired)**

Two Intrepid Marines

It was in the battle of the Koromokina River that Sergeant Herbert J. Thomas, South Charleston, West Virginia led his squad to destroy two machine gun positions ... and was attacking a hidden third when the grenade he threw rebounded from vines to fall back among the squad. Thomas smothered its explosion with his body. He attended Virginia Polytechnic Institute and the State University, was active in sports and just about everything else.

Private First Class Henry Gurke, at a 3rd Marine Raider roadblock near Empress Augusta Bay, was in a foxhole barraged by Japanese grenades. Certain the enemy could annihilate the position, Private First Class Gurke threw himself on a grenade to spare the automatic rifleman whose life could have made the difference ... and it did.

Thank God For A Watchful Dog

We had war dogs with us on Bougainville. One night one of the dogs growled and Slim Livesay, a squad leader from Montana, shot and hit a Jap right between the eyes. We found the Jap the next morning, three feet in front of the hole. **Sergeant William Otho McDaniel, Paragould, AR**

Alfred E. Lewis seems to leap as gleefully as a spring bird from his subterranean foxhole on the edge of the Torokina strip on Bougainville.

Hunkered down away from mortars, a communications team passed the word. (U. S. Marine Corps Photo)

Though known only by a few, the feats of the Coastwatchers were tremendous. This is one of them, Paul Mason

The Coastwatchers

"From Perches On Bougainville They Watched Us, And For Us"

"In the history of war there is no finer chapter than the heroism of these men," says the Oxford University Press, first to ring out the post-World War II story of the Coastwatchers after years of secrecy when they received no more official attention than wraiths on the jungle landscape.

The Coastwatchers were landowners, plantation managers, shopkeepers, public servants, roustabouts, and strayed Europeans who knew the land, its people and how to live and hide.

They operated with bulky tele-radios (16 men to carry) from the far borders of New Guinea through Papua, Solomons, Admiralties, to the far New Hebrides. They were formed under Australian Intelligence Director, Commander R.B.M. Long, and commanded by Lieutenant Commander Eric Feldt. General Douglas MacArthur called their heroism "decisive." Their code name was "Ferdinand" for the friendly bull.

Bougainville was a critical watch post for the Coastwatchers, notably Paul Mason and Jack Read. Their mountaintop observations of Japanese aircraft and warships headed for Guadalcanal and Tulagi provided the alerts that kept the Cactus (Navy) Air Force of Gregory "Pappy" Boyington aloft.

Once Mason and Read reported 24 bombers headed to the Canal in time to allow the defenders to shoot down 23. The watchers could even report the return of the one lone aircraft.

Paul Mason, particularly, was hardly cast in the hero image. Glasses, self-effacing, soft spoken, he was described by Walter Lord in "Lonely Vigil" (Viking Press) as a likely bank clerk who wandered into the brush.

But looking the part or not ... Mason and Read were often only steps ahead of torture and death. The Japanese were once at Mason's front door when he fled out the back.

Mason's famous report, "Forty bombers headed yours," called from his mountain aerie, is now classic World War II history. The Marines accounted for 26 of them. Admiral William F. Halsey said succinctly: "Mason and Read saved Guadalcanal, and Guadalcanal saved the Pacific.

Knew The Territory

Mason had worked and managed a plantation north of Empress Augusta Bay, and he camped and traveled as a Coastwatcher through the haunts we knew: the Laruma River, Torokina, Numa-Numa Trail. He operated south near Buin while Read was in the northern Buka area.

Fortunately the Australians thought to give the Coastwatchers military rank which would protect them by Geneva conventions. As civilians they would be spies and shot on the spot. The military rank protection helped little for any Coastwatcher who fell into the hands of the Japanese where they were tortured, beheaded, or otherwise dispatched.

Historians now say that the Japanese command at Rabaul would have been wiser to put even more emphasis on catching Mason and Read on Bougainville—though the Japanese did their best. The Coastwatchers' cryptic notes often spelled disaster for unwary Japanese air and sea skippers.

Coastwatchers were landed again on Bougainville before our assault (3rd Marine Division) to help guide our ways and warn of Japanese movements. (On Guadalcanal Martin Clemens worked with 1st Division Commanding Officer General Archer Vandegrift. Among Clemens' assistants was the legendary Sergeant Major Jacob Vouza, a local who scouted for the Americans, was captured, bayoneted, and left for dead by Japanese ... but who made it back to our lines.)

With a large team necessary to carry the radios, provide security, and scout against the Japanese, Mason depended heavily on the Bougainvillians—although a few did go over to the side of the Japanese.

Shortly before Bougainville was recently stricken by a revolution of landowners, which still paralyzes its economy and government, Mrs. Noelle Mason returned to the island and to the Inus Plantation which she and Paul operated after the war.

There, in a little memorial park cleared and landscaped by the Lions Club of Bougainville, she visited the grave of a Bougainville native son, named Boros. It was Boros who refused to reveal the hiding place of Mason. For this reluctance he was beheaded and buried on the spot.

More Than Just Coast Watchers

It was Coastwatchers who drew the young skipper of PT-109, John F. Kennedy, out of Japanese territory near Blackett Strait, Munda, New Georgia. They rescued a group of nuns and whisked them off to a submarine. And, they aided the getaway of some 165 USS *Helena* survivors, hidden by the Reverend A.W.E. Silvester, a Methodist missionary on Vella Lavella.

Actually only a few of top rank or G-2 personnel in the 3rd Division ever chanced to meet the phantoms called Coastwatchers. From perches on Bougainville they watched us, and for us ... and some of us are just learning of that now.

Mrs. Noelle Mason dusts off the inscription on the memorial to the Bougainville man named Boros who lost his life when he refused to reveal the whereabouts of her husband Paul Mason.

Christmas In The Marine Corps

Marines don't always enjoy Christmas, but they never forget it, even if their foe does. This is clear from the collection of stories that follow: anger, terror, frustration, irony and wild humor have all been components of Christmas, Marine style.

(My own three Christmases in the Marine Corps remain memorable: 1956, with the 6th Marines, drinking Aqua Velva in the barracks at Camp Lejeune; 1957, walking guard duty on Okinawa with the 9th Marines; and especially 1958, again with the 6th Marines, restricted to ship on the French Riviera, when my buddies smuggled aboard everything I needed for a Christmas celebration!) **Paul Colaizzi**

A Message From Pappy

This entry was not written about a 3rd Division Marine. But Marines of the 3rd Division fought and died to get him and other Marine pilots an airstrip to fly from on Bouganville.

There are some Marines whom every Marine has heard of, and there are some who are known to all Marines but also to all civilian citizens of the United States. Colonel Gregory "Pappy" Boyington, a native American of Pima Indian origins, is one of those known to Marines and civilians alike. His exploits are legendary.

He shot down Japanese planes over Burma and China serving with the Flying Tigers. He shot them down all over the Solomon Islands from one end to the other, from Guadalcanal to Rabaul, while leading his Marine Black Sheep Squadron. He shot down more enemy planes than any other Marine pilot in World War II, 28 to be precise. For this he won the Medal of Honor.

But he didn't receive his medal until after he returned from the dead, so to speak. He himself was downed over Rabaul, declared missing in action and presumed dead, then endured the harsh captivity of a POW in Japan for nearly two years. (The story we heard in Japan in the 1950s was that our 9th Marines barracks on the slopes of Mt. Fuji were the site of his POW camp.) From his accounts of his POW experiences, however, it was just as much his Japanese tormentors who were his captives as he was theirs (reminiscent of another Marine POW, Martin Boyle, author of *Yanks Don't Cry*).

All of this is graphically described in Colonel Boyington's own book, *Baa Baa Black Sheep*. In the entry that follows, he describes what it was like to be a combat pilot in the air fights over Rabaul on Christmas Day 1943, in Chapter 12, "Good will to all men ..." **Paul F. Colaizzi**

"Good Will To All Men..."

... On Christmas on the peace-on-earth-good-will-to-men day, I went around the skies slaughtering people. Don't ask me why it had to be on a Christmas Day, for he who can answer such a question can also answer why there have to be wars, and who starts them, and why men in machines kill other men in machines. I had not started this war, and if it were possible to write a different sort of Christmas story I would prefer to record it, or at least to have it occur on a different day.

I was leading a fighter patrol that was intended to intercept any enemy fighters that followed our bombers, which had preceded us to Rabaul. We saw them returning from their strike at a distance, and saw that Major Marion Carl's squadron was very capably warding off some Zeros; before we got within range, I witnessed three go up in flames from the .50-calibers triggered by Carl's pilots.

We caught a dozen or so of these fighters that had been heckling our bombers, B-24s. The Nips dove away and ran for home, Rabaul, for they must have been short of gasoline. They had been fighting some distance from their base, with no extra fuel because they wore no belly tanks. They had not expected us to follow, but we were not escort planes and didn't have to stay with our bombers.

Nosing over after one of these homebound Nips I closed the distance between us gradually, keeping directly behind his tail, first a thousand yards, then five hundred, finally closing in directly behind to fifty feet. Knowing the little rascal couldn't have any idea he was being followed, I was going to make certain this one didn't get away. Never before had I been as deliberate and cold about what I was doing. He was on his way home, but already I knew he would not get there.

Nonchalantly I trimmed my rudder and stabilizer tabs. Nonchalantly, I checked my gun chargers. As long as he could not see me, as long as he didn't even know I was following him, I was going to take my time. I knew that my shot would be no deflection and slowly wavered my gun sight until it rested directly upon the cross formed by his vertical tail and horizontal wings. The little Nip was a doomed man even before I fired. I knew it and could feel it, and it was I who condemned him from ever reaching home—and it was Christmas.

One short burst was all that was needed. With this short burst flames flew from the cockpit, a yellow chute opened and down the pilot glided into the Pacific. I saw the splash.

Using my diving speed with additional power, I climbed and as I climbed I could see off to my right two more enemy planes heading for Rabaul. One was throwing smoke. I closed in on the wounded plane, and it

dove. His mate pulled off to one side to maneuver against me, but I let the smoker have it—one burst that set the plane on fire—and again the pilot bailed out.

His mate then dove in from above and to the side upon my own tail to get me, but it was simple to nose down and dive away temporarily from him. From a new position I watched the pilot from the burning plane drift slowly down to the water, the same as the other had done. This time his flying mate slowly circled him as he descended, possibly as a needless protection. I remember the whole picture with a harsh distinction—and on Christmas—one Japanese pilot descending while his pal kept circling him. And then, after the pilot landed in the water, I went after the circling pal. I closed in on him from the sun side and nailed him about a hundred feet over the water. His Zero made a half roll and plunked out of sight into the sea. No doubt his swimming comrade saw me coming but could only watch.

This low altitude certainly was no place for me to be in enemy territory, so I climbed, but after searching for a half-hour, I saw no more of the little fellows in this vicinity.

I next decided, since I was so close, to circle the harbor of Rabaul so that I could make a report on our recent bombings there. Smoke was coming from two ships. Another had only the bow protruding from the water, and there were numerous circles all around that had been created by exploding bombs.

While I was looking at all this, and preparing mental notes, I happened to see far below a nine-plane Nip patrol coming up in sections of threes. Maneuvering my plane so that I would be flying them from the sun side again, I eased toward the rear and fired at the tail-end-Charlie in the third "V". The fire chopped him to bits, and apparently the surprise was so great in the rest of the patrol that the eight planes appeared to jump all over the sky. They happened to be Tony's, the only Nip planes that could outdive us. One of them started after my tail and began closing in on it slowly, but he gave up the chase after a few minutes. The others had gotten reorganized, and it was time for me to be getting home.

On the way back I saw something on the surface of the water that made me curious. At first I thought it was one boat towing another but it wasn't. It was a Japanese submarine surfacing. Nosing my Corsair over a little steeper, I made a run at the submarine, and sent a long burst into her conning tower. Almost immediately it disappeared, but I saw no oil streaks or anything else that is supposed to happen when one is destroyed, so I knew I had not sunk her.

My only thought at this time was what a hell of a thing for one guy to do to another guy on Christmas. **Colonel Gregory S. Boyington**

A Gift For A Chapel

In the midst of being shot at by Japanese snipers on Bouganville, on Christmas Eve, 1943, one Marine tanker found something spiritual in it all. The author, Thomas F. Murphy, a lifetime member of the 3rd Marine Division Association, was that Marine.

Christmas Day 1943, on Bougainville was approaching. Action had begun to lighten up for our 3rd Tank Battalion ... and the Army was soon to relieve us.

Many of us had assisted Father Conway or other battalion or division chaplains at Sunday services. So Corporal John Siracusa of East Boston, now a retired longshoreman, and I decided to set up a chapel for Christmas.

What to use?

We were now in a rear jungle area, and so scouted the area for a location and anything to use for a chapel. We headed for a clearing—not far from the intersection of two fine roads built by the Seabees. Slightly up a hill we could look down on the road going toward the bomber strip.

As we approached the clearing, there lay a brand new tarpaulin, all tied up in regulation style ... and the lines attached to the canvas were new and unused—like everything had just come out of a Navy warehouse. Alongside the giant-sized tarp was 50 to 75 feet of brand new one-half or three-fourths inch line.

These were sitting in the middle of the clearing as though gently placed there by some unknown persons ... in an area where there were no signs of any encampment, and not a soul in sight.

I said, "Johnny, I do believe this was forgotten by a Seabee unit. Let's borrow it for our chapel."

John was a pretty rugged fellow and we had no problem hoisting up the tarp and taking it back to our area. We strung the line between trees and hoisted that fine tarpaulin up and over, tied it down, and secured the side lines. We now had a fine roof for Christmas.

Next was an altar of sorts, which was not too difficult.

We needed flowers ... and I remembered a place down the hill near the river below—where we had been dug in a few weeks before. Called E Company Grove, it had been kind of a sniper's playground ... but I remembered there were giant beautiful red flowers in that area.

John and I went over and cut some of those flowers and brought them back to the chapel ... arranged them with some large green leaves and tree branches for a background.

We had a beautiful and very peaceful Christmas service. I still don't know ... or can I forget the whole affair ... but who left that tarpaulin in the clearing? **Thomas F. Murphy**

Admiral Arleigh Burke
"31 Knot Burke" Protected Marines At Bougainville

So close to the beach at Empress Augusta Bay, 3rd Marine Division troops mistook its flashes and thunder for a "summer storm" this fast and hectic sea battle stayed an angry Japanese fleet from chewing up the transports and beachhead of the Bougainville invasion. Dispatched at high dudgeon from Rabaul, the red-ball fleet was intercepted by Rear Admiral A. Stanton Merrill's Task Force 39, notably Destroyer Squadron 23, the Little Beavers of Captain Arleigh A. Burke. It was in turning the Japanese around that Burke, later the prominent chief of Naval Operations, won himself the title of "31-knot Burke." All considerations paled said Admiral William F. Halsey, South Pacific fleet commander ... "beside (the) clear duty of protecting the Marines at Torokina."

**With permission of
Random House, Inc. New York, NY**

Saved by A Curtain of Fire

When we suspected pretty substantial Japanese installations inland from the Bougainville beach at Koiari, 3rd Marine Division headquarters decided on ending all speculation. Major Richard Fagan had just arrived at Empress Augusta Bay on November 29 with his First Parachute Battalion. They got the job of making a destructive raid on the Koiari Beach area, tearing up everything they saw, and getting out.

To play safe, we made test landings; and reconnaissance spot-checks found no Japanese at all. The coast was clear. We went ashore and landed plumb in the middle of a Japanese supply dump. A Japanese officer was as surprised as we, came out to meet us, and was killed.

In no time we were 350 yards across, and 200 yards inland. The welcome from the Japanese was light. They just hadn't realized what had happened. They learned of the presence of the "Chutes" very soon, however, and came on strong. Even with our reserve in the skirmish, it appeared the Marines could, by nightfall, be pushed into the sea. For sure, any objective inland was kiboshed.

So Major Fagan needed a rescue. It came, and skillfully. Naval craft appeared off shore and put down a curtain of fire around the tangible perimeter. The curtain of fire around the Marines, effectively boxed them in from the angry enemy right there in the front.

The fire enabled the Marines to get aboard the rescue craft, then aboard ship and taken back Empress Augusta Bay. The group had been hardly able to get off the Koiari Beach to go inland...and even luckier to get off to go home. A total of 290 Japanese died compared to 15 Marines.

A Walk Underwater

With new replacements after Bougainville, our new lieutenant (right out of school) wanted everything by the book.

We were on an exercise that called for "A" Company, 21st Marines to advance in a combat patrol formation. As lead scout I was out in front by about 150 feet as we advanced through grass four to six feet high. About half a mile from the jump-off point, I came upon a bend in the river. There was a 20 foot, steep bank to the water, which was deep and fast at this point. To my right, about 200 feet from the bend, the water was shallow. To my left, about 300 feet away, the water was also shallow.

I went down the embankment, and pushed off into the river, with the second scout right behind me. The water was about eight to ten feet deep for about ten to 12 feet out, then it shallowed for the next 30 to 35 feet.

Upon reaching the other side, we checked the area and then signaled for the rest of the men to cross. First man down the bank was our BAR man, Jack R. Watson from Sanford, North Carolina. He secured the BAR to his left arm so as not to drop it in the river; then with all the rest of his equipment, he entered the water and promptly went under. We waited, but he didn't come to the surface—not even to pop up for air.

I pulled off my pack and ammo belt and ran back to the river, arriving just in time to see Watson walking out of the water on our side.

He had been unable to surface due to the wight of his BAR and all of his equipment .. so he had walked across, under water, on the bottom of the river.

Needless to say, our new lieutenant had a change of plans and no one else had to enter the river at that position.

Jack R. Watson made it through the rest of our campaigns, but to the best of my knowledge ... has not gone swimming to this day. **John F. Pelletier, Warner, NH**

Peaches For The Captain

In 1943 on Guadalcanal there was a food supply dump called the "A & P Store" where outfits could pick up canned goods. My outfit (K-4-12) sent over a truck and loaded it with canned food. Needing someone to guard the food that night, I drew the watch duty.

The procedure was for the person on duty to just lock himself inside with the food. During the night I decided our tent needed some peaches, so I took a case of peaches over to our tent. The next morning the mess sergeant discovered the case of peaches missing and reported it to the captain.

Since I had been on duty, the captain came to my tent to ask about it, and I admitted that I had taken the peaches. He asked what I was going to do with them, so I told him my tent mates and I were going to eat them.

The captain responded, "Open a can and let's get started." After eating the peaches, he thanked me and left. Nothing more was ever said about the case of missing peaches. **Richard M. Coulter, Daingerfield, TX**

Harbor Of Siege

As the morning of August 1 grew we approached Guadalcanal of which we had heard so much. At dawn the mountains of San Christobal had been off to port and by 0930, we were entering Indispensible Strait, the eastern entrance to Iron Bottom Sound. By 1230 we were off Tetere Beach, and with the other ships took up our assigned position. The anchor chain roared out of the hawser pipe and in the light breeze the bows swung toward the shore.

The Sound in which we lay had become the graveyard of many American, Australian, and Japanese fighting ships and sailors in the past year, but on that sunny morning the calm, blue surface gave no hint of the deadly battles fought there.

Astern 20 miles, Florida Island with its Tulagi base rested on the horizon.To starboard, a bit farther off the cone of Savo rose a blue cloud ... there the year before, Quincy, Vincennes, Astoria, and Canberra were sent to the bottom in the brief and deadly fight with Admiral Mikawa's seven cruisers.

Tetere was about the halfway point on the Canal's 85 mile northern shore, well outside the besieged perimeter of the original landing.

From the ship we could see the fringing beach, backed by coconut palms, and the thin screen of tropical hardwood in the middle distance—bare brown ridges covered with kuni grass (the fall coarse sawtooth grass of the tropics). In the distance were the rainforest—covered mountains which reached to 8,000 feet.

Happy Days At Tetere ...

Bill McAllister, a telephone lineman with H-2-3 talks about the life at Tetere.

The trees were in rows, about 25 feet apart. Tents (pyramidal) were in neat rows, between trees. Six men lived in a tent, two to a side, two across the back and one on each side of the flap.

The center pole had boards nailed across to hold weapons by their slings ... and any food (packages from home) was suspended with kerosene-soaked string to keep away ever present ants, rats, and lizards.

Down the company street about every 20 tents was a Lyster bag. This contained purified water, drawn from a nearby river. Near the bottom of the bag wee several small push button faucets for filling canteens.

The mess tent had tables like picnic tables. Food generally ranged from terrible to inedible, although the cooks did a credible job with Spam—fried, baked, boiled, minced, sliced or chopped. And there were powdered eggs, powdered milk, good coffee and good bread—bread that looked liked cracked wheat with maybe weevils baked in.

On one occasion I was part of a "work party" who staged an armed holdup of an Army supply dump and liberated a truckload of Ten-in-One rations (one man, 10 days) ... which contained such delicacies as canned peaches.

We did visit the Seabees who had great chow ... and would let Marines go to the head of the chow line, and boot out any dogfaces who tried to take advantage of their hospitality ... and on, and on, and on ... **Bill McAllister, Churchville, NY**

Saw The Rescued Jack Kennedy

I saw the future President Jack Kennedy after his rescue by the Navy and return to Guadalcanal.

Raymond Spiers from Waukegan, Illinois, Warrant Officer Danis, and I were on the beach on the Canal at the end of Henderson Field where the P-38s took off.

Raymond and I saw a Navy destroyer sending in a small boat. Five men were already on the beach and two were facing the water. There was one man to their right in a metal type Navy stretcher, bandaged up and unconscious.

Two civilians were standing at the officer's left. One was an Australian coast watcher, and the other an island native. I called some Marines nearby and asked, "What's going on?"

They said the Navy had rescued the son of the ambassador to Great Britain. I knew I was looking at somebody important, and I got an uneasy feeling ... so we left. I did not realize for years whom we had seen ... until his story was told. **Richard Walker, Monroe, LA**

A Close Call For A Star

On the "Canal," before and after Bougainville, morale of the troops was considered essential to offset the vigorous training sessions.

Various recreational activities were offered on the battalion level: movies and boxing tournaments, always popular, however, could not compare with the enthusiasm generated by select members of the 1st Battalion, 21st Regiment's basketball squad. Each company in the battalion had a team and from these came members of this special group, all of whom had varsity high school and or college experience.

They were all good, but the one who stood out by far was Private First Class Goebel F. "Tex" Ritter of Able Company.

Goebel F. "Tex" Ritter

The Richmond, Kentucky native specialized in the one hand set shot, almost never seen in those days, and his range was from "down town!" Over six foot in height, and not yet 20 years of age, "Tex," upon receiving the ball, would delight the troops in their anticipation of a sure basket; and he did not often let them down in sinking the long one. "Swish" was the sound, as the ball cleanly settled through the net, time and time again, followed by tremendous cheering of the appreciative onlookers. This squad was not a one-man unit, however, as the team went on to compete for Regimental honors.

On Guam, in July 1944, Able Company with other battalion units, was fac-
(Continued on page 71)

Her Mind Was Made Up ...
We Fly Mrs. FDR To Guadalcanal

In the early days of World War II Eleanor Roosevelt was determined to visit the South Pacific war zone and comfort the wounded.

This proposed desire on her part met with stiff opposition from the President down the ranks of senior naval commanders in whose commands she would require inordinate attention for her safety and logistic support. The loudest complainer was Admiral Bull Halsey, Commander, South Pacific, who would be charged with her well being during this female invasion of his sacrosanct domain!

But Eleanor, a charming, mild mannered, well loved lady most of the time, could be a feisty tough fighter if her will was denied. The first to fall was the President who practically told her—a pox on your house—you're on your own. The rest was easy despite the anguished howls of King, Nimitz and Halsey. The buck fell on the latter to arrange the trip.

The flight to Noumea, New Caledonia, Halsey's headquarters, was easily arranged on a VIP flight from Australia as was the flight to the burgeoning base at Espiritu Santo, jumping off spot to the war zone. Eleanor, by this time, had been joined by Miss Collette Ryan, a big, friendly Australian attached to the International Red Cross; she was to be her constant companion on her dangerous mission.

How to get the First Lady and her companion to the war zone—Guadalcanal—was the next problem. Surface transportation was out of the question due to the risk involved.The Japs were still making nightly forays with the heavily gunned Tokyo Express running down "the slot."

Air was the only solution and Halsey thought of his heavily gunned Navy Liberators (PF-4Ys) and Sears, commanding officer of VB-104 and the Navy Search Group operating out of Carney Field.

Having flown the Admiral several times in Car Div Two and Anacostia, his next choice was simple—send Sears and his crew to Espiritu Santo (our rear base in the New Hebrides) and fly the ladies to Guadalcanal. Make it at night to arrive just before dawn. (This was the safest procedure travelwise; however, Henderson airfield was severely bombed the next day!)

We flew down to Espiritu next day in routine fashion, having negotiated the 625 nautical miles several times before. In the meantime, we had prepared old "48" as best we could to accommodate our distinguished guests. A plywood cargo liner was installed in the bomb bay upon which two passenger type seats were securely fastened. Flight jackets and several thick Navy blankets completed the accouterments.

A point of interest: our old workhorse PB-4Y had a most luscious nude young woman painted on the nose of the plane, under which was her indelicate pseudo name. The latter and the plane's nickname were painted over—but not the lady!

First Women ...

... on Guadalcanal are greeted by 3rd Marines in formation. Colonel William C. Hall is left. Although most of us never saw a skirt go by, their presence puritanized the area—no nakedness to showers, screen the urinals, and they even threatened to put pants on the Lyster bags.

Visit Of The First ... And Other Ladies

Eleanor Roosevelt, wife of President Franklin Delano Roosevelt and the First Lady in those days, and a very busy one indeed, visited Guadalcanal one day, recalls Colonel Austin P.Gattis, USMC, who was there with the 12th Marines.

"We got the word that she stopped her convoy of vehicles in the middle of one of those 'rivers,' stood up in her jeep and waved to a laundry bathing party alongside. It didn't seem to faze her that many of the Marines were totally nude. She simply waved, said 'Yoo hoo boys,' and drove off."

"After we'd been on the Canal for some months, we learned that six Army nurses had arrived. To this day I haven't figured out why. Surely they had little time for nursing, being entertained at officers' messes on the island.

"... We (12th) were one of the first units to be honored. Our commanding officer 12/3 assigned me to pick up the ladies ... Let me tell you when I settled my gal in that hot sedan and sat beside her, the smell of her makeup, powder and perfume was enough to take my breath away .. That plus seeing an American female after so long a time ... was traumatic ... haunted me a long time ...

"Just the fact that six American women were on the island generated an order from Island Command to the effect that we would immediately cover piss tubes, heads, showers etc. ... with cotton bunting available for that purpose from division dumps ... all 65,000 ashore were to comply. What price modesty!"

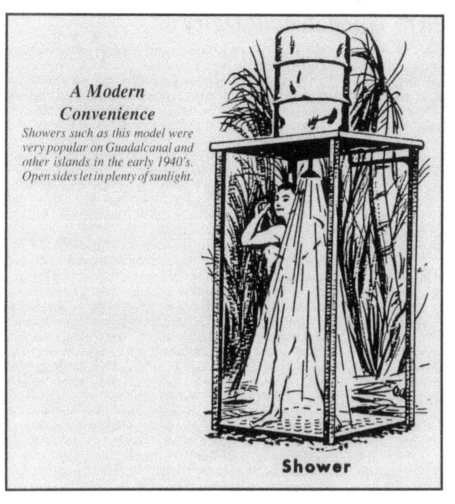

A Modern Convenience

Showers such as this model were very popular on Guadalcanal and other islands in the early 1940's. Open sides let in plenty of sunlight.

Shower

Jap Air Attack
Backfires With Downed Bombers

Red alert sounded after dark.

Normally we'd jump into slit trenches and stand up until we could hear aircraft approaching ... then get below ground. When standing we could still get a good view of the sky through the coconut trees.

One pitch black night—no moon—the search lights came swinging back and forth through the black sky. Suddenly three of the search lights locked on a Jap bomber. Short orange traces then streaked the sky and into the bomber. It burst into flames and started falling like a flare, but not as bright. It was like an orange ball, trailing smoke.

The lights switched to another Jap bomber. Tracers flamed the second. Then the lights locked on a third ... red and orange tracers came from above the Japanese plane. The second plane didn't hit the water near Savo Island before a third started to go down in flames.

In a matter of minutes, a fighter from the Cactus Air Force had shot down those three bombers before they were able to drop their bombs. I often wondered who the fighter was who accomplished this feat. **George J. Green, Webster Groves, MO**

ing the enemy. Private First Class Ritter was hit in the top of his helmet by a Jap bullet which creased his scalp. At the battalion aid station it was determined the wound was superficial but because of its near fatal nature, it was recommended he be evacuated. Ritter declined, stating he wanted to rejoin his comrades. Reluctantly the corpsmen agreed. However, about 30 minutes later he returned, asking to be sent back. Those 30 minutes were wrenching for him as he understood so vividly how close death came. He was directed to the Division Field Hospital.

Several years after the war, a corpsman who had treated Ritter on Guam, was watching a TV basketball game between the New York Knicks and the Washington Capitols. The corpsman became nonplussed in suddenly recognizing Ritter was playing for the Knicks. He mentioned this to a group of his high school students, citing the unusual wound Ritter received.

Several weeks later, one of the students attended another game between the Capitols and Knicks. At the end, he spoke to Ritter and asked if he remembered the name of a corpsman on Guam. He didn't recognize the name, but when asked about his head wound, Ritter replied, bending over showing his scalp: "I don't remember him but he sure remembers me!"

For three years, from 1948-1951, Ritter played guard for the New York Knicks. Understandably, Tex's point production in the NBA was not the high caliber he attained while starring with the unit of 1/21. Even so, for the three seasons with the Knicks he had an average of five points per game.

In reviewing Ritter's basketball background, one can readily understand how he made a professional team. At Eastern Kentucky State College, for example, he was twice named "all conference" by the Kentucky Intercollegiate Athletic Conference (KIAC), first in 1942-1943, and again in 1947-1948, after spending 16 months with the Marines in the Pacific. In his last year at college he scored 328 points in 25 games.

Basketball was not his only sport, however, as he was mult-talented in a variety of them. While at Eastern Kentucky, he lettered in basketball as well as baseball, golf, and track, twice winning honors in the state broad jump.

So if a Japanese rifleman on Guam had not been slightly off-target, the American sporting audience would have missed seeing a great athlete perform. **Carroll M. Garnett**

"Another" Guadalcanal Dairy

Carroll Garnett of Chester, Virginia is a free lance writer who served as a corpsman with the 21st Marines during World War II. These are some notes from his diary during those days.

- Harbor of Noumea, New Caledonia, aboard the transport, USS *American Legion*. Today, July 29, 1943, we sailed out of the harbor, passing several moored American battlewagons, and without mishap landed on Guadalcanal on July 30. There, we moved into a large coconut grove and spent a miserable night in the rain. The following day we dug large foxholes, erected tents, and got squared away. Several days later, we moved again; this time to Lunga Point, where earlier the Marines and Japs had a terrific battle; few evidences were still to be found. We are now right on the ocean; take two swims a day and are really having it rather swell at the moment except for the last two nights.

- Tojo (General Hideki Tojo, Premier of Japan) announced that August 13, also being Friday the 13th, he would raid the "Canal." The night was light as day with a full moon and low hanging clouds. Sure enough, about 8 p.m., here they came! The official recording said 12 planes came over, but I doubt that. Somehow one of their torpedo planes sneaked up on the transport, USS *John Penn*, and sank her. We were all standing over our foxholes when a hell of an explosion rent the air. The whole sky was one mass of flames. We all dove for our foxholes. The ack-ack guns put up a terrific barrage, but officially no Japanese plane was shot down; 150 lives were lost on the John Penn. Those Japs are sneaky bastards.

We all figured they would be back the following night. Sure enough, about 9 p.m., they returned. None of us knew how many planes, but the enemy had no success. One Jap plane was caught between two search lights; the ack-ack was hounding the hell out of him when all of a sudden the guns were shut off as one of our planes caught up with the enemy from the rear. Then the whole air was a mass of fire as the American plane fired a burst. I never imagined our planes had so much firepower. The Jap must have gone into a steep dive as the United States plane gave him another burst, straight

down. We never heard the outcome of this air battle. We were secured at 0200.

- September 25. Much has occurred since last writing. We have had so many air raids, I can't keep account of them all. It gets pretty tiresome—getting up in the middle of the night and staying in one's foxhole for several hours. We are now back in Coconut Grove.

- At a regimental parade the other Sunday, the regiment was given its colors. Also, Captain Carlson, C Company Commander and son of noted Colonel Evan Carlson of the Marine Raiders, was awarded the Silver Star Medal (3rd ranking medal for gallantry in action) for outstanding bravery on Guadalcanal in August 1942.

- Several days ago we went into the jungle for a two day problem. What a mess ... soaking wet with sweat, dirty, and exposed to "jungle rot." Still, there's something very mysterious and interesting about the jungle and all the queer noises. I had the 2 to 6 a.m. watch. Along about 5 a.m. the birds began to call, as the early rays filtered through the dense overhead. Along with the bad parts of the jungle, there are also interesting parts. Beautiful flowers are to be found everywhere; there are large and mysterious trees and ferns, and brilliant colored birds. However, the variety of insects, ants, and other ever present arthropods are a nuisance.

Marines always seem to cope with their environment. On the march, individual water supply in the tropics is always a problem. We learned that a certain vine growing around trees contained a limited amount of water; so, we would simply cut a section of vine with our machetes, hold the cut section over our mouths and let the water run in. We called them—what else—"water vines."

- October 31, 1943. Well, it looks like it's here at last. We are all prepared to go into action on the 4th of November—to Bougainville—the dreaded "Bogey" as it is called in lower military circles. B-24s have been bombing the hell out of the island for the last two weeks. Morale is high and, after 15 months of strenuous training, everyone is glad to be finally going into combat. **Carroll M. Garnett**

Prayer In The Jungle

The 12th Marines had this chapel on Guadalcanal in early 1944. Austin Gattis recalls helping Chaplain Duplissey construct the altar, in its entirety from the platform up.

Times Of Laughter Helped Overcome Times Of Stress

Dear ...

...Remember things like playing bridge at Guadalcanal ... Minitree Foulkes (Major) and I against you and Lieutenant Rogers? When you two were vulnerable, Minitree or I would say to Rogers, "Say, are you venereal?" And he'd get just flustered enough—obviously overawed by our ranks—to make a mistake. I think Minitree and I had as much fun out of seeing your frustration as we have had from upsetting Rogers. (You, never being appropriately respectful of our exalted rank, didn't think it was as funny as we did.)

Poor Rogers. He was killed during the Jap counter-attack on you at Guam. As I recall, he had a son born just before you all landed. You named the regimental recreation field after him.

And, Minitree lost his leg at Iwo. After the war, I used to see him in Richmond when there on business. He had some great stories to tell about predicaments he got into with his prosthetic leg.

Do you remember in the fall of 1944—on Guam—when some of us made up a presidential slate of Minitree and Jack West (from the 1st Battalion, 12th)? Ran them on a platform of "Free Love and Nickel Beer." Would go around to various officers' messes, barge in and give a campaign spiel. Was always good for a free round of drinks for us.

Jack West (Joshua C. West III) was from Suffolk, Virginia, a graduate of VMI. He and Minitree were our selection primarily because they were both old enough to be eligible to run.

In 1946 on one of my trips to Richmond I spotted a newspaper item ... Minitree was running for the Democratic nomination for the United States House of Representatives, against the Byrd machine incumbent. So, I looked up his phone number and gave him a call. When Minitree came to the phone, I asked, "Mr. Foulkes, for the record, Sir, are you still advocating your 1944 platform of Free Love and Nickel Beer?"

There was absolute silence. Then, almost in a whisper he said, "Who's this?"

Fun and games. I think that while all of us had times of stress, et al. during the war, some of the high points were the funny things that happened ... times when laughter prevailed ... Semper Fi" **Lytle G. (Bud) Williams, S-4 12th Marines, Monterey, CA**

Heads Up ...

Early on the "Canal", and it may have been on the truck ride from the transport dock to the Coconut Grove, home of the 3rd Marine Division, we were on the road of sorts on the inland side of Henderson field, when I spied a Jap helmet.

I yelled for the driver to stop, jumped out and ran back a few yards to lay claim by lifting the helmet by its camouflage net. A head came up with that helmet. I dropped the helmet ... a little wiser and embarrassed by guffaws. Well, you hadda learn. Austin Gattis, Washington, DC

World War II Marines frequently had field problems near and through the Balesuma River which ran near the 3rd Marine Division's camp at Tetere.

On a return trip to Guadalcanal and New Zealand in the 1980s, Bob Lowry of Hillsboro, California, formerly of the 12th Marines found this sign about an oil palm project among the coconut trees which were so prolific during the World War II days on Guadalcanal.

The Colonel Was Sniped

Some of us were having a few drinks in the back tent (on Guadalcanal). As we looked across, we were in a direct line with the tent of Colonel Smoak, 21st Marines. He had a tent with a big coleman light and was living it up pretty good while writing a letter. He had a fifth of whiskey, down by his feet ... would reach down, take a swig on the whiskey, write some more on the letter.

We had a few guys from the 3rd Platoon—Gayge, Buck, and I don't know whether Bozikis was there or not. But we were all having a pretty good time—McDowell, Sabatinelli, and a few of the others.

One of them who had been to Scout and Sniper's School, said, "You know I think I can sneak across and get that fifth of booze."

"Naw you can't do that." Well he did. Slipped across, carefully eased the booze from behind, and soon we were sitting there drinking it.

In the meantime Colonel Smoak is writing the letter. He reaches down for his bottle, can't find it. Reaches down again, and can't find it. So he takes down the Coleman lamp, looks all around. No booze.

If you remember, we were assigned to cut down all the grass in that area next day. **Jim McKearn, A Company, 9th Marines, Beloit, WI**

An Outstanding Marine

Lieutenant Colonel Eustace Smoak of High Point, North Carolina was a battalion commander who was awarded the Legion of Merit for his conduct during the Guam Campaign. He had enlisted in the Corps as a private in 1929, and was commissioned a second lieutenant in 1933. He served as executive officer of the 21st Marine Regiment on Iwo Jima.

A Marine Corps combat correspondent wrote:

"Marines who bore the brunt of repeated Japanese attacks on Banzai Ridge on Guam will long remember the cool battalion commander, trim, tall and erect, moving along the forward lines of the 3rd Marine Division checking position and asking questions of the most vital importance in the same tone with which he might ask about the weather. They will not forget the clean shaven figure, an M-1 rifle slung on his shoulder, directing the unit unmindful of mortar, machine gun and rifle fire bursts around him during the first 24 hours of the assault against the island. During that period, he lost two company commanders (one killed and one wounded) and 40 percent of his combat personnel.

"The Nips launched a coordinated attack the night of July 24-25, 1944. Some of them overran Colonel Smoak's positions and threatened his battalion command post. Calmly, Colonel Smoak maintained his units on the line and, employing headquarters and mortar personnel, retook his positions and restored the situation in the rear."

The Chaplain Couldn't Turn The Other Ear!

I recall a lecture by the chaplain to all men of the battery back in base camp on Guadalcanal in 1943.

He said our language had become intolerable ... He pointed out that no matter what the subject—whether it be Japs, rifle, helmet, chow, guard duty, or whatever—it was preceded by this most "versatile word." And, just recently it had been called to his attention that the Colonel (C.O.) had been referred to in this manner. This was the last straw and he proceeded to say if we didn't clean up our language of foul words, there would be some extra duty handed down.

Breaking ourselves of this habit was about as difficult as not eating. But after sometimes controlling ourselves a little better I recall the Chaplain being on a more intimate basis with us ... with a wave or a "Hi" from the more assured. **E.B. Thompson, Chillicothe, MO**

Wrong Territory?

Colonel Robert "Bob" Blake, Chief of Staff, came bouncing back to headquarters, 3rd Marine Division, Rein., on Guadalcanal.

"Hancock. Take a note." he ordered.

It was directed to Colonel Alton Gladden, commander, 3rd Service Battalion. Seems Colonel Blake was upset with the condition of the area he visited.

Your area reminds me of a National Guard encampment," he dictated. "No, that would be an insult to the National Guard!" he continued.

Dispatched through channels.

Wasn't long until Colonel Gladden appeared. A striking officer with gray goatee and a big walking stick, the colonel was senior to the chief of staff.

In addition to that fact, Colonel Gladden was dealing from a loaded deck.

"Bob," He said to Colonel Blake. "That is not my area!"

Chagrin. The chief of staff. We always compared the learned Colonel Blake (he read profusely) with a Methodist preacher, what with the the wavy gray hair and the command of the language. But he could cuss with the best of em! **Brent I. Hancock, Hershey, PA**

Ship-To-Shore Fire Control

A little known segment of Marine activity in the South Pacific was the use of 3rd Division personnel as ship-to-shore fire control teams in support of beachhead landings.

In late 1942 I was a member of an eight man team sent to Camp Niland, California for training in pre-invasion landings and observation of enemy beach installations ... in order to direct destroyer fire during the main landing.

This unit was eventually utilized in August of 1943 when we were attached to the 12th Marines on Guadalcanal.

The fire party was headed by Navy Lieutenant Ellis, and the Marines were Lieutenant Ira Elsham, Private First Class Charles Becker, Corporal Earl Honeyager, Corporal Charles Allum, Corporal William Cramer, Sergeant John Hennion, and myself, Corporal Charles Felix. Our group departed from Guadalcanal on August 14, 1943 aboard APD-16 and disembarked at Vella Lavella Island in the pre-dawn of August 15.

The main landing force was the 35th Regimental Combat Team of the 25th Army Division under Brigadier General Robert B. McClure, and our team advanced inland to higher ground to direct destroyer support fire on Barakoma Beach for the Army unit. We maintained communication during the attack with the destroyer USS *O'Bannon*.

Our fire unit remained on Vella Lavella until evening August 17 with only light enemy contact ... about 300 Japanese were located in the beach area. Enemy air strikes were frequent, at times quite heavy, but the Japanese lost 17 planes on D-Day in dive bombing and strafing attacks over the beach.

On evening August 17, our party of eight went to Barakoma Beach to catch passage back to Guadalcanal on one of the ships that was unloading.

The beach area was a litter of decaying bodies. On August 6, four Japanese transport destroyers had attempted to run from the Shortlands and land reinforcements on Kolombangara. Carrying some 1,000 troops they were intercepted by an American task group and three of the vessels were sunk with some 1,500 enemy killed. Many of the bodies washed up on the Barakoma Beach, and the hard coral prevented immediate burial. Although scores of decaying bodies had been covered with lime, the stench was terrible.

We boarded LST-396 to return to Guadalcanal after the ship had unloaded its cargo of ammunition and aviation fuel. However, near dark the beach came under extremely heavy air attack, and on one occasion eight bombers and 45 fighters bombed and strafed the beached LST's. Eventually the attacks became so intense that the vessels aborted unloading and put out into Vella Gulf. About midnight our LST was hit by enemy bombers off Barakoma. The ship with its volatile cargo exploded ... and sank quickly. It was abandoned by all personnel, and we took to life rafts.

All members of our ship-to-shore fire team were picked up by American destroyers. I was rescued from the water about daybreak by the destroyer USS *Saufley*. Throughout that day of August 18, the Saufley engaged enemy bombers and fighters, and shot down several Japanese planes on its run south to Guadalcanal.

We arrived at Guadalcanal and disembarked late August 19 ... subsequently rejoining our 12th Marines unit ... all intact and glad to be back. **Charles Felix**

Bathing Canal Style

Often we'd pile into a 3/4 ton truck for a laundry/bath party in a river. The driver would stop the truck in the middle of the river and we'd jump out, clothes and all.

We would take one piece of clothing off at a time, wash it and toss it on the bank. When all the clothing was washed, we'd soap up ourselves, rinse off, get out of the water, put on the wet clothes we'd just washed.

When all were ready we climbed back on the truck and headed for camp. I swear, dust would coat us and dungarees just as bad and thoroughly as before we'd washed. Back at camp we'd be pretty dried off, but dust covered again as well.

Oh well, it felt good, for a few minutes, to be cleaner.

Sweep The Sky

We had a gunnery sergeant named Bookout. He had picked up a Tommy gun somewhere and was in the habit of running out and firing it off at Jap bombing planes when they came over.

One night the alarm went and Bookout grabbed his trusty Tommy gun, and ran out to do battle. A sergeant named Murphy followed him out and picked up a palm frond. "Look, Gunny," said Murphy, "you get the high ones and I'll get the low ones." **Roland B. Abbott**

The smoking lamp is lit as 1st Lieutenant Al Jenson, 21st Marines meets with two pipe smoking natives.

Movie

Trials Of A Working Party

When we were on Guadalcanal, we used to do work party duty down on Lunga Beach. The whole battalion was there. On this particular night Lieutenant Nolan and I took the better part of the company—three truckloads—down to the beach to unload supplies in an Army Supply Depot. Supplies included clothing, writing material ... and about everything you cold think of. We started work at 6 p.m. and they brought us sandwiches at midnight, and we were due to quit at six the next morning.

Sometime between midnight and six in the morning, the Army captain who was in charge came around and said, "Sergeant, pass the word to your men that we're going to shake them down when they leave this depot."

So I go around and make a special point to talk to each group, telling them they were going to be shaken down at the end and not to be packing anything off. They heard me, but they didn't listen. So, when we loaded up in our trucks at the depot in the morning, we went to the gate about half a mile away. When we got there, here is the Army captain and half a dozen MP's.

The captain said, "Sergeant, fall your men out right here" and he indicated an area about 100 feet wide between the trucks and the jungle. So, I fell them in about 30 or so feet from the jungle. When the captain and MP's looked in the trucks, there was gear about a foot deep in each of those trucks. Nobody in

the world needed all that stuff, but they were packing it out anyway. So the captain had his men, each with a pad, itemizing the things these guys were packing off. There was a lot of it, and they were very intent on the itemizing ... and not paying any attention to us guys. Suddenly the captain looked up and said, "Sergeant, where's your men?"

I said, "Right behind me," and I pointed over my shoulder, and says, "Right here." With that I looked around, and I had 13 guys left. The rest of them just infiltrated into the jungle and left. I've often wondered why those 13 stayed ... but they did.

The captain took us down to a little barbed wire enclosure that they used for prisoners, and threw us in there. He didn't throw the lieutenant in there with the rest of us. The lieutenant called Captain Fowler back at the camp, and by and by Captain Fowler came down to the enclosure.

Captain Fowler said, "Captain, have these men had breakfast?"

"No breakfast, chow's been over a long time," he said.

"That makes no difference," said Captain Fowler, "I want these men fed breakfast, and I want them fed right now."

"We'll see what we can do." The Army group disappeared and about 15 or 20 minutes later, someone came and got us and took us to the mess hall and fed us breakfast. Captain Fowler had a truck with him, we loaded up and went back to camp.

Incidentally, the 13 guys got off with a warning; and I got a summary court martial. Fortunately, I was cleared. **Jay Strode, Jerome, ID**

Auckland To The "Canal"

Arriving in Auckland Harbor, the city was very impressive and reminded everyone of San Francisco because of all the hills. It was their summertime, and the weather very pleasant.

Time in New Zealand was spent training in all combat activities. This included a 50-mile hike where somebody got the idea that we should all live like Japs for three days—on rice cooked ourselves.

Quarters was a wooden hut accommodating two to four. The mess hall was enclosed and, except for the mutton, we were able to survive. There was a recreation building where ladies served tea and pound cake.

Liberty was rare and we took a train to Auckland for the movies, drinking, and sundry. If you missed the 11 p.m. train you were out of luck.

Our numerous field exercises included amphibious landings ... which required digging gun pits. I always said, "In New Zealand we have really learned to dig."

On July 16, 1943 we sailed on the USS Crescent City and arrived in Noumea, New Caledonia 11 days later ... tropical climate, high temperature, high humidity, just like New river, North Carolina.

We left Noumea for Guadalcanal, arriving July 30 at Tetere beach. The first thing people did was to get a coconut for milk. Food was always of interest ... and we were fortunate ... with excellent cooks who did the best they could with dehydrated food and Spam, Spam, and Spam.

At Guadalcanal the Army was pretty much in charge of supplying food and were always short of dessert—such as fruit cocktail.

One day Sergeant C.A. Rouse had an idea to take a 3/4 ton truck to Lunga Point Supply Depot to see if he could pick up some extra fruit. He secured a set of captain's bars and we took off ... our driver (a private first class), "Captain" Rouse sitting next to him, and myself— with a three man working party in the rear. We proceeded to Lunga Point, bluffed our way into the store area, loaded the truck with fruit cocktail and canned peaches, and made our way back to camp.

Needless to say we enjoyed the dessert for the next few weeks.

Bathing was a problem. We used our helmet for almost everything ... shaving and washing socks ... but on occasion we'd go to the Balesuma River for bathing and washing clothes. **George J. Green, St. Louis, MO**

And Now Meet Corporal Robert Dwight Peart

Every outfit has a comic or character, and we had more than our share in the Tankers. One in particular in B Company Tanks ... very inventive, a fine brain on electronics, a poet (we called him the Professor) ... was Corporal Robert Dwight Peart.

He was not your typical gung-ho Rambo Marine. He was quiet, I assume a man of the fine arts, always smoked with a long cigarette holder, and had a small moustache above his lip. I understand that today he is in the jungles of New York City tuning pianos.

Robert was a man who could, if given all the equipment and plenty of space, put together a great space ship or perhaps a bomb of sorts. He would invent the unknown, or blow us all to kingdom come.

In New Zealand he had rigged up an alarm clock that would turn on a radio in the A.M. Also, he was working on a system of wiring to turn off the lights at the same time ... but then we departed Warkworth for Guadalcanal.

One night in the Tank Battalion's coconut grove area, Robert had come on guard sentry duty. We were still patrolling and mopping up resistance on the island, and sometimes Japs did come into our tent area at night to secure food and water. One Jap did come in one evening to watch the movie, was sitting on an oil drum enjoying the show when taken prisoner by sentries.

When Peart came on duty that night, he had an old Marine Corps issue raincoat ... which came down to his ankles, since he was not a tall Marine (perhaps five foot seven inches). He also had a belt full of M-1 ammo, but nothing in his rifle.

At about 2130 hours we heard hollering—"Corporal of the guard, help, Jap, Jap, Jap, help!" It seems three Japs came down the trail from where they had been tapping the water barrels. Now they were heading for the galley food supplies and cooks' tent.

R. Dwight challenged them, flashlight down by his side. Two of them made a quick exit, but the third one fired a pistol shot at Dwight's light, missing him. If he had the light to his front he would have got it in the guts.

Peart drops the light, and even with 10 yards of raincoat and all that ammo, tackles the Jap, and it's ass-over teakettle in the mud. Somehow he had got the pistol away from the Nip ... who now realized he had a tiger by the tail.

About this time, the company commander, Captain Quentin Joy, a very tall, rugged Marine and former wrestler, got the Nip in an armlock and settled things down.

Someone said to Robert, "Take the pistol. It's yours."

He said, "I don't want it." If it were a screwdriver or pliers, I guess he would have said, "I'll take it." However, the Nip had a beautiful silver watch which had a gear type face on it. The sergeant gave it to Peart. The Nip was taken into the mess hall ... and it developed he was a warrant officer in the Japanese naval forces construction engineers.

I really would love to see that old genius today. New York is a big area; he could be anywhere. But if anyone knows Robert's whereabouts, the Tankers of B Company, 3rd Tank Battalion of World War II would love to have his address.

Thomas F. Murphy, who lives in Andover, Massachusetts with his wife of 16 years, was a tanker in the 3rd Marine Division for three years, including the heavy action along Motoyama Airfield No. 2 and other parts of Iwo Jima. He went to North China with 1st Marines in 1946, joined the Army in 1950, later served four years in a religious order as a working brother with aged and infirm and doing mission work, and still later was superintendent of buildings at a college of the Augustinian Fathers.

"You Must Remember This ..."

...Words of a song of the period must trigger your recall of less than elegant surroundings, with tropic skies and running water. The big camp in the coconut grove sometimes became very wet. Here Private First Class Herberta A. Eannacone of H&S, 21st Marines squats for a better look at the situation to see if it is worthwhile wading in. **Photo by Dugo, submitted by Al Jenson**

It's Called Being Resourceful

We were on Guadalcanal, on the beach unloading ships. One evening they sent a working party with me as sergeant in charge. Howell Heflin (now the senator from Alabama) was the lieutenant in charge. We were put on a fishing boat, loaded with beef from Australia. We hadn't had any beef, except for canned rations and such. On the boat, the cook said, "Anybody here who can cook?"

A couple of the guys said, "Yes, we can cook." So he took them to the galley to show them where they could cook some midnight supper for us. Had steaks, and everything else. We really had a good meal.

Along the way someplace, Harry Lee Gilliam said, "Strode, don't you think I oughta ride along to the dump, and see if this meat gets there all right?"

I said, "Nah, it will get there all right."

He persisted, "I believe I should ride there with it." I looked at him and I saw he had something in mind. I didn't know what it was, so I said, "yeah, I think you better."

He took off, and I never saw him again that night. The ducks would take a load of beef ashore and then come back for a load. Gilliam just stayed on the duck all night. Everytime it went under a big banyan tree before getting to the camp where the beef was being unloaded, he would throw a quarter of beef off.

Next morning we figured we would pick up the quarters of beef when our trucks came and got us. Our trucks didn't come and get us. One of these guys in a duck took us home .. so we were scared to stop and pick up the beef then. As soon as we got back to camp, I said, "Lieutenant, we've got four quarters of beef laying back there under a banyan tree. We can't pick it up with this outfit. Do you suppose we can get a truck and go back and pick it up?"

"Yes, you boys go get a truck, go back down there and pick up the beef, and I'll tell the captain that the officer on that boat gave it to us."

So, we had some fresh beef. Along with this, I don't suppose we stole it ... appropriated probably ... we got a sack of potatoes and some fresh onions off this boat to go with our beef. **Jay Strode, A-1-9, Jerome, ID**

Not Sorry For "Borrowed" Beef

This ties in with Jay Strode's story about beef "borrowed" from the ship we were unloading. We hadn't had anything that tasted like a fresh vegetable in a month of Sundays, and that beef was mighty tasty compared to those rations we wuz getting.

We had Spam every way you could think of ... boiled, baked, fried, you name it—we had it that way. That beef went over very good with our group ... and I never felt a bit sorry that I helped steal it. **Harry Lee Gilliam, Sunnyvale, CA**

The Monster And The Doc

Our regimental doctor, 12th Marines was a real screwball ... Doc Anderson. In private practice he was an obstetrician and, not much good to us. His favorite prescription, regardless of what ailed you was ... "soak it ..." He kept this up for all of our stay on the Canal and even most of the time on Guam.

One day he got diarrhea and every man Jack of us in our battery hounded him with "soak it, Doc."

He loved to go hunting. One day on the Canal (along with a warrant officer buddy), he came back with a Gila monster with a rope around its neck. They put it in the sack of a red-headed quartermaster, who didn't have a noticeable sense of humor. Well, when the quartermaster climbed in that evening and found the monster, and that it crapped all over his sack, he was one mad redhead.

We ate it (the monster) ... kinda tasted like chicken, rather bland. **Austin Gattis, Washington, DC**

Rear Echelon E-X-P-L-O-S-I-O-N-S

Most Gung-Ho Marines never get to know what is done by the rear echelon left behind while a campaign is going on. It happened to me on Guadalcanal when the 3rd Marine Division went to Guam. Most of us on this dubious duty were sick, lame, or lazy. My excuse was a bad knee.

We worked hard on all kinds of clean up details: strike tents, assemble various gear, clean up grounds. We also got into more than a little raisin jack, or anything we could find to ferment, to pass the evening boredom.

One inexperienced brewmaker mixed a batch of raisins, sugar, water, and yeast. He mixed the stuff about right, but put it in a five gallon can and sealed it up. It was one of our usual hot days, the sun poured down on the can all day. That night, after all had quieted down, there was a mighty explosion. The mixture had blown up. Next morning we found the can, top blown off. Raisins were imbedded in the nearby coconut trees ... looked like freckles on the bark.

Two French 75 shells were used on the altar in the chapel for decoration, or flowers, or whatever. On examination, the shells were found to have the detonators still in them.

It seemed unsafe to try to ship them on up to Guam. Maybe it would be safe to send them home if I could detonate the primers.

The company repair shop had a fair sized vise. Why not put the shells one at a time in the vise? With the vise on ground outside the tent, one could crawl up to the shell with punch and a hammer and set it off. It worked. Boy, did it work.

There was a giant smoke ring that encompassed most of the coconut grove. As quick as the primer went off, I jumped up and ran back into the tent, stuck my head out and looked up and down the row of tents. Each tent had a head or two sticking out, looking to see what had happened.

It was a few days later before I dared to try it again. Both made it home okay. One made an ashtray. The other is in a small museum. **Clarence Brookes, Cortland, NY**

"Is Relished By The Best Of Men"

Members of Able Company, 1st Battalion, 21st Marines were a special unit.

While working loading ships at Guadalcanal, the men obtained a wooden keg from a life raft. A few night patrols were conducted around Divisional Officers Mess, the proper ingredients were assembled and some apricot brandy was brewed.

When the proper mixture was poured into the wooden keg, the question arose on where to store it until it aged. The center pole and boards were removed in one of our tents, a hole dug large enough to store the keg and the boards and tent pole replaced. Sea bags were placed around the pole and all was well.

A surprise inspection was called for the day before the brandy was to be ready.

Our area was spotless, everything in order, stolen Army jeeps were hidden in the elephant grass near the motor pool and we were standing outside our tents in clean khaki uniforms

Suddenly there came a hissing noise and the pungent odor of apricot. There was nothing we could do as the General and his inspection team were in our company area. All we could see was court martial.

When the General and his inspection team came through our area, no one moved to enter the tent or said a thing about the smell.

When the team left our area to go to B Company, our lieutenant came to our tents. We all felt we were in deep trouble until we saw the lieutenant was carrying his canteen cup.

How fortunate we were to have such excellent combat officers who were so understanding in the rear areas. **John F. Pelletier**

Such Stupid Officers!

When 3rd Marine Division, Reinf., was in the field in New Zealand, three officers headed by Colonel Bowser (12th Marines), were ordered to Guadalcanal to select a site for the 3rd.

They returned to 3rd Marine Division Headquarters, and presented their report to Colonel Alfred N. Noble, "The Little Chief," chief of staff, Major General Charles D. Barrett was commanding general; Brigadier General Allen H. Turnage was assistant commanding general.

In any event, "The Little Chief" was displeased with the findings.

He had a manner of starting with a low voice, then increasing the crescendo until he hit: "Never, never, in my God damned life, have I seen such stupid officers!"

Things ended well. The 3rd was quite comfortable on Guadalcanal in the coconut grove. **Brent I. Hancock, Hershey, PA**

Smuggling Booze In His Socks

It was a pontoon barge section, as I remember it. And, I was on the rear echelon in Guadalcanal. One day, about 115 in the shade, they decided we ought to load that stuff (special whiskey consignment) in a pontoon barge. There was a big tent where we had been guarding it for some time. Rain got into some of the boxes—some rum in cardboard cartons.

I got the dubious honor of going down inside that pontoon thing and stacking some of this whiskey. It occurred to me that a wet carton would be quite easy to get into. So, I got a pint and put it in my sock (tied it in my sock with my handkerchief). Some of the boys on the outside saw what was going on, so they quickly relieved me. I went over and stashed the bottle of rum under a bunk and went back to work. Before I got through, I had five pints under that bunk. And, of course had to split it up with the guys who were in on it.

I started back down the company street and had a pint in each sock—scared to death that I would clunk them. When I was down, probably a couple hundred feet, one of the officers came up and said, "Hey Brookes," and motioned me back.

"Judas Priest," and I knew the way I hesitated about turning around that he was probably suspicious. But he came up with a pitcher of orange drink of some kind, and he said, "Here you fellows might need this." **Clarence Brookes A-1-9, Cortland, NY**

The Band ...

... They did much more than stand parade, wake you up and toot their horns. Here were the stretcher bearers, who lifted their comrades—often from the firelanes where they had been hit—and carried them back to the aid stations. Many lost their lives in the attempt to save wounded comrades. Here Tech Sergeant Richard A. Linden, then of North Hollywood, California musters a beat as this jiving group rocks the coconut grove.

The 21st Marines undergo a final inspection in 1944 on Guadalcanal before embarking on the recapture of Guam Campaign. (Submitted by Corporal Allen L. Robbins, H&S, 21st Marines).

Anecdotes From The Solomons

Colonel Henry Aplington II, Warner, New Hampshire wrote prolifically about a long and combat-active career in the Marine Corps which included Bougainville, Guam, Iwo Jima, China reconstruction and Japanese repatriation, and Vietnam. On Bougainville he served with the 3rd Marines and commanded the 1st Battalion from January to September 1944. Here are some of his anecdotes from the Solomons.

Something Beats Nothing ... All To Hell

When we had our USO show of the Pacific War, it was not the Hollywood glamor girls of whom we had read entertained the Army.

... But Ray Bolger and Jack Little putting on their show on the division boxing ring to which we had hoisted a small piano. Instead of glamorous costumes we could have expected from Dorothy Lamour or Betty Grable, they were dressed in rumpled khaki.

The show was appreciated, but the only jingle we remembered was:

They sent for the Army
To come to Tulagi,
But Douglas MacArthur said NO!
He gave us as the reason
It wasn't the season,
Besides there was no USO.

Once A Thief

One day there arrived on my desk a great sheaf of endorsements on a basic letter from the FBI saying that one of my men was a fraudulent enlistee. He had concealed the fact that he had done time in Atlanta for forgery.

I sent for him, told the story, and asked if it were true. He was incensed: "No sir, I did time in Atlanta but I am a patriotic American and when war started I went down to enlist. The recruiter said, "Do you have a criminal record?" I said, "Yes."

The recruiter again said ... "Goddamit, I asked do you have a criminal record?" And I figured I was supposed to say "NO."

That had a ring of truth so I told him to forget about it, and put on an endorsement saying, in effect, that I didn't care what his civilian profession was, he was a good BAR man and I needed him. That was the last I heard of it.

After all, he wasn't that good a forger. And I had a good lesson on how the troopers can louse up a supply system.

One day we received word that a box of requisitioned machine gun parts, which we badly needed, was at Lunga Point and a sergeant was sent to get it. That evening he reported to my tent with the disgusted announcement that ... "Major, some son-of-a-bitch stole our machine gun parts."

Then he brightened up and said, "But you ought to see what I got."

Dunkin' The Chief

Moving from New Zealand to Guadalcanal, Major General Charles D. Barrett and chief of staff "The Little Chief" Colonel Alfred N. Noble, were with the advanced echelon.

General Barrett decided to go ashore. Colonel Noble asked of the general's whereabouts and was informed that the general had gone ashore.

With that "The Little Chief" went into full action. Landing craft came alongside, one with twin props.

Ashore we went. Coxwain ran into a sand bar. "Drop the gate" demanded The Little Chief.

"I'll back her off" suggested the coxwain. "Drop the gate" demanded The Little Chief.

He stepped off the gate, and disappeared in a deep hole! You know we couldn't laugh. But later, did we ever.

Nice man, "The Little Chief." **Brent I. Hancock, Hershey, PA**

None Had Our Name

On Guadalcanal we worked at unloading around the clock with different shifts.

One particular night our group was unloading 500 pound bombs, and it happened to be the night when a single Japanese bomber came over. He dropped three bombs, two on the outside of the ship towards the bay, and the other between the barge we were on and the ship. There was absolutely no shrapnel, and no one hurt. But you can imagine how we felt riding on a barge with about 50 or 60 500-pound bombs, and just waiting for that next one to hit right dead center of that sucker. But that Jap bomber only dropped three bombs and then left for home. **Harry Lee Gilliam, Sunnyvale, CA**

One Tough Sergeant Major

CWO T.O. Kelly USMC (Retired) was a no-nonsense sergeant major of the 21st Marines and later division sergeant major. His was a name remembered, respected, and often feared—and his prestige extended to officers as well as the troops.

"T.O.," the titular singularity to which he was respectfully referred, became the sixth president of the 3rd Marine Division Association (1962-1964).

Tom lived much of his retired life in Annandale, Virginia, near Washington, DC, and in later life moved to La Plata, Missouri.

Third Marine Division, reinforced, occupied the vast coconut grove on Guadalcanal. When word came to division headquarters that Sergeant Major T.O. Kelly was coming from the 21st Marines to become division sergeant major, his reputation as "one tough sergeant major" was well documented.

As a matter of fact, Sergeants Major McKnight and McKenna quietly requested transfers. Master Sergeant Brent I. Hancock, chief clerk to General Allen H. Turnage, was informed: "Don't laugh, Kelly is being assigned to your tent."

And thereby was formed a lasting friendship. Kelly would put up with the antics of "The Hershey Kid," would really gripe when taps were sounded and guys would not freeze. Officers of less than field rank felt the wrath of Sergeant Major Kelly.

Seagoing for 14 years, Kelly was a creature of habit: exercises in the morning, chow on time, after chow—washed socks and skivvies. Insisted that "The Hershey Kid" cover his belly with a blanket at night "to keep from getting sick."

General Turnage enjoyed sharing the attempted diversion with Kelly concerning the 3rd's next move: "Going to work with MacArthur this time," the Hersey Kid would tell Kelly. Actually, plans were for the Marianas Campaign.

"Saw something today, Hershey," reported Kelly one evening. "Working with MacArthur," he growled. "How about Guam!: **Brent I. Hancock, Hershey, PA**

No Mistaken Identity

Preparing for the Marianas Campaign, a group from the naval service established a headquarters with the 3rd Marine Division on Guadalcanal.

On an evening, General A.H. Turnage, commanding general, 3rd Marine Division, reinforced, and some of the key staff officers of both commands went to the nearby headquarters of General Geiger, commanding general, 1st Amphibious Corps.

That resulted in a little beer party at the naval headquarters, which was in officers country and near the general's headquarters.

One of the chiefs drank a bit too much, and got unearthly sick. Next day around 1700, the general summoned me to his back door. "Hancock," he said. "You were at the party last night. I heard your voice when I came back from Geiger's party." He continued, with a twinkle in those blue eyes. "That little chief was pretty sick all night." "I don't want anyone to think that the General can't hold his bourbon!" "Aye, Aye, Sir!" **Brent I. Hancock, Hershey, PA**

Rain Again Today

"We had enemy raids (Guadalcanal) ... and watched aircraft fight in the sky at night and return in the day time after the battles ... They would come in over the air strip and roll for every enemy they had shotdown ...

"Then we landed on Bougainville and put up with 19 days of rain and foxholes full of rain and the ground was full of other small holes ..." **Eugene J. Rosplock, Elmira, NY**

Sanitation From Heaven

On the Canal we were obsessed with an effort to maintain a modicum of cleanliness, our bodies as well as clothing.

It was a good thing, too. If we didn't exercise care, jungle crud would get a foothold ... and you'd pay hell getting rid of it. We rigged 55 gallon drums to catch rainwater ... and if the rainfall was insufficient, we hauled water from the streams (called rivers on the maps.)

We often soaped up right outside our tents during tropical downpours ... except sometimes the rain stopped ... and left you soaped all over. **Austin Gattis, Washington, DC**

USS Windsor APA 55 is one of the ships that took elements of the 3rd Marine Division from Tetere Beach, Guadalcanal on June 2, 1944 and landed troops on Guam, Mariana Islands on D-Day, July 21. Veterans of the Guam landing will remember there was a slight delay in the landing, and considerable time was spent in the Marshall Islands. The Windsor earned five battle stars for her World War II service.

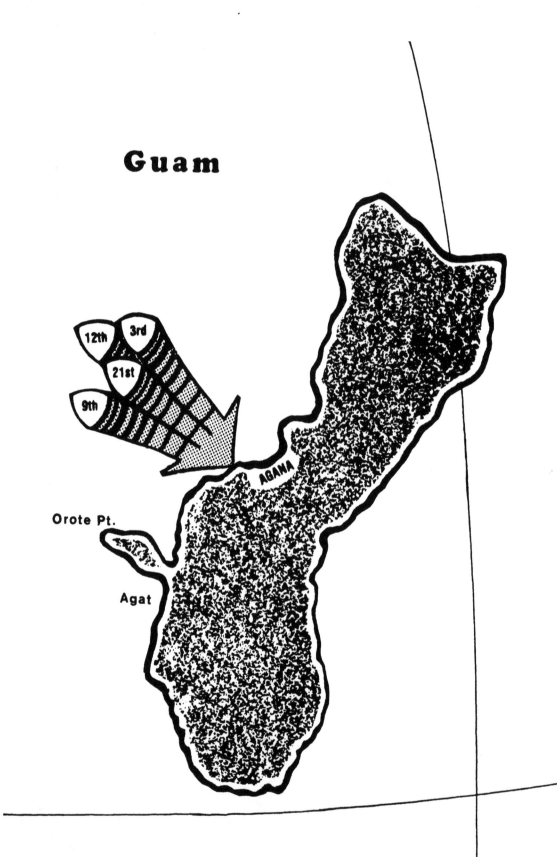

Exploring Agana—To St. Louis Blues

Agana had been taken earlier on the day that our outfit moved through its rubble strewn streets. For some reason, our company halted temporarily at the end of a long street of bombed out dwellings. Our company commander told us to take a break, but not to wander off too far.

I had passed about halfway down the narrow thoroughfare when I stopped in front of a stone house with attractive porch and entranceway, and which had been spared the complete destruction dealt most of the houses. Its roof was partly down and portions of the back and side walls were gone, but through the open windows I could see that its furnishings were still there, though mostly in helter skelter condition.

I climbed through one of the windows into what must have been the living room. With the removal of a few pieces of plaster and rubble I uncovered an old upright piano, which was barely recognizable but still in one piece. After exploring the other rooms, risking having my head bashed in from falling framework and ceilings, I returned to the piano, and—using my field jacket—cleaned it off the best I could.

Seating myself at the dusty keyboard on a broken three-legged chair, I began playing my best three fingered version of the *Saint Louis Blues*, the only piece I felt proficient at. I don't know if it was the beautiful music I was making (I really doubt that) but I soon had Marines climbing through the windows to stand around the old upright, where we joined voices to the strains of that grand old piece. We were momentarily transported away from the war and the awful destruction round about us.

A short time later, remembering where I was, I reluctantly—and a little sadly—climbed back through the broken window and rejoined my company at the end of the street. **Maury T. Williams Jr., Dayton, OH**

First ... As A Boy ...

He was a boy when the Marines came ashore in July 1944, and the 9th Marines were the first he met. Later he was to join the Marines, and in Vietnam command that very regiment. Ben Blaz became a brigadier general in the United States Marine Corps, and now represents his island in the United States House of Representatives.

"There were the United States Marines, who, after hopping from island to island, eventually liberated one of their own and seemed almost as glad as we were that they came back to Guam...

"The years have dimmed the sight but not the memory. To me the war years were the most precious of my life. They provided me with a reservoir of wisdom and strength from which I have drawn abundantly over the years.

"Si Yuus Maase, Guam, for enriching my life so much.

"Si Yuus Maase, Guam, for making me the person I am today." **Ben Blaz**

Ben Blaz, Congressman from Guam and former Brig. Gen. USMC and CO of 9th Marines.

Good Instincts

It happened on the Guam Campaign. We were dug in along this line, and a man named W.L. Livesay was either in my foxhole, or the one next to me. Suddenly, I heard this old 45 he had go "Bang" and I reared up and asked, "What's going on?"

He said, "I don't know, but something just told me to stick this thing out over the edge of the foxhole and pull the trigger." In the morning we found he had got his game ... within about 10 feet of the hole there was a Jap laying out there. **Jay Strode, Jerome, Idaho**

The Liberation

If you had placed yourself in harm's way to liberate Guam, no matter what fate has meted you since, you will always be a hero out there. Even if you fit only the age and girth and have that stateside look, you will be greeted on the street for what you are—one of the liberators.

The day you came ashore, July 21, 1944, is the island's big holiday, superseding by far the 4th of July. The places you fought are marked with bronze plaques and people, old and new, learn of you again and again in history books.

Among Marines that same affection is exchanged. Returning veterans have met Chamorros and cried. Marines who knew they'd never return have asked sons and daughters to make the mission for them.

General Louis H. Wilson, USMC (Retired) whose heroism as a company commander and captain on Fonte Ridge merited him the Medal of Honor, put it best in a recent return: "We, who got to know you under battlefield conditions, have never forgotten ... We have come back to the arena of our greatest challenge ..." Guamanians suffer visitors gladly ... Marines they embrace. Vital to Japan—also to the United States.

Located 1,200 miles from Japan, Guam was eyed as essential to Nippon expansionists. American sachems gave the loyal American island no such weight. Congress refused the money to fortify it, and when hostilities threatened, gave Guam up as a lost cause. Even as the actual island war expanded, Guam was the back forty. General Douglas MacArthur's avowed principal interest was the liberation of the people in the Philippines.

Scarce was said of the rapes, beheadings, holding camps, forced labor on that little piece of America ... except on Guam itself. There were people like Father Jesus Baza Duenas who spoke out against them and was beheaded barely 12 days before the Marines came back.

When The Japs Came In 1941

It was a short hop for the bombers from Saipan (half an hour) and they

(Continued on page 85)

Even Truck Drivers Got To Use Cargo Nets

When the Big Brass decided it was time to recapture Guam someone in that group dreamed up the idea that DUKW's were just the thing for hauling any and everything from ship to shore and even to the front lines. Now I guess that looked real good to the Brass, sitting around the table and looking real dignified, but that landing was a far cry from what they dreamed about.

The plan was to get some "professional" truck drivers, such as us, and we were all "Gung Ho" to go on a campaign so we could get some Jap souvenirs. The thought never occurred to us that we might just get shot. As stock holding members in this great organization, we should have had the opportunity to take a vote on this matter. This did not come to pass and we got the DUKW's and started to learn all about them.

Being a member, in good standing of C Company, III Corps Motor Transport Battalion, I just happened to have a berth on the USS *Zelin* that just happened to be scheduled for the recapture of Guam. I do not remember what compartment I was "supposed" to be assigned to because I slept topside every night from the time we left Guadalcanal. I did not much like the idea of being below decks in case we did get a torpedo below the waterline.

As we remember, there were poker and dice games most of the time and having a little gambling blood in my veins, I played poker a little. Well, that last night aboard ship was no exception and as the night wore on, the poker decks either wore out or were torn up by an unhappy loser and the only cards left were pinochle decks. Yes, you guessed it, we played poker with a pinochle deck. Now if you like a wild gambling game this is the way to go. Anyway, in the wee hours of the morning, 0400 I think, somebody bellowed out on the PA system to saddle up and get to your station and be ready to disembark. Now if there are any of you guys were in the 4th and 22nd Marines that were on the USS *Zelin* and just happened to lose a few shekels in a game with me, I thank you for your contribution to this old country boy. What concerned me was that I had over $700 in my money belt and no place to spend it.

As my status was a mechanic, I assumed that the Big Brass figured that we mechanics should get ashore and find us a DUKW to ride in just in case those gear grinders tore one of them up. Climbing down that cargo net with full field pack and rifle would not have been so bad if those boat jockeys had known how to drive one of those things.

Before I went into the Corps I was just a country boy without a care in the world who had worked on farms and ranches most of my life and I could hold my own when it came to riding a bucking calf or horse; of course I had been thrown a time or two and have seen the tops of more trees than most riders but I managed. Getting into a Higgins boat from that landing net was a different experience.

Burnell Focks, Guymon, Oklahoma manages to flavor with humor even some of the most dangerous and exasperating of situations on Guam, giving very little credit to the boatsmen who rescued us from our descent down the cargo nets. He was with III Corps Motor Transport Battalion, III Amphibious Corps.

CHONITO CLIFF
Hill Of Hell And Heartbreaks

This story was written by Sergeant Cyril O'Brien. He had fought in an earlier campaign with the men he describes. But for the accident of a recent transfer, he would have been with them in the fight for Chonito Cliff. The cliff was taken, with the help of another Marine unit, but not until almost all the men in O'Brien's company had been killed or wounded.

Nearly half my old company lies dead on the barren slopes of Chonito Cliff. Four times they tried to reach the top. Four times they were thrown back. They had to break out of a 20 yard beachhead to make way for later landing waves. They attacked up a 60-degree slope, protected only by sword grass, and were met by a storm of grenades and heavy rifle, machine gun and mortar fire.

The physical act of forward motion required the use of both hands. As a consequence they were unable to return the enemy fire effectively. Most of the casualties were at the bottom of the slope. They had been hit as they left cover.

There was Pappy, his name stenciled on his canteen cover. A bullet had ripped away the first "P" in his name.

My former assistant squad leader was beside him. He often had me on his working parties. I had seen those arms of his, which in death still clutched a splintered rifle, throw full ammunition cases about as if they were empty.

There was the first scout of my squad. We had shared the same tent for months, when I was second scout. He was always promising himself a "white Christmas in 45." He was facing the sky, his hands at his sides. You'd think he was dreaming.

Willie, who volunteered for mess duty so he "wouldn't have to stand inspections," was lying halfway up the slope. His feet were dug into the dirt. His arms were bent as if he were ready to charge again. But his Garand rifle was empty and thrown aside. The pistol in his hand was empty too. Perhaps he intended to club the enemy with the empty weapon.

The "Beast"—we called him that because he was so big—had charged his big frame to within five yards of an enemy machine gun nest. He caught a blast in the chest. The fancy lettering he always placed on the back of his dungaree blouse was torn by bullets.

There was Frankie, who had received a shiny, chrome plated pistol from home. He had boasted he would get many Japs with it. Now the sun's rays from over the ridge glinted on the handle. The pistol was still in its holster.

Peter had a strong voice in camp. He had it in the face of death. He was halfway up the ridge, yelling something about the "bastards" on the top, when their fire cut him down.

The lieutenant we called "Chicken" because he seemed so young, was the only one to reach the crest. A grenade smashed in the side of his head. Those skinny legs which had led me so often to exhaustion were white in the dried grass. Two Japs, five feet in front of him, had holes in their heads. An American grenade which the Chicken evidently had thrown was lying between the Japs. How often the lieutenant had drawled in his slow hesitant voice: "Now a grenade— it explodes five seconds after you heave it." His hadn't.

Eddie was lying in a bed of mountain flowers. He was fond of flowers. He used to put them in his helmet for camouflage when others used grass. The hand that was not on a rifle was crushing a flower.

Behind the lieutenant, his face anxious as if awaiting an order, was Angelo. He loved to sing—but couldn't. He and I were penalized once for singing *Put on your Old Gray Bonnet* after taps.

The company was still under fire when, on the ridge, I talked to the men who had made it. Private First Class Leon Slicner of Perth Amboy, New Jersey tried to tell me how "Smokey" could have been saved had they been able to pull him out of the fire lane in time. His words came slow. Finally he stopped in the middle of a sentence, leaving the story in mid-air. He really didn't want to talk. Besides, he was pressing low into a foxhole, and he couldn't breath well, for a machine gun was spitting fire over our heads. **Sergeant Cyril O'Brien, 3rd Marines**

Reprinted from *Semper Fidelis,* a collection of stories of Marine Corps Combat Correspondents (Published by William Sloan Associates Inc., New York, New York.)

came in the morning of December 8, 1941, smashing first Apra Harbor and the docks and sinking the Navy minesweeper USS *Penguin.* At 0400 December 10 flares above the beach north of Agana signaled the approach of Major General Tomitara Horii's South Seas Detachment of over 5,500.

A Chamorro-manned machine gun stuttered and some of the Japanese fell but there was little more resistance and Captain George J. McMillin, USN, commander of the Apra Naval Station, was not about to throw lives away against such odds.

From the outset the Japanese brought their arrogance and disdain for all that was American and Chamorro. Lorenzo Blaz, a school boy then, remembers it well. His father carrying a truckload of local people was struggling in second gear up a hill when willy-nilly a Japanese machine gun opened fire, killing them all.

The Guamanians will tell you how in two and a half years under the Rising Sun there were friendly Japanese, civil Japanese, arrogant and bully Japanese, and even sadist Japanese, but because of the superiority accorded them by their flag there was little recourse but to bite your lip or cry. While Guamanians hid American flags, buried some—for later hoisting—in cans among the banana fronds, the Japanese changed the name of the island to Omiyajima or Great Shrine Island.

It was with the immediacy of the Marine Invasion that the Japanese became vengeful, cruel, sour—even to rounding up young men and beheading them.

Devastating Bombardment

The Japanese had reason to be sour. The greatest bombardment by air and sea ever to descend on a Pacific island centered on Guam. It leveled Agana, and destroyed every coastal gun. Day and night, bombs and the shells rained under command of the crusty Rear Admiral C. Turner Joy, Task Group 58.18. Under Turner-Joy's flag were the New Mexico, Pennsylvania, Tennessee, Idaho and California.

One young Japanese out of his head with the "pounding, pounding" said ... "no matter where I went the shells followed." Some Japanese officers, seeing it was all over, committed suicide even before the first Marine came ashore.

(Continued on page 87)

An Explosive Performance ... By A Guamanian Freedom Fighter

Move over John Wayne ... take note Tom Clancy. The story of Francisco J. Cruz matches the best of any action movie; and to this day Mr. Cruz has not fully received his due honor.

In the early 1940s Francisco Cruz was a young dynamiter at an Asan rock quarry on Guam. With the arrival of the Japanese in late 1941, Cruz hid 700 sticks of dynamite, and kept them hidden for two and one-half years.

Then in mid-1944, Alberto Tenorio, A Saipanese Chamorro working as a Japanese interpreter on Guam, advised Francisco that the Americans were expected soon ... and informed him of Japanese machine gun locations in the hills above Asan.

When the Marines landed at Asan they did, indeed, encounter heavy enemy fire ... and the advance temporarily halted in some sectors. The Japanese, some 700 or 800 of them, became jubilant ... prematurely sensed they had repulsed the enemy ... and commenced to toast each other with sake.

Francisco Cruz saw the situation ... that the Japanese force of 700 plus (many now intoxicated) had left only one sentry to guard 10 machine gun bunkers. He overpowered and killed the lone Japanese sentry, disguised himself as a Jap soldier by donning the dead sentry's uniform.

With some of the hidden explosives, he placed dynamite charges at each of 10 machine gun emplacements and in an underground cave headquarters nearby.

Within a couple hours after the charges had been placed, 3rd Marine Division troops resumed the advance in that sector causing the Japanese to cut short their celebration and return to positions in the bunkers and cave.

Boom! Boom! Boom! Boom!

Cruz set off explosive charges at each location—effectively annihilating some of the enemy and its positions ... thus probably saving hundreds of Americans and Guamanians from later injury at the hands of these Japanese soldiers.

Throughout the operation of the American liberators Cruz continued to assist in finding and eliminating Jap forces—especially with the Guam Combat Patrol formed in late 1944.

As a member of the Combat Patrol, he was awarded a Bronze Star from Secretary of Navy James Forrestal ... but was never decorated for his bravery in destroying the Japanese positions above Asan. Francisco had been verbally promised the Medal of Honor by Admiral Nimitz himself ... but the award has never been carried out.

The Guam Legislature has named Francisco Cruz for inclusion into Guam's Guma Onra ... and is attempting to have him nominated for either the Medal of Honor or Medal of Freedom.

Like many other Guamanians, Francisco Cruz proved himself a brave and resourceful fighter for freedom.

The Listening Post

The 12th Marine's Regimental Fire Direction Center was dug into a hill side on the west bank of the Asan River, about 2,000 yards inland from Green Beach. Enemy resistance had been bitter when we landed, and the beaches were still being heavily pounded. Now in the early morning hours of D+2, we were expecting a counter attack at any time.

Anticipating this the evening before, Lieutenant Colonel William T. Fairbourn placed two men on listening posts in the draw, about 200 yards in front of our position. One was dug in on each side of the narrow river valley, down which such an attack would likely come. They were in direct line, telephone contact with each other and with us at the fire direction center.

Since midnight, Japanese small arms, mortar, and artillery fire had gradually built up, and the beachhead was continually lighted by the eerie, green star shells put up from our ships off shore. The situation was rather tense, as it had been throughout the day. Due to the heavy shelling on our landing beaches, some of our units were unable to make it into shore at all, and our supplies were critically short.

One of the radio messages I sent from Colonel Fairbourn to each of the firing batteries stipulated that if they fired a preparation, they were not to exceed 20 rounds per gun. While in contact with our 4th Battalion, I learned that their commanding officer, Lieutenant Colonel Bernard Kirk had ben hit.

"Find out how he is," ordered Colonel Fairbourn, immediately concerned.

"Is that you, Bill?" came the gruff voice in answer to my query.

"Yes. How bad is it?" asked Fairbourn, when I handed him the mike.

"My head may be bloody, but it's far from bowed," grumbled Kirk.

Our direct line telephone, to the two men up in the draw, was next to my radio. Periodically Colonel Fairbourn would place it to his ear and listen. At one point, he nudged me, tilting the phone so we could both hear.

"Hey, Charlie, how you doin'?" came a muffled whisper from one of the listening posts.

"OK, I guess," was the hushed answer. "How about you?"

"Jeeze, I'm scared. I keep hearing noises."

•

"Yeah, me too. Why do you suppose the friggin' colonel has us up here anyway?"

"Who knows why a friggin' colonel does anything?"

Bill Fairbourn nudged me again, and in the ghostly light of an overhead flare, I detected a grin on his face.

"This is the friggin' Colonel," he said into the phone. "Now when you two are relieved up there in the morning, you are to report to me. Over and out."

Sometime after daylight, Charlie and Mike were relieved, and they reported to the fire direction center as ordered.

"The next time I place you two on listening posts, it means just that," said Colonel Fairbourn. "There will be no talking unless you have an urgent message, understand?"

"Yes, Sir," they answered in unison.

Jack Kerins, Terre Haute, IN

You'd Bite Too!

On Guam, 1945, 1st Battalion 21st Marines were issued warm beer when it was available. To maintain control we were issued two cans, twice a week and were not allowed to save more than a week's allowance.

Anyone wanting more beer could hitch-hike down to one of the United States Navy camps on Saturdays and buy all they wanted in the Navy "Slop Chute."

One Saturday, several of us went to one of the United States Navy areas and I watched a young seaman sitting near us pour beer on the table for his pet monkey he'd acquired in the Philippines. It was a wild sight. The monkey was soon drunk, staggered all over the table entertaining the crowd. Finally the poor animal fell over in a deep sleep. The seaman told his buddies not to go near the pet the next day, "Cause he'll have a hell of a hangover and will bite anyone who gets close." **William I. Pierce, Merrillville, IN**

But that was not all the story.

There were plenty of Japs left even though that intense 15 minute immediate invasion barrage of everything just before H-Hour would have indicated otherwise.

Elements of the 3rd Marine Division's three regiments, 21st, 3rd, 9th, landing three abreast, touched ground at 8:39 a.m., July 21 between Asan and Adelup points, General A.H. Turnage commanding.

Audacious UDT (underwater demolition) teams did make things a little easier—blowing up 640 obstacles at Asan and another 300 at Adelup—then had the temerity to erect signs near the surf welcoming Marines to the USO.

Skilled and slippery, they lost but one man. The Navy and Marine Air blew the Jap airfleet out of the sky with kills ... like 150 one day, 50 the next, and so on.

"I couldn't believe that even grass could live after that bombardment," recalled Jimmy Helt of Pittsburgh, radioman with the 1st Battalion, 3rd Marines, who came in with assault waves ... He was recalling that final quarter hour bombardment of the beach.

Eyes Of The Nation

It cost the Japanese, but not all that much. Many crouched in relative safety behind reverse slopes, waiting the bidding of the 29th Division Commander, Lieutenant General Takeshi Takashina. Takeshi followed almost by rote the tactical bromide: kill them all at the water's edge.

Not all Japanese soldiers were sure even the top seeded 29th could do that. A note found on one sure casualty: "I will not lose my courage ... but now is the time to prepare to die."

Roy E. Geieger III AmphibForce Commander and boss of the invasion (called Stevedore) emboldened his own troops (us) with the solemn knowledge ... "The eyes of the nation watch you as you go into battle..."

On the beach, that island war became a series of little battles, not Patton at Bastogne or Tomischenko on the Russian steppes. The Marines swarmed up the high ground. One observer with the 12th Marines, down at the beach, said they looked like dots that moved. Elements of the 9th got to the ridgeline early. The 21st seemed at first to have easy going, until they faced the sheer

(Continued on page 89)

THE CUP THAT CHEERS

The heavy drinkers in the outfit didn't have any trouble finding something to drink. There was always some kind of alcoholic beverage to be had from some source.

Several of the men had learned to make raisin Jack; and the natives made a drink from the coconut tree sap that would knock a fellow on his rear end. The natives would climb to the top of the tree and put a tap tube into the green portion of the frond. It worked just like tapping a maple tree for syrup.

The natives called the stuff, "Tooba" or "Tuba". After collecting it in a bamboo section container they let the juice ferment for a period of time, sorta' like making wine, and ended up with a beverage that had about the same alcoholic content as wine. It had a milky look, with a kind of sour taste—almost beer flavored.

This stuff was rather milky and for a heavy imbiber it took a lot to get a buzz on, but the natives carried the process one step further. They would distill the Tooba, and make what they called "Aigee" or "Aggi" (spellings were subject to much imagination). Now that stuff was something else ... it was about 90 proof.

There was another source for booze, from the Navy ... "torpedo juice"—a propellant for torpedoes. I believe that stuff was 180 proof and was almost always cut in strength by mixing it with some kind of juice. I have heard some pretty hairy tales about it and raisin Jack making some of the men go blind.

You fella, wan' buy "tuba"?

Ole Dick Everhart took on a load of that stuff one night and came within a spate of dying. It was close enough so that Dick gave up drinking anything with alcohol in it after that night. **George E. Lyon, Florence, AL, Motor Transport**

He Rode That Cowboy

Thinking about the Seabees, brings to mind an incident that took place not long after we made our move to Agana. The military, mainly the Army, had introduced a new type of vehicle during this Guam Campaign. Called the "Duck," (or DUKW), it was a huge amphibious thing that had four wheels on it and could either float in the water or be driven on land.

They were massive outfits and took up more than their half of the road. The roads on Guam weren't extremely wide to start with, and these cowboys seemed to take delight in hogging. They had been known to crowd other vehicles completely off the road, and most all the vehicle operators on the island were hossed off at 'em. The "Duck" drivers were going to have to straighten up their act, or the battle with the Japs was going to be secondary.

One day a convoy of these monsters were speeding down a narrow road and met a convoy of Seabee equipment, going in the opposite direction. Well the

"Ducks" had hogged the road for so long and intimidated the other drivers, because of their size, that I reckon they felt pretty good about it and didn't even slow down.

The first "Duck" in line was traveling at a pretty fast clip and was about 20 feet away from a Seabee operated bulldozer going in the opposite direction, when the Seabee raised the bulldozer blade and, just as they met, swung the point of that blade right into the side of the "Duck."

He laid the side of that thing wide open, from one end to the other, and about six more "Ducks" piled in on top of the first one. The impact swung both of them around so as to block all traffic. It brought the Army driver and Seabee just about face to face.

The Seabee climbed up over the guard plate and out onto the radiator of the dozer. Here, assuming his best John Wayne stance, with hands on hips, said, "Now try hogging the damn road, cowboy."

I would imagine the six hours or

so of bottled-up traffic, before the vehicles could be extricated, gave the Army commanders time enough to figure out a way to explain to the "Duck" drivers the much needed virtues of driver courtesy. We never had hogging problems after that. Bless our little old Seabees. **George E. Lyon, Florence, AL, Motor Transport**

Didn't Know Password

One night on Guam we dug in and went back to eat at the chow line which had caught up with us. Earl Richmond turned around and asked a fellow the password. It was a Jap and he took off on the run. So dark we couldn't see to shoot him.

That night we filled our canteens with water at a stream. Next morning we went back there and the stream was full of dead Japs—all swollen up. **Sergeant Otho McDaniel, A-1-9, Paragould, AR**

The Regimental C.O. Viewpoint

Colonel George Watts Carr who landed with Colonel E.A. Craig (CO 9th) gives a brief, and clear picture of one officer's view of the assault on Guam.

Colonel (Lieutenant General, Retired) Edward A. Craig ... was anxious to get his feet on the shore ... and take personal command ... A pall of smoke and dust hung over the landing beaches and Navy guns continued to pour shells over our heads, and beyond, as our troops moved inland.

As we began the slow ride to the beach ... I stood behind Colonel Craig who was directing our amtrac driver where to go. Our enlisted personnel were crouched down in the cockpit behind us.

We were not proceeding as an organized wave, but independently. On our left and to the front we could see other amtracs heading shoreward and some returning to get a second load of Marines. Big splashes in the water among the tanks indicated the enemy was still throwing mortar and artillery rounds.

Suddenly off to our port an amtrac took a direct hit and exploded violently and virtually disappeared. We were now approaching the reef and with a sudden lurch mounted the coral reef which was covered with three to four feet of water ... The amtrac treads now had traction and we moved swiftly to the beach ...

"All out" the coxswain yelled and we poured over the side taking our equipment.

At the spot we landed was a small grove of trees and we gathered in it to regroup before moving out. About that time the Japs began to drop mortars around us and we heard cries to the left for a Corpsman. It turned out that one of the wounded was Lieutenant Colonel Sabater, regimental executive officer. He was evacuated and never came back.

When Colonel Craig found we had landed behind his battalion he ordered us to move 500 yards to the right so we might be nearer the center of the regimental beachhead. This we proceeded to do, walking along the edge of the water, dropping to the ground whenever a Jap barrage would start ... Soon we reached a spot the colonel selected and proceeded to establish our command post ... Our troops began to land and the spot the colonel had picked turned out to be as attractive to them as it had to him ... Of course, we were only about 50 yards from the water's edge.

At about this juncture Colonel Craig called a small staff meeting and assembled five of us to a circle, lying on the ground heads and feet out like the spokes of a wheel. Hardly had he gotten the first words out before a Jap sniper began firing at us and his first round struck the barrel of the R-1's carbine. We scattered like a covey of quail. Our R-1 was bleeding about the face, though he was not hurt seriously (splinters from carbine) ... but he was evacuated and never came back.

It was now mid-afternoon and rather still. Jap shelling had become very sporadic. Supplies were beginning to come ashore and pile up. Off to our left, a bulldozer scraped out a big hole 30 feet wide and 100 feet long which soon was filled with dead Marines, brought in by Graves Registration.

It was becoming almost impossible to function on the beach and the front lines had advanced well beyond the first day's objective. Colonel Craig ordered me to find a command post forward for the night.

Some of my wire teams had located a small cave on the reverse side of the cliff to our immediate front. After a quick inspection, we leap frogged the command post forward, just before dark, and had a relatively peaceful night in the Jap cave overlooking a crowded beachhead. **Colonel George Watts Carr Jr., Durham, NC, H&S Ninth Marines.**

No Sword Play

On Guam's Mount Alifan, Marine Corporal Roger Spaulding, a farmer from Sheridan, Indiana, plopped to the ground and found himself face to face with a Jap officer.

The Hoosier turned to his platoon leader."Hey, sarge," he shouted, "there's a Nip here staring me in the eye."

"Shoot him!" the sergeant hollered.

"I can't, my gun's jammed," the farmer answered.

At this, the Jap rose from the ground and charged Spaulding with a large samurai sword. Gripping his rifle with both hands the Marine rose to meet the charge. A shot rang out. The Jap fell, a victim of another Marine rifleman.

"Why'd you do that?" Spaulding called to the Marine. "I was just fixin' to club him to death." **Technical Sergeant George R. Voight**

cliffs and the enemy mortars. By late afternoon the 21st was on high ground, and just after noonday the 12th was set up and firing.

Up above the 3rd Marines—as close as 100 yards away—the Japanese defenses were intact, and the act of forward motion required the hands as well as the feet. Worse, there above the 1st Battalion, the Japanese were in a position for ideal enfilading fire. When the 3rd called in big guns, the Japanese hid on reverse slopes to come out again when the storm died down.

One Marine recalled the doggedness of the Japanese. Directly up from Red One, a Japanese machine gun position was continually reconstituted. As the gunner was shot, his body was dumped down the slope and another took over.

They Rolled Grenades

Able Company 1st Battalion 3rd will be remembered by anyone who watched them in that defilade on the face of Chonito Ridge. Commanded by Captain Geary Bundschu, the Marines were so close to the Japanese on the top that the enemy had to roll grenades. To throw them would put the grenades out of range ... impossible!, so it seemed and Captain Bundschu asked to sever contact. Permission was refused. He assaulted again with Lieutenant James A. Gallo Jr. the only officer left. Bundschu died on the way up. Eventually with supporting fire Gallo got to the top.

What he found were spider trenches wound around the top like the tracks of some great worm. They were shallow, the width of a man's body ... but not enough to shield from the splinters ... particularly of the 81's which sliced off limbs and parts of a man's body with the smoothness of a scalpel.

An attack by the 2nd Battalion 3rd (led by Lieutenant Colonel Hector de Zayas, later killed by a sniper), on the flank of the Able Company Hill, secured the crest for good.

An old survivor, Takashina was quick to conclude that once the 3rd broke through the Chonito ring, the Americans could spread out. No longer confined, they'd be hard to push back into the sea. He called in his people from Tumon, Pago, Sinajana, anywhere near—even though it weakened defenses elsewhere. He would interlock and then initiate the

(Continued on page 91)

The Reef ... Beach ... And Inland

No sooner than we unloaded the last equipment, the sailor was winding up the ramp, and the coxswain proceeded to back the LCVP off the reef. I don't think he went 20 feet when a mortar landed next to him, sinking the LCVP. All three sailors floated to the reef and, as fast as they could, got aboard an LCVP to the right of us ...

With mortars continuing to fall, we had to carry our guns and equipment ashore. We were ready to fire missions that afternoon.

On the left we could see the 3rd Marines attacking Chonito Ridge. They would go so far, dig in, and try again to advance. They appeared as little dots all over the face of the cliff. It took them several days to clear the ridge. To assist, a Navy destroyer was practically sitting at the edge of the reef, firing their 20mm and 40mm guns.

Late that afternoon I decided to nab something to eat, found what I thought was a comfortable position, and pulled out a "K" ration. After a few bites, and to be more comfortable, I scraped away some of the dirt ... and exposed the left hand of a Jap. I covered it back up and moved somewhere else to eat.

Lieutenant R. Talles, our battery executive officer, was controlling the fire and had little rest since we landed. Toward evening he told me to take over, so he could get some rest, and wake him around midnight. At midnight, I woke him as directed, crawled into the foxhole that he left, and fell asleep.

I was awakened at the gray of dawn by a tremendous amount of firing. I could even hear small arms fire ... this surprised me and woke me in a hurry. As I crawled out of the foxhole, I could see our shells striking the cliffs in front of us that the 21st Marines had captured the first day.

At the battery exec's position, Lieutenant Talles told me the Japs had broken through the 9th and 21st Marines trying to get to the artillery positions and beaches.

I immediately went out to insure that Sergeant English's local security, 50 caliber machine guns, were awake and all guns manned. Sergeant English had everyone on the alert, but no Japs had gotten as far as our position.

Lieutenant Talles said battalion had called and ordered all men to assist the infantry in stopping the attack and to restore our lines. We took everybody except four men per gun. Colonel Wharton was not enthused about leaving his gun position, but said he must go ... and did. He never returned to our position, because he was wounded badly.

Our forward observer teams were with the 3rd Battalion, 21st Marines who were hit hardest and we received word that Private First Class "Cowboy" Rowland and Walters had been killed. Walters was being evacuated by stretcher, when he was hit again ... in the head.

Later in the day, Sergeant English and I scouted the flank of our position but found only dead Japs.

Shortly after, the infantry advance was fairly rapid and we relocated our positions near Tumon Bay (in 1990, Guam's principal resort area). Sergeant English was checking local security when a Jap in a cave threw a hand grenade that exploded on Sergeant English's leg ... and he died before we could get medical attention.

On another forward relocation of our position, we heard several rifle shots. To our pleasure the object attacking was a nice fat pig. We were fortunate that several of the men had been raised on farms, and it wasn't long before the pig was gutted, and in a 55 gallon drum of boiling water. It was the first fresh meat. **CWO George J. Green, Webster Groves, MO**

If It Stays White Drink It

We had secured the island, and were on squad patrol through the jungle and surrounding fields, when we came upon a cow that had four or five pieces of shrapnel ... and the cow was mooing.

Gieseke, a farm boy from Illinois, said that cow needs milking ... so we chased down the cow ... and Max Brotherton and I held the cow's head so that Gieseke could milk her.

Gieseke said, "Throw me your helmet." So, I handed him the metal part of my helmet, and we used that as a milk bucket. In a short time, Gieseke had the helmet filled with milk, and passed it to me. I took the first few swigs of milk, and passed it to Gieseke. About that time I remembered what I had used that metal portion for the night before in the foxhole. I didn't say anything, but when it came back to me, I passed it on to the others. I didn't tell Gieske about that for several years, but he seems to have survived. **George J. Briede, La Belle, FL**

Hi, Buddy!!

We were on Guam, working our way across the island, and had encountered a little difficulty that day. That night we dug in along a little paved road ... it wasn't very wide, maybe 10 or 12 feet wide ... got back among the tall grass and dug in. It rained that night, and it was my turn to stand watch. I was sitting there, in the rain, with my poncho over the top of me, and my rifle underneath the poncho to keep it dry. I had a field of fire out across the road, maybe two feet wide.

Despite the rain, it was light that evening, and I could see pretty well. Suddenly, standing right in the road was a Jap looking at me. I threw the poncho off, up over my head, got it tangled a bit. As I was doing this, the Jap took off, running down the road. I hollered, "This is Strode. There's a Jap coming down the road."

Just before he took off on his run, the Jap had said, "Hi, Buddy." So for several days after that, the rest of the men in the company were calling me, "Hi, Buddy" Strode.

As for the Jap, he hit just right. Nobody seemed to have a weapon ready to fire ... because as he ran down the road, not a shot was fired. **Jay Strode, Jerome, Idaho**

Ready To Leap From The Truck

Prior to the Guam invasion our trucks and jeeps had been prepared for driving through salt water at depths as high as the vehicles ... because of coral reefs surrounding the island.

On D+1 my truck and I were transferred from the Navy ship into a Navy landing craft since the tide was higher and allowing the boats to reach closer to the beach.

When the landing craft made contact with the coral reef, we stopped, letting the front ramp of the landing craft down into the water. Then I drove the truck into the water, which came as high as my waist while sitting. Now, you'll have to keep in mind that all kinds of shelling, shooting, and flares were all around the area. I got over the coral reefs with the truck okay, but I was prepared to jump out at any moment because reefs are like hills under water.

I finally got to the beach, but little did I know that was as far as I was going. Our infantry was pinned down by Japs firing from the large hills in front. The main part of our 75 howitzer outfit was in front of me, so I made a foxhole on the beach.

Because our troops were pinned down, Navy destroyers were called in for some direct fire on the hills near the beach. The shelling worked with success because the infantry was able to forge ahead, our unit moved ahead, and I got that truck off the beach. **Henry J. Klimp, St. Simons Island, GA**

Floating Gasoline

Corporal Everad F. Horton now of Johnson City, Tennessee won't forget (especially after 50 years) how his platoon floated 50 gallon drums of gasoline over the reef on Guam, then rolled them ashore, against Japanese protesting with small arms and mortars. Tough enough to come ashore with pack and rifle, how about accompanying drums of gasoline to get inshore attention? Still, Guam never matched Iwo for sheer violence, he adds.

Respected Chamorros

William T. Hewitt recalls well the mop-up operation across Guam, when just about everybody—except perhaps chaplains—joined in the sweep ... Chamorros found Japs on the island 20 years later, so some were missed ... Hewitt would speak warmly for Chamorros anytime. They are now home in his picture album. **William T. Hewitt, Binghamton, NY**

This Japanese flag, captured on Bougainville, is the center of attraction for this rifle platoon squad of A Company, Ninth Marines after they had arrived on Guam. When squad leader Otho McDaniel of Paragould, Arkansas had sent the photo home during the war, the inscription on the back said, "Dad, this is the toughest bunch of boys the Marines ever put out—what I mean, they're rough, rugged and ready for a fight any time, place, or situation." Pictured are, first row from left—H.A. Harper of Wisconsin, E.L. Whittemore of Iowa, C.F. Price of Pennsylvania, McDaniel, J.P. Marenda of California, and R.B. Dusek of Illinois. Standing are H.E. Dunsworth of Kansas, E.J. Fuller of Massachusetts, G.A. Lincoln of Montana, S.W. Easly of Texas, F.L. Ferreira of California, and E.L. Richmond of Iowa.

counterattack in depth—one he was not in a position to muster.

Over Chonito, Bundschu Ridge, and on the plateau, later to be named for Admiral Chester W. Nimitz—Takashia had organized his repost. In an encroachment almost in the lap of the Japs, F Company (under Captain Louis H. Wilson) with 2nd Lieutenant John J. "Jack" Eddy, bore the brunt of these last ditch banzai attacks, almost as they kicked off. One machine gunner in the F Company defense had 80 Japanese piled in front of him. Jack Eddy recalls calling the 105's to tear up the brush all around him.

Moment of Banzai ...

Down at Agat near the Orote Peninsula, and not far from Apra Harbor where Major General Tomitara Horii's brigade had come ashore in 1941, the 1st Marine Provisional Brigade, under Brigadier Lemuel C. Shepherd Jr., and the 77th Army Division (305th Reinforced Combat Team) under Major General Andrew D. Bruce, had no cakewalk.

They faced the Japanese 38th Infantry Regiment under the zealous and bushido Colonel Tsunetaro Suenago, with much heavier defense fire than at Adelup and a nasty obstructing reef which pilloried landing draft and cost lives.

From Gaan Point, with concrete many feet thick and fire ports like picture windows, pillboxes poured on fire before Marines could touch the sand. The 22nd Marines, forced to commit reserves early, saw battalion momentums sag. General Shepherd ashore called the situation "critical." Had, in fact, the Japanese stuck to procedure and generated one of those wild banzais, some Marines indeed would have been pushed into the sea.

When it came the Japanese threw themselves into whooping, yelling onslaughts ... sometimes resembling a second rate Indian western, with officers who should know better slashing tanks with swords.

Not all of the Japanese were so senseless, bushido or not. Harry Gailey in *The Liberation of Guam, Presidio Press*, tells how young Japanese soldiers cried as they tore up pictures of loved ones before their fatal rally into American gunfire.

The Guam Campaign lasted 21 days, July 21 to August 10. The cost:

(Continued on page 93)

Guam ...
The Detail ...
The Background
Color

Colonel Henry Aplington II, who commanded 1/3 from January to September 1944, through the Guam Campaign, now of Lower Warner, New Hampshire, enjoyed a long career in the Marine Corps with a second tour with the 3rd Division (1955-1956) and service in China.

To Guam

As it turned out, we were to be aboard the LSTs for much longer than the anticipated nine days. The crowded conditions under the Central Pacific sun were to go on for six weeks.

After passing through Kwajalein's reef, course and speed were set to bring us over the horizon off Saipan on June 15, D-Day. The Pacific weather was perfect, day after beautiful day passed without event ... The formation seemingly motionless on the calm sea. Under the crowded conditions there was little that the troopers could do other than lounge in the make-shift shelters, clean their weapons, play cards, or look at the ocean and the other ships. After a few days the uniforms began to get pretty ripe and the troops did their laundry by streaming it over the side on the end of a line.

The 15th was another beautiful day and we cruised over the horizon as the 2nd and 4th Divisions hit the Saipan beaches. We had been provided with the maps and frequencies in the event of being committed and on the conn of 449 my communicators had the radios up so we could follow the action ashore. Fighting came through as heavy and the progress slow but by later afternoon, it was evident that "the good guys" were winning. So too did it appear to Admiral Spruance for he confirmed W-Day for the 18th and our Guam forces hauled off for the run south.

During the night, however, the situation changed. The next morning heavy casualty reports led to the decision to land the 27th Division USA on Saipan, Guam was postponed, and the 3rd Division and the 1st Provisional Brigade were retained in reserve. We were sent to the east where we were to steam 150 to 300 miles off shore, reversing course every 12 hours, while awaiting the development of the situation.

On the 17th, the monotony was broken, when at 1800, we were hit by four Japanese torpedo planes. When the general alarm went, I bolted down the passageway and out into the starboard deck to see all AA firing and, low in the water, a Japanese releasing his torpedo, seemingly headed directly for 449. At that moment LCI (G) 468, a bone in her teeth, guns firing, and signal flags streaming, surged between us and the torpedo. There was a flash and 468 was dead in the water, nothing forward of her superstructure, the flags burned remnants, and dead sailors hanging over the gun tubs.

It was all over in three minutes and 468 was taken in tow, stern-to, by another LCI and the formation resumed its course. Three of the Japanese planes were splashed.

On the 18th and 19th as we cruised back and forth, the Battle of the Philippine Sea was fought between Spruance's and Ozawa's carriers resulting in the destruction of the remaining carrier air strength of the Imperial Japanese Navy from which they never recovered.

By the 25th Admiral Spruance felt that Saipan was under control and the 3rd Division was released from floating reserve and directed to Eniwetok.

Even though Japanese naval offensive power had been destroyed, the bitter resistance on Saipan convinced the high command that caution was the watchword and it was decided to take more time to work Guam with air and naval gunfire; and the 77th Division USA in Hawaii was added to the Guam order of battle.

We Gather At Eniwetok

We entered Eniwetok Lagoon on the afternoon of June 30 and the ever efficient Navy Postal System had mail aboard the ships by evening. Gathered there was the whole Guam assault force as well as the supply ships and oilers of the fleet train and the myriad craft of the service force, the mobile, floating, facility capable of repairing anything not requiring major yard repair.

The might of the United States Navy was impressive but of more import to us was that we had now been aboard for over three weeks, and to get the men off the ships and stretch their legs ashore. All of the islands not being developed, and within reasonable small boat range, were assigned in rotation to the battalions and other units. The islands were not large enough for any sort of training or maintain the physical edge which we had when we left Guadalcanal. Calisthenics, some minor sports, and just the relief of being on dry ground had to suffice.

Guam was rescheduled for July 21, the 77th Division caught up with us, and we sortied on July 15, this time for real.

Two distinct memories of the next five days are of reading in the chart room a news flash announcing that the Army Air Forces had made a strike (originating on the China mainland) on Yawata, the large Japanese steel-producing city; and even more immediate for us, a message from Admiral Conolly in his flagship *Appalachian* that the underwater demolition teams had destroyed all of the obstacles planted on the invasion beaches of Guam, and the *Appalachian* was operating with impunity off the coast.

Steak And Eggs

Reveille on Friday July 21 was around 0230 and breakfast itself was the traditional pre-invasion steak and eggs. As dawn came we rounded Ritidian Point, the northern end of Guam.

Off Asan, *California* and destroyers were bombarding the shore. *California's* 14" guns seemed almost depressed and every few minutes she gave a belch of flame, and twelve shells tunneled ashore.

In preparation for the 0730 traditional "Land the landing force," the troopers—laden with fighting gear—went down to the tank deck and squeezed between the overhead and the sides of the LVTS. Motors were roaring and the exhaust fans whining. It was claustrophobic, particularly with the knowledge that within range were Japanese coast defense guns. When the bow doors opened and the ramp dropped, we gratefully rattled out and into the swell of the sea.

The tractors formed and we started the run for the beach bracketed by the splashes of Japanese mortar and artillery rounds, following LCI (R)s which fired sheaves of rockets to supplement the naval gunfire still pulverizing the beach. As we approached the reef, the LCIs peeled off to each flank, the gunfire lifted, and fighters strafed the beach area. The LVTs climbed the reef, crossed the low water shallows to the beach and after crossing the debris cluttered coast road deposited us amid the wrecked houses and shattered palms of Asan village.

The Asan Beachhead was indeed a little Anzio. The whole 1/3 zone was heavily defended, on the right by the terrain, and on the left by Captain Nakamura's professional 320th Independent Infantry Battalion. With their interlocking bands of fire and pre-planned mortar and artillery concentrations, they knew what they were about and died hard. We did not know at the time that the 3rd Marines were knocking on the front door of the command post of the commanding general of the 29th Infantry Division in the hill mass in front of us.

The terrain was horrendous with its impenetrable brush and cliff-like outcropping of rocks. The battalion had extreme difficulty making forward progress, as well as a most difficult job of maintaining contact with units on the left and right.

On the fourth day, the battalion was occupying ground on what was to be later called Nimitz Hill after the Admiral moved his headquarters forward from Hawaii ... and the officers club was built

1,350 Americans killed, 6,450 wounded. Japanese losses: 17,300 killed, 485 prisoners.

Pathway To The Heartland

Capture of Guam and the Marianas (Saipan and Tinian) opened the way to the heartland of Japan and the awful product of that liberty was the burning of Tokyo, and the bombing of Nagasaki and Hiroshima.

With the fall of the Marianas, Hideki Tojo, prime minister under whom the war began, was forced to retire with his cabinet. Admiral Isoroku Yamamoto, Japan's military icon, commander of the Combined Fleet, planner of the attack on Pearl Harbor (who, ironically, opposed war with the United States) was dead ... shot down by United States' P-38s over Bougainville. His influence and mind-set might, indeed, have meant a shorter war, and may well have averted the atomic bomb.

On that same August 10 the United States VII Corps in the Avranches-Mortain sector in France was forcing German divisions to withdraw to the east. The United States VIII Corps on that same day entered Nantes and reached the Loire. Before August 15, the Allies would attack southern France.

In Europe the war was winding down. In the Pacific a long bloody road was ahead.

on the site of the battalion's command post. Thirteen years later I had lunch there, and it was eerie to sit on the terrace drinking martinis, looking over the neat lawns, officers quarters, barracks, and office building where once was scrub brush covered by my machine guns and 81 mortars.

In a morning regimental briefing, I learned we were to be relieved by 2/9, which would be attached to the 3rd and with the rest of the division, would attack at 0930. We would assist by fire until masked, and then move back into a designated division reserve position.

Alerted by the activity, the Japanese reacted with machine gun and mortar fire and ... in the middle of a burst of shells ... Bob Cushman, the 2/9 Battalion commander, slid into my dug-in command post. It was the first time that I had encountered him. Twenty-three years later he would be commandant. During his tour as CMC, after Lois and I were married and living in Annapolis, we were invited to sit on the reviewing stand while Bob took a midshipman parade. As we sat under the canopy on a lovely Maryland spring afternoon, I looked at Bob with his sparkling whites and four stars and wondered to myself whether it could be the same person who slipped into my command post on that long ago morning.

In the next 24 hours, Lou Wilson, one of his company commanders, would win the Medal of Honor and be on his way, too, to being commandant.

... There had been intermittent showers through the day, but with the dark came heavy rain. On the line, Marines huddled under ponchos in their wet foxholes trying to figure out the meaning of the obvious activity on the part of the opposing Japanese.

Around midnight there was enemy probing. At about 0430 my three companies on their hills erupted into fire and called for mortar support. Company commanders told me there were Japanese all around them. When I said to Dave Zeitlin, from whose area came the most noise, that I wanted to see some dead ones in the morning, he told me not to worry. The firing continued sporadically until light when I went up to see—there were dead Japanese all right—some apparently blown up by their own demolitions which they had been carrying, some by our mortars, and the rest by rifle and automatic weapons fire. The demolitions carriers had been sent to infiltrate the American dumps and artillery positions and destroy them.

We didn't stop them all. Some got into the division field hospital area where they were killed by corpsmen and patients, and in other rear installations by cooks, communications, administrators and cannoneers. The Japanese had been close ... three of my four dead were killed by bayonet thrusts.

While General Takashina had bet the store and lost with his night attack ... chewing up seven of his best battalions in the process ... the Japanese were so reduced that they could never again mount a major offensive. We didn't know it, and division anticipated a resumption of the attack the following night.

Forty years later one of my young academic friends, well read in military history, but without personal experience borrowed my *1/3 On Guam* in which a scene of numerous Japanese casualties was described. He told that his reaction was sadness for "all of those young boys from little Japanese villages lying there dead." I suggested that his sympathy was misplaced, that I was a young boy (27 seems so young today) from a little Vermont village and that if one of those Japanese "young boys" had gotten me in the sights of his Model 99 light machine gun, he would have squeezed off a burst without shedding a tear.

As one day passed, we successively displaced forward along a dusty ditch-lined road flanked by pineapple fields. It was a clear, hot afternoon and at one of the halts one of my runners climbed the fence and with his "knife, utility and fighting," the famous Ka-Bar, chopped off several of the pineapples which he brought back. They were juicy and delicious.

As A Company came slogging by one day, Pat Patterson, a heavy man, grinned and as he always did, said, "Major, this is a hard war on a fat man."

On the morning of August 2, shortly after the companies moved, the command post personnel were startled by the appearance of an enemy light tank which charged by them on the road ... an officer in the open turret was shooting wildly with a pistol. A hundred yards or so toward Agana, it ran into a ditch and was abandoned. There were no casualties resulting from this little "Banzai."

While advancing northward up the island, a road block had halted the lead company, and I went forward to find out what was happening. A Japanese 20mm gun, protected by riflemen, was firing and the advance party had taken cover in the ditches. I got them moving and,

seeing a mustard-colored uniform break cover in the woods ahead, I cranked off a round from the '03 which I habitually carried in something of a Pattonesque gesture to give my troopers the comfort of believing that their commander was out of the "Old Corps." Also, I didn't like the carbine.

There was a roar behind me. It was our new regimental commander and the two of us stood in the middle of the road, the Japanese blasting away, while we complained about the war stopping while the "battalion commander gets himself a Jap."

Telling me that I was too far forward, he stomped off to the rear. It was a silly exercise, but the story lost nothing in the telling throughout 1/3.

Later in the day, the column again stopped and the word was passed, "Battalion commander to the front." This time it was an American flag waving from the brush ahead. Beckoned forward, it turned out to be carried by a group of Guamanian men, women, and children, overjoyed and shaking hands with and patting the Marines.

The children stood at attention and bowed, but were told by their parents, "You don't have to do that. They are Americans."

One night a typhoon rain pelted down and the troopers huddled under their ponchos in the waterlogged fox holes. At about 2200 flares went up from the woodline to our front when we received a full fledged infantry attack. The .81 mortar crews had draped ponchos over the muzzles against the rain, and one crewman didn't quite get his off in time. Before he got it clear, a round went through it but fortunately, the time was too short for the fuze to arm, and it went right through the poncho.

On one rainy morning at the end of the campaign, the head of the 1/3 column turned off the road into the location where we were relieving a battalion of the 77th Army Division. They were formed up under their helmets and ponchos ready to move out.

They appeared to be smaller than my youngsters and older, many wearing glasses. In their fatigues so different from our herringbone utilities and their olive drab ponchos—contrasted with our camouflage ones—they looked strange. But there was no doubt in our minds that the 77th were good people to have alongside in a fight; and as a result, we referred to them as "The 77th Marine Division." We considered it a compliment, what they thought of it is not recorded.

A DOCTOR OF COURAGE

Fred Schlumberger joined the 2nd Battalion, 12th Marines sometime in the spring of 1944 after the division had completed the landings on Bougainville. We were encamped in a giant coconut palm grove at Tetere on the island of Guadalcanal.

Schlumberger was a specialist (urology) and was older than most battalion "docs" who for the most part were just out of medical school. It was unusual for a specialist to be assigned to a line battalion but he was happy with the assignment.

I suppose he had little opportunity to practice much of his extensive knowledge of medicine for most of the battalion was made up of enlisted men who were 19-20 years of age and junior officers. Schlumberger was probably 10 years older than me which I thought was incredibly old.

The doctor became famous when he made a major, who was unpopular, get his boils lanced. This major was a notorious malingerer who suffered from skin eruptions which were unpleasant looking. Schlumberger was uncompromising in treating Marines who just wanted to "play off." However, he did send home one of my noncommissioned officers who would get sick when the time for combat approached. He had a heroic mind but cowardly legs. I think the doctor realized that it was not the fault of the young man and that it would be useless to punish him for what he could not help.

In May of 1944 the transports arrived in Sea Lark Channel to lift the 3rd Marine Division 5,000 miles into the Marianas operations.

I was on the transport Wayne, the APA 54 carrying the battalion landing team of the 1st Battalion 21st Marines. Fred Schlumberger travelled with the battalion headquarters which was on another transport. Artillery was landed right behind the infantry and about two hours after H-Hour, a 105 howitzer of the 3rd Battalion 12th Marines fired the first adjusting round.

I was with the 21st Marines and in mid-morning I had an 81mm mortar round explode about three feet in front of me, a shard slashed me across the ribs on the right side of my chest but did not penetrate the chest cavity. It did cut in two a cigar and a pencil.

I was the reconnaissance officer responsible for laying the wire and putting in a switchboard and I got this job done and went back to the beach to be evacuated for treatment. I went out with a Higgins boat and we had difficulty getting taken aboard because of the flood of casualties. Finally, we went aboard the APA-54, the Wayne, from which I had landed in the morning.

Shortly after I went aboard, Fred Schlumberger arrived. He told me that he had been standing in front of his aid station on the beachhead when a Japanese artillery shell exploded about 100 yards away. A fragment hit him right on the belt of his trousers and opened up a gash in his stomach.

He asked if I would like to accompany him to the operating room where he would get treatment. We went below but found that there was no doctor available. So Fred calmly cleaned his wound, pointing out to me the large intestine which was trying to poke its way out of the opening. Then he got some gut and a needle and sewed himself up without any anesthetic. After this was done, he went to work treating the wounded.

The "Wayne" must have received 300-400 wounded men, many of them stretcher cases, and the staff consisted of two medical doctors and a dentist. Frederick Schlumberger operated on the wounded all through the night despite his own injuries.

Morning came and he said he was going back ashore as he felt that he was needed at his own unit. The next day I returned to duty and went up to the 1st Battalion 21st Marines who were holding a hill line above the beach.

During the rest of the Guam Campaign, he regularly dressed the wound in my side and cauterized it. He had a dry, sometimes biting sense of humor, but all of us in the battalion respected and trusted our battalion surgeon.

He left the battalion after Guam and did not go with us to Iwo Jima. After the war, he entered practice in Beverly Hills as a urologist. I visited him once in 1950s in his office. He seemed to be prospering. I think he was a solo practitioner.

In 1970 I returned from Australia and went to visit Fred and his wife, Holly, in Beverly Hills. We talked over old times and he spoke with deep affection and regard for the young enlisted Marines.

He told me the story of one of my men, Sergeant Homer Cornelius Wright of the state of Texas. "Stud" Wright was a rangy young man of about 20, who was an expert rifleman. Like many of our artillerymen, he really wanted to be in the infantry where he could get at the Japanese at close quarters. Sergeant Wright had contracted malaria on Bougainville and had several attacks. His illness was severe and Fred Schlumberger told Wright that he would have to go back to the United States for treatment. "Stud" Wright and another man remonstrated with the doctor telling him that they could not be spared from the coming operation.

Frederick Schlumberger did not send Sergeant Wright back to the States. On the night of June 24, 1944, "Stud" Wright was with the 1st Battalion 21st Marines when we were overrun by two battalions of Japanese infantry in a night attack. It was a chaotic situation in the night with the battlefield being illuminated by the yellow lights of shells fired from the destroyers offshore. Sergeant Wright did heroic deeds that night.

I never saw Frederick Schlumberger again, and we lost track of him. I finally located him with the help of a private detective agency who gave me an address and a telephone number. When I called, I was told that he had died on June 11, 1983. I wish that I had been able to talk to him once more to express my respect and affection for him. **John McKinnon, Oakland, MS**

About The Gunboats

Colonel Newport E. Hayden USMC (Retired), a mustang and former commanding officer of the 3rd Regimental HQ Company, lived with his wife, Barbara, in Piedmont, California before he died in 1990. He joined the 3rd Division on Guadalcanal, just before Bougainville. Guam was his most memorable campaign.

My duty on the Guam landing was as assistant operations officer for the 3rd Regiment. I landed with the second wave on Red Beach, near Adelup Point. I remember well the shelling we took on the beach for three days and nights as we were in front of the Chonito cliffs.

I was assigned as a fire control officer for the first three nights for three rocket LCI gunboats. I went forward in late afternoon to locate our forward positions on the left flank and was sent out to the lead gunboat before dark, staying aboard all night with a backpack radio man. Our mission—to fire rockets inland from the reef line and to close the beach above Adelup Point. I was to see that the boats didn't fire on our troops ashore.

We were to close the beach every few hours and be prepared to close and fire on call many a rocket into Agana. On the second or third night our gunboat rammed the reef and hung up. At night the two other gunboats tried to pull us off, but it wasn't until dawn that we broke loose.

I think I was the first Marine sitting on the reef for six hours with a frontal view of the whole town of Agana ... and you know, they never fired a shot at us.

The gunboat commander was pretty nervous and I had to tell him a few things to do—like breaking out rifles and posting a guard on the bow with orders to shoot anyone approaching the ship. That was my connection with the Navy the first three nights on Guam.

Colonel Hayden later returned to Guam, where he stayed with Captain Peter Siguenza, a native of Guam, who returned under Hayden's command to help liberate his homeland. A business official on Guam, "Pete" helped organize the 3rd Division veterans into a unit there. Pete lives on the tourist haven, Tumon Bay.

The Hitchhiker

When we were in the Northern part of Guam near Finegayan, Marine Jesse Barber, a Scout of 3rd Battalion, 9th Marines was sent back to division headquarters.

Jesse followed the trail through the jungle and, as he reached the main road, heard a rumble coming around the bend. Thinking he would have a ride back to division he slung his rifle over his left shoulder, tilted his helmet back on his head and assumed the hitchhiker position, right hand up, thumb facing south.

Jesse's smile froze on his face when a heavy Japanese tank came down the road with its tank officer standing in the turret with blood streaming down his forehead, and a pistol in his right hand. Jesse froze in the hitchhiker position. He looked up and the Japanese tank officer looked down ... with their eyes meeting.

The tank continued on but was hit by a Regimental Weapons roadblock further down the road.

Until this day I think of what must have gone through the mind of the Japanese officer ... some crazy Marine. **James J. Carvino, 3rd Battalion, 9th Marines.**

Guam Folk-Hero Was Underground Resistance Song Author

"Sam, Sam, My Dear Old Uncle Sam"—it was a favorite underground song on Guam—and the man who probably sang it most was Pete Rosario at far right. At left are two Guamanians who knew the war as youngsters, and next to Pete is World War II 1st Lieutenant John J. "Jack" Eddy who received the Silver Star for action on Fonte Ridge. Rosario died in 1986.

An ever vigilant Japanese occupation force never quite understood the enemy it had in a simple underground song often sung by Guamanians within earshot of the hated enemy during Guam's dark days from December 1941 to July 1944.

The song was written by Pete Rosario and had the Japanese known the hope it expressed for the Guamanians—and the derision for the Japanese—they certainly would have administered Rosario much more than the beating he received, his penalty for singing it in public. He often sang the song with Louis Futado, an Hawaiian on Guam, who helped Pete with the original version of *Uncle Sam, Please Come Back to Guam.* Others, too, could claim a piece of the song as it was readily adaptable to impromptu verses.

The Japanese put Pete to work in a hospital mess after they occupied the island. Hoping to gain the cooperation of the population, the Japanese were somewhat lenient once they secured the island. However, that all changed immediately before and during the American assault that began July 21, 1944.

The Japanese became vengeful and brutal, and stories of their cruelty—and Chamorro courage—were immortalized in songs often attached to Pete Rosario's underground song, *Uncle Sam ...*

After the American landing, Pete Rosario was one of the Chamorros assigned to the security patrol by United States forces. The patrol's volunteer role was to protect the homes and villages from the dispersed and dangerous Japanese stragglers, a duty he took very seriously.

Pete was bitter, and with good reason. Near Merizo, the enemy assembled the leading and most formidable Chamorros, shot them and beheaded 40, even as the 3rd Marine Division approached over the nearby hills.

After the war, Pete zealously served the war-weary youth of Guam. He organized clubs and sports leagues, obtained playing fields and maintained them.

Pete Rosario died August 10, 1986, just a few weeks before his 66th birthday at Barrigada, Guam. **Cy O'Brien, Silver Spring, MD**

After the island of Guam had been secured, various companies and battalions were assigned areas in which to patrol daily to clean up the Japanese remnants. A Company, 9th Marines was assigned the southeast portion of the island, centered around the Village of Merizo. We were there for about a week. The people of the village were very friendly and hospitable, and glad to see us. We had patrols on a daily basis, and the chief of the village suggested we have a party for the Marines and the natives the coming Saturday night.

Marines saved candy and cigarettes out of their C-rations, and the natives had some things to eat ... and we had a nice party. The natives did their stick and pole dances, which were quite intricate, and interesting to us who had not seen them before.

About the time to secure for the night, the chief suggested we sing *Old Lang Syne.* So we all stood to sing. The first verse was carried off real well. The next four verses, only the people of Guam were singing. We Marines didn't know the words.

While in Merizo, we often heard the kids sing a song, which little Jose—sort of a mascot around the camp—had sung for us before. They sang:

**Sam, Sam, my dear old Uncle Sam
Won't you please come back to Guam?
I don't like saki, I like Canadian ...
I don't like Japanese, I like American.
Sam, Sam, my dear old Uncle Sam
Won't you please come back to Guam?**

That was the first verse. It seemed like there were 20 or 30 verses, and they knew them all. **Conrad M. Fowler, Lanett, AL**

Bold Prophecy

"In the Pacific the United States has many rivals, but of them all, Japan, by reason of position and power, is the greatest. She is the only purely Pacific world power and her very existence depends upon the place which she makes for herself there.

"We may conclude then that the powers with whom the vital policies of the United States are more likely to conflict are Germany, in the Atlantic, and Japan, in the Pacific ..." **Captain E.H. Ellis USMC, 1921**

The untimely and mysterious death of Major Ellis in 1923 on Japanese held Palau deprived the Corps of one of its most brilliant minds ... **Isely and Crowl, Princeton University Press, 1951**

The "Human Flies" Incident

Privates First Class Jack Evans and Joe Young, both scouts for H&S Company, 2nd Battalion, 21st Marines, had the experiences of human flies in trying to scale the heights of X-Ray Hill on Guam.

They were on a reconnaissance mission to the top of the 500-foot hill. They reached the base of the hill, really a plateau, without accident and were gradually working their way to the top, hand over hand on knotted ropes. They were tossing the ropes above them to lasso crags when a group of Jap machine gunners on a rocky bluff nearby spotted them.

Machine gun bullets bounced off the rock surfaces of the cliff all around the Marines and in the excitement, Private First Class Young dropped his rifle.

Unencumbered by the rifle, Young worked his way to the top while Evans dangled in mid-air, but with great difficulty Young pulled Evans up the cliff and they hid behind rocks on the summit for about an hour.

Eventually the Japs turned their fire on other Marines and the scouts continued on their mission. Another Jap sniper later took a few shots at them before Evans put him out of commission. The Marines stripped the Jap of his rifle and ammunition which they took back to their commanding officer.

Joe Young, who was from Baltimore, Maryland, and a real Marine's Marine, was later killed on Iwo Jima. Jack Evans lives in St. Albans, West Virginia.

Joe Young and Jack Evans are shown with a Jap flag. The holes in the flag were caused by the Jap who blew himself up in front of them when they cornered him below Fonte Ridge, Guam.

A Cook Who Cooks Against His Will' Will Never Quite A Belly Fill

August Fopiano (private first class that is) was summarily transferred to Headquarters, 2nd Battalion, 12th Marines and told he was a cook ... with hardly any experience even in boiling water. But those in authority didn't care about that ... And he was to cook for the officers, no less ...

Let Private First Class Fopiano tell it:

"After a few days of learning, the mess officer secured a piece of veal (fresh meat, no less!) and the master sergeant gives me instructions on how to proceed with it ... place veal in pan ... hot oven ... etc ... two hours ... golden brown etc ..."

"So at 2 o'clock (1400 to the salts), I placed the veal in the hot oven ... But why stay and wait ... and watch that veal ... since the major league players were playing baseball across the road ...

"After an hour and a quarter, I returned to the galley ...

"Wow! the entire area was engulfed in black smoke ... and I was able to find my way to the stove ... cleared the area of smoke ... opened the oven to find ... charcoal ...

"The mess officer rushes into the galley ...

"Arrest that man for destroying government property."

The mess sergeant never told me to baste the darned veal ...

All of the officers had a gourmet dinner of S.O.S. **August Fopiano, Brooklyn, NY**

"Oh, Hell Another Day Another Dollar."

On patrol on Guam, I passed out from the heat, or whatever. Pappy Horton, our platoon leader, told Tom Chambers and me to go back to our area which was going through dense jungle.

As we are walking back, we heard voices talking—Japs? We couldn't make out the words. Maybe American? We didn't know. We decided to split, one going to the right and the other to the left. We were sitting beside the trail to see who was there. No one came down the trail. We waited a few more minutes, then we decided we couldn't stay there all day. We'd go see what was in store for us, rifles at the ready.

Well—there they were—15 to 20 Pogey Bait Marines sitting there talking like they were on the front porch at home. When asked what they were doing there, they said they were lost and had no idea where they were. They were surely lost—way out in front of our lines. They didn't know that. So they all followed Tom and me back to our lines. At least for the time being they were safe. **Ben Byrer, Valencia, PA**

Wish The Boys Goodbye

We were on Guam, all organized resistance had ended, it was getting dark and raining ... and it was the first night we had been able to sleep on the ground without digging a hole. I grabbed my poncho, and put it on. I had fallen asleep, and someone stepped on me. "I said get off me, you S.O.B." The gentleman tapped me on the shoulder and said, "That's all right son, that's a colonel." MY GOD! I slept next to the colonel all night.

Next morning, Dan P. Bozikis, Hacienda Heights, California was wounded and I had his BAR, and there I am early in the morning, nervous as heck. I cleaned my BAR, wiped it off, put the magazine in the chamber, lifted it in the air, pulled the trigger ... and off went a round. The colonel says, "Watch it, son, someone can get killed that way." I took off. I didn't get to stay near him.

Then on Iwo, we were near the Second Airfield. I was in the mortar section, in a hole with my gunner (in Hank Meyer's squad). We lost quite a few officers by that time, and here comes Captain Conrad Fowler, Lanett, Alabama on a stretcher—four men carrying him—and he asked them to stop, so he could say good-bye to me. And I thought this was amazing. If I were him, I would want to get the heck going. "Captain, you're the best." **Edward Duch, Chicago, IL**

Who's Nervous? ... I'm Not Nervous

About six or seven days into the Guam campaign, we had run into some fire fights during the day, were kinda nervous, dug in for the night. We had an area out in front that was cleared for quite a few yards. With three of us in a foxhole I had the watch from 12 at night to 2 in the morning. I was half asleep when I took over the watch, looked out in front to see where the shadows were and what the terrain looked like.

As I sat there thinking to myself, and the other two guys in the foxhole sleeping away, I looked out and thought I saw something move out there. Looked away, looked back, and saw one of the shadows move again. So I stared at it ... it didn't move ... so I looked away and then back, and thought it moved again.

As was the custom, I got a grenade, pulled the pin on the thing and laid down the pin. Next time that shadow moved, I was going to throw the grenade. I looked and looked, and finally realized that as the moon went across the sky, the shadow moved. Since I still had the grenade in my hand, I looked around for the pin ... dug around for it ... couldn't find it ... probably the more I dug, the more I covered it up. By then my hand was getting numb, and I was going to have to do something.

I decided that since everyone was sleeping, I would throw the thing just as damned far as I could, but at an angle so it would be in front of someone else, some other platoon area. I did that. Sure enough, with the explosion everyone grabbed their weapons, and there was shooting all up and down the line.

Mercado was in the hole with me, and when he rose up, he said, "What happened." I said, "I don't know. Damn thing went off down there some place." I felt kinda stupid, and never told that story until now. **George Briede, LaBelle, FL**

"A Paper Doll" They Could Call Their Own

The "12th Marine Buckaneers" swing band performed on Guam for various units of the 3rd Marine Division... and even for the natives. One major problem: the vocal soloist knew only one song "Paper Doll", and the band wearied a bit of that selection.

Courtesy of Japan

Stateside mail was distributed to troops preparing to assault the beaches to liberate Guam. We were watching the softening of Jap defenses. Dick Rodgers opened his letter and let out a whoop of joy. He'd just learned of the birth of his first, a boy who was named a junior! I sincerely congratulated Dick. He was my friend, a great guy and a fine officer.

A couple of days later, he was killed by Japs who had infiltrated our lines to blow up our artillery and trucks. The Japs were stopped but we paid a price in casualties. Dick, the Battery Commander was leading a patrol of Marines to find and destroy the enemy who were still sniping around the command post of Regimental H&S 12th when he was hit. The rest of the patrol was pinned down by hostile fire and returned to the command post.

A subsequent squad was formed and dispatched to retrieve Rodgers. His body, uniform, and equipment had been severely riddled by a heavy charge designed to blow up an armored vehicle.

There was no difficulty in identifying Rodgers's body due to his size, a 6' 5" former UCLA basketball player, and as he had visited regimental sickbay the evening before. A big toe was infected and Doc Anderson packed cotton under the nail to relieve the pain. The cotton was still in place.

Dick was interred in the cemetery on Guam.

I have often wondered if his family accepted the offer of the Corps to return his remains to the States. **Austin Gattis, Washington, DC**

Lord Of The Flies

George T. Walden of Hicksville, New York had a young puppy which grew into a mutt, then became mascot of H&S, 3rd Marines. The commanding officer didn't like the dog because he "barked" for colors.

George was in Graves Registration, not usually the repository of cheery stories. But he sure did come up with a spooky one.

He buried two Japanese soldiers, and every morning the dirt was removed from them. The third day the lieutenant was really upset ... "The dirt is off the Japs again."

So Walden decided to stand by to watch the bodies, to see what really was happening. The villains were toads, imported to the island of Guam to eat flies. As Walden watched, the toads scratched the dirt off the Japanese bodies, so the flies would accumulate again on the corpses ... and the toads would have a feast. **George (Shorty) Walden now lives in Hicksville, NY**

My Favorite War Story ... God Bless America

This is the story of how Navy pharmacist's mate, Virgil Warren, of Oakland, California, came to sing God Bless America—not very well, he admits—on a hot beachhead at Guam.

It happened during the fiercest fighting there. Marine Master Gunnery Sergeant Israel Margolis of Los Angeles, had been hit trying to man a machine gun. No one knows how Margolis got there. He was a weapons expert. A man with that job usually is to be found somewhere in the rear.

But Margolis must have seen the wounded streaming back. He knew what was going on. All of a sudden, he was on top of the 100-foot cliff above the beach, looking for action with "H" Company, 21st Marines.

Margolis was 48 years old, not young for a front-line fighter. But, like many Marines, he was a professional. Born in Volkovisk, Russia, he had fought as an officer of the Czar in the first war. After the Russian Revolution, he had joined the American Army in France, then had come to the United States, been naturalized and joined the Marine Corps.

Margolis was short and wiry, with a square cut head. He had been all over with the Marine Corps. His record book showed service in the Caribbean, China, Iceland, New Zealand, the Solomon Islands. He was in action on Bougainville before Guam.

The Japs had that cliff top on Guam well marked. It was flat and open. The Marines were trying to bring enough strength together there to launch an attack. Machine guns were holding them up, sweeping the position with murderous fire.

Just after Margolis was first seen there, Jap machine gunners spotted one of our guns. In succession, five Marines tried to man that gun and each was hit. The fifth, though wounded, tried to drag the gun to a new position.

Margolis went to his aid and began to man the gun himself. The Japs caught him immediately. Bullets cracked his legs and hips and he fell. Warren pulled him to the cliff edge, but there was no way to get him down. He dressed the wounds. It was useless. Margolis was dying. He knew it and pleaded with Warren to leave him.

The corpsman looked around helplessly. If there were only block and tackle, a sling to lower the wounded man down the cliff. But men were just beginning to fashion a sling. Margolis' life wouldn't wait.

Final thoughts must have crowded the man's mind as he lay there, the St. Petersburg Cadet School, the Czar's Uhlans, the A.E.F., his 18 years in the Marine Corps, his adopted land he would never see again.

Suddenly he opened his eyes. Warren asked if he wanted anything. Margolis nodded. Warren knelt by him "Please," Margolis whispered, "please sing God Bless America."

The corpsman swallowed. He tried. "God Bless America, land that I love ..." Warren sobbed. Muddy Marines, shaken by battle, crowded around. Warren tried again. "Stand beside her and guide her ..." Lumps came to the throats of the Marines. One, then others, knelt. "Through the night with the light from above ..." Margolis' eyelids fluttered.

"From the mountains and the prairies ..." On the cliff top above the beachhead there was a hush. "From the oceans white with foam ..." Now the corpsman lifted his voice; "God bless America, my home sweet home."

In the silence that followed, the tough little soldier, United States Marine, and naturalized American, died. **Staff Sergeant Alvin M. Josephy, Marine Combat Correspondent (submitted by James M. Galbraith, Lorain, OH)**

He Who Fights And Runs Away

Private James Ganopulos, Homestead, Pennsylvania had just come in from a patrol when a few Guamanians reported Japs. Barely willing, on request of 2nd Lieutenant John H. Leims, he volunteered to go out again to scare the Nips up if he had to.

"All of a sudden we heard chopping and we all froze in our tracks. With silent movement, we followed the sounds ... came upon a river which we had to cross, being the scout. The river was about 40 feet wide with a high bank on the other side. I got to the other side and was ready to crawl up and all hell broke out. I froze against the bank and watched tracer bullets from our side going over the bank and all I can hear is shooting from the Jap side.

"In the meantime, I noticed Lieutenant Leims motioning to me with a grenade in his hand, for me to toss mine over the bank ... I got mine off my belt and started throwing them over my head at wherever I thought the Japs were. Some of the others were throwing over the bank.

"All of a sudden there was no more shooting and now it was almost completely dark. The men came across from the other side and we all went over the bank and found eight Japs sprawled out and we heard the movements in the jungle where the rest took off.

"It was too dark for us to try to seek them out. But, we found the bamboo raft they were building which wasn't finished. They had ample food and ammunition, and plenty of weapons. We did figure there were more than 15. None of our men were hurt ... I've had more on Guam, but this one I can't forget."

Some Things ... You Never Get Over

As we were pushing north of Agana in the wrap-up phase of the Guam campaign, we approached a small Jap encampment—and there, stacked six feet high in 10 x 10 storage cases, were 12 oz cans of crabmeat.

Our HQ 81mm mortar platoon took a break and gorged ourselves. What super respite after having nothing but "C" and "D" rations since we disembarked—15 days ago. Of course, we stuffed what we could carry in our packs so we could gorge ourselves again later.

We ate so much crab we got sick.

So years later, attending college in Wisconsin, I was again treated to a feast of shrimp and crab. My memory of Guam came back—and I just couldn't eat them. My stomach for seafood is just returning. **Walt Wittman, 1st Battalion, 9th Marines. Now in Florida, Walt made his home in Merrill, WI before and after the war.**

One Last Road ... For The Japs

Art Cassaretto, El Capitan Way, Delhi, California—with H&S, 19th Marines—was quick to note on the way to New Zealand that there was not another ship in sight, and the only tug could go only 11 knots an hour. The *Flying Yoke*, or USS *Robert Fulton*, traveled lonely because she was so full of explosives and high octane gasoline that no other ship was permitted near her.

"... With all these explosives and aviation gas ... anything that explodes within 150 yards of this ship is going to blow us too ... the Navy couldn't afford to have us around other ships ... so Marines have the run of the ship," the skipper said. "No movies, no radio ... but you can get a cup of coffee anytime you want."

Art recalls these adventurous engineers and Seabees doing things their way.

"On Guam one of your jobs as engineers was to build roads. George Jensen, small as he was, drove a big TD-18 tractor which is pretty good size in those days. He left one morning on his 18 and came back seven hours later. The colonel was concerned."

When he brought the rig back they asked him where he had been and and he said, "Well you told me to build this road through the jungle, which I did. So I got it built and turned around and came back ..."

"I don"t think you know," says the colonel, "But you were four miles behind enemy lines." Sure enough, the tractor had a lot of holes in the side. Later one night the Japs retreated down the road that George built that day.

Yes, Humor Did Get Weird

The 1st Battalion, 21st Marine first tent area on Guam was set up in a coconut grove, surrounded by jungle on the north end of the island. Our guard posts were around the tent area in a perimeter just inside the edge of the jungle. Often Japs still not captured would try to slip into camp at night for food. We were all nervous on guard duty, but my buddy Ed was always more so.

I used to play a cruel trick on him at night when he came around to my post with the corporal of the guard to relieve me. Standing quietly I could hear people coming down the small jungle path. When I knew they were very close but still could not see me I would flip off the safety in the trigger guard of my M-1. It made a loud "click" and hearing it Ed would holler, "Its me, Ed, your relief." I would never answer till I could see him. Then he'd say "You promised not to do that again." I guess our sense of humor got weird under combat conditions.

Some Recollections Come With Pain

A history such as this where those on the front are asked to relive it—though O so briefly—brings back pain with memories. Take K. Sterling Felton of Long Beach, California. He was with Baker Company, 1st Battalion, 3rd, remembers well the Japanese suicide attack on Guam and the loss of so many like Sergeant Bonaccio ...

"... That night when the Japanese suicide squad broke through our front lines (land mines in hand) intending to blow up our artillery ... but as fate would have it, they ran smack into our 1st Platoon position (Baker Company) along the hillside between the Japanese and the beach ... Did you know though that even USMC history books say it was the 9th Marines they attacked and not us?

You probably know the events ... and then there was the time when as we were prepared to attack (about July 23) there came "friendly fire" as a Navy/Marine dive bomber dropped a bomb on our lines, killing half of the Weapons Platoon ... It was the following night the 9th was overrun on Fonte Plateau and we later went up to replace Captain Wilson's people in their foxholes ... It rained in the same water filled foxholes ... **K. Sterling Felton, Long Beach, CA**

Chamorros Escape Japanese

Eleven Chamorros, native residents of Guam, had just escaped from the Japs following the Marine landing on the island. They had nothing good to say for the Jap conception of the new order.

Zippo ... and They Were Off

As 2nd Battalion scouts, we followed the front lines closely to check caves for maps and pertinent information left by the Japanese. It was my turn to enter the cave with Sergeant Fernwood behind me followed by Marines Benson and Worrell.

After yelling the Japanese phrase for "surrender and come out with your hands up," I tossed two grenades into the cave. As I started down a number of steps, I could see that it ran for about 20 feet with steps going up the other side of the ridge. There were also four carved-out sleeping quarters along the side.

Sergeant Fernwood and Marine Benson went past me as I had stopped to search the first sleeping area with Marine Worrell. Sergeant Fernwood was checking one area and Marine Benson another. Marine Benson could not see, so he took out his Zippo lighter. When he lit it, he shoved it right into the face of a Nip who was waiting in the dark with two grenades. The Jap struck the two grenades to his helmet and threw them at Marine Benson; both bouncing off his chest. At this moment all Olympic records were broken by four Marines for as we know Japanese grenades have only three second fuses.

I had only heard the pop and didn't find out until later just what had happened. Marine Benson dropped his lighter in the Jap's Face and I had left behind my BAR. I guess I needed both hands to push Marine Worrell up through the entrance. It must be the Marine instinct and training when a grenade goes off behind you to take off without speaking a word.

The only casualty was Sergeant Fernwood. He was hit where no one wants to be hit. Marine Benson lost his Zippo, something which was irreplaceable. I had to replace my BAR with another. **Submitted by James J. Carvino, 3rd Marines, Brooklyn, NY**

First Impressions

Ernest Baals of Erial, New Jersey met his first Marines casually. He only saw them at the Marine Quartermaster Depot in Philadelphia where his father worked. He was so impressed by the caliber of the men that he later joined them ... and, of course, wound up in the 3rd Marine Division.

From Bar To Ka-Bar

Private First Class Charles L. Moore lives in Chickamauga, Georgia. He was a BAR man with Easy Company, 2nd Battalion, 3rd Marines on Guam ... This is his account of personal combat with a Japanese soldier on the high ground over the Adelup invasion beach.

"Our 81 mortars pretty much cleaned out their defenses, but there were still plenty of Japanese left in the spider trenches and foxholes. Our job was to flush them out.

"It was shortly after 1300 on D plus 2 and I ran up a bank and sprayed into the trenches. One hole was sheltered by a big boulder; and a soldier (tall for a Japanese) lunged at me with his bayonet. I knocked away his piece with the barrel of my BAR and his weapon fell on the ground. He grabbed a knife from his belt and attacked me again. My clip was empty so I turned around the BAR to club him with it, but the barrel was red hot, burned my hands and I had to drop the weapon. The Jap struck at me again and got my other arm. Then I pulled my Ka-Bar and slashed him across the head. It was a deep gash and the skin fell off his brow and down to cover his eyes. It was easy to dispose of him after that."

Charles usually returns to the 3rd Marine Division Association reunions with his wife, daughter, and three grandchildren.

A Find For Drinking Men

While going through the jungle one day in campaign, we came upon a small open area with several small buildings which was used as headquarters for a Japanese unit. We had quite a fire fight with the Japs that were there, using several tanks, anti-tank guns and usual company weapons. The open area with the small buildings was in area of 1st Battalion, 9th Marines.

After the fire fight was over and the Japs had returned to the hills beyond, we discovered one of the buildings to be either an officer's club, or club of some kind, as it was loaded with Japanese beer, sake, and some hard liquor. Naturally the men and officers of 1st Battalion consumed the beer and it being a very hot day it tasted pretty good.

Semper Paratus

While the men were enjoying our find and were relaxing, Colonel Randall, came up to me and said that we are a little bit lax and maybe inviting a counter attack from the Japanese. He thought we had better go down a little road to our right and tie in with adjoining battalion.

So Colonel Randall and I started down the road. After going 50 to 100 yards down the trail, we encountered rifle fire from the Japs we had just flushed out of their positions.

Colonel Randall looked at me, and I at him, while crouching in a ditch out of line of fire. He asked if I had a carbine and I answered all I had was a 45 pistol and I didn't think I could hit anything with it. My rifle was with my pack, and I don't know where his was. After being pinned down for 10 minutes or so, some of our company men heard the firing, came to our rescue, and chased the Japs further back. We did tie in with the adjoining battalion, but we both were caught in an embarrassing position in not being able to fight back. Thank God the Marines came to our rescue ... **Gaylon Souvignier, Canton, SD**

Third Marine Division Leathernecks are shown entering Agana, the wrecked capital which was full of mines planted by Jap forces as they retreated. (Defense Department Photo, Marine Corps).

A Marine wounded in the fighting at Guam receives first aid from Navy doctors and corpsmen at a dressing station right on the firing line. Litter cases are treated and prepared for evacuation to ships standing offshore. (U.S. Marine Corps Photo)

In The Middle Of Screaming Japs

George Walters attended Western Michigan University ... was a member of Buchanan Michigan Police Department ... and later enforcement officer for Michigan Secretary of State. He died in 1988 in Galien, Michigan.

Frank Brandemihl, Livonia, Michigan was with the police department for 33 years, specializing in homicide investigation. He is a graduate in graphic arts, and associated with Detroit Society of Arts and Crafts.

The strafing attacks are somewhat dim, but I remember being on the beach at Bougainville, hearing machine gun fire and Marines yelling "Zeros!!!" Then I saw two aircraft flash by strafing the swamped landing boats and those still coming in with more Marines.

L Company was about two weeks into the Guam Campaign. We broke out of the jungle and onto the high ground with very little cover ... feeling we were being observed every step of the way. As we climbed the mountains, Japanese snipers were able to "zero in" their weapons, so our lines were thin and spread out.

As we dug in for the night, the word was passed to be "extra alert" as a possible counter-attack might be launched. The weather was getting nasty, and heavy monsoon rain was in the air. We felt the presence of the enemy, and L Company was "spooked" to say the least. My partner, George Walters from Galien, Michigan, and I dug in on the top of the red clay outcropping ... right in the area of a probable approach if the Japs tried to over-run us.

Just before dark the Japs fired air bursts right over our heads, indicating that they knew where we were, but the airbursts stopped as darkness fell. We set up trip wires and other approach warning devices ... then put on a 50 percent watch.

Some time after 10 p.m. a very heavy rain and windstorm struck our lines ... limiting visibility, distorting sound, and filling the two-man foxholes with water.

We thought conditions could get no worse ... but then Japanese troops ... screaming, throwing grenades, and firing weapons—hit our lines from the rear. We surmised they were Japs we had passed in the jungles on previous days and were attempting to rejoin their main force.

Navy fire was called for and they fired shells with parachute flares that lighted the whole battle area. We had a panoramic view of the action as figures moved about in the driving rain throwing grenades and firing weapons. Both Japanese and American voices could be heard through the sounds of battle. We were forced to hold our fire for fear of hitting our Marines. Grenades arched and sparkled towards our water-filled foxholes ... sometimes we had to duck into the water as grenades came close to us.

The surrealistic scene seemed to last for hours. As the sounds of grenades, machine gun and rifle fire died off, the sounds of wounded men and Marines yelling for corpsmen prevailed.

At the first light of dawn we knew some of the men from L Company had become casualties. One rifleman, whose position was over-run, was attempting to join Marines in another foxhole. He was mistaken for a Jap soldier and was shot by one of his friends, dying from the wounds. The emotional impact on his friend was so severe that he had to be evacuated.

After the usual "scrumptious" breakfast, L Company was forced to deal with another thorny problem. Up to this time, enemy kills were claimed by as many as four or five marksmen. Word came down that the marksmen had to bury the enemy soldiers they had killed. Because of the hard digging in the coral ground, burying the dead became a problem.

Suddenly, the number of over-zealous marksmen dried up.

There was a lot of conjecture on how the Japanese troops had met their demise. Some of the theories were: heat exhaustion ... heart attack ... or finding and eating Marine rations.

I would like to say something about the Japanese soldier. In my humble opinion, and the opinion of most Marines who met him face to face ... he was an excellent soldier—tenacious and courageous. **Frank A. Brandemihl, Farmington Hills, MI**

Handful of Jap

In establishing the first command post after going ashore on Guam, I was going to install a phone in a lone thatch shack ... The doorway made for poor vision, so on grabbing into the palm fronds, I got a handful of Jap that was concealed under the palms.

Armed only with a telephone, I shouted "Japs" ... I cut and run from the shack and jumped behind a fallen tree ... From the protection of the tree the armed Jap appeared in the doorway ... where he directed the fire of other Marines that took care of the problem in short order ... **E.B. Thompson, 4th Battalion, 12th, Chillicothe, MO**

U D T And The Welcome Mat

"Welcome Marines, USO That Way ..."

The Japanese didn't erect the sign the first assault waves to Guam were so surprised to see. It was put there by the Navy's super daring Underwater Demolition Teams (UDT).

James R. Chittum, Las Vegas, Nevada was one of them and he tells the story as if it were all so matter-of-fact. The only trouble was the Japs were so close, they could see the Swabbies' mustaches. Frank F. Laht, Stevensville, Maryland recalled how the Japanese had barred the harbors with cement, palm logs, and coral.

In three days, more than 1,000 obstacles were blasted away in front of Guam by UDT personnel. Many of them were Seabees. One UDT man was killed in preparing the "laying the carpet" for the Marine buddies.

In some instances, obstacles were less than 50 yards from shore and the reef was completely dry. It was necessary for the men to run across 150 yards of exposed reef carrying 40 pounds of powder to get to the obstacles. In all cases the obstacles were completely removed. In daylight the average time for a platoon to remove 30 obstacles ... from the time the rubber boats left the LCPR (ship) until detonation ... was 16 minutes. **James R. Chittum, Las Vegas, NE, commander, USNR**

Heart Of Gold

As a correspondent I was a little freer to roam than most, and one day my wanderings took me into a Jap quartermaster supply. What did I see there but socks, socks, socks, and socks.

Over there ... wet, dirty, you had no hope of a shower, or clean clothes ... but just think how heavenly. You, and only you, O'Brien, could have a clean change of socks every day. And nobody, nobody, knew you had all those socks, but you. So I dumped everything out of my second pack, even chow—and filled that second, the lower pack, with socks, socks, socks.

And came morning. Quietly, and most selfishly, I reached into my pack and withdrew my first pair of clean, unused, never even washed, Japanese socks. And I died ... the socks were made for Tabis (that Jap big-toed sneaker), and I couldn't even walk with one of those on. No non-Jap could stand that chafing. I shared the socks then with everybody— but I always was a nice guy. **C.J. O'Brien**

Beer Cooler

THE SCHOOL SOLUTION

When the Island of Guam was secured, we moved to the coconut grove on the east side of the island. You will recall that the Japanese had very substantial defenses over there. There were trenches, sandbags, dugouts, etc.

The story behind that, is that in the 1930s (and presumably in the 1950s and 1960s), Marine Corps Schools, Quantico had an advance base problem team. They would dream tactical situation problems 15 years down the road. They would dream what equipment would be available. And they would determine a plan or program to overcome the problem, based on tactics they would anticipate, and the equipment available.

In the 1930s, they had an advance base problem, entitled "Recapture of Guam after it had been seized by hostile forces." The school solution was: there would be a landing on the eastern shore, about the midpoint of the island where our billet was, in those coconut groves. And so, they were ready for the landing at that point. But as you know, we landed on the center of the island, western side.

Now at the Marine Corps Schools there was a poem that is sort of appropriate:

Here lie the bones of Lieutenant Jones
A graduate of this institution
In the thick of the din,
He died with a grin.
He had used the school solution.
                Conrad Fowler, Lanette, AL

Fighting Aftermath

Harold H. Schwerr, North Mankato, Minnesota had several experiences with Japs on Guam after they had quit fighting, but were still active.

One day the people of C Company, 21st Marines were going about their camp duties when into camp walked a rather imposing soldier in a Japanese uniform. He was quite tall and regal looking ... and a surprise to the members of C Company when they realized that this was the enemy walking among them. There was no trouble, however, for the man had come in to give himself up.

Late one night the company members were in their tents when noises were heard outside, and someone got up to investigate. Several Japs were trying to steal their clothes off the line. The thieves were routed, and took off with machine gunfire chasing them. No one was killed, but Schwerr thinks the road was wet with Jap urine.

C Company experienced enemy soldiers trying to steal food from the galley; and there were even a few who crept in at dusk to watch the American movies on the outdoor screens.

Cold Hand Luke

On the second night on Guam, I was in a hole with a corporal named Breckinridge. There was a dead Jap within a few feet and he was liberally covered with toads attracted by the flies on him.

The moon was fairly bright and Breckinridge had taken his helmet off and he was pretty bald. Suddenly I saw him stiffen up and roll his eyes back in his head.

It turned out that one of the toads had hopped on his bald head ... and he thought the Jap had reached out and touched him.
Richard B. Abbott, Sanger, CA

A Party In The Village

After the Guam campaign, A Company 9th Marines was detailed to the community of Merizo to conduct a series of patrols to find Japanese stragglers over a two-weeks period. At this village on the southeast part of the island, the Marines and villagers became good friends—and it was decided to have a party on the Marines' last night there. The Marines would have some rations, and perhaps some candy and soft drinks for the children. The villagers would prepare some of their native dishes.

On the day of the party I went out with some of the natives to get food for the party that night. We went out in a dugout canoe that they had made themselves. Part of our expedition consisted of the use of a few hand grenades. After dropping grenades over the sides of the boat we had an abundance of fish ... plenty of fish ... brought them back ... and that was part of our meal for that night.

It was quite an experience going with those folks. The goggles they wore, they made themselves. I can't remember what they were made of, but they had coconut shell lining on the outside. **Harry Lee Gilliam, Sunnyvale, CA**

Dukw's On The Pond And Dry Land

As we got closer to the reef on Agat Yellow Beach, I decided the coxswain was lost and going in the wrong direction, so I told him so. He told me that if I did not get my head down this could be a one way trip for me. As we got closer to the reef, I decided he might be pretty smart, even though he was a swab jockey.

When we hit the beach Warrant Officer Louis E. LaBohn told me to catch a ride on the first DUKW that came by and keep an eye open for any DUKWs that might be having trouble. I got on one that was heading out to sea. After we dropped off the reef into deep water, I was getting a shower from a stream of water, coming from a bullet hole in the side of the DUKW. My first thought was to get my toolbox which had wooden plugs for such an occasion. But, you guessed it, the toolbox was still aboard ship.

My next thought was to tear up T-shirts and use the strips for plugs. Neither the driver or I had a T-shirt. Only other things were skivvies and socks, and the driver said "no way" on the skivvies ... socks if he had to. Those sock plugs held for at least a week.

Tricky Unloading

Along with ammo and supplies some of the DUKWs had been loaded with 75mm pack howitzers before we loaded aboard the LSTs. Coming off the ramps of those LSTs was an experience ... some went off a little too fast and went Deep Six.

After Orote Peninsula was secured, we started using the Old Agat and Sumay Roads. If we did not have wounded to haul out to hospital ships, occasionally we had time to pull off the road and go sightseeing. On one such occasion, we found a stash of Japanese sake. We were all as happy as a six-year old at Christmas. Just one small problem though ... who was going to take the first swig? Our imagination ran away with us, and after some discussion, we got mad and broke every bottle.

A Good Mascot

There was a young boy, probably 10, that we adopted as a mascot. He was real good at keeping us supplied with Tuba and Aggie. He was always on a DUKW hitching a ride, and was real personable. I have thought of him occasionally, and wondered if he still is alive.

The carabao cattle on the island turned out to be a costly experience to some of us country boys. One day we saw a little calf sucking a cow ... and we country boys could attest to the fact that a sucking calf makes for the most tender eating of any beef. A shot rang out, the calf hit the deck; and in short order it was skinned and butchered and on the its way to the cook's shack.

As the cook started frying that steak, we fellers gathered like a bunch of vultures with our mess kits for a feast. We each latched onto a piece of steak, tried cutting it with our mess kit knives—but to no avail—so used our Ka-Bars.

After chewing on the first bite for about 30 minutes we gave it up as a lost battle. Shoe leather would have been like an angel food cake compared to that meat.

Three days passed, and the captain had us fall out. He had a native with him. In a loud voice he said, "Does anyone know who killed this farmer's sucking calf?" All is quiet. Then he proceeded to the tall end of the platoon (my six foot, three inch end), very calmly removed his hat, and holding it towards me, said, "A $2 donation from every man should pay for that calf." We did.

On Parade

You all remember how much trouble we often had with swab-jockeys? On one trip out to the ship, to pick up a load, a bearing on the propeller shaft froze up and would not turn ... leaving us helpless in the water. There was no way to fix it at sea, and wouldn't you know, there was not another DUKW in sight.

There were some tank lighters circling pretty close to us and one decided to check us out. I told him our plight. He looked at his buddy and said he would tow us to shore ... and threw us a line so we could tie on. The beach was in one direction, and those other tank lighters were in another. So what does he do but go past all the other tank lighters to show what he had caught.

Days Of Barter

Several days after the landing, another driver and I had each picked up two Jap rifles and had them laying on the dashboard, just behind the windshield. On one trip, alongside a ship, a sailor hung over the rail and asked if we would sell those rifles. We said we had no use for money under our present conditions. So, he said how about a pack of spuds ... and I agreed.

He disappeared and in short order was lowering a hundred pounds of spuds over the side. I was about to tie the rifle to the line, and he said, "Wait." He came back with a case of eggs, which he lowered to us. Then I tied the rifle to the line. We had one good meal on Guam.

No Target Practice

Army MPs would have had a happier and more peaceful life if not for some country boys who were driving DUKWs. By the end of the second week, the Asan-Agana Beach was secured. Offshore in this area of the beach there were some tankmines that had not been detonated, and at low tide those "bananas" would show.

Those "bananas" made good target practice, so that's what we country boys did. We could always manage to hit at least two, and sometimes three, before those Army MPs would swarm that beach like flies on a cow pile. Don't know why they had to get so upset.

Difficult Chore

To me one of the most nerve racking things was transferring wounded from our DUKWs to the tank lighters. These transfers were no easy task and I know some of the wounded were injured in addition to their wounds. But I can assure you, we tried to be as gentle as possible. **Burnell Focks, III Corps Motor Transport**

Banzai And Snipers ... 9 Years Later

On July 25, 1944, I was executive officer of A Company, 1st Battalion, 21st Marines. The company was dug in on the crest of the ridge that defined our beachhead on Guam. To our front the terrain was rolling and fairly open. About 400 yards away was a low cliff. A long range Japanese sniper had taken up a position somewhere on that cliff, and fired a single round every hour or so with devastating effect.

On our right, B Company had been given the formidable job of defending the only ravine that led straight down toward the beach below, now littered with supplies.

That night the Japanese struck with what was perhaps the biggest banzai attack of the war. It was a totally black night until eventually the sky came alive with flares. The Japanese spearhead was aimed directly at the ravine. The assault was so overwhelming that it succeeded. The enemy poured down the ravine and fanned out onto the beachhead.

In the chaos of the next morning, rear echelon units found themselves in vicious firefights. The area within the perimeter finally was retaken.

A Chance Meeting

Nine years later I attended a press conference at Argonne National Laboratory on the outskirts of Chicago. I was then science editor of *Popular Mechanics* magazine.

The purpose of the news conference was to show off a new and radically different type of research reactor. In attendance were science writers from many of the larger American newspapers and magazines, along with some representatives of the foreign press.

At mid-morning, the director of the news conference announced a coffee break, and directed us outdoors, where small tables had been set up in the warm spring sun.

I took my coffee to one of the tables. A moment later a Japanese man, about my age, asked if he might join me. We exchanged names, and he identified himself as a science reporter for *Asahi*, the Tokyo newspaper that is the largest paper in Japan. He was on temporary assignment to Washington, DC, and had been sent to Argonne for the press conference.

We made small talk, carefully avoiding the subject of the war. Finally I couldn't resist. I asked him if he had been in the military. Yes, he had been in the Navy, as a meteorologist. I asked where he had been stationed. He mentioned a couple of places, and then said that he had been on Guam when the Americans landed there.

I told him that I was one of those Americans.

I asked if he had participated in any way in the banzai attack. He replied that every man on the island had participated. His unit, not normally a combat unit, had formed up that afternoon, at the base of a small cliff some distance from the Marine positions. I asked if there was a sniper somewhere on that cliff. He gave me a startled look, and said, indeed, there was a sniper in a small cave almost directly above his head. I told him that the sniper had killed one of my men that same afternoon while the man was stringing barbed wire.

Of course I had to ask him what happened to him during the banzai. As soon as it was totally dark, the Japanese

He gave me a startled look and said, indeed, there was a sniper in a small cave almost directly above his head.

had surged forward. My newfound friend had scrambled down the ravine. He must have passed within 30 yards of me. He made his way through the chaos of the firefight, and out across the beachhead. When dawn came he observed the carnage. The banzai attack had succeeded in penetrating the line, but now the Japanese were trapped within the Marine perimeter.

He managed to make his way back through the Marine positions, and ultimately to a beach on the far end of the island. There a Japanese submarine picked him up along with a handful of other survivors.

Sitting over coffee that day, we were both somewhat shaken by the course of our conversation. We instinctively turned to the present. We learned that we were precisely the same age, we were both science writers, we were married the same year, I had two sons and he had two daughters, and in each case our children had been born within a few weeks of each other.

We were mirror images.

Twice our lives had crossed, once during this half-hour in the warm Illinois sun; the other time, a desperate moment nine years before and half a world away.

Clifford B. Hicks, Brevard, NC

Sgt. Carroll J. Williford of Mooresville, NC stands amid the debris of a Catholic Chapel near Asan Beach, Guam. The chapel, walls two feet thick, was wrecked by the pre-invasion bombardment. The statue of Christ, however, remained upright on the altar, though it was nicked and one hand was gone. (Official U.S. Marine Corps Photo)

It Was High Priced Ammo!

On Guam (after the campaign), I was near battalion headquarters and heard our colonel talking to another officer. He mentioned this native guide who had been assigned to the battalion. The colonel said, "I dismissed him because we didn't need him anymore. He's really a nut ... his name is Joe Gurito. I told him, "Joe, we won't be needing you any more, so you can give me your rifle.'"

"Colonel," Joe says, "Will you let me keep the rifle for a while. I know where these Japs are hiding out. Give me some ammunition and I'll get some Japs for you."

The colonel says, "Joe, you will bring this rifle back?"

"I promise."

The colonel gave him a handful of ammunition and jokingly said, "Now for each round of ammunition I want you to replace it with two Japanese ears."

"Yes, sir," and he left.

The colonel said in about a week or so, Joe came back and said, "Could I get some more ammunition, colonel?"

The colonel said, "Joe, I thought you were going to turn your rifle in. Don't you think it's about time?"

"No, I did pretty good." And, he handed him a brown paper bag, and said, "I need some more ammunition."

The colonel opened the bag and saw some gray and white objects. "What the hell is this, Joe?"

"That's some Japanese ears, colonel."

He threw the sack back at him and said, "Joe, get the hell out of here." The colonel later reaffirmed, "That guy's nuts."

Later on, while on Guam, I was visiting the quartermaster and he had bags full of old clothing and shoes ... surveyed items that had been turned in by our people in order to get new clothing and shoes. I asked, "What are you going to do with those things?"

"They're going out to the dump."

"I said, "Why don't you give them to the island people?" I had been to mass the weekend before, and most of them were barefooted, and clothing all patched up.

"Well I can't give them to you. I'm busy here, and if I don't see you, that's up to you. But I can't give you permission to take them."

So I got a few bags, and gave them to the natives ... who accepted them like they were brand-new clothes. **William Parrie, Benicia, CA**

Fruits of An Inspection

On Guam after the Japanese organized resistance was over, my platoon, B Company, 21st Marines, went out on jungle patrols each morning for several months. Jack, one of the men who had been part of the outfit before the Guam Campaigns, always walked down the trails with his bayonet attached to his M1 rifle. He was considered a little "Asiatic" since the others never did that.

Later on during an inspection by the battalion commander and the usual groups of officers including the doctor, they happened to go into his tent which was next to mine on the platoon street. One of the inspecting groups saw under a bunk a one-gallon glass jar full of liquid and some brown curled objects. Thinking he had discovered home-made Raisin Jack liquor or whiskey, the officer asked whose bunk it was. Jack responded and was asked if those curly things were apricots. "No, sir," he replied, "those are my Jap ears." A hush fell over the inspecting officers, then some excited conversation. The next day Jack packed his gear and was shipped out to some rear area. Now some of his buddies, instead of laughing at him, envied him. He found a way to get home. **William I. Pierce, Merrillville, IN**

Really, Really Chicken

On Guam near Agana, there was a chicken yard with a 15-foot high wire fence around it. With so much corn willie and Spam who would not go for the nostalgia of a real chicken dinner. Howard Fix, who still lives on a rural route in Harrisonville, Pennsylvania, tells this story.

First we went to the cook and asked for some lard. He said 'sure' and we also got a five gallon can for a pot.

So one Saturday night, without authority ... Hell, we weren't going to tell anybody about this ... The village was about three-quarter mile through jungle and being a bit leery of the dark, we went at dusk. We got our chickens, cooked them and had a feast.

About the third night—and also my buddy always took his rifle—I go over the fence and by golly by the time I hit the ground, a Jap goes over the fence on the other side of the pen.

My buddy wanted to shoot him ... I said "Hell no, he's only in here after chickens like we are ... besides, if anybody hears the shooting, we'll be on guard duty day and night.'" Well, some of the officers got word of the chickens and asked about it. We didn't know anything.

Many years later at home on the farm, I was coming down off the hill and there was my buddy from the Corps ... "How are you, you old chicken thief?" he says ... I answered indignantly ... "I never stealed no chickens." **Howard Fix, Harrisonville, PA**

Souvenir Hunting Can Be Dangerous

I didn't have any specific duties, so another Marine and I decided to go souvenir hunting on Guam that summer's day in 1944. Off into No-Mans Land we went, following a trail through the brush.

Then, across the trail we spotted a taut rope, about a foot above the ground ... couldn't tell where it started or ended ... and immediately figured it was a booby trap.

There was a hill to our right, about ten feet high, covered by brush, so decided to leave the trail and go up and around the hill. At the top, hidden by a bush, was a hole three feet in diameter. I lost my footing, slid into the bush, and then into the hole, feet first.

Falling in the darkness, I heard Japanese music playing ... was I falling into a cave full of Japs? The floor of the cave was about nine feet from the top where I had fallen through ... landing on my tail-bone, I was badly shaken up ... I could hear my buddy at the top of the hole screaming my name.

In the cave were two shadowy figures, with rifles pointed at me. I pointed my rifle in their direction. Thank God nobody was trigger happy as we all were Marines, out souvenir hunting. The other two Marines had found the cave opening at ground level, and I had fallen into an air-shaft at the top.

Before I arrived, the other two Marines had found a Jap victrola, some Jap records, and decided to make some music. This was the music to which I had made my sudden entry to the cave. **Lou Hum, Santee, CA, was formerly with M-4-12**

From Crap Games ... To Replacements

After being transferred from the 2nd Marine Division to 3-G-12 of the 3rd Division, we sailed from the States to the South Pacific aboard the USS *Mount Vernon*.

... On board with us was a contingent of Seebees. One of them could smell a crap game from stem to stern. He was known as "C-1-X-6," his favorite expression when throwing the dice.

... Bougainville—the swamp where we couldn't dig a foxhole. One of our gun crew had a theory that anything GI—that was not attached or in use— was available for his use. He and another of our crew, scrounging at the beach, "found" an enormous tarpaulin which they brought back to our battery. It was put to good use as a ground cover mattress on which eight or 10 men could lie side by side.

... Another of his acquisitions were two cases of Thanksgiving turkeys. President Roosevelt may have promised every serviceman turkey for Thanksgiving, but there may have been some who were a little short.

... One of our exciting fire missions on Bougainville was a rapid fire mission with our 75mm pack howitzer. We were receiving incoming "mail" from the Japs so it became an artillery "duel" in our minds even though we had no idea what our target was. By the time of "cease fire" our adrenalin was taking effect, so that we were all very pumped up. The sight of our rear trail man covered with mud, caused by the recoil of the gun, created more laughter.

... Guam was quite a different landing for me, as now I knew a person could get hurt—or worse. As the landing craft headed toward shore I was aware of the pot holes in the coral reef and the thought came to my mind that if I were to step into one of them just right I could break a leg and get a return trip to the ship without having to endure the upcoming danger. When the ramp went down I was more careful than ever! Cowardice was not in my blood that day.

... When we arrived at Iwo Jima the one redeeming action that I saw was the first B-29 come in for an emergency landing. Then we could readily see the value of that pile of sand. But at what cost!

... After returning to Guam, we awaited replacements so that we could be rotated home. One of the new "90-day wonders," a 2nd lieutenant, was given orders to take us out for combat instructions. After getting us away from the camp, he seated us around him and proceeded, "If you think that I'm going to tell you men who have been through three campaigns how to fight, you are nuts!" Our estimation of this officer took a decided turn for the better. **Ira M. Shelton, Brawley, CA**

"Extending" Was In The Cards

After awhile we sorta' caught up on the movie going as only a few were available each month, and were circulated among the many different units on the island of Guam. Got to where it was like television is today ... lot of reruns.

So, the main source of entertainment was card games. Most every night there were several good poker games going on ... the limit being determined by how much money a fellow had, or how good or lucky he was.

My friend, Lee Cox, got so engrossed in a good poker game one night, that he let his better judgement get the best of him. He had been on a fairly nice winning streak for several days and had won about $200. Now that may not seem like much this day, but for 1945 when you're making $55 a month, it's a bunch.

Seems like the game was only a quarter ante and ten dollar limit on any one bet; and most of the time stayed pretty mild as far as the betting went. Well, Lee was drawing some really nice hands and was getting more aggressive as the game progressed.

One of the other fellows was also drawing some pretty nice hands, but he wasn't being as aggressive as Lee. He was just biding his time, sucking ole Lee right in. He bluffed a couple of sorry hands and let Lee catch him at it.

Next time around, he loaded up on Lee and by the time they had finished bumping the final bet, Lee's $200 was in the pot, plus an IOU for about $200 more. Needless to say, Lee lost the pot.

I was the winner of the pot and I sure didn't feel good about it because Lee was my friend and I knew what $200 meant to a private first class, drawing $55 per month. Next morning, I told him to forget the IOU, but Lee was too bullheaded to listen. It was close to time for him to either extend his enlistment, which was worth $400 or let it ride and be discharged when the war was over. He elected to extend his enlistment so he could pay off his gambling debt. The extension caused him to have to stay in the Marine Corps until 1947 or 1948, at which time he got out and joined the Air Force. **George E. Lyon, Florence, AL, 3rd Motor Transport Bn.**

"Fortune Teller ..."

I joined the rear echelon of the 3rd Division on Guadalcanal while the 3rd was still fighting on Bougainville. When the 3rd returned to the Canal, I was assigned to the machine gun section and was taken under the wing of B. White of Virginia. White was with the 3rd from the onset and spent time in New Zealand at $21 per month. While on New Zealand, White went to a fortune teller and was told that he wouldn't live until his 21st birthday.

During invasion of Guam, White turned 21 years old (Guam time) and we were all kidding about the fortune teller's prediction being wrong. That night White was on outpost when a Jap came through and killed White and Corporal Barkley. After it was all over we realized that White hadn't turned 21 United States time.

We buried White and Barkley in graves 17 and 18 on Guam. I don't believe in fortune tellers but this makes one stop and think. Of course, not reaching 21 in the Marines Corps in the Pacific wasn't too unusual. **Ernest Baals, Erial, NJ**

His rifle converted into an instrument of mercy, a Marine rests his head on his pack while receiving blood. (Official U.S. Marine Corps Photo).

Marines blast Japanese pillboxes as they crouch for cover in this action in Guam. (Defense Dept. Photo - Marine Corps)

Hot chow was served to this unit on Guam, two days after the July 21 landing. The food was of course a welcome change from C-Rations. (Defense Dept. Photo - Marine Corps)

A Sleeping Sentry

Captain Conrad Fowler, A Company commanding officer, ordered me to take a patrol in the area and see if we could locate something. After walking the area for some time, we came upon one of the native shacks with a Japanese soldier sitting, sleeping at the front door. (He was supposedly on guard duty).

As we approached the shack, the Jap woke up and started running. I motioned the men not to shoot and to let him go and we would get the group that was inside the shack. We then proceeded to make a semi-circle around the front door of the shack, and after we were all in position, opened up with our fire power. We counted 12 dead Japs in the shack and we did not have a casualty. We believe we would have had a real fire fight if the Jap sleeping at the front door had not run away and left his fellow soldiers. We did try to find the Jap that got away, but to no avail. **Gaylon Souvignier, Canton, SD**

Mine Kills Member Of Grave Detail

I was a member of the Graves Registration Team of the 3rd Division during the liberation of Guam.

We were beyond our lines searching for isolated burials, when we spotted the body of a Navy man. Our lieutenant ordered us to retrieve the body. (The Japanese had mined the area, and we had already detected several detonators). I took a rope and approached the Navy man's body, attempting to secure it and pull it from the mined area.

One of the men accompanying us on this mission was my good friend, Sergeant Rawls. He had been guarding our flanks.

The sergeant stepped in to assist me, and what followed imprinted my memory forever. My friend stepped on a mine detonator ... there was a terrible explosion ... both of his legs were blown off at the knees ... and the lower part of his jaw was blown away.

I was slammed to the ground, dislocating my jaw and was disoriented for a period of time. Our lieutenant was also injured, but recovered.

Unfortunately, my friend's injuries were too great. Sergeant Rawls died the next day aboard a hospital ship from the wounds received while honoring a fallen comrade in arms. I was pleased to find that there is a street on Guam named after this good Marine and friend. Semper Fi! **Edgar Neer Jr., Independence, MO**

Where There's Firing — — There's Smoke

On Bouganville or Guam—I'm not sure. All of those battles seemed to run together for me. We had stopped for the night, and the lines were grouped into three men in a foxhole. Dale Griffin and Goober Wilson and myself were in one of them. These foxholes were on the jungle trails. As always we stood watch 1/3 of the time, each man. Just as it was starting to get light, I was on guard, Goober woke up and was going to light a cigarette. I said "don't light that match."

Sure enough, there were Japs coming up the trail toward us. We both laid back down and woke Dale. We all got our rifles ready. We laid quiet and when we saw they were several feet from us, we started shooting. The air was filled with smoke. Don't know how many we killed or how many got away. When we started shooting everyone on the line was shooting too. We were the only ones to see the Japs, but everyone was shooting with us. There was so much gun smoke in the air, we couldn't see in front of us. **Ben J. Byrer, Valencia, PA**

A Visit By The Commandant

The commandant, Lieutenant General Alexander A. Vandegrift, at left, and Major General Allen Hal Turnage, commander of the 3rd Marine Division shown as the commandant was leaving General Turnage's command post, after congratulating him on his part in the recapture of Guam. (Defense Department Photo, Marine Corps).

Welcome To The "Big E"

General Graves B. Erskine "the big E" shortly after taking command of the 3rd Division is greeted by Colonel J.A. Stuart, commanding officer, 3rd Regiment. Colonel Stuart relieved Colonel William Carvel Hall, who became the "Four" of the division and retired as a Brigadier General.

And The General Did See A New Enlisted Mess Hall

I recall this incident from my service with General G.B. Erskine.

While on Guam in our rest camp, the officers of Headquarters Battalion and the general's staff secured work from the Seabees to build an officers' wine mess in the officer's country of Headquarters Battalion. Finished in about two weeks, the wine mess was situated on a cliff overlooking a valley with a tremendous view.

At the grand opening of the wine mess, the officers invited General Erskine for an evening of celebration. The officers also invited the nurses stationed at the hospital in Agana. Music for dancing was provided by the division band. The opening was on a Saturday night and everyone had a great time.

On the next day General Erskine called his officers together and asked if the enlisted men had anything comparable to the officers' wine mess and, of course, they said, "No."

General Erskine then said that one week from that Sunday he wanted to attend church services with the enlisted men in a new chapel and afterwards to have breakfast with the enlisted men in

a new concrete-floored, screened—in mess hall. If these two items were not completed by that Sunday, he personally would gather all the enlisted men from the battalion, lead them up to officers' country, and shove the officers' wine mess over the cliff.

During the following week construction went on almost around the clock. The Seebees, not having enough troops to build the chapel and mess hall, requested the officers to get some volunteer enlisted men from Headquarters Battalion to help them with the construction. The officers supplied us with some "liquid" refreshments and the construction began.

True to form, the following Sunday, bright and early, General Erskine led his officers down the road through the battalion street and went to church services with us. After that, he led them and us to our new mess hall ... all screened in and with concrete floor and wooden tables.

We all had a great day, thanks to a great leader in General Graves B. Erskine. **John D. Guilfoyle, New Hampton, NY. Retired gunnery sergeant.**

Mightier Than The Sword

In "A" Company Command Post ... runner, Kelso, was on watch. The command post was awakened by his shout ... "Gimme some support ... Gimme some support!" It was just break of day and we saw a Jap running to Kelso's foxhole with a samurai sword, grasped with both hands and raised over his head.

A Marine quickly shot the Jap and saved Kelso. Kelso explained that he by mistake pressed the magazine release instead of the safety on his carbine. His magazine had fallen to the bottom of his foxhole.

Explanation came years later: A passage in the novel *Shogun* explained that it was a greater honor for a Japanese to kill an enemy with a sword. The Japanese had a chance to complete his attack in two ways. The Japanese did neither. **Conrad M. Fowler, Lanett, AL.**

A Word From Sonny

One of the saddest moments I remember: On the third night after the Guam landing, a sergeant in the Command Post Security Guard came to me and said, "Gunny, Sonny is on a stretcher down on the beach asking for you." (There were several Marines who remember Jenson as the "old Gunny" in the 6th Marines.)

Sonny, I don't remember his real name, joined my company in a replacement group on Guadalcanal after the Bougainville operation. His record book showed his age as 21, but I doubt if he was even 18 or 19. In that training period I gave much instruction, especially to beware of booby traps.

Since the night wasn't too dark I found my way to the beach, and the first stretcher case I found was Sonny. He called out to me when I was still 10 feet away. How he knew me that far away I'll never know. There was no time to talk. Stretcher bearers arrived at that moment to carry him to an amtrac, which would get him across the reef and aboard the transport hospital ship *Rixie*.

But there was just time for him to say, "Oh, captain, I'm so glad you came. I just want to thank you for all the training you gave me. But I also want to apologize for making the worst mistake you always told me to look out for. I crawled into a cave over a booby trap."

There wasn't time for more. He was being lifted up over the high gunnel of the early model amtrac which had no ramp. I didn't get to shake his hand or even say a word. **Albert L. Jenson**

Point Of View ... Was A Sweeping One

It all depends on your point of view. Anyone who ever looked down from Guam's Nimitz Plateau to the invasion beach below is struck by the visibility. Marines in 1944 feared the Japs could look right down their throats. A return to the Japanese positions now proves they were right.

The view wasn't lost to General Louis Wilson, who received the Medal of Honor for his action on the Nimitz Plateau ...

"I don"t think there has ever been such an opportunity to see here exactly what happened. You can see where you were, how it was" (he told returning Marines).

"... And I think you'll agree with me that, had I ever been up here before, I would never have been down there ..."

This photo shows the commanding view of the landing beaches from Nimitz Plateau, then known as Fonte Ridge, when the 3rd Marine Division made its landing on July 21, 1944. This photo was taken by Karl Appel of Baywood Park, CA, formerly of A Company, 9th Marines, on July 30, 1989.

NO ROSE GARDEN

Jack Eddy has come back to Guam to live. He gave up all that rat race in St. Louis and Chicago to get back where things are paced as they should be and where the "pretty Chamoritas still are running down the beach."

But Jack had long ago made a pretty good investment in Guam, three or four days after he and the rest of the 3rd Division landed between Asan and Adelup Points facing Chonito Ridge.

Eddy, a 1st lieutenant, led the second platoon of F Company, 2nd Battalion, 9th Marines in the battle for Fonte Ridge. It was the engagement in which Captain (later General and Commandant) Louis H. Wilson earned the Medal of Honor. Jack recalls being with Lieutenant W.A. O'Bannon of Maripola, CA, Joseph W. Bell of Wimberly, Texas, and Louis Machala of Dallas in a fight that was a real "slugfest." It was where General Takeshi Tkashina, Island Defense Commander, died ... where the back of the Japanese resistance was largely broken.

Later on Iwo Jima, near Cushman's Pocket, Jack was shot through the arm. Planning a renewed assault with his commanding officer (then O'Bannon) Eddy—"Obie" discovered—also had a hole in his chest. He was taken from the scene and spent two years in United States hospitals.

Dangerous business, Eddy recalls now, as Iwo Jima ended, every platoon leader who started with the 9th Marines had fallen or moved up to more responsibility. Eddy received the Silver Star for his role in that Iwo engagement.

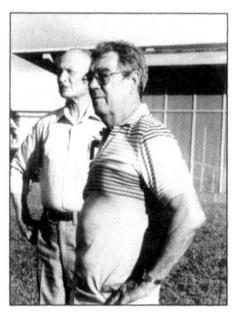

Return To Fonte

It was just about where a swimming pool is now located on Guam that General (then Captain) Louis H. Wilson had the command post of F Company, 2nd Battalion, 9th Marines. The bitter engagement on that spot, with Japanese close enough to touch, was the critical Guam battle of Fonte Ridge. Captain Wilson received the Medal of Honor for his role in the engagement.

Above, returning to the spot is, at right, Jack Eddy platoon leader in F Company in that battle, and James Paponis of San Francisco, who recalled his part in that battle as "the proudest thing I've done in my life."

Just Hopping Along

The division made a "sweep" of the Island of Guam a few months after the campaign was officially terminated, in order to round up a large number of Japanese who still remained on the island.

On this particular evening, we had completed the day's work, set up our lines for the night, and were going to get some rest. We were exhausted, and everyone had bedded down on the ground for the night. Private Russell Gieseke and myself had a guard duty watch, and were patrolling along the line of sleeping men. While patrolling, we heard a couple of thumps on the ground, and whirled about, and saw this giant toad hopping along. He probably weighed about a pound and a half anyway, maybe two pounds.

He was headed toward Lieutenant Gaylon Souvignier, who was laying flat on his back, mouth open, snoring, and really enjoying the rest. That toad made about three hops, and on the fourth hop landed right in the middle of Souvignier's chest. Souvignier jumped off the ground, about three feet in the air, landed on his feet like a cat, ready to fight. He saw the two of us laughing; and I think he felt we probably tossed the toad in the middle of his chest. I think he feels we did that, to this day. **George Briede, LaBelle, FL**

113

Some Guam Memories ... Close To The Heart

Guam will always be particularly close to any Marine who served there, especially if (or she) was among the Liberators, those who hit the beaches at Adelup-Asan or down at Agat that July 21 morning in 1944.

Marines do return to battlegrounds with consistency, as do the GIs who once scurried across Omaha Beach. These battle sites are where they fought, lost companions, brothers—and the return is bitter sweet. On Guam, veterans return to a land where now they belong, of which they are a part, where they are saviors, liberators to loving, grateful people and find an embrace reserved for only the dearest of heroes.

Among returnees to Guam are many who had gone, and will go to no other battlefield ... although they are veterans of many. They return with children, grandchildren. A few sons and daughters have returned only as proxy for dead or disabled fathers who had promised themselves and eternity that some day they would go back.

General Louis H. Wilson USMC (Ret), who became Commandant of the Marine Corps and carried the Medal of Honor for action on Fonte Ridge, returned to hail the Guamanians "for your indomitable spirit.

"No other American community since the Civil War has paid more dearly in terms of suffering and citizens killed and wounded in action. We who got to know you under battlefield conditions have never forgotten ... and because we have never forgotten we have come back to the arena of our greatest challenge.

General Craig Remembers

General Edward E. Craig USMC, who as a colonel commanded the 9th Marines (now lives in El Cajon, California) has long, fond memories of Guam, preceding even the invasion to rescue it from the Japanese.

Many years ago at the age of nine, I was on my way to the Philippines on an Army transport with my father. We stopped at Guam, and I went ashore with him. We landed from a native canoe on Asan Beach. It was the same beach we landed on in World War II.

Later, at Marine Corps Senior School in Quantico, one of our map problems involved the capture and defense of Guam. Since then I have passed through Guam on many occasions.

In 1947, I took the 1st Marine Brigade from Tientsin to Guam and stayed there for two years. The brigade cantonment, built entirely by Marines was right next to Yona. Most of the land we built on belonged to a Mr. Cruz, and one time his daughter competed for the "Queen of Yona" title.

The old Quonset where Marion and I lived for two years was the same one used by General Erskine when he had the 3rd Division. We updated it and finally cleared it of rats. When the heavy rains and typhoon hit, we slept on our cots under ponchos. What a beautiful view we had from that Quonset. So, you see from the above that Guam has meant much in my life. Reading *Juan Alta* put me right back there.

With some 5,000 Marines in the Brigade, and no town to go to on liberty, morale was a problem. Agana had not been rebuilt. We solved it by keeping the men busy with construction, heavy and intensive training, and erection of clubs, bowling alleys, soda fountains, beaches where we had mobile hamburger stands on trucks and plenty of cold beer. We had very few problems and morale was high. Officers and staff NCOs who wanted to have their wives come out had to build modified Quonsets themselves. Our Engineer Company assisted with plumbing and electricity. Everyone had their problems and solved them.

The Beauty Contest

General Edward Craig confirmed a story I had written about a native of Guam beauty contest. Some guys thought I had not fully recovered from rock fever. Good old Eddie made my list of honest officers, I will follow to the valley of death—if required. **F.J. Hoban** (Miss Cruz was named the "Queen of Yona" in 1949. It was her father who offered the land where the 3rd Marine Division camped, as did the 1st Marine Brigade when they came there from China in 1947).

For Her Dad

Paula Wilkes, a school teacher from Alta Vista, Virginia, returned to Guam as a tribute to her father, Warren Y. Wilkes (C/1/3) who died in 1987 ... He always talked of Guam, Paula said, and she promised she'd go back to represent him.

Return To Guam

They Come Back ... again ... and again ... says Governor Joseph F. Ada of Guam. Here the old warriors and Liberators form up before the Asan-Adelup Memorial on the beach. General Louis H. Wilson, center foreground, is here flanked by Harry Noble of Sarasota, FL, left, and Harvey Tennant of San Diego, CA.

Horse Trading Goes A Long Way ...

When we returned to Guam, and commenced training in earnest, we had five new officers.

It was also the era of free enterprise, because since we didn't get too many luxuries (the comforts of life) we had to obtain them as best we could. It developed we had a corporal Hite from Tennessee in the company—a man who was not too well suited to military drill and procedures, but he could really "trade."

He came over one day and said, "How would it be if we had some ice cream for the company?"

"Great!"

"I think that for a bottle of whiskey, I can do some trading." The bottle was obtained, and he went trading.

At an Army Air Force group, he put the bottle of whiskey in the right place. Result, twice a week, we got 20 gallons of ice cream. Our timing was that at 4 p.m., a jeep would leave the Air Force unit with the 20 gallons of ice cream on board, and come barreling down the road. We had the company fall out at 4:30 ... and messmen started ladling ice cream into mess gear. Regular chow was at 5 o'clock ... which meant everyone had dessert first, but nobody minded.

Another example of free enterprise was borne out as we tried to build wooden frames and decks for individual tents. There were company carpenters who could build the frames and decks, as long as we could get lumber. The corporal could get lumber, as long as there was an occasional bottle of whiskey available.

The company also had a pretty good basketball team, when it could find time to practice, or dodge rain showers. So, our enterprising corporal said, "How would it be if we got a cement basketball court?"

So, again, a bottle of whiskey put in the right place resulted in one morning ... truckloads of cascajo (coral), a cement mixer, and some knowledgeable people ... and the result was a basketball court that was superb.

As training went on (in between ice cream and basketball, which was really not the major part of our days) we couldn't help but reflect on the difference with our group now as when we went to Iwo Jima. Going to Iwo, we were pretty confident about how good an outfit we had.

We were experienced. But we weren't as confident this time as before ... because of many new men and new officers.

We had just started sending some of our gear down to the ships. We knew we were going to some place like Japan, when the first of the atomic bombs was dropped ... and the word came down shortly. "All right, you can unload your ships." There have been a lot of pros and cons over the years about the atomic bombs, but our group was certainly glad to unload our ships. **Robert Cudworth, Camillus, NY, A Company, 9th Marines**

Advice From The Doctor

In the sweep northward on Guam—August 1944—A Company was in reserve position. Two Marines came through A Company with a small Japanese prisoner. We were not moving at the time but we were hot, thirsty and it had been weeks since we had rest.

As the Marines moved to the rear somebody in A Company (9th) asked, "Why doesn't somebody shoot the son of a bitch?" The Japanese turned and said in clipped but perfect English: "That's exactly what I wish someone would do."

Turned out that he was a medical doctor, educated in the United States and gave valuable information about the Japanese on north end of island. **Conrad M. Fowler, Lanett, AL**

Return To Beautiful Guam

When we left Iwo what little bit of equipment we had left was loaded on landing crafts for the return to home base on Guam. With a final salute to our comrades, who would remain forever on this rock, we boarded one of the transports for the return trip. Having been hung up on C and K rations, and no way to take a real sho'nuff bath for the past month or so, it was pure heaven to get back aboard that ship.

The sailors really gave us the royal treatment. After washing Iwo Jima off our bodies—and it took a while to do so—we were presented with a feast of steak and fresh sunny side-up, fried eggs. One of those all-you-can-eat deals, you know.

The return trip was restful and allowed much time for contemplating the awesome experience that had transpired. A fitful night's sleep, without the nightmares riding roughshod over you, was many, many nights away. There were no body wounds, but the soul and mind wounds could only be healed by the passage of time.

The sight of good ole battle scarred Guam, coming up out of the horizon, was enough to gladden the heart of any weary Marine. Our rear echelon men were there to pick us up and carry us back to camp. A lot of hand-shaking and back-patting went on as the welcoming committee assisted us with our gear. They didn't roll out the red carpet, have the band meet us, or anything like that, but it was real warm family type welcome. **George Lyon, Florence, AL, 3rd Transport Battalion**

This is the Chapel Regimental Headquarters of the 12th Marines near Ylig Bay on Guam ... in a photo taken shortly after our return from Iwo Jima in 1945. (Photo submitted by Jack Kerins).

Telephone Reunion
After 45 Years

Private First Class Jack Kerins of Regimental H&S, 12th Marines with a young Guamanian boy, Jesus Dydasco Lazama. The photo was taken in September of 1944.

An interesting side note: In September of 1989, Kerins corresponded with Peter Siguenza, President of the Guam chapter of the 3rd Marine Division Association. Kerins sent a copy of this photo, along with others, and asked Siguenza if he knew Lazama.

Although Siguenza did not know him, he obtained Lazama's address from a cousin. Lazama is now retired from the United States Army and lives in California. Kerins then sent a copy of the photo to the California address, and a short time later received a phone call from Jesus Lazama.

After 45 years, the two talked again.

Jesus was eight years old when the photo was taken. He is now 53.

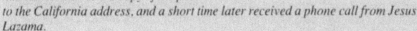

Trade-Up To
A Banquet

Things being in a quiescent state after Guam was secured, the PX started passing out free beer. Each man could get two cans every other day. Well I didn't drink the stuff, but it had a good trading value, so I accumulated mine and used it for trade. Our food supplies were still pretty much on combat rations, but were beginning to improve somewhat.

A lot of the time I was the driver that was dispatched to pick up our daily rations from the commissary stores and got to be pretty well acquainted with the supply sergeant. He just happened to be one of the old timers with a beer gut that had been on a starvation diet since we'd left Guadalcanal, and Man, was he ever a prime target for a swap-out. His two beers a day just couldn't quench that thirst.

On this particular day, I had loaded up and was signing for the groceries, when he casually mentioned how hot it was, and wouldn't cold beer go good about now. Just as casual, I asked him what he reckoned two cases of beer would be worth.

"You got beer?" he said. "Look, guy, you get me two cases of beer and I'll give you this whole damn supply dump. No kiddin' now. If you've got it, we'll work a deal."

I asked if they had gotten fresh eggs in. He said, "Some, but they were for the officers' mess."

"If I had some beer, could I be the proud owner of some those eggs?"

"Man, they'll court-martial my ass," he said. "That's private stock."

We kept dickering around until I traded the two cases of beer for 12 dozen eggs and two 20-pound hams, and a five gallon can of fresh ground coffee. After I delivered the supplies to our mess hall, I got my beer and went back. I hid all of it out in the jungle beside our tent for right then, but that night, one of the guys borrowed a huge skillet from the mess hall for the ham and eggs, and a pot for the coffee. Out there in the jungle we had feast like you wouldn't believe.

The ham and egg supper worked out so neat that it got to where it became almost a weekly affair of one type meal or another. The little jungle clearing was beginning to take on the appearance of an outdoor restaurant. Real home-like. That bunch of truck drivers sure knew how to scrounge too. If there were any tidbits to be had, they managed to figure out how to obtain them. **George E. Lyon, Florence, AL, Motor Transport**

Inconsiderate Japanese Officer

After Guam was declared secured, two men and I were on a night outpost. I had promoted a cot somewhere and was asleep ... A Marine named Johnson was sitting on the foot of it. Suddenly he let off the M-1 right over my head.

I bailed off the cot and almost drove my pistol which was in my waist band through my ribs.

What the hell happened? I asked. Johnson said a Jap just ran past my head. Just about then we heard him hit a grenade on something. We heard the primer go and I think we could even hear the fuze burn. It was no use to run, so we just humped and laid there. There was a muffled blast and crap flew all over us.

In an hour or so it was light and we found the Jap officer lying a few feet away, with his head gone and a broken leg. Johnson rolled him over and began to curse. "The dirty son-of-a-bitch. Look what the bastard has done to my sword! He held up a samurai sword with half the hilt blown off. He was so mad at the Jap, he wanted to kill him all over again. **Richard B. Abbott, K-4-12, Sanger, CA**

The Condom Caper

After returning to Guam from Iwo Jima, the 9th Regiment settled into their new camp inland, instead of on the beach where we had been before.

The humidity in the new area caused increased problems with rust and mildew, which affected almost everything, until someone discovered that some smaller items—such as candy bars, cigars, etc.—could be protected by placing them in condoms and tying a knot in the open end.

This idea really caught on, and PX was deluged with requests for condoms. This sudden increase in demand, for an item that was seldom requested, created quite a stir in higher quarters ... and an investigation conducted to determine the reason.

I never did hear what the investigation showed, but I'm sure it was discovered what was being done with the condoms. I've always wondered whether the investigation was conducted for the good of the troops, or if someone was afraid they were missing out on something. **Karl Appel, Baywood Park, CA**

The Day The Earth Shook

Shortly after being released from the hospital and rejoining A-1-9, I developed a coral infection in the scar tissue on my right leg, which resulted in my being given light duty.

While on this assignment I was honored by being given the title of "Captain of the Officers' head" and the responsibility of burning out the "pit." This was usually accomplished by getting a jerry can of diesel fuel from the motor pool, pouring it into the pit and then igniting it.

On one occasion there was no diesel fuel, so gasoline was given as a substitute. I was apprehensive about using the gasoline, and wanting to protect my posterior anatomy, I checked with the lieutenant in charge of the detail ... and was told to proceed.

After pouring the gasoline into the pit, I moved back, ignited some paper which I threw into the pit, and ran like hell.

Violent reaction ... is putting it mildly. There was a ground shaking explosion, and a ball of flame resembling a small atomic bomb blast. Contents of the pit were blown sky-high, resulting in an odoriferous fallout. Geiger counters were not needed to enter the area of contamination, but a gas mask would have been helpful.

When things settled down, I looked for the lieutenant but he had made a strategic withdrawal and was no place to be found. Gasoline was never used again in the officers' or enlisted mens' pits. **Karl Appel, Baywood Park, CA**

From Fishing To Prisoner Detail

While in training on Guam to go to Iwo, three of us (Parrie, Kinniard, and I) went up the Talofofo River to try to get some fish. We had our rifles, some hand grenades, and a native guide.

Along the river we ran into more and more signs of recent Jap life, and down at the mouth of the river, we could see some sailors swimming. Suddenly a sailor hurried up, shouting, "Bring your rifles. I got a Nip." We followed to a nearby spot, and he pointed toward large rocks near the foot of a cliff.

Seeing several Japs behind a big rock, we opened fire on them and Parrie was pointing and yelling, "Over there, right there, two of them." He was out of ammunition and had lost the pistol he carried; and since I couldn't see them from my position, I handed him my carbine. He shot several times. I saw two Japs fall.

By this time there were several servicemen looking on. A Navy doctor rushed to the two fallen Japs, and said, "Don't shoot, don't shoot. This one is alive and is O.K."

We stripped off his clothing and he had not a mark on him ... but pretended to be unconscious. But when two men put his arms over their shoulders and carried him to a jeep, he didn't drag his feet ... he raised them to prevent injury. Parrie, Kinnaird, myself, the native guide, his son, and the doctor all got in the jeep with our prisoner seated on the floor in the center. I was holding him by the hair of the head as precaution and noticed his eyelids move. Just as I said, "Watch this S.O.B., he is not as sleepy as he pretends," He kicked with both feet ... hitting the doc in the left hip so hard that the boy almost fell off the doc's lap. The Jap then grabbed the sheath knife on the

doctor's hip and the muzzle of Kinniard's rifle.

I put a headlock on the Jap while others struck the Jap's hand to force him to let go of the knife and rifle. I tried to use some of the two Japanese expressions I had learned when I was in the 4th Raiders. One was "Lay down your arms and surrender." The other was, "Now Japanese soldier you must die." I goofed, and said the latter. This made the Jap resist all the more.

As the doctor and Parrie hit the Jap in the temple and then full in the mouth, I was applying a choke hold. The Jap relaxed and there was no more fight left in him.

At division headquarters, he was laid out on the ground. The doctor sent for a pair of his shorts, and it developed the jap was so small, he could use one leg of the shorts for a skirt.

When the Jap asked for water, a corpsman brought some in a bottle marked "Poison." The Jap pointed to the skull and crossbones and shook his head, no. The corpsman sampled the water to show it was okay, and then the Jap drank even though he said his throat hurt.

When we A Company men left, we talked it over and agreed not to say anything about our adventure back at the camp. The next morning there was a story around of a native who killed four Japs and one escaped. Another tale was of a native killing four Japs and letting one be killed by his son.

A day later Captain Fowler questioned me. At the time I denied everything, but I did not fool the captain. He said, "I wish you had told us about it. We could have got credit for it." **Clarence Brookes, Cortland, NY, A Company, 9th Marines**

Jap Flag Trophies

Snyder, a replacement who had joined A Company, 9th Marines after our return to Guam from Iwo Jima, was unusually busy around his tent one day, ripping up a sheet into several pieces. On each piece he inked or painted a large solid red circle, and added some black lettering much like Japanese writing.

It was his version of Japanese writing. He was making Jap flags.

The Jap flags were not for flying—they were for selling. After the ink and paint had dried, each flag was dirtied a bit, folded several times, and made to look like the flags often found on Japanese soldiers.

On his next weekend liberty, Snyder took two of the flags down to one of the air fields on the island where there were always new pilots from the States. A pilot looking for souvenirs would gladly spend several dollars—or a bottle of whiskey—for a Jap flag.

A bottle of whiskey was a much sought after commodity in the ranks of Army, Navy—even Marine—personnel on the island and was usually worth about $60.

So, from each bedsheet and inking materials, Snyder could expect to take in $300 to $400 or more.

But even more interesting would be the number of game rooms or trophy cases back in the States proudly bearing a Japanese flag—a pseudo "Japanese Flag" made by Snyder. And, oh the stories that were probably told about how those flags were "captured." **Robert Cudworth, Camillus, NY**

Man At The Well

Chamorros will not forget the amiable Abe's job to distribute the water and guard the precious source at the well head. In the role, Abe, big, portly, waring his heart on his sleeve, got to know every woman and child from Inarajan to Dededo.

Just a year ago, Abe went back and searched them out—the old Chamorros of his own generation. They embraced him. They exclaimed his name. They cried—and Abe cried too. **Abraham Jacobson, "The beloved Seabee," Bayside, NY**

Vignettes of Guam

Photo Return

Dominic J. Savino of Oceanside, New York came on quite a cache, some 100 or more pictures taken by a professional Japanese photographer. He brought them home and it told a lot to the people of Oceanside. Then Dom took them, with his wife Mary, to Guam and turned them over to Governor Ada for his archives. The Savinos were treated royally by the Governor. **Dom J. Savino, Oceanside, NY**

All Expenses Paid

Charles W. Ford of Ocala, Florida with 12th Marines, remembers like yesterday Roy W. Beard, Peter Batna, William Utley ... and the pictures of them in front of tent on the canal (after Bougainville) looks more like a resort than a training ground for Guam.

Careful Cargo

Lieutenant Colonel C.L. Gutherie of Johnson City, Tennessee will not forget now for sure—after all these years—how his platoon floated 55-gallon drums of gasoline over the reef on Guam (D-Day) then rolled them up on the shore to the all around pop of imperial mortars and small arms.

It was a shaky business coming ashore with pack and rifle, but the 2nd Separate Engineers (and others of their trade) confronted the inhospitable beach with a lot that was heavier, and drew more Japanese attention. Gutherie retired from the United States Army.

Old Faithful

Jim Freeman saw them carrying everything from Texas doglegs to 45's and carbines, but will still take an M-1.

"I saw several old vets carrying .03 Springfields, and .44 Texas doglegs. Not one to leave a mystery unsolved, I asked why the switch from the carbine to the Colt .45. I was told that in a combat situation, it is not always possible to field strip your weapon and keep it clean. They were right, so I threw away my carbine and carried an M-1 for good ..."

No Longer The Same

Henry Jennings Krueger, McPherson, Kansas got to the airstrip, later called Harmon Field, on Guam when it still was a few scars in the red dirt. Hank worked on the B-29's as a mechanic, and usually at night for the big bombers mostly took off for Japan at six or seven in the morning.

Hank greeted hosts of Marines ... joined many in the chow lines as they came sightseeing to view these aerial behemoths of the 62nd Bombardment Squadron, 20th Air Force. Then it was the center of Pacific, American, and world attention.

Hank revisited with his wife, Myrt, the heart of the 20th on Guam. All he could find was a stone wall and a bare field.

Maintaining True Spirits

On Guadalcanal we had a big poker game one night. Someone came up with the idea you could put in a whiskey order back to the States. Someone had received a flyer that if you ordered so much, they would sent you glasses and everything with it. So we cut the pot. People put in their order for a case of this or a case of that. We sent the damn money off. And wasn't too long until, sure enough almost by return mail, came this tremendous pile of whiskey.

We had already loaded (to go to Guam), struck our tents and everything ... and here we have about 100 cases of whiskey. You're getting it for like 10 or 12 dollars per case in those days, no freight, no nothing.

What the hell you going to do with this stuff? We decided we could take a bottle apiece. Then a boy in quartermaster got a hold of a great big pontoon ... Remember, those big 10 foot tubes that they use for all sorts of things? He cut a hole in it and stashed all this whiskey in it, with palm fronds, padding, and welded the cover back on. Labeled "First Battalion, 9th Marines Showers."

After that we went on up to Guam. We had a deal that if anyone got shot, he just rolled over his share to the rest of the group. By the time we got through Guam Campaign, just about everyone's share was doubled. Along with our seabags and other gear, came this receptacle labeled, "First Battalion, 9th Marines Showers ..." not a damn bottle of it broken. You know, you say that is one of those things that can't happen in a war. Hell, it happened. **Carey Randall, Jackson, MS, 1st Battalion, 9th Marines.**

A native Chamorro girl meets a United States Marine. (Photo submitted by George Green, Webster Groves, MO)

A few minutes after their first landing on newly-captured Orote Peninsula Airfield on Guam, Marine pilots were seen eating their noonday C-rations aboard derelict Japanese torpedo. All three of the pilots were veterans of air combat over Bougainville. (Defense Department Photo-Marine Corps.)

Led by their mascot, a stray dog, 60 Japs walk to the trucks which will take them to the Island Command stockade. This photo was taken September 12, 1945, more than a year after the Island of Guam had been recaptured from the Japanese. (Defense Department Photo-Marine Corps.)

Guam .... Crossroads of the Pacific

"The 3rd Marine Division's actions on Guam had, without question, contributed greatly to the ultimate victory over Japan," points out Prof. Harry A. Gailey, author, teacher, and military history authority.

In this treatise on the campaign, he notes Guam's pivotal importance in 1944 as a base for B-29 bombers and a forward command post for the Air Corps and Navy.

A professor at San Jose State University, California, he has received numerous honors and grants including SJSU Outstanding Professor and President's Scholar.

Among the numerous books he has written, four are about World War II actions in the Pacific. They include: Peleliu: 1944; Howlin' Mad vs the Army; The Liberation of Guam; and Bougainville, 1935-1945.

Here is his thorough report and analysis of the Guam Campaign as a part of "Two Score and Ten."

3rd Marine Division Actions on Guam June 1944-February 1945

by Professor Harry A. Gailey

Guam was the first United States territory seized by the Japanese in World War II. On December 10, 1941 an overwhelming Japanese force landed and forced the surrender of the government and small garrison with the only resistance coming from a small half-trained Guamanian Insular Guard.

Until late 1943 the island was governed by the Japanese Navy whose rule over the native people, although not beneficent, was not as harsh as it would become after the Imperial High Command decided to fortify the island.

Despite the difficulties in running the gauntlet of American naval vessels, the Japanese poured men into Guam and its neighboring Mariana Islands—Saipan, Rota, and Tinian. Ultimately there would be over 18,000 Army and Navy personnel on Guam. Frantic efforts by the Japanese commanders, utilizing their own troops and local Chamorro forced labor, and taking advantage of the natural terrain, made the island very defensible.

The Japanese, who at first overran huge areas of southeast Asia and captured most American, British, and Dutch possessions in the Pacific, had seen their fortunes reversed by the end of 1943.

Their southern advance had been halted in New Guinea and their Navy had been severely damaged at Midway and in the Guadalcanal battles. The Solomons Campaign, highlighted by the defense of Guadalcanal, represented a turning point in the war.

Neutralizing Rabaul and Truk

In conjunction with MacArthur's offensives in New Guinea, the later conquests of New Georgia, Vella Lavella, and the landings on Bougainville in November 1943 signalled the end of usefulness of the Japanese air and naval bastion of Rabaul. Meanwhile, Admiral Chester Nimitz (CINCPAC) had built a huge naval striking force which by early 1944 dominated the Central Pacific. A portion of that fleet struck Truk, the major Japanese naval base and command headquarters, in February 1944 and rendered it useless.

The Orange Plan

By late 1943, Nimitz felt strong enough to begin to implement the major features of the pre-war Orange Plan, the drive across the Central Pacific. This called for the step-by-step occupation of Japanese air and naval bases first in the Gilberts, then Marshalls, and finally the Mariana Islands.

The spearhead for these attacks was to be Marine Corps units, backed by Army divisions, all supported by overwhelming might of either the 3rd or 5th Fleet.

The first objectives, Tarawa and Makin in the Gilberts, were seized in September 1943 and then by Nimitz's direct order the pace was stepped up. Kwajalein and Roi Namur were occupied in January 1944 and the following month Eniwetok was taken.

Once these bases had been taken, the main targets for Nimitz and the Army Air Corps commander, General Arnold, were the three main islands in the Marianas group—Saipan, Guam, and Tinian. Possession of these would give the Navy the forward bases for more efficient attacks north and west. More important, they would provide the air bases for the superbomber, the B-29, to launch bombing attacks on the Japanese home islands.

The Marianas Invasion Plans

Initially the plan for the Marianas called for the utilization of four divisions and one brigade. The overall commander of the operation was Vice Admiral Richmond Kelly Turner who would also be directly in charge of the Saipan invasion.

The commander of ground force operations was Lieutenant General Holland M. Smith who would also supervise the invasion of Saipan with the 2nd and 4th Marine Divisions in the assault and the Army's 27th Division in reserve for use either on Saipan or Guam.

The Guam operation was commanded by Rear Admiral Richard Conolly with Major General Roy Geiger commanding the ground forces—the III Amphibious Corps. This corps had the 1st Marine Provisional Brigade and the 3rd Marine Division for the assault.

The invasion date for Saipan was June 15 and the very ambitious, but short-sighted, planners expected to invade Guam three days later. The Japanese resistance on Saipan proved greater than expected and the only reserve for Guam, the Army's 27th Division, was committed there. The Guam Operation was therefore postponed until the Army 77th Division could arrive from Hawaii to act as the reserve force.

When the warning order for the Marianas reached Geiger on Noumea, the III Corps was hardly more than the 3rd Division. The elements of the Brigade were scattered over the South Pacific. The 3rd Division, commanded by Major General A.H. Turnage, had been formed in September 1942 and reached the South Pacific in January 1943.

Some of its special troops were used against the Japanese on Vella Lavella and Rendova, but the first action involving the entire division was Bougainville. On November 1, the division landed at Empress Augusta Bay. There, in conjunction with the Army's 37th Division, the 3rd, 9th, and 21st Marines (along with supporting tank and artillery units) fought through swamps and almost impassable jungle against the Japanese, relatively few in number but extremely tenacious.

By the time the 3rd Division left Bougainville in December it had inflicted heavy loses on the enemy and helped establish a relatively secure perimeter within which the Army engineers and Seebees would construct three airfields.

The Kavieng By-Pass

Soon after returning to Guadalcanal for rest and reorganization, the Divisional staff was informed that its next objective would be Kavieng Island west of New Ireland. The success of air and naval actions, however, convinced General MacArthur that he could bypass Kavieng. The 3rd Division, already on its way, was called back. Later elements of the division were used to occupy Emirau without meeting resistance.

Meanwhile, Admiral Nimitz's Central Pacific campaigns had wrested Kwajalein and Eniwetok in the Marshall Islands from the Japanese in early 1944. His next objective was the Marianas with emphasis on the island of Saipan.

General Turnage had been notified in early May that the 3rd Division would have the main assault role against Guam, an operation scheduled to begin three days after the initial landings on Saipan. The plans for the invasion were formalized with Geiger and Conolly's staffs and practice loadings and landings were conducted by the division off Cape Esperance.

The Seaway to Guam

On June 1 the first elements left Guadalcanal for Kwajalein; the entire Amphibious Corps with the heavy naval escort left the Marshalls on June 11. Softening up of Guam's defenses began on the 16th when a task force of battleships and destroyers pounded the landing sites. This was accompanied by air strikes. Once it was determined that the

Saipan Operation would need the Army 27th Division, the Guam Invasion was postponed indefinitely. The invasion fleet which was orbiting from 150 to 300 miles west of Kwajalein on June 25 was ordered to Eniwetok to await the arrival of the Army's 77th Division from Hawaii.

Marines of the 3rd Division would spend 52 days cooped up aboard the APAs and LSTs before hitting the Asan beaches. Finally on July 15 the armada left the Eniwetok anchorage bound for Guam with D Day scheduled for the 21st. Fortunately for the Marines, the Navy had worked over the beaches and adjacent areas very well. Admiral Turner Joy with four cruisers, 12 destroyers, and two light carriers bombarded the island for 13 days and air groups from TF-58 cooperated with almost continual air strikes. Admiral Conolly brought up three more battleships on the 12th, and by D Day two more battleships joined the firing line. This heavy bombardment did what was seldom accomplished during the Pacific conflict. It destroyed most of the heavy guns and permanent defenses of the Japanese in addition to the heavy toll of lives taken on the enemy's forward units.

Landing On Guam

The landings went as planned although the Japanese from their cliff positions and at Adelup Point poured fire onto the DUKWs. By the end of the day, 36 of these were out of action.

The 3rd Marines at 0829 on July 21 landed on the northern Red Beaches 1 and 2.

The 21st Marines in a line of two battalions in the center came on to Green Beach while the 9th also landed in a column of battalions on Blue Beach.

The objective of the 3rd Marines was to secure Adelup Point while other elements drove straight ahead to secure the cliff facing them. The objective of the 21st Marines also was to move off the beach and take the bluffs overlooking Green Beach. The 9th was to secure Asan Point while one battalion moved southwest to join with elements of the 1st Marine Brigade.

By evening of D Day all the major objectives for the first day had been secured. The 9th Marines advancing across dry rice paddies took the ridgeline fronting them by nightfall. At first there was considerable enfilade fire from Asan Point, but tanks moved up to support 3/

9 and this key area was taken before evening. Meanwhile, the 21st had been halted briefly by hostile fire from the cliffs, but its commander, deciding against a frontal attack, sent one group up a defile on the south and another to follow the Asan River. Both 2/21 and 3/21 had reached the top of the bluff line and secured their positions by late afternoon.

The most difficult fighting of D Day was in the sector of the 3rd Marines. Immediately in front of their narrow beach was a high ridgeline, Chonito Cliff, and to the left was Adelup Point. Supported by tanks, Marines of 3/3 clawed their way up the 60° slope of Chonito Cliff and after taking heavy casualties reached the top by noon. The Japanese on Adelup Point were finally driven off when LCI gunships and destroyers added their fire power to the tanks supporting 3/3.

The most spirited Japanese defense was for a ridge to the right of Chonito Cliff named Bundschu Ridge after the commanding officer of A Company 1/3. Despite taking part of the ridge, there were too few Marines left to hold the gains, and before evening of D Day they had retired to their jump-off lines.

The Bluff Line Taken

Once established along the bluff line, the 3rd Division was in a commanding position. With each day, General Takashina's (the Japanese commander) position became more desperate. Bundschu Ridge was taken by the 24th after days of heavy fighting. The 9th Marines took the Piti Navy Yard on W + 1 and made an amphibious landing on Cabras Island. The Japanese were routed from there the following day.

Meanwhile, to the south, General Shepherd's 1st Brigade had expanded its beachhead and by the 24th began its assault on the well-defended lines of the Orote Peninsula.

General Geiger had also brought the Army 77th Division ashore beginning with the 305th RCT on the night of the 21st. Except for local defense, sometimes desperate as along Bundschu Ridge, and with probing night assaults, the Japanese had not seriously counter-attacked. General Takashina and his subordinates, having passed up the best opportunities to smash the Marines, on July 24 concluded that they could not hold the island with such tactics and made the fateful decision to order **banzai** attacks against both the 3rd Division on the plateau and against the 1st Brigade on the Orote Peninsula.

Banzai! Banzai!

The **banzai** attack against the 3rd Division was well organized and was coordinated with that to be launched against the 1st Brigade on Orote on the night of July 25-26.

The Marines were aware that something was happening in the Japanese areas before they were attacked. The weather that night was foul; heavy rain made the Marines in their foxholes even more miserable. The Japanese sent out probing attacks before the first main attacks began at 0300. One Japanese battalion flung itself against the 21st Marines in the center of the line. Despite huge losses, some Japanese broke through to the beach area where they were destroyed by tank fire. Elsewhere, the 21st line had disintegrated into individuals fighting to stay alive. Some of the attackers even carried demolition charges and blew themselves up, hoping to kill as many Marines as possible. Before the final attacks ceased just before dawn, the Japanese dead were piled like cordwood in front of the Marines' foxholes.

Meanwhile another large Japanese force had exploited the half mile gap between the 21st and 9th Marines. The Japanese commander seized a ridge parallel to the beach and set up machine guns there. This position was not retaken until mid-afternoon of the next day. Others in his command reached the beach where they were finally destroyed by ad hoc groups of non-combatants after penetrating within 10 yards of the Fire Direction Center of the 12th Marines.

The heaviest fighting of that morning was on the Marine left, held originally by the 3rd Marines who had later been reinforced by 2/9 pulled from the right flank because of the heavy losses sustained by the 3rd. The 3rd sustained seven separate charges by the Japanese 48th Brigade. Supported by direct gun-

Here the Japanese "banzai" losses showed up since the Marines had little difficulty in taking the crest despite the natural defenses of the mountain.

fire from naval support ships, the Marines cut each Japanese attack to pieces. In the morning, 950 Japanese dead were found in front of the lines.

Defeat ... Inevitable

The failure of Takashina's attack, coupled with that on the Orote Peninsula, confirmed the Japanese defeat on Guam. No exact count of Japanese dead in the 3rd Division area could ever be made. The figure arrived at by Marine Intelligence of 3500 might have been conservative. After the disastrous attacks, the Japanese were never able to conduct any aggressive offensive in force, and the way was left open for the Marine and Army units to move to the north coast. A fact expediting this movement was the death of so many officers who might have organized a more logical, concerted resistance.

General Geiger resisted the criticism of his superior, Holland Smith, and did not hurl the battered 3rd Division immediately into action. Rather, he waited for the 1st Brigade to gain the upper hand in taking the Orote Peninsula.

Finally on the 27th, the 3rd Marines began their attack with the main objective the Fonte Plateau. Resistance to the advance was sporadic, heaviest in front of 2/3 and 2/9, particularly along the depression known as the "pit." This strong defensive position was not reduced until the 29th. Two of the companies of 2/9 had suffered 75% casualties by that time.

There was also fighting in the sector of the 9th Marines who assaulted Mt. Chachao. Here the Japanese banzai losses showed up since the Marines had little difficulty in taking the crest despite the natural defenses of the mountain. By the 30th, two RCTs of the Army 77th Division had taken position on the right of the 3rd Division and the line now stretched across the narrow neck of the island.

Final Phase

In the realignment for the final phase of the operation, General Shepherd's 1st Brigade searched the southern part of the island for enemy units. Although difficult due to terrain, the 4th and 22nd Marines found few Japanese in southern Guam.

All during July 30-31, corps and division artillery augmented by guns from the naval flotilla battered road junctions and possible concentration points. Then on the 31st a general advance began.

The 3rd Marines advancing along the coast road occupied the wrecked capital, Agana, within four hours.

The 21st in the center and the 9th on the right of the division line also moved rapidly ahead against sporadic resistance. The division advanced three miles that day. The main enemy to rapid progress was the hilly, overgrown terrain of northern Guam with few well-defined trails through it.

Geiger, knowing that time was not a factor, continued to resist pressure to move more rapidly. Instead, he brought up the 1st Brigade on August 4. Three days later he positioned Shepherd's force on the extreme left, having the 3rd Division move to the right to assume the center for the final drive to the north coast.

On August 3 the drive was resumed against weakened Japanese defenses. General Obata, commander of all Japanese forces in the Marianas, had assumed direct command after Takashina was killed. He ordered a fall back toward Mt. Santa Rosa with strong blocking positions established at two points. One of these, the Finegayan area, was in the 3rd Marine sector. Finegayan was the last major action for the division. The Japanese made a strong stand there, but were soon overcome. One reason for the success was that the Navy provided the additional firepower of a battleship, a cruiser, and five destroyers. By the 5th the division forward lines were far beyond Dededo

After the 1st Brigade joined in the northern offensive, the 3rd Division advanced on a northeastern axis along the roads and trails and met only scattered pockets of resistance. The favorite Japanese tactic was to defend road or trail junctions by siting light artillery pieces and machine guns backed by a few tanks to ambush the advancing Marines. One of the most serious actions of this type was in the sector of 2/3 on August 8 where enemy tanks actually broke through the lines before being destroyed.

The cautious nature of the final advance can be seen on that same day where the infantry was halted while Corps and Division artillery blasted an area suspected of sheltering hundreds of Japanese for two and one-half hours. By the evening of the 9th, Marine and Army units had reached the northern end of the island, and on the following day General Geiger declared all organized resistance to be at an end.

The previous day, scouts from the 9th Marines had discovered a small concentration of Japanese near Mt. Mataguac. General Bruce's headquarters were notified since this was in the sector of the 77th. An army patrol was fired on by the Japanese on the 10th. The following day the area was surrounded and artillery and machine guns destroyed the Japanese defenders at General Obata's last command post. By the time that Obata had written his apology to the Emperor, the Americans had counted over 11,000 Japanese dead. Losses by the III Amphibious Corps by that time were 1190 killed in action, 377 dead of wounds, and 5,308 wounded

Aftermath

The declared official end of resistance did not mean an end to the fighting. There were still thousands of Japanese in pockets in both north and south who had been bypassed.

It became the task of the 3rd Division to handle this serious problem since the 1st Brigade left Guam in Augsut and the 77th Division in early November. The 21st Marines located in the northern part of the island were given the main task of ending Japanese resistance al-

> *The declared official end of resistance did not mean an end to the fighting. There were still thousands of Japanese in pockets ... who had been by-passed.*

though patrols from the other regiments as well as station troops were involved in these combat actions. The task of the Marines was made much easier because of the cooperation of Guamanian volunteers.

The largest, most complex sweep was organized by the new Division commander, Graves Erskine, in late October. He utilized the 21st on the left and the 3rd on the right to move slowly north, killing more than 100 Japanese before reaching the coast. By than, many Japanese still holding out were starving and over 1200 had been induced to surrender. However, the body count of Japanese as late as February 1945 was 145 killed by patrols that month. Major Sato in the south surrendered 34 men in June 1945 and Lieutenant Colonel Takeda in the north gave up with over 100 followers, only when he heard the Emperor's message ending the war. Thus although the island was secure enough for the construction of three B-29 airfields and Nimitz's new headquarters, certain areas of Guam remained dangerous for the continued patrols by the 3rd Division.

In late fall, General Erskine was notified that the Division would be a part of

the assault on Iwo Jima in the Bonin Islands, and planning and training for this operation were put into effect immediately. On February 17, the Division left Guam to act as a reserve force for the two assault divisions.

An Important Role

There were probably very few Marines at that juncture who understood what an important role they had played in winning the war. Although generally regarded as secondary to the Saipan Operation, the liberation of Guam provided the United States with a forward command post for the Air Corps and Navy.

By the time the division left the island, B-29s were already utilizing the airfields to strike at the Japanese homeland. The successful Marianas Operation helped bring about a complete reorganization of the Japanese government with General Tojo being replaced as premier. Many senior Japanese commanders recognized after the American conquest of Saipan and Guam that they had lost the war.

The 3rd Marine Division's actions on Guam had without question contributed greatly to the ultimate victory over Japan.

The Longest Night
Recollections of PFC Dale Whaley

When we left the Higgins boats on November 1, 1943 on Bougainville, the Navy was short of machine gunners so I was selected to man the machine gun on the landing craft. This gave me a splendid view of the landing which you didn't get when you were crowded in the bottom of the boat.

... The only opposition our unit encountered (H/2/9) was strafing which did us no harm. I sprayed the shore line as did the other boats, but all we hit were coconut trees.

This time (Guam) it was different. We took heavy fire ... and my most vivid recollection was a direct hit on the "alligator" next to us ... metal and bodies flew through the air, splashed all around us ... but we landed okay.

... On the fourth day we moved to Fonte Ridge ... we got cut off ... I was with the second machine gun squad ... We were hit very hard and unable to fall back ... Our squad leader and No. 1 gunner were hit ... One by one, the others were hit ... I was out front by myself, and the company was not able to get to me ... it was the longest night of my life ...

... The ground in front and on both sides was littered with bodies of Marines and enemy dead ... I was the only one left in our squad ... When I found the bodies of my two buddies, I felt more alone than I had ever been before or since that day. We reorganized under Lieutenant Jack Eddy ... I was promoted to corporal for this action.

Private First Class Dale W. Whaley, USMCR was awarded the Navy Cross for action on Fonte Ridge July 25-26, 1944. He received the gold star in lieu of second Purple Heart, and the Meritorious Service citation for similar action on Iwo Jima. Whaley lives in Loomis, CA

123

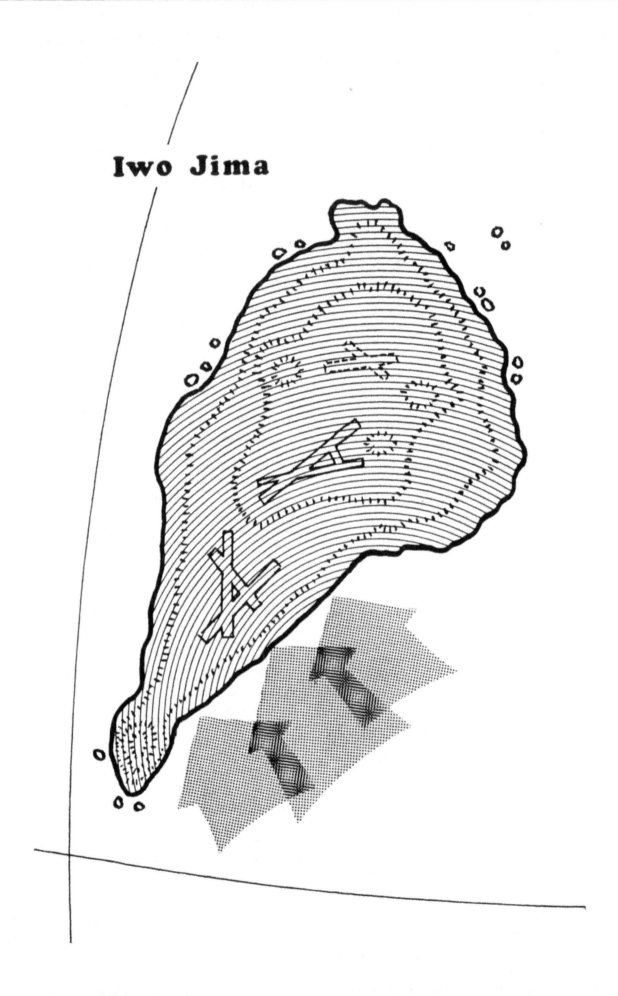

Iwo Jima

The Briefing

On the ship's open deck, some sitting, some standing, the men of the 3rd Marine Division crowded closely around a large, three dimensional map of a strange looking island. Comments ran from, "Another damned island" to "It looks like a pork chop." The chattering dwindled to silence, as a lieutenant from Division Intelligence stepped into the center of the circle.

"OK, now listen up. You're about to get the word," he announced.

"This is our objective—Iwo Jima," he said, waving a long, wooden pointer over the model. "It's five miles long and about two and one-half miles wide, and it's in Japan's front yard, only 700 miles from Tokyo. Our landing beaches are along the east coast, here. The initial assault will be made by the 4th Marine Division, on the right and the 5th Marine Division, on the left. Our own 3rd Division will be floating reserve."

"Maybe they won't need us," muttered an optimistic private first class.

"You can bet your ass they will," said someone else.

The lieutenant continued, "Enemy resistance will come from Mount Suribachi, here on the left flank, from the high ground and cliffs on the right flank, and from bunkers, pillboxes and entrenchments to the front, along the terraces, overlooking the landing beaches. We estimate that the island is defended by twenty to thirty thousand heavily armed and well concealed Japanese troops."

There was another noticeable murmur among the gathering of Marines.

"We are not sure of what all they will have waiting for us, but they've had more than twenty years to put it there, so we should be prepared for anything. For example, photographs of the landing beaches, taken by one of our reconnaissance submarines, have revealed something we have never seen before."

The lieutenant held up an enlarged photograph of what looked like a concrete obelisk about three or four feet tall with short pipes jutting out on the right and left sides.

"These structures are spaced at intervals all along the landing beaches," he said. "We are not certain, but we believe these are stationary flame throwers that can be fired from a central control base hidden somewhere on the island. In other words, we may have to go through a wall of flame upon reaching the beach. If our theory is correct, and you find yourself confronted by these structures, we suggest that you use your entrenching tools to bury them with sand and smother the flames."

Again, murmurs of nervous uneasiness and mounting tensions moved through the group of concerned Marines.

"Now, a little bit about the island itself," continued the lieutenant, in an effort to quell the increasing anxiety. "Iwo Jima is one of the Volcano Islands. As the name implies, these islands are formed by build-up from submerged volcanic activity. The islands are actually mountain tops protruding above the surface of the sea. Mount Suribachi, for example, is a smoldering volcano.

"Historically, it is not uncommon," he said, "for one of these mountains to blow its top and the island to disappear beneath the sea only to reappear again, in some future era, following another eruption.

"Are there any questions?" he asked.

Private First Class Clay W. Jordan, of the 12th Marines, raised his hand.

"Yes, what is it?" asked the lieutenant.

"Lieutenant, how many Japs did ya say are on this island?" asked Jordan in his slow Texas drawl.

"We are estimating about 20,000 to 30,000, but of course our pre-invasion bombardments should reduce those numbers considerably."

"Yes, Sir, we've heard that before. Did yaw'll say they'll be lookin' down our throats from three sides when we hit the beach?"

"Yes. Actually, there are but two possible landing sites, but both are faced with that same situation. We have chosen the eastern beaches because of better surf conditions," explained the lieutenant.

"Uh huh, and ya say we may have to go through a wall of fire on that beach?" pressed Jordan.

"That could be a possibility, if those beach structures are what we believe them to be."

"Well then, lieutenant, I have one more question for you. Why in the hell don't we just cruise around out here 'til that damned island sinks; then when it comes back up, we'll just get on it before the Japs do?"

Needless to say, all tensions, resulting from the briefing, were immediately relieved. **Jack Kerins, Terre Haute, IN**

Honing In On The Homeland

Now, the enemy was at the edge of the home islands.

The enemy was you.

And the battle you made to take that last Japanese bastion, and hasten the end of a terrible war has gone down beside Thermopylae, Gettysburg, and Verdun. America lost 6,821 of her sons there. Casualties for United States numbered 25,851. But victory for Japan and even the dignity of negotiated peace was lost to Japan forever.

Only 750 miles from the Japanese heartland, Iwo Jima in the Bonin Islands—made of sand, sulphur and scraggly vegetation—had not, as the war began, been considered a fortress or much of anything. In 1940 ground was broken for an airport and the next year a naval detachment set up some big gun defenses. It was not until the United Stated forces hit the Marianas that the Japanese began the construction of the major surface and labyrinth defenses that our forces would come to know so well.

So dense were fortifications that virtually anywhere in the immediate shade of Suribachi a dozen bunkers or such were within stone's throw. Even tanks were imbedded in the dusky ash as mini pillboxes.

"The length of the island meant that the enemy artillery could reach any place on it," recalls Brigadier General Wendell Duplantis, USMC (Retired) who commanded the 3rd Battalion, 21st Marines. "There was no possibility of drawing back out of range.

Lieutenant General Tadamichi Kuribayashi, commanding Iwo's defenses, knew that also. A samurai of the warrior caste he eschewed the bloody tactics of banzais and "killing them at the water's edge." He would let the invader ashore, and kill him there.

Like Admiral Isoroku Yamamota (who planned the Pearl Harbor attack) General Kuribayashi had spent much time in the United States. He understood its might, and despaired of a war with the sleeping giant. Yet, when his country called, he fought as a true samurai.

(Continued on page 127)

LIKE A SCENE FROM HOLLYWOOD

We had made the top of Hill Peter after a hectic day of being pinned down by Japanese machine gun and mortar fire, getting hit by our own naval gunfire, and then storming up the hill after a heavy artillery barrage followed by smoke cover.

Spread out across the south and east sides of the hill, it was obvious we were going to have to neutralize that large gun emplacement just ahead of us to the north.

It housed a large anti-aircraft weapon (probably 155mm in size) which stood like a sentinel atop Hill Peter. The big gun was not firing at us, but the emplacement had tunnels feeding into it and was serving as operation center for Jap soldiers firing at us with knee mortars and small arms.

We had called for flame throwers and Private First Class Fransko joined us. Fransko was formerly an A Company, 9th Marines man and had gone to battalion for flame thrower and bazooka training a few months previously.

Our problem was how to get Fransko across 90 to 100 yards of flat terrain, close enough so he could use the flame thrower on the gun emplacement. We had to do it quickly, our mortars had not moved up yet; and besides some of our own men were too close to the emplacement to chance mortar fire.

Smoke might help, but the only smoke we had was a red smoke grenade in the lieutenant's pack—a grenade used to mark front lines for aerial or artillery shelling. It would have to do.

A big machine gunner from Texas was instructed to "Throw this baby as far out toward that emplacement as you can. There's a light wind behind us and it may blow the smoke toward the emplacement. We'll put as much fire on the emplacement as we can."

He gave the canister a tremendous heave. It landed and commenced spewing out red smoke. Fransko and a fire team took off, using the single grenade's smoke as much as possible for concealment. When Fransko reached the emplacement, he gave a few bursts of his flame thrower ... bounded to the top of the parapet and fired more bursts.

As he did this, his helmet fell off. Fransko had a red bandana tied around his head—pirate style—presumably to help cushion the weight of his steel helmet.

Here was pure Hollywood, in color. A fighting Marine with red bandana on his head, standing on the parapet angrily sending bright orange bursts from the flame thrower into the emplacement, and lacy fingers of red smoke wafting across the whole tableau.

After Fransko's coverage with the flame thrower, one of the other Marines carried a pole charge into the emplacement, rammed it in the largest of the openings from where Japs had been coming, and pulled the igniter wire.

Several seconds later—and it seemed longer—the whole emplacement erupted in a tremendous explosion. Debris shot upward and it seemed the whole top of the hill had blown up. The pole charge evidently ignited a Japanese ammunition cache within the hill ... a hill which later proved to house a network of tunnels and various levels of Japanese activity—sleeping, eating, storage, and communication center. One of the major Japanese strongholds on the island had been badly damaged, if not neutralized.

But even though Japanese activity had been slowed down from within part of Hill Peter, the Japanese were still very much active in openings on the extreme north side of the hill and further north.

Suddenly, a Japanese tank came from behind a smaller hill in front of us ... firing effectively with his machine gun and causing numerous casualties on the hill.

Fransko had now picked up a bazooka; and after firing a few rounds, hit one of the tank's treads. The tank could no longer go forward, but could still pivot and fire effectively.

Attempts were made to further neutralize the tank by hitting it again with bazooka fire, but during this action Fransko was wounded in the arm and had to be sent to the rear.

Fransko had a great day, and he was one of the many who helped A Company take Hill Peter and allow other units to pass through and continue the assault the following day. He was later recommended for and awarded the Navy Cross. **Robert Cudworth, Camillus, NY**

A Walk To The Front

Iwo Jima, February 26, 1945 (Delayed)—you leave the 3rd Marine Division command post on Iwo Jima shortly after a breakfast of bouillon, hard biscuits, and vegetable hash.

Wire is criss-crossed along the soft, loamy soil. The ground is pock-marked with foxholes that makes it impossible to walk on a straight line for more than a few feet. Tin cans, paper, boxes, pieces of wood litter the ground. A Japanese sailor's blouse decorates a hole that formerly housed a Nip mortar.

Suddenly the whir of a mortar shell sends you to the deck. You lie there 10 or 15 minutes, pick yourself up and start off again, this time keeping a sharp lookout for foxholes to jump into. Carrier planes circle and dive several hundred yards ahead, spitting machine gun bullets and rocket shells at the enemy.

Another mortar shell lands 100 yards or so to your right. Somebody screams in pain. Corpsmen rush out to his aid.

A Nip sniper bullet whistles overhead and you zig-zag to a hill. You are now several feet higher than when you left the regimental command post. You look around and see jeep-ambulances bounce along their trails. Over to your left, an intelligence party noses around pillboxes and dugouts, hunting for documents.

Suddenly you are aware that the Marine lying a couple of feet above you hasn't moved or said anything. You look at him closely and see a bullet hole in his neck. His name is stamped in bright black ink on the front of his jacket.

Walking further up in a sparsely wooded area, you find yourself among deep ravines and wrecked artillery emplacements. You take your carbine down from your shoulder and carry it at the ready. Too many Marines have been killed by snipers hiding in the holes and cement caves.

Artillery and mortar fire from Marine guns in the rear whish through the air like tissue paper tearing. It makes you feel more comfortable. You arrive at the top of a terrace just as a Nambu machine gun spits, and you see a company of Marines up ahead hit the dirt.

Lying there, you look at your watch and discover you've been walking and dodging for two hours. Twenty minutes later, you decide to move up again. You bend low and half-run, half-walk to a shell hole near a column of tanks. You sit down, take out a cigarette and light up. You've come 800 or 900 yards, and are exhausted. Your nerves are on edge.

Mortar shells begin landing on and around the tanks. You scrounge into the earth, almost digging a cave with your knees and elbows. Later, you peep over the hole. A dozen Marines lie around the tanks. They will never get up.

Moving to the right, you leap from foxhole to foxhole, then make one final dash to a company area. The riflemen and automatic riflemen lie motionless, except for their trigger fingers and arms that vibrate when their weapons fire.

The Japs are 10 yards away. You are at the front.

Sergeant Dick Dashiell, a Marine Corps Combat Correspondent, was formerly with the Associated Press in Charlotte, NC.

Thinking Of You Margie!

When I was 3rd Battalion radio chief on Iwo, I had a radioman at the forward observer post by the name of Ken Curry. (Because I had been overseas longer than him, I used to tease him that I was going to call his girlfriend when I got back to the States, because I knew her phone number. Other times, I said I didn't know it.)

One day while firing, Sweeton, the radioman at headquarters, called and said, "You know we change codes at noon?" I said, "Yes." He said, "Curry has a different code than I do, and we can't fire the guns until we both have the same code numbers to use. Curry wants to talk to you."

I called Curry and he said, "Crerand let's use Margie's (Ken's girl friend) telephone number as the code." I said, "Fine" and told Sweeton. So her number was the code we used to fire the guns.

A few days later, Curry, who was always smiling, came back from the front and with a big grin on his face, said: "Don't you ever tell me again you don't have Margie's phone number."

... By the way, Curry is married to Margie now, and they live in Los Angeles. **Richard B. Crerand, York, PA**

Assault Was Inevitable

General Kuribayashi had 21,000 troops, many hunkered down in bunkers with five foot walls of steel and concrete. They knew they were the front line of their nation's defense, and prepared to die there. Each promised to take at least 10 Americans with them. Eventually, even the General's runner knew that the invasion of Iwo Jima was inevitable.

That inevitable assault was planned for February 19, 1944 to commence at 0900. Occupation of Iwo Jima would eliminate air threats on the Marianas, provide a base for our Navy in Japanese home waters, offer a rescue haven for crippled bombers returning from Japan and give fighter escort for the same planes on their way to target.

Chosen for the role was the Fifth Marine Amphibious Corps under command of Major General Harry Schmidt. It was composed of the 3rd Division under Major General Graves B. Erskine, the 4th Division under Major General Clifton B. Cates and the 5th Division under Major General Keller E. Rockey.

Richard Wheeler in "Iwo Jima" (Lippincott & Crowell, New York) observes how the Japanese were aware that the Marines always got the toughest jobs, hence the formidable United States respect for Iwo Jima ... but the Japanese were hardly intimidated by the Marines. Some even considered the Marines "In spite of their reputation, rather cowardly because they'd been known to surrender when the honorable choice was death."

The 3rd Marine Division with 21,000 men was still 80 miles at sea when Vice Admiral Richmond Kelly Turner in Navy command of the invasion at 0630 ordered: "Land the landing force."

That force was the 4th and 5th Marine Divisions, numbering 42,000 men. The assault called for landing on a 3,500 yard stretch of sandy ash on the east side of the island. The 5th was to occupy the southern half of the beach, nearest Mount Suribachi, and the 4th was to hit the strand beginning at the first airfield (there were three) and extending the hellish distance to where the island gets fatter at the boat basin.

(Continued on page 129)

Remembering The Flag

Weapons Company, 21st Marines, was temporarily ensconced on the north edge of Motoyama Airfield No. 1 awaiting a call to bring up our anti-tank guns on the day the world would remember forever, because of Joe Rosenthal's famous photograph.

Things had not been easy for us because of enemy fire being received from the mountain heights to our rear. Enemy guns on Suribachi had an unobstructed bead on every inch of the airfield and most of the rest of the island.

"There's a flag up there," one of our guys shouted, pointing to the top of the volcano.

We strained to see what flag was flying where none had been before. Since the distance was upwards of a mile, no one could make it out.

As usually happens when someone points, others took it up and soon there were several hundred Marines looking up on those lofty heights. While I strained to make it out, one of the 75mm gun captains went for his field glasses. Our interest was not casual; we'd received a lot of fire from up there during the past several days.

The man with field glasses focused in on the fluttering object and let out a yell ... "It's ours!" Others along the taxiway made the discovery at about the same time, and I've never heard such a shout as then erupted up and down the entire length of the strip. Suribachi had been taken!

We could not know that the most famous photograph of the war had been snapped that day. When I finally got a look at the flag through the borrowed glasses, I couldn't anticipate the emotional reaction I would have upon viewing it again more than 40 years later.

Others, who would later share in the pride of that tattered old flag when they saw it hanging on the wall of the Marine Corps Museum in Washington, DC, most probably had a picture of Rosenthal's photograph in mind. For me, however, it brought back a vision of my comrades who were scattered along the runway of that airfield in 1945, many of whom would not live to see that grand symbol of victory again.

Every Marine who saw the flag atop the mountain that day viewed it from a perspective the rest of the world could never really understand. **Maury T. Williams Jr., Dayton, OH**

If only we could have knocked the top off that Mount Suribachi...and the rest of the hilltops as well. (U.S. Marine Corps Photo)

The Flag Was Inspiration

From our location we were able to see the first flag raising on Mt. Suribachi and it was an inspiring sight that did much for our morale.

We were finally able to cross Motoyama Airfield No. 1 and turned toward Airfield No. 2. Another night was spent in foxholes during which we could hear shrapnel falling all around us from air raids.

As we moved up toward the second airfield, I was in a foxhole looking for my next place to land. As I stuck my head up and laid my left arm across the front edge of the hole to lift myself up, a sniper's bullet went up my left sleeve and out the elbow of my jacket, making only a red streak but not drawing blood. Needless to say, I did not wait for him to take another shot.

Shortly after we came to a small incline. As I started up I suddenly found myself looking into the muzzle of a rifle. I momentarily froze. I could see the uniform of a Japanese soldier behind it, and as my gaze went up the barrel of that rifle, I felt a sudden surge of relief as I saw he had no head.

When we arrived at the edge of the second airfield. I was one of two scouts and two BAR men chosen to lead the way across the field. The other scout was Bill Speiers and the two Bar men were Charles Mains and Kenneth Cross. This was on February 24, 1945.

As we ran across the field, there was hardly any cover, just a few holes made by our Navy's shells. We made it all the way across before we found holes large enough for cover. The two BAR men were in a long trench, one at each end and Bill and I were in a round shell hole about eight feet deep at the bottom with sides tapering enough that we could lay against them and watch over the top.

While we were waiting for our company, just ahead of us on a ridge, we saw a squad setting up a machine gun. They had on dungarees that looked just like ours and we thought at first they were Marines until they swung the gun around and opened up toward our lines. We were able to take the whole squad out and as they sent up new men to man the gun we were able to pick them off. Our BAR men, with one in each end of the trench would fire alternately enabling them to keep the Japs pretty well pinned down. There was a cave just behind the gun. We had a good view of the entrance, and could pick them off when they came out. They finally spotted us and began firing a pattern of mortar shells, each one getting closer. During this time, Bill and I both had a nature's call. He went first in the bottom of the hole while I watched and fired, then I went. Just as I was about finished, a mortar shell struck at the top rear

edge of our hole and showered me with dirt. I pulled up my pants, dirt and all and said let's get the hell out of here. We had no sooner left the hole when a mortar shell struck it dead center.

Since things were getting out of hand, and our company still had not caught up, we pulled back to rejoin them. However, when we reached the ridge where they had been, they were nowhere in sight. As we covered the last few yards, we could see sand kicking up around our feet from the Jap machine gun.

As we went over the ridge and were descending, a machine gun blast from a spider hole caught Charles Mains across the chest and neck, killing him instantly. Before we could figure out what to do next, mortar shells began coming from the Japs and a rocket barrage came from our side catching us right in the middle.

When things let up, Ken Cross was hit in the rear end and my right leg was nearly blown off. My rifle that I had been holding in front of me was blown apart and my pack was ripped but thanks to my mess kit the fragment didn't go through. Bill Speiers was the only one not hit but he did not know how to stop the bleeding of my leg. I was able to get my first aid kit, dump the whole packet of sulfanilamide powder into the wound and apply the bandage with enough pressure to stop the bleeding. Bill went back for help and Ken was able to walk ... so he went back to the aid station.

For the first time, I really felt fear. Before then, there had not been time for that. But now, unarmed and helpless, I was afraid of who would reach me first, the Japs or our own men. A corpsman finally arrived and gave me a shot for the pain and left. Not long after a big husky Marine came up and seeing I could not walk, carried me to a safer place to wait for a stretcher. After about four hours, four stretcher bearers arrived and began to take me back. Every few steps a machine gun would open up and they would drop me and hit the deck. We finally reached an ambulance and they loaded me in. I was lucky to be on the bottom (there were two stretchers on bottom and two hung above), as the man above me was hit by machine gun fire on the way back to the beach and killed.

When I finally was aboard the hospital ship, men were in stretchers lined up in the passageways waiting to get into the operating room. It was around 3 a.m. when they finally got to me. The next thing I knew, it was daylight and the doctor was standing over me with a glass of bourbon. He said "here, drink this, it will make a new man of you." He had been in the operating room all night and after taking that short break, he went right back. **John E. McKinnon**

Holding On By Our Fingertips

By mid-afternoon on D-Day, the base of Motoyama #1 Airfield had been reached by the 23rd Marines (4th Division) and the 25th was already commanding the critical heights above the boat basin. But the 25th reported 45 percent casualties by 1600. All tanks were ashore and the 4th—in the words of Combat Correspondent David Dempsey was "holding on by our fingertips."

More than 1,000 4th Division Marines—that same day—were evacuated to hospital ships and an undetermined number lay dead.

The 5th Division faced Suribachi, "the cone shaped mountain. The ashen terrain on and below it, in the words of Marine Combat Correspondent Keyes Beech ...": as thick with guns as a porcupine has quills."

Here General Kuribayashi decided the Japanese would offer minimum fire on the landing, but open with everything when the 5th made a crowded beach. Death and casualties were immediate. Among them was twice-honored Medal of Honor recipient John Basilone (NY) killed by mortar splinters as he came ashore.

The platoon of Easy Company, 2nd Battalion, 28th Marines led in scaling Suribachi. At 10:31, February 23 it raised the first flag, photographed by Leatherneck Sergeant Louis R. Lowery.

The second flag was raised that afternoon after Lieutenant Colonel Chandler W. Johnson, commanding officer 28th, complained that the first flag was too small. He wanted a bigger one for all to see, and got it from an LST, that flag raising was pictured by Joe Rosenthal.

Would the 3rd Marine Division be needed on Iwo? Advance bombardment of the island had been so heavy and thorough from air and sea, perhaps the 4th and 5th Divisions could handle the whole job.

During first hours of battle on shore, men and officers of the 3rd Division crowded around loudspeakers on board their transports. Radios tuned to frequencies being used by combat teams and aerial observers of the two divisions landed showed there was action aplenty.

On the next day it was verified the 21st would be going ashore. On D + 2, troops of the 21st Marines went over the side. But they were not landed immedi-

(Continued on page 131)

Marines on the sand of Iwo Jima. (Defense Department Photo - Photographer Lou Lowery)

Even Land Sickness On Iwo ... Especially Land Sickness

On February 16, we pulled anchor and headed out to sea from Guam. Shortly after sailing we were advised of our destination and shown maps of the island. Tokyo Rose had been right, we were headed to Iwo.

We arrived the day the invasion was scheduled. We awoke that morning and the ship's guns were shelling the island. The 4th and 5th Divisions went ashore. Our unit went aboard our landing crafts and stayed out in the water all day. We heard there was not enough room on the beach to land or that the tide was too high. We circled in the water from daylight to dark, and that evening returned to our ship and waited.

On the morning of the 20th, we again went out in our landing craft. The water was so rough and the waves so high, you couldn't even see from one craft to the next at times. Men were getting sick and vomiting in the crafts. We kept circling and returned to our ship after dark.

The following morning, we went aboard the landing craft, we circled the entire day, and finally that evening they put us ashore.

I remember the sand was like volcanic ash—dark grayish in color. We started digging in the beach and by the time we got the hole dug and got in, it would cave in. It was just like it was filling up from underneath us. It seemed we were constantly digging.

That first night on the beach I got sick after finally getting off the ship. It was caused from being on solid ground after getting used to the constant rocking aboard ship. They call it land sickness and I was sick all night long.

The next day was spent digging in on the beach and waiting. I can remember the next day because it was the day they took Mt. Suribachi—the volcano at the south end of the island. The day they raised the flag.

I can remember being down below and some of the comments that were made: "Looks like they've taken Mt. Suribachi and the high ground so maybe we're on our way." It didn't turn out that way. After being on the beach for two or two and one-half days we got our orders to move out and take the second airstrip. It was right in the middle of the island.

I remember going up to the airstrip to get ready to make our assault. As the corporal and I were going along he was telling me what to do. He was experienced. He had been in other campaigns.

Shells were coming in and he would say, "Let's get in this hole and get in quick!" We would both jump in and stay a while and he'd say, "Now, let's keep down, keep low. We'll listen to the shells, and when the sound is real short, we'll move out because they'll have us zeroed in."

He'd keep listening and I'd keep listening, not really knowing what I was listening for, and he would say, "OK, let's go! Let's get to another hole!" We jumped and ran to another hole closer to the airstrip. We crawled along until we got in position at the edge of the airstrip. All the men were lined up on this little embankment just at the edge of the airfield.

The word finally came down to move out. It was passed from man to man. The corporal told me, "Now, when you start out across that airfield, don't run straight, run just as hard as you can and zig-zag back and forth. This will make you more difficult for them to zero in on."

The entire line of men on that embankment started running at the same time. I ran just as fast as I could run and the shells were popping all around me and the machine guns were bursting. It was just wild; everything was going on at once.

We learned later our Lieutenant Lane was killed crossing the airstrip. He was hit by a mortar shell and never really located.

After we crossed the airfield, I ran up a little embankment on the other side and came face-to-face with a Jap machine gun nest. We were all surprised. I began firing my weapon, a BAR (Browning automatic rifle), and I fell back down the embankment into a hole. I looked over my shoulder and saw hand grenades coming over the top, so I just laid there as still as I could, hoping I could get loose after they exploded.

I know they went off, but I don't know what really happened. It was kind of like a dream. I laid there for a while, then began to wiggle to see if I was all together. I jumped out of that hole and found another one where a friend of mine was digging in.

We laid there quite a while because the Japs had us pinned down pretty good. I looked off to the right and there was some hand-to-hand combat going on in another hole. A Jap was swinging a saber at a Marine. The Marine reached out and grabbed it with his bare hand and took it away from the Jap. He then killed the Jap and came over and fell in our hole. His hand was badly damaged so we patched it up the best we could and sent him back to the corpsmen. Later, when we got back to Guam, this same Marine found our camp to see if anybody had picked up the saber. One of our men had brought it back and the boy got to take it home with him.

Still in the hole, I looked around and saw a bayonet fight going on just above us. This Jap had put his bayonet through this Marine's collar bone, but the Marine managed to kill the Jap. The boy then came over and got in our hole and we patched him up and sent him back to the aid station.

That evening we were told to hold our positions at all costs. We dug in and there was fighting off and on all night. A lot of our men were getting wounded, pulled out, and new troops brought in. They brought us ammunition all night long. We held the hill that night and fighting continued for several more days. After securing the airstrip, we were then to try and advance forward. It was inch by inch, foot by foot. In a whole day we would make only a few feet and then be pinned down for some time. Our assault plan was a different objective each day.

During one of these assaults, five of us got in front of everyone and were pinned down by two or three machine guns. Three Marines, a sergeant and I were pinned in a shell hole. I remember a Jap charging at us. We were all firing at him and he just kept coming. Our rounds were going right through him and hitting the ground on the other side. But he kept running and running. Finally, we got him down.

We were pinned down until late that evening and not knowing if anyone would find us or not; we made a plan. We decided to wait until after dark, then try to work our way back to our own lines, hoping we wouldn't be mistaken for the enemy and shot. As luck would have it, we were discovered by a runner who went and got a BAR man—a good friend of mine—and brought him to help.

He told us to use a narrow path he knew about and he would pin down one of the machine guns while we ran down it, one at a time. We would break and run, as he kept one of the machine gunners halfway pinned down. But, he couldn't keep them all busy, so they were all shooting and we were running like crazy. Luckily, we all got out alive.

The days sort of run together, but I remember one very sad day for me. I had this close friend I had known a good while. He and I were in separate holes about five yards apart and we were talking back and forth. He would holler something and I'd holler back. I yelled at him once and he didn't say anything. I yelled again and he still didn't answer, so I crawled over to where he was and found him—hit by a sniper. He was just laying there—already gone.

It was a very hard experience and when we did get orders to move out, it was hard leaving him laying in that hole. To remember him, I still have one of our mess kit forks. I'll hold onto and cherish that fork forever.

One area on the island had underground sulphur mines and the surface was very hot. We had to go through those mines on one of our assaults. We were pinned down for several hours in ately because of heavy surf and the heavy Japanese artillery and mortar fire on the beach. Landing boats circled and circled for much of the day ... men got sick ... icy rain and salt spray added discomfort.

Late in the afternoon, the troops were landed on the black sand amidst the junk heap of twisted equipment, weapons, and bodies. Marines of the 21st dug into the slope just beneath the first airfield which the 23rd and 27th Regiments had seized the previous day.

It was into that maelstrom the 21st Marines (commanded by Colonel Harnold J. Withers) were thrust on D + 3 (February 22, Washington's Birthday) to relieve the 23rd Marines of the 4th Division in the center of the line.

First in the attack northward were the 1st Battalion under Lieutenant Colonel Marlowe C. Williams and 2nd Battalion under Lieutenant Colonel Lowell E. English.

The Crisis For The 3rd

Sergeant Alvin Josephy, also a Marine correspondent, described the engagement: "They no sooner relieved the battered 23rd when a storm of enemy fire broke loose around them. For a moment they were stunned. Artillery and mortar fire thundered into their ranks. Machine gun fire cut them down. There was nothing to see but dry, powdery sand, hillocks and humps of it stretching in all directions. The 21st was nearly surrounded and pinned down—out in the open. Every hump hid Japs, firing as fast as they could."

By the end of the second day of the attack, one battalion had made 50 yards, the other a hundred ... "Actually the whole operation was in peril ... and they couldn't flank the enemy. The line ran from coast to coast. Somebody had to go straight through."

The somebody was Lieutenant Colonel Wendell Duplantis and his 3rd Battalion who had just relieved the tattered remnants of the 1st Battalion on the right. His orders were to go through at all costs.

Thunder paved the way from our ships, artillery and aircraft. Then 3/9 moved out. "In the first hour and 45 minutes, Duplantis recalls, "I lost over 500 men, almost half of my command. Other front line battalions were in even worse shape."

(Continued on page 133)

the area and had to dig foxholes and get down in the steaming hot ground. It was nearly unbearable, but there was nothing we could do but just stay put.

We were supposed to be and, in fact, were going north, but in my mind we were going south and I couldn't get my directions straight. We were heading for the enemy, but in my mind I was going back to our lines. It was very confusing and I remember being very careful about where I would shoot. I couldn't tell if I was shooting at my own men or at the enemy. I finally got my bearings after daylight, but that was an awful feeling not knowing what direction to point and shoot.

Men were constantly moved to the front, then back again so they could have breaks. We were down behind a ledge one day and a new lieutenant came in that morning. I never knew his name.

We were getting ready to try to take another position and he kept raising his head and we'd tell him to keep his head down.

He just kept looking up until finally one of the Jap snipers zeroed in and hit him right in the top of the head. He had been with us only a few hours.

We laid there for a little while, and knew we had to get out. There was a little cave to the right that needed to be knocked out before we could go anywhere. I volunteered to take a couple of grenades and knock it out. I crawled on my stomach and skirted around to the mouth of the cave, threw the grenades in and took care of the machine gun nest.

No Mission Impossible

Our captain was Captain Raoul Archembault. One day he was talking on the phone and I was close enough to hear his conversation. We were on assault and I could only hear his side of the conversation, but I heard him say, "Yes, Sir. Yes, Sir. Yes, Sir. All right, Sir. Now, I'll tell you something, Sir. You're back there, and I'm up here, and if you want that done, you come up here where I am, and you and I will do what you want done; but I won't send one man of mine out to do what you are telling me to do."

I don't know what that officer was telling us to do, but he didn't show up to do it; I know that for sure. This made our whole company respect that captain even more because he had our interest at heart. He would not send his men on a mission he didn't feel he could accomplish.

Ka-Bar
WW II

One day we were in the back on a rest period when orders came down that every man would be clean shaven. Well, there were only one or two razors among the whole group, so each of us took our turn and used a dull, dry razor with no soap or anything. I assume the purpose was to make the Japanese think we were fresh troops when we went back to combat. Sounded like a good idea to me.

One of our daily experiences was the "Screaming Willie"—a rocket fired by the Japs about the same time every day at evening time. It had no particular direction that it was aimed, but every evening you would sure start thinking about it.

There was a lot of "combat fatigue" when you get to where your nerves are at their very end. One fellow in our unit hid back behind the lines one day when we were not really in assault. Shells were going over and he would just sit in his foxhole, shivering and shaking. He would say, "Is that one of ours? Is that one of theirs?" We knew he was about to crack. A couple of us went to the platoon sergeant and told him we had better move him out because he was getting ready to crack and if he went back to combat he was liable to get himself, and us, killed.

The sergeant talked to him awhile and sent him to the aid station. From there he went aboard ship and back to Guam. It was nothing to be ashamed of. It's just that sometimes a person has a cracking point they reach.

I also recall the unloading and loading of the ships. The Beachmaster was in charge of directing the Seabees on how and where to load and unload the ship. He had some kind of PA system you could hear a long way off. He hollered all night long. I don't know when he slept.

You sure had to give those Seabees credit for doing what they needed to do with their dozers and equipment to get the airstrip ready for planes. It wasn't too long after we had the airstrip secured that they had it ready for airplanes to come in and land.

After the island was secured and all the fighting completed, our company was put on a burial detail to find, bring in and bury the dead. If we found any of our men who had been left behind, we had to get their personal belongings and anything else they may have had on them. We wrapped the dead in a poncho or shelter half and carried them to the nearest road to be picked up. They were then taken to the cemetery. It was quite a detail to have to go through.

When we found enemy dead, we buried them right where we found them. We did dig holes and we did bury them.

A memorial service was held for the dead that had been buried on the island. I remember going up to the cemetery, listening to the services, and going down through the rows of crosses trying to pick out the ones that were in our particular unit—the ones we may have known and that might have fallen at our sides. **Cecil "Bub" Harvey, 21st Marines**

The Guys Were Proud Of Colonel Carey Randall ... With Guts Enough To Nix A Bad Command Decision

One day when we were supposed to move back on line with other units, Captain Harper sent me to find B Company, determine their location and area so that we could move right in and occupy or pass through their position.

I was returning back to the company when I happened to look over my head and saw one of our Hellcats approaching. It looked like he was coming right at me ... and apparently he was ... because I took off and headed for a big rock, which gave me just time to miss the six 50's that he opened up on me. I was really out-gunned ... one little carbine against six 50-caliber machine guns, so I stayed where I was ... in the cover of that rock.

After we moved into our new position, the high command was planning on launching a night attack, and Colonel Randall was opposed to this. He came up to our A Company lines, looked over the situation. We saw him just as he was leaving ... and he was leaving on the run ... which was a good way to move. Just as he was about half-way out, he lost his canteen. We all hollered at him to tell him he had lost his canteen, but he wasn't interested in canteens, and he made it.

That was a hot spot for anyone to come into, let alone a colonel. I always wanted to thank him for sticking with his decision (no night attack), even though it did cost him his command. He did the right thing, as it turned out. This position was under fire at all times ... every time you moved and exposed yourself, you drew fire.

Our communications wire was always getting cut by mortar fire ... and a young Marine—I don't remember his name—ran wire out of that position a number of times, replacing wire that had been cut. He drew fire every time he strung wire.

One day he came diving into a foxhole about the time enemy fire was closing in on him. He disappeared, and I thought he had been hit and 1st Sergeant Samuels thought he was hit as well. We both crawled out of our holes, much as we didn't like to, and both stuck

our heads over his foxhole about the same time. There he was, sitting down there, eating a can of C-Rations ... unconcerned about the whole war. Sam really jumped on him.

And, the young Marine said, "Aw those SOB's can't hit me," and they never did.

I remember Luke Cantrell. He looked 16 years old ... but I remember he bawled me out one day for setting off a pole charge. "Let us do that, Pappy, you might get yourself hurt." And, I remember him waving at a Jap fire lane, to direct me across ... things that you really, really remember.

I think of our whole A Company group ... man after man, who enter my thoughts at times. People, who I don't think of ordinarily will enter my thoughts as I think about our group as a whole. **Carl Horton, gunnery sergeant, A Company, 9th Marines, Pioneer, CA**

On The Beach

I went ashore on Iwo Jima late on D-Day plus two with the 34th Replacement Draft attached to the 3rd Marine Division.

The weather was rainy and cold; about 45 degrees. The night was relatively quiet and two other Marines and I shared a miserable time in a Jap communication trench. Dawn finally arrived, so the three of us dug a "deluxe foxhole."

Just as we completed it, our platoon leader told me and another Marine to dig in up on the ridge at the south end of Airfield No. 1. With resentment, we complied with this order and left our friend, Jim, behind on the beach.

That night the Japs shelled the beach and Jim was one of those Marines killed by an airburst.

Sometimes what we dislike doing turns out for the best. **Eric S. Ruark, Baltimore, MD**

Among the Americans who served on Iwo Island, **Uncommon Valor was a Common Virtue"**
Admiral Chester W. Nimitz

The combat which continued was described in the earlier history of the division (Infantry Journal Press, 1948) ... "That fight in ankle-deep sand which clutched at men, tripped them and clogged their weapons, will be recalled as one of the most freakish nightmares of the Iwo battle. Led by Archambault (1st Lieutenant Raoul), K Company battled with bayonets, knives, clubbed rifles and entrenching shovels in a screaming, hacking melee that was over in a few minutes with nearly 50 Japanese killed in hand-to-hand combat."

With the numerous casualties encountered it was obvious more units of the 3rd Marine Division would be needed. On D + 4 the 9th Marines (commanded by Colonel Howard Kenyon) were landed. (The 3rd Regiment was to be retained in floating reserve until it was certain it would not be needed ashore).

The 9th Marines' first assignment was to pass through the lines of the 21st Marines at the 2nd Airfield and continue the drive northward.

Progress ... Then Isolation

Bolstered by fire support of the 3rd Tank Battalion and artillery of the 12th Marine Regiment, battalions and companies of the 9th fought for and took hills surrounding the 2nd Airfield—despite sometimes overwhelming fire from Japanese infantry and artillery. The artillery included everything from the highly mobile knee mortars to anti-tank guns, field artillery, and the harassing "Screaming Mimi" shells with devastating explosive power.

The 3rd Division, maintaining its position in the middle of the three division drive northward, frequently leapfrogged the 9th and 21st Regiments. But advances were seldom regiment or even battalion wide. They were company wide ... sometimes even platoon or squad advances ... knocking out pillboxes, machine gun nests, deadly snipers.

In one early morning maneuver, units of both regiments made a surprise night attack. There was progress as company after company moved silently through the maze of Japanese defenses.

But progress caused problems too. When it turned light, some companies found there were enemy machine gunners and riflemen in front of them ... and behind them. They were surrounded.

(Continued on page 135)

133

Uncommon Valor

The morning sun on February 27, 1945 was just beginning to light up the gray skies with golden shafts of sunlight that bounced off the clouds above us. The rough choppy water surrounding the island of Iwo Jima looked almost pleasant.

Four Marines from the 2nd Platoon of A Company, 9th Regiment, 3rd Marine Division shared a gigantic shell hole in the cinder-like volcanic sand: Private First Class Russ Gieseke, Private First Class Dean Cooney, Private First Class T. Clark, and myself, Private First Class George Briede. The crater was 25 feet across and 10 feet deep and it had provided us with a haven for the night. As a further precaution, we had dug individual foxholes in the bottom of the crater.

We had failed to take a heavily fortified high ridge the day before, and were now in various stages of "waking up" before we attacked the ridge again. Sleep had been fitful because of intermittent mortar shells. We were dirty, gritty, smelly, and dog-tired.

The light of day grew brighter and the exchange of shellfire became more intense. Shells were bursting all around us. You could hear the shrapnel whistling through the air and hear it "thud" when hitting something solid. My BAR was set up at the top of the crater and I had to retrieve, test fire and clean it before we attacked that ridge again. I waited until the exchange of shellfire seemed to subside before I scrambled up the sides of the shell hole in ankle deep shifting sand. I reached the top of the crater and cautiously peered over the rim and scanned the terrain in front of me.

Mortar shells were landing some distance away from my position, far enough to make it seem safe to pick up my BAR. I reached over the edge of the shell hole. My hands closed around its blue steel barrel and wooden stock. There was a blinding flash a few yards from where I was kneeling. A mortar shell exploded sending white hot pieces of metal flying in every direction. One of the pieces passed through my face between my eyes and mouth going from the left to right side of my face. It tore away part of my nose, my cheek bones and sinuses. The impact was like getting hit in the face with a baseball bat. I was stunned and tumbled back down to the bottom of the crater.

When I stopped tumbling, I sat up and grabbed my face with both hands. I could feel the shredded flesh, the gaping hole and what was left of what was once my nose and cheekbones. My face was racked with pain, a dull pain, the kind you experience when you are really hurt bad. I could hear Russ Gieske and Dean Cooney yell for a corpsman as they rushed to my side. I could feel the warm blood trickle through my fingers and down my arms as I waited, but I could not see a thing.

The corpsman knelt by my side, I took my hands from my face. The cool air on my wound gave me an idea of just how large it was. The corpsman said "tilt your head back." I did as he ordered and the blood gushed down my throat choking me. I sat straight up, trying not to panic, and let the blood drip down the front of my dungaree blouse. He gave me a shot in the arm, put a pressure bandage over the wound and tied it around my head. He asked if I could walk back to the battalion aid station or needed a stretcher. I decided I could walk the 100 yards.

An aid station had been set up at the airstrip we had taken two days earlier. I needed the medical attention badly, but the 100 yards that lay between me and the aid station was flat, wide open and under heavy rocket and mortar fire. I could stand, I could walk, but I couldn't see. Dean Cooney volunteered to take me back to the aid station. So we started, Cooney leading the way, with one of my hands on his cartridge belt and the other holding my bandaged face.

We ran from one shell hole to the next, in short dashes, bending low to make as small a target as possible. Each step jarred my head and face. Shells fell close to us and bullets whined past us as we ran from cover to cover. Dean would jump into a shell hole when the shelling got close and I would tumble in after him. I was bleeding despite the pressure dressing and I was getting weaker and weaker. We went on running, stumbling, stopping until I could no longer run but just stagger in his wake. I could no longer hold my head up or speak above a whisper. Those 100 yards were turning into the longest in my life.

Cooney took one of my arms and draped it over his shoulder and behind his neck. Holding to that arm he put his other arm around my waist to support my weakening legs. I rested my head on his chest and we staggered on. It seemed like forever. He half dragged and half

carried me the last 25 yards but we finally arrived at the aid station.

As we stood there, trying to catch our breath from our perilous journey across the battlefield, the sounds of the aid station penetrated my consciousness. The moans of the wounded and dying mingled with the frustrated curses of the medics. I could feel people hurrying past us going from one place to another. I stood there being ignored as my blood seeped slowly through my fingers covering both myself and Dean Cooney. I was classified as "walking wounded" and stretcher cases had first priority.

I was too weak to care, but Dean wasn't. In desperation and with what Admiral Nimitz would have called "uncommon valor" Private First Class Cooney, the Iowa farm boy, pulled his .45 revolver from his holster pointed it at the next medic that came near us and said "he's next."

And I was.

I was put on a stretcher and I felt a needle slide into my arm. As the plasma flowed into my veins, I would feel the warmth spread through my cold body. I was put on a stretcher and taken to the beach where I was ferried to a hospital ship.

I was to spend the next 22 months in Navy hospitals in Oakland, California, and Philadelphia, Pennsylvania. Surgeons rebuilt my face and therapists taught me how to read and write braille and touch-type on a regular typewriter. During this period I was too concerned with adjusting to my new life in the shadows and dark to seek out and thank the man who saved my life.

It was many years later, at a Marine reunion in Madison, Wisconsin, that Dean Cooney and I were reunited and I had an opportunity to thank him for his act of "uncommon valor." **Corporal George Briede, A Company, 9th Marines, LaBelle, FL**

134

The Scene:

A Medical Aid Station On Iwo

Iwo Jima, February 26—(Delayed)—One of the busiest spots on Iwo Jima is the battalion aid station. At one 3rd Marine Division station, Chief Pharmacist's Mate William M. Carey of Long Beach, California directs a staff of corpsmen which hardly has a minute's rest as it brings in the wounded on stretchers, administers blood plasma and albumen, soothes the shell-shock cases, probes for bullets and shell fragments.

The station is situated in a revetment a couple of hundred yards behind the front lines. As the doctors and corpsmen go about their tasks, mortar shells and sniper bullets land nearby and whistle shrilly overhead. Occasionally, a fragment strikes a stretcher bearer just as he climbs out of the revetment to go to the aid of the wounded.

Packs of gauze, tape, splints, drugs and antiseptics lie in readiness against sandy hillsides. Ambulance drivers highball their jeeps up to the stretchers and hurry away in a cloud of dust, often with mortar shells pocking the road behind or ahead.

A shell lands on a weapons company command post 100 yards to the rear. Corpsmen dash out, give first aid on the spot, carry or lead in the wounded, help the medical officers give further treatment, and put tape on the wounded which show the nature of the injury, the patient's name and rank and his next of kin. Then the wounded are taken to the regimental aid station.

Coffee and cigarettes comfort the shock cases, who lie huddled up in blankets, cotton wadded in their ears to shut out the noise of gunfire. A corpsman bends over and talks to them in a normal conversational voice.

And still the wounded come—and from all directions—not merely the front.

"I've never seen so many wounded men come in from the rear areas," says Carey. Most of them are from the front lines, of course, but this morning almost half came in from behind us here." **This story was filed by Sergeant Dick Dashiell, a Marine Corps Combat Correspondent of Chapel Hill, NC, formerly of The Associated Press at Charlotte, NC.**

The Long "Mile"

"No liberty, no R&R," those were among the memories of H.P. Garnett, Captain, USMCR, as he made the shuttle from States ... to New Zealand ... to Guadalcanal ... to Bougainville ... back to Guadalcanal ... to Guam ... to Iwo Jima ... back to Guam ... and then to the States—the same route a few other 3rd Marine Division veterans also made.

He joined the division at New River and was assigned to Loco Battery, 4th Battalion under Colonel Bernard Hope Kirk, a real "old China" Marine. **At the war's end, "H.P." studied medicine and is a board certified surgeon, now retired in Kittery Point, ME**

B Company, 9th, now commanded by Lieutenant Jack Leims, was one of the companies who found themselves to be sitting ducks ... every movement bringing deadly fire. Orders were received to withdraw, but movement was impossible. Finally, under cover of darkness, the survivors of B Company were able to draw back, carrying their wounded ... and making return trips for wounded. Sixty-five men had walked up the hill ... only 22 walked down.

Fox Company, 9th, under 1st Lieutenant Wilcie O'Bannon was in an even worse predicament. They too were pinned down by deadly fire from all directions but were not able to withdraw that first night. Star-shell illumination kept the Japs from sneaking closer, but it also prohibited movement by Fox Company as well. Tank support, laboriously brought into the area, eventually helped the beleaguered Marines reach safety.

No Such Thing As Secured Positions

Because of the terrain and the Japanese network of caves and tunnels, organized fighting often broke out well behind Marine lines ... in locations that were considered "secured" and had to be retaken, usually by riflemen because terrain was sometimes impossible for tanks to traverse.

As casualties mounted among front line troops, some volunteers streamed northward to help fill the ranks. Clerks, drivers, and headquarters men moved up from service elements. Almost 2,000 replacements (many not much beyond boot camp) had been brought along by the division to help unload ships. They were sent to rifle platoons and given on-the-job training in how to fire a BAR, throw grenades, and how to "keep your head down and still move forward."

Early in the day of D + 18, a platoon of Able Company, 21st Marines reached the water at the northern end of the island. They experienced casualties at the point but did send a canteen of water back to Lieutenant Colonel Eustace B. Smoak, executive officer of the 21st Regiment. He sent the canteen thru channels to General Erskine and then to General Harry Schmidt with the famous note, "Forwarded for approval, not for consumption."

(Continued on page 137)

Lieutenant Ira Shaw of 2nd Battalion, 9th Marines is treated for a gunshot wound which he received in the 9th Marines Command Post on Iwo. (Official U.S. Marine Corps Photo - taken by S/Sgt. R.R. Robbins).

Twenty Shower Spigots At The Iwo Hilton

The Seabees were ingenious—we all knew that. They built roads, put up buildings and other installations, and performed hundreds of do-it-now miracles—often times under fire. Here is a story of one of their more unusual accomplishments on Iwo Jima.

Our company (A-1-9) was called off the line one day during the fighting on the northern part of the island ... and brought to a point near the center, above the second airfield.

"What's that contraption up ahead?" someone asked. It turned out to be a maxi-shower complex, designed by the Seabees to take advantage of the hot sulfur springs which were in evidence by the steaming vapors coming through the black sand in many spots on the island.

With the Seabee rig, water was pumped from the ground, aerated over a rack arrangement, and dumped into an overhead tank. There were about 20 shower spigots so that a good sized group could shower at one time.

As we reached the location word came through ...

"Okay, men, you can all take a hot shower ... and then we have clean skivvies and socks for you when you're done."

So we stashed our weapons and other gear, took off our clothes and started showering.

Oh, that water was hot ... and being sulfur water, it was hard too. There was hard water soap available, but it didn't help much. In fact, if you soaped yourself too generously, much rinsing and rubbing was required to get it all off. But it was refreshing.

Actually the whole bit was unbelievable. After the past days of combat, fatigue, shock, fear, and misery, that shower did feel good. And so did the clean skivvies and socks.

Where to now? Back to positions on the line, of course. We had been away from the action for less than two hours, but that hot shower and clean clothing provided some much appreciated rejuvenation.

Thanks to the Seabees, a new wrinkle was added to 1945 combat. **Robert Cudworth, Camillus, NY**

Perils Of Night Attack

Attack on Hill 200-Peter: On the morning of February 25, 1945, Captain John B. Clapp, commanding officer of B Company, was issued orders to pass through I Company, 21st Marines and continue the attack to seize Hill 200-Peter.

At about 0830, after a preparatory artillery barrage, "B" Company passed through the lines with the 1st Platoon (commanded by Lieutenant John H. Leims) and 2nd Platoon (commanded by Lieutenant Robert F. Hagamen) in attack and the 3rd Platoon (commanded by Lieutenant Stanley A. Yoders) in reserve. C Company of the 1st Battalion was on the left and a 4th Division unit was on the right.

After about 150 yards advance the company came under intense enemy mortar and machine gun fire, and was forced to hold up the attack and reorganize. Several casualties were suffered, including four key men, Lieutenant John A. Fitzpatrick of Mortars, Lieutenant Yoders of 3rd Platoon, and two machine gun section leaders.

B Company fought back with all the power it had and made very slow—but steady—gains. The Hill was an outstanding piece of commanding ground, having observation of the entire 2nd Airstrip and the beach.

Enemy artillery air bursts made advance terrifically slow and by nightfall only about another 100 yards was gained.

The next morning (February 26) a new plan of attack was in order. The 1st Platoon with two platoons of tanks was

to flank the side of the Hill under the protecting fire of the 2nd Platoon. The hour was 0800 after an artillery preparation. The drive was on ... and under the excellent work of the tanks, the 1st Platoon succeeded in flanking the right side of the Hill.

A total of three caves, four pillboxes, and one anti-aircraft gun were wiped out. A Jap field piece, some 300 to 400 yards to the front, took its toll of six tanks and forced the 1st Platoon to a position some 75 to 100 yards from the base of the hill. For two complete days and one night, the company threw everything it

had on the top of the hill. Resistance was strong but showed signs of weakening.

Orders were issued to have the 2nd Platoon advance up the hill under the protection of flanking fire from the 1st Platoon. Sergeant Albert M. Villa, now in command of the 2nd Platoon, started up the hill under enemy grenade and small arms fire. They (the platoon) succeeded in reaching the top and destroying about four emplacements and one anti-aircraft gun. This hill was a beehive of connecting trenches and caves on top with a number of tunnels descending some three stories down. It was no doubt a forward observer's headquarters for enemy artillery. Some 70 casualties, minor and fatal, were suffered in the assault of Hill Peter. Jap dead were numerous ... and even after lines had passed on, quite a few prisoners were captured—hiding deep in their unbelievable tunnels and caves. Firepower was terrific from both sides and uncountable amounts of high explosives were used by both enemy and Marines in a 3-day battle of stamina and courage.

Night Attack: During the days of the 3rd, 4th, and 5th of March (1945) on the island of Iwo Jima, the Japs were making a fanatical stand against the 1st Battalion, 9th Marines in a defense pocket around Position 201-Easy. Day movement was held almost at a standstill. On the morning of the 6th, at about 0230, orders were issued to C and B Companies to execute a night attack starting at 0500.

C Company was on the left and G Company 23rd Marines was on the right (of B Company, 9th). B Company under 2nd Lieutenant Leims prepared to move out with the 1st Platoon (commanded by Lieutenant Donohue) on the right and 2nd Platoon (commanded by Lieutenant Harry L. Mertens) on the left. Orders were to advance about 250 yards, set up a line, and be prepared to work small groups back to destroy enemy caves and emplacements by-passed during the night.

At 0500 the attack was underway and no resistance was encountered for 50 yards. When daylight came, small groups with demolitions worked back over the ground passed thru. Numerous emplacements and large caves were blown up. About six casualties were suffered during this period.

At 1430 B Company was issued orders to advance with G Company, 23rd Marines due east in an endeavor to contact the 21st Marines. C Company was to hold fast, which left B Company's left flank exposed during this assault.

Attack moved rapidly with minor casualties until the company reached Hill 202 Able where it ran into a strong enemy defense. Heavy mortar and machine gun fire was delivered by the enemy, and Lieutenant Donohue and Sergeant David H. Hackett of the 1st Platoon became casualties.

For three hours, B Company fought back with small arms, artillery, and mortars in a desperate attempt to hold the hill in order to establish contact with the 21st Marines. At about 1800, Lieutenant Leims was notified that G Company, 23rd Marines was pulling back to positions previously occupied before jumpoff of this attack. This left the company's right flank exposed. At 1830, B Company marked its flanks with WP (white

phosphorus) grenades and found that the 21st would be unable to reach our left flank. At 1845 the enemy launched a mortar barrage which caused numerous casualties. The situation was desperate and no reinforcements were available.

Orders were issued to B Company to withdraw to positions about 200 yards in rear of Hill 202 Able. This withdrawal was covered by a platoon of Able Company, commanded by Lieutenant William Zimmer. Under cover of darkness some 15 wounded were rescued from the hill and the remainder led back to safety. Though the withdrawal was an unfortunate situation, B Company's 500 yard drive and the blow dealt the enemy on Hill 202-Able enabled the 21st Marines ... who passed through B Company the following morning at 0800 ... to hammer their way into that strongly fortified hill and reach the long sought-for beach.

A total of 23 dead and 20 wounded were suffered by B Company from 0500 to 1930 (on March 6), but a considerable amount of distance was punctured into enemy territory. Within one hour after withdrawal, some 15 wounded were rescued from the hill in total darkness and rushed to the rear for evacuation.

Key men who became casualties in this drive were Lieutenant Owen R. Beede, executive officer; Lieutenant Donohue, 1st Platoon; Sergeant Harold Burns, platoon sergeant, 2nd Platoon; Sergeant Hackett, 1st Platoon; Sergeant Melton, 1st Platoon machine gun section leader; Corporal Charles N. Harris, 1st Platoon squad leader; and Sergeant Robert Pavlovich, 1st Platoon, platoon sergeant. **Signed by Lieutenant John Leims (now deceased)**

This was not the end. There were still pockets of savage Japanese fighting ... but the end was nearer. The 5th Division to the west was experiencing tremendous resistance in its area, so units of the 3rd Division moved there to help.

One reason for the fanatic resistance was because General Kuribayashi had his headquarters in this area. Efforts were made to offer him the opportunity to surrender ... but this was not to happen.

The End Of Resistance

It is not clear how General Kuribayashi died, but according to Bill Ross in "Iwo Jima" (Vanguard Press, NY) the most likely explanation brings up the draconian code of 'Bushido.'

Gathered from fragments of information, interviews and volunteered comments, largely from Japanese enlisted men: "General Kuribayashi died shortly after dawn on Sunday, March 18, after issuing the last order to the remaining trickle of troops to 'go out simultaneously at midnight and attack the enemy until the last.' At the mouth of what the American Japanese personnel stationed on Iwo still call the general's cave, Kuribayashi knelt and bowed three times, facing north toward the Imperial Palace ... begging forgiveness from Emperor Hirohito in a final prayer, the general plunged a hari-kari knife into his abdomen. At that moment a trusted aide, standing over him with a sword, brought it down across his neck. The traditional coup-de-grace severed Kuribayashi's head."

Admiral Nimitz announced the end of organized resistance on March 16, 25 days after the Marines had landed. But sniping and harassing fire continued for several days ... including an attack one night on Army fighter pilots and ground service troops who lay asleep in their tents near Airfield No. 2. Nearly 200 Japs were killed after they were stopped by members of the 5th Division's Pioneer Battalion.

Within three months after the securing of Iwo Jima more than 850 B-29s were saved by emergency landings there. This meant more than 9,000 crew members were saved in 90 days because of the sacrifices of our Marines.

In addition, the fighter planes based on Iwo could now fly support for bombers coming from Guam and Saipan on their runs to Japan.

In summary, Secretary of Navy James Forrestal stated, "Iwo—when we got to it—was the most heavily fortified island in the world.

King For A Day

Gaylon Souvignier of Canton, SD and I had the shortest terms of company commander on Iwo, you ever imagined.

One afternoon, our third day on Iwo, Bulley Fowler was shot in the leg, and Souvignier became company commander, A Company, 9th Marines. We were assigned to take a large hill with gun emplacements ... an objective for two or three days. We took off in an attack the following morning, but did not get far. Half of Japan was in that hillside shooting at us. We were pinned down, tried to treat the wounded, and had trouble ... everytime you wiggled back to help someone, snipers were taking potshots.

Finally, came the attack order again for the hill, and it included a lot of artillery, followed by smoke. Assumption being that we could move up on the hill during the smoke.

Suddenly, a message came over the walkie-talkie, "You are Drum Red Six (which meant company commander)." "Who says so? Where's Souvignier?" "He's been shot." I knew there were at least two other lieutenants who should be there ahead of me, so I asked, "Where are Lindlbad and Glass?" "They've both been killed. What do you want to do?" I said, "I'll call you right back."

Didn't take me to long to realize what to do ... I wasn't calling the battalion commander to say, "We aren't ready yet." So, called the company command post, saying, "Follow original orders. We'll move out after the artillery has finished and the smoke starts to hit the hill."

And it did work. We got up on the hill before the smoke had lifted, and surprised the Japs. When they came out of their holes, there we were. We had a pretty good little skirmish that afternoon ... several good skirmishes, in fact.

I remember Blas was up on that hill, and I thought, "Where did Blas go?" I went around a rock looking for him, and there he was—coming back. Talk about a red-blooded Indian boy ... he was. He had been hit by grenades, and was bleeding from head to toe. I don't know how he made it out of there on his feet. Shock, I guess. Because they carried him back by stretcher from there.

Jap tanks fired at us ... kept us busy all night. But by morning, things were pretty quiet. We had the hill and another outfit (from the 21st Marines) passed through us ... and moved forward.

The hill we had taken was honeycombed with caves and several levels or stories of Japanese activity. About five days after we left there, the regimental headquarters had moved up to that location and the regimental command post

was established near there. At that time, the Japanese decided to come out again from within the hill ... and made things a bit unpleasant for regimental headquarters for a time. **Robert Cudworth, Camillus, NY**

Every Day Was Hell

Iwo was something else ... the mess on the beach with equipment of all kinds, the deep black sand so difficult to walk in, the many causalities on the way to the front. That first night, Baranowski and I were beside each other, and a red ball of fire appeared at almost eye level. We both ducked. It was a shell and it hit a tank at a lower level behind us at the first airstrip.

Two days later as I jumped into a shell hole with three others, Baranowski came running. He thought the hole too small for one more. He hit the deck, and enemy fire hit him almost as quick as he hit the ground. We pulled him into the hole with us. He died in my arms. We put his dog tag in his mouth, and put his body up on the rim of the shell hole.

At about that same time, the next hole to us took a direct hit. Hennessy, Downey, Meek, and Lieutenant Lindblad were in that hole, and body parts fell all around us. Downey was the only one not killed instantly ... and he died on the way back to the Aid Station. It was our own naval gunfire that hit them. I actually saw a shell come from behind us. Sometimes you see things in the corner of your eye that you might otherwise never see.

A short time later, we were able to move from our pinned down position when an artillery barrage, followed by smoke, was plastered on Hill Peter.

As we charged up the hill, we were getting mortar fire from the large Jap gun revetment ahead of us. It was out of reach for hand grenades, so we were firing at the undercurve or bottom of the gun to provide ricochets down into the revetment. This allowed us to get closer so that our grenades and a flamethrower could go to work.

The skirmishes continued as we received fire from machine guns and occasional grenades. Inside the hill, ammunition in the big gun hole was grumbling and several small explosions occurred. I thought the whole hill was going to blow up, so I moved back into a hole with Russ Gieseke. Right then, one hell of an explosion took place as the ammo in the big gun revetment went up. About two inches of dirt fell on us.

Everything was very quiet for a moment. Then we heard Japs talking very close to us over the rim of the hole we were in. We pulled pins on hand grenades and let the spoons fly, counted three, and rolled the grenades over the edge. There were no more Jap voices.

After we had secured the hill, word came to straighten up our lines. We dug in at the foot of a hill, saving top soil to put in the bottom of the foxhole for insulation. Sulfur was burning underground, and you could heat your C-rations inside your foxhole.

During our skirmishing on the top of the hill, our bazookaman had knocked out one tank and disabled another out in front of us. That disabled one was able to go around in circles on one track, and he seemed to do that all night long.

Wounding Paralleled

When we went into attack on the morning of March 6, I had hardly left my foxhole when I was hit in the neck with shrapnel from a mortar shell. It knocked me out, and when I came to I was bleeding fairly good. Another piece of shrapnel hit my canteen on my right hip. This was one of the older galvanized canteens. I went to the aid station and was sent to B Med where a corpsman pointed at and asked for my canteen. This was the first I knew I had been hit on the hip.

Ten hours later I was back in the hospital at Guam between clean sheets. I had heard that Cork had been hit in the neck also by a Jap bullet and was paralyzed from the neck down.

We stood at the door of that Guam hospital when a new load of wounded came in. Can you believe, here came Cork on a stretcher and able to lift his arm when I asked, "Is that you, Cork?" (Charles Cork was a machine gunner in A Company from Birmingham, Alabama).

Since we had been two in a foxhole all the time in combat, it seemed okay while visiting to stretch out on a guy's bunk. No more had I put my feet up on Cork's bunk and lay back than the nurse ward supervisor came in. She took a quick second look and yelled, "What are you doing two in a bunk? Don't you know that is a court martial offense?" A good laugh for most of the other patients. I was more than a little teed off.

Because of a knee injury, as well as the neck, I was sent to the States. Cork bet he would beat me back to the United States. He was at San Diego when I got there. **Clarence Brookes, Cortland, NY**

That Dad-Gummed Black Sand

We landed on Iwo Jima in the afternoon, nearly evening. A lot of our supplies were covered with that dadgummed sand (black as tar). We were used to the white sand ... the coral sand of the South Pacific ... the beautiful islands.

About the first thing we did was to go through the 27th Marines, to take up their positions.

The next morning each platoon had a tank that it was supposed to support ... and the tanks were commencing to draw rocket fire. Some of our men were getting wounded, because of the fire attracted to the dad-gummed tank. The tankers kept hollering, "stay up close to us, stay up close." And, we were getting as far away from them as we could because of the rocket and mortar shells falling around the tanks. That was the only time they attached a tank to each platoon ... a detriment to us rather than being a help.

We were scattered across the airfield that night (Airfield No. 1) ... and the Japs were shelling us with rockets. First squad of our platoon was practically all wiped out that night. Sergeant Benedetto, and maybe one other, were about all that were left. Then when he had to be evacuated, it meant as a platoon sergeant I was down to two squads.

Next day as we moved out, we encountered intense machine gun fire from our front. We were constantly pinned down, making little progress. That evening, a little before sundown, I looked back about 300 yards and saw our company commander (Bulley Fowler) running. A machine gun raked across, and I saw him just fold up and roll into a shell hole. I grabbed a corpsman and we ran back, right quick. He was lying there with his pack off, under his head. While the corpsman worked on his leg, we visited. "It looks like you'll be going back, captain."

And he said, "Yes Strode, but I'd rather stay and go back the normal way with all the rest of you." He didn't know at the time that was the way all the rest of us were coming back.

We spent the night under cover, and the next day it was more of the same ... couldn't move. We advanced a little bit—with the mortars and machine guns trying to give us cover—but every time we moved those Jap machine guns were spraying constantly.

Somebody said, "If they could put some smoke on that hill so they couldn't see us coming, maybe we could get there." So Lieutenant Cudworth got on his radio (walkie-talkie) and requested a barrage be laid down with a few minutes of smoke. The request was confirmed, and when we knew the last shell had fallen, we were up on our feet and a-running.

Oh, For a BAR!

We got to the top of one little sand pile ... I happened to be in the lead ... and a Jap popped up out of a hole with a hand grenade. The other guys were right square behind me. If I had my old Browning Automatic Rifle, or M-1, like I had in the other campaigns, that would have been the last of him. But I had that little old carbine ... might just as well had a pocketful of rocks. I whirled around, and told the guys to "spread out, there was a grenade coming." About that time, that grenade went off on my back. He had just counted his time, tossed it out there, and it knocked me off the that hill. That gave the other guys time to get scattered.

Blas (Corporal Cipriano Blas) told me later, when on board ship together after both being wounded, that "J.R. knocked him out, and as soon as you were out of the way, I smoked him out, Jay."

I have thought since that time ... if I had folded up and thrown the carbine at that guy I could have knocked him over. I was a pretty good shot with my rifle ... but I felt completely helpless with that carbine.

When the grenade hit me, it just knocked the stuffing out of me, and I fell on my face. I got up on my hands and knees trying to get my breath, and told myself you better get in that shell hole there. So I kind of rolled and fell, and got into the shell hole, and passed out—I guess.

The next thing I knew someone was packing me. (Back in training, we always taught people not to walk in step when carrying a guy on a stretcher ... stay out of step ... that way it was easier on the person being carried. These guys carrying me were a-pounding me, I was really hurting. I woke up and said, "Get out of step back there, you guys." Old Silver Bill Brady spoke up, "Strode, I can't, I'm by myself."

So, I said, "All right, keep a-going." When we got back to company headquarters ... and why I did it, I don't know ... I said, "I want to stop and talk to Carl Horton." They set me down on the ground and Carl came over and I said, "I guess it's all yours, I'm all done."

The Journey Back

They shoved me into the back end of a jeep, but I don"t remember another thing until I woke up and was on one of the barges, laying in the middle of about 50 men. Next time I remember anything, I felt like I was a-swinging. I woke up, and I was swinging ... I was in a wire basket being lifted from the barge onto the deck of a ship.

Eventually I woke up and found myself in a bed in sick bay - only half a dozen of us in this little hospital operating room. This little corpsman, a black-haired fellow who took real good care of me, said, "Dad, I'm glad to see you back. I didn't know whether you were going to be with us or not."

I don't know how long I was there ... time didn't mean much ... when they moved me into the chief's quarters, where somebody said something. Right next to me, here laid Blas. We started talking, and the doctor said the wounds on Blas' hands and legs were full of sand ... and he started pulling off the scabs, and literally peeled that man alive. Blas just laid there and looked at me, and we visited like he was going to church. I never saw the like. And, actually when he peeled those scabs, there was just the raw flesh underneath ... that was all you could see ... but he never uttered a sound. We just went to visiting like we were back in camp.

Sometime along here, Sergeant Albright from the 1st Platoon came to see Blas and I. Albright was just as touchy as a man could be. He hadn't been wounded, but had lost control of his nerves (combat fatigue). He acted like someone was going to step around a bunk and shoot at him. He was watching everything. I never felt as sorry for anyone in my life. I think it was easier for Blas and myself than it was for guys wounded like that.

Eventually we landed in the Marshalls, stayed there all night, and the next day they loaded me on one of those little Marine gooney bird mail planes, in one of those wire baskets. They packed someone else right alongside—head to foot—and who should it be but Parrie ... W.J. Parrie. We were taken to an Army Hospital, someplace else in the Marshalls, for a night or two ... and I never saw Parrie again for another 40 years.

God bless all of you ... I love you just the same as I did the day we landed on Iwo. We all know how that feeling is, and we'll never get over it. God willing, I'll see you all again. **Jay Strode, Jerome, ID, A Company, 9th Marines**

Marines and tanks advance on the Japs during one of the March 1945 battles for possession of Iwo Jima. (Defense Department-Marine Corps Photo)

Two Marines in hole hastily dug from smoking sulphur rock (note mine and refinery in background) stand ready to repel Japanese snipers, many of them in Marine uniform. (l. to r.) Pvt. Robert K. Marshall and Cpl. Allen L. Griffin. (Defense Department-Marine Corps Photo)

One Little Epic On Iwo Jima

I recall our landing, and was so amazed to find the coral ash was so much different than I had imagined. We pushed forward in reserve, but a short time later were on the front lines.

The evening we landed I got word that an officer of B Company was killed ... so Bill McCrory (our executive officer) was sent to replace him. A short time later Captain Bill had also been killed.

On the next day, February 26, we were moving our line forward when word came on the radio from Colonel Carey Randall that Bulley Fowler (our company commander) had been injured, and that I was to take charge of A Company. I felt a load of hay had fallen on my head. That was a very long day, and a very long night.

Our orders from Battalion Commander Randall on the following morning were to keep moving forward, and prepare to attack ... while receiving sporadic fire from the fortified hill ahead.

We had some artillery support, and kept mortar fire and machine guns firing on the hill to keep the enemy pinned down, and to forestall their attacking. But, we were getting a lot of machine gun fire from our right flank, and losing many men. One of our platoons in the advance was pinned down by that machine gun from the right flank.

I was trying to direct some of our men to work around and knock out that gun (or guns) ... using field glasses to pick out the fire from this gun, and directing by hand signals. I believe when I was wounded I got in the way of that same gun.

That evening, I was evacuated on a hospital ship, sent back to Guam, then Pearl Harbor, and later on Great Lakes Hospital.

Two or three years later, after I was out of the service, I received a letter from the Marine Corps asking me to return my field glasses. I wrote back that they were on Hill Peter, Iwo Jima ... That is the last I heard about my field glasses. Gaylon Souvignier, former A company, 9th Marines commander on Iwo Jima, lives with his wife Dorothy in Canton, South Dakota, where he runs a cattle auctioneer business and livestock yard.

"Extra Metal" For A Jasco Man

I was shore beach party attached to the 3rd Regiment in the Guam Campaign and we landed right at the base of Chonito Cliff.

All hell had begun—mortars, direct fire from small arms—and we had to set up our TBX radio (the hand generator type for transmission) to establish communications with the floating contingents. The beachmaster also used our facilities when he got ashore. Fortunately, we had learned from Tarawa and were put ashore in "alligators."

The first several waves were pinned on the beach and we had several casualties. An unforgettable moment was when a colonel (later identified as Colonel DeZayas) came charging up the beach to rally the troops to "move out, get off the beach."

For three days we manned our TBX radio as the communication link to the ships off-shore, and direct Japanese artillery and mortar rained down on our beach parties. On the third night a banzai attack reached clear to the beach, and all hands had to revert to riflemen.

I was ordered to the 1st Battalion under Lieutenant Colonel Marlowe Williams to provide the command post with a wire party to keep telephone communications moving forward. At one point we "borrowed" a reel of telephone wire from an army unit which joined our right flank. (Today Chonito Cliff is gone. It was bulldozed for the road from Agana to the Naval base.)

At Iwo when our shore beach party mission was completed, I was assigned to replace the battalion commander's radio operator at 1st Battalion, 21st Marines ... now commanded by Major Robert Houser.

One unforgettable night in March, with only a handful of troops left, I had not had time to dig much of a hole, but I secured my radio and spread my blanket. Suddenly, a charge with much small arms and Jap grenades—looking like lighted cigarettes—came at us. In close combat the 30 caliber carbine was not as effective as the .45 Colt revolver which I had bought from a Seabee ... a most propitious investment.

The battery on my radio was holed by grenade fragments, my blanket shredded, and my right knee was injured so I cold hardly walk. At daylight, the commanding officer sent a runner to the rear command post for a replacement battery.

We started to receive replacements one day—50 to 60 new people. All of a sudden heavy shelling began almost on top of our troops. Fortunately, I had the frequency of Corps Artillery—there was no mistaking it was our own artillery—and passed the word about what was happening. The shelling ceased; I don't think we had casualties.

The commanding officer sent me back to the rear command post as the injury on the knee reduced my mobility. When we returned to Guam, it was off to the Naval Hospital for knee repair—removal of grenade fragments and damaged cartilage.

After I retired in 1983, the knee was replaced. It wasn't until recently that modern X-rays showed a Japanese grenade pellet has resided in my back since that night. I imagine I'm one of many old Marines from those days with "extra metal." **Robert M. Hayter, Port Ludlow, WA**

THE GODS OF WAR

The gods of war have few friends;
but the gods of liberty must have their heroes,
tho soon forgot by lesser men.

So while the Japs raged, the bands played.
The politicians brayed, and the brass prayed.
Old Mac strayed, and Wainwright paid.
But the grunts stayed and won the damned war.

Wm. D. Corathers

Communication—Offshore And In The Air

Some line Marines even to this day are not aware that the Marines who dragged ashore the big tarp-wrapped electronic and radio consoles, while machine gun bullets splattered in the water, were specialists who brought down the thunder of heavy sea guns and aircraft on positions the Japs were foolish enough to maintain.

This is the excerpted story of one such Marine with JASCO (Joint Assault Signal Company), Lieutenant Fred M. Renfro, Fort Wingate, New Mexico.

When a salvo from USS *Utah* struck Suribachi, smoke would ensue and rocks would erupt ... but overall the effect seemed like breaking a window with a peashooter. Assault battalions had clearly landed ... on the beaches ... but a pall or apathy seemed to have occurred. Nothing seemed to be happening. A small white plane flew haphazardly about. This was the aerial observer, and he was shot down as we watched. Another plane soon replaced it. Little talk or noise generated from the troops aboard our LSP (Landing Ship Personnel). I was reminded of a song my mother used to play, *Just Before the Battle* from Civil War days. I resolved not to think of the subject anymore.

When the ramp lowered and I stepped onto the island, my maps, briefcase, and the other six thousand things I carried rendered me as agile as the slowest of turtles.

"Puff, whiff, whiff, puff, psst, psst ... at two second intervals a speeding trajectory was cleaving the identical space I was in ... the constant enemy fire was ineffective as long as I kept down. I was certainly willing.

I laboriously topped the terrace, ran toward a monstrous shell hole and dived in. A mass of humanity occupied the 15 foot deep crater (made by a 500 pound bomb). Before I had time to right myself, an explosion roared and dirty ash and smoke splattered blindly into the hole. That the last shell missed the hole was no proof that the next shell would follow suit.

So, began a journey across 500 yards separating the beach from the airfield. The sprint from shell hole to shell hole rarely exceeded 15 yards ... rapid movement in the deep, fine black sand at every step was legion and automatic. Life assumed a more permanent perspective below ground level than above it. Rarely was a living thing seen during the short intervals on the surface.

The beach was havoc ... broken, overturned, uplifted, crushed landing boats ... no living thing was discernable. Occasionally, a deadly stalk and kill duel was staged between the boats and shell splashes seeking a target.

For whatever reason, Halp (unidentified) stuck to me on that perilous, tragic, terror stricken jaunt toward the airstrip. Wherever you are, old Halp—thanks from the bottom of my heart for going with me.

Into the smoking hole we dived to find two dead men and one badly wounded. "Corpsman," I yelled. "When their eyes look like that, Lieutenant, nobody can do them any good." "Goddamit," I yelled, "Get the corpsman, pronto," and I cradled the fellow's head in my arms. That damn Halp actually came back with the corpsman.

"He's gone ... there's nothing I can do for him ... it's just a matter of minutes or seconds," the corpsman said, " ... see, lieutenant, he's already dead." sure enough, the Marine had imperceptibly given way to his life there in my arms. We gently laid him in the bottom of the hole and figured our next move.

We enacted a similar drama that day in a cave past the airstrip. I did not need a corpsman to corroborate the helplessness of a person who stared peaceably, unseeingly after having been badly wounded. Both of those fellows died looking peaceful.

The going was not as bad as that damn beach trip had been. We could tell places that were not under direct fire by the calmness of the people there. In one area, the major was talking to five men. "You men can't stay here indefinitely. If it gets dark and we haven't a perimeter, the Japs will kill us all tonight."

The major went on with his reasoning, but his audience refused to look him in the eye. He acknowledged my arrival gratefully ... a junior officer jumps and runs to his death when a superior snaps his finger. Enlisted men can feign deafness.

In another area, I crawled over to the major and asked if they could use an air strike. Such a possibility had not occurred to them. Finally, he said, "You're damn tootin' we could use one! How soon? ..."

"Five or ten minutes," I said, "if everything works right." "Go back and set up, and I'll send you an order..."

A rocket strike had been specified at a rise 60 yards in front of us, a little to the left. I checked my map's grid coordinates and told the sergeant to send the message. One has faith in one's pilots, but knowing those potent rockets would be bounding around shortly, one is more interested in cover than an observation point.

We heard the planes roar over us, but the sound of the rockets blended completely into the ever constant noise of battle. So I didn't know how the strike went ... I went over the taciturn major and he was jubilant. "I'll send you another message," he said.

The formal message written with pencil on a message center sheet came from the major, calling for a rocket strike but directly to our front and closer. I was busy finding firm protection between me and the direction from which the planes would come, when the sergeant waved. "Some difficulty in making contact," I thought as I moved toward him.

"We are talking to Charley Baker, but they won't send an air strike ..." "Able Dog Three calling Charlie Baker One ... Over," I yelled into the mike I had grabbed from the sergeant ...

"Charley Baker One" came the faint answer. "Proceed with rocket strike immediately ... Over, " I said, loudly and distinctly. "Planes cannot fly after sundown, Over ..." "You goddam bastards. Japs are all over, shooting hell out of us. Their banzai attack will start any time now. Get those son-ov-a-bitching planes up here. You're all we've got to throw at the bastards. Pronto."

After a brief pause ... "Sorry, Charley Baker is securing ... Over and out."

Lieutenant Fred Renfro died in May 1990. His son Stanley found this account among his papers and sent it to us.

JASCO was a "service" force that supported all divisions and all campaigns. This is printed to give a typical insight into the critical role of these sea-air-ground communications specialists.

Gunboats send up a barrage of rockets in support of the landing of the Marines. (Official U.S. Navy Photo)

Once Marines had reached the top of Mount Suribachi, it was obvious why Suribachi and the high ground at the northern end of the island made the beaches such hot-spots for troops landing on Iwo Jima. The lone Marine in this photo covers the flank of a patrol. (Defense Department Photo-Photographer Lou Lowery, Leatherneck Magazine).

What A Place For Memories!!

After 45 years I still retain some of my memories of Iwo as a platoon sergeant in A Company, 9th Marines. I remember the day we went in, and the beach was so littered with landing craft, LCI's, LST's, all kinds of gear, and everything that had been bombed and beached. I remember the destroyers and rocket boats laying up close to shore, just pounding different targets. When we got ashore, I remember the sand that you could hardly walk in.

Remember the morning we went up across Airfield No. 2, and seeing seven or eight tanks there, in a row, that had been knocked out ... seemed like they were all in a straight line ... wondered why they had ever got knocked out like that.

I remember up past Airfield No. 2, the three gun emplacements up there on a hill in a half circle. They looked to me like they were 155s. Every once in a while you would see Japs in there with them. Had spider traps everywhere, and you wondered why the guns had never been knocked out. I don't believe the pounding by the Navy and aircraft ever touched them.

As we went across the field toward Hill Peter ... and all the naval and artillery fire that was supposed to precede us ... I remember the SBD that evidently got in the line of fire and was shot down. That's where we lost Lieutenant Glass. I believe it was from that shelling. We went on then up to the Ampitheater where we lost Lieutenant Zimmer and Brother Neal from Machine Gun Squad. That night we lost Sergeant Widener. That was some night, I know. The Japs were coming out of the holes everywhere ... a lot of good men were lost there.

Like Pappy (Sergeant Horton, 2nd Platoon) said, "We'd wind up with five or six men in each platoon."

Later on they relieved us, and we came back down for a day or two of rest. We were on Airport No. 1 ... I believe it was No. 1 ... and the Japs got a lucky shot in and got the ammunition dump. I remember the island shook like it was just going to go ahead and sink.

We were still eating our C-rations. We had plenty of good water, and we never were short of food. Had plenty to eat, such as it was.

We all remember the Buzz Bombs ... like 50 gallon drums that the Japs shot out into the air and would go all the way down to Airfield No. 1. What a noise they made.

We must remember all of our lieutenants and we can't say enough about Conrad (Captain Conrad Fowler) until he got hit and had to go on. We'll always remember Lieutenant Cudworth, Strode, Pappy, Sergeant Allen, and all the men that were left ... how they looked when we got aboard ship and left the island. Everybody was very thin, eyes sunk back in their head. There's so many things we could talk about, if we remember. **Woodrow Easterling, Moselle, Mississippi**

Heartbreak Patrol ...

After the 4th Marine Division secured their objective and left for Hawaii, units of the 3rd Marine Division moved into a large bivouac area north of the ground vacated. I was with K Company, 21st Regiment.

One night I was sitting on an improvised head located in a large shell hole. As I used this facility a roving patrol, consisting of two Marines, challenged me. I gave the password but, unfortunately, it was the password for the next day. This created a lot of suspicion, and I ended up with an M-1 carbine pointed at my head. Finally I convinced the patrol that I was one of them. We "shot the bull" for a few minutes and one member of the patrol said he had just found out that his brother, a cook, had volunteered to participate in a "mop up operation" and had been killed that day. I expressed my sympathy and the patrol moved on.

The next day I was the second man in a column which I think was operating at the mouth of the gorge where General Kuribayashi made his last stand. We came upon a dead Marine who apparently was responsible for the death of four or five Japs. The Marine in front of me turned around and said, "That's my brother; get him out of there."

Little did I know that the Marine who nearly shot me the night before would be the one to find his brother! **Eric S. Ruark, Baltimore, MD**

Night Attack And Medal Performance

So the 3rd was being employed in its first night attack which was held secret. We were told about the dawn attack and if you wanted to stay behind you could. But, I was more scared to stay behind, so, we stripped off anything that would make any sound. We had complete silence, no artillery, no shots etc. We moved up at night and waited for dawn.

When dawn broke we were mixed in with the Japs, and all hell broke loose. I was shot in the chest and Rabbit Pavlovich was also wounded. All you could hear was a shot and silence. Japs were killing off our wounded. I kept a .38 revolver pointed to my heart in case they came my way. The Japs were taking what they could from the Marines.

That night Lieutenant Jack Leims (who received the Medal of Honor) came about 200 yards up from the rear with a roll of telephone wire on his back so he could find his way back. He was also wounded ... but he grabbed Pavlovich and said he would come back for me. I said, you'll never make it ... Let me put my arm around your cartridge belt ... I'll drag along ..."

When we got back to our lines, I couldn't move my body, 'cause I was also shot in the arm and leg but managed to crawl back with Lieutenant Jack Leims and he with Bob Pavlovich on his shoulders.

Included in the citation of 2nd Lieutenant John H. Leims for action with the 9th Marines on Iwo Jima:" Ordered to withdraw his command after he had joined his forward platoons, he immediately complied, adroitly effecting the withdrawal of this troops without incident. Upon arriving in the rear, he was informed that several casualties had been left at the abandoned ridge position beyond the front lines. Although suffering acutely from the strain and exhaustion of battle, he instantly went forward despite darkness and the slashing fury of hostile machine gun fire, located and carried to safety one seriously wounded Marine, and running the gauntlet of enemy fire for the third time that night, again made his tortuous way into the bullet riddled death trap and rescued another of his wounded men ..."

Tankers To The Rescue ...
The Night Battle In Cushman's Pocket

This is a story of a night battle for high ground on Iwo Jima near the second airfield, and later to be know as Cushman's Pocket. The battle was carried on by the 2nd Battalion, 9th Marines, led by Easy and Fox Companies. The attack was not successful and elements of the companies were pinned down, remaining beside the Japanese for some 36 hours. Lieutenant Wilcie O'Bannon, F Company, commanding officer, received the Navy Cross for his role in this attack and withdrawal.

This material was not prepared originally for the 50th anniversary book, but as a memoir requested by an old friend, Roger Radabaugh of Willmar, Minnesota who came up with C Company 3rd Tank Battalion to rescue the Marines in and around Wilcie. Wilcie now lives in Mariposa, California.

Some of the people whom I wonder about are—a wounded radio operator that was in the same shell hole with me during the whole 36 hour ordeal. Did he live? Where is he today? Been so long that I cannot remember his name. (Radio operators were assigned to assault companies from the battalion communication section, so sometimes we did not know them by name.)

He was shot in the belly at belt level ... could see his diaphragm move up and down as he breathed. I dusted him with sulfa powder two or three times from his first aid kit as well as from mine. At the end of the episode I could smell the infected wound ... guess gangrene had begun to set in. He was carried out on a shelter half at the time you guys in those wonderful tanks arrived and furnished the covering fire for our "strategic withdrawal" ... a Marine unit never retreats, ye know.

Also, I remember an unknown man who exposed himself several times in directing me where to throw hand grenades to silence a couple of machine guns, as well as some snipers. He was in a position a little higher than me and concealed behind a large rock so he could view the situation quite well. However, he did have to expose himself to determine where the grenade targets were. I can't remember his name; what a man!

Then another ... happened just as day was breaking and the Nips had discovered we were among them. Guess they were asleep at that time. Anyway, my runner and I (a rifle company commander always has one or two men along to carry messages for him) had just taken cover behind a large rock 'bout the size of a overstuffed chair, we both heard a thud ... a splat sound. Guess we both knew it was the impact of a bullet in our vicinity. He says, "Are you okay, lieutenant?"

I says, "Yep, how about you?" After a very short delay—guess he was sort of taking a quick inventory—he said, "I'm okay, but my canteen is missing." I immediately recalled sort of a tug of my dungaree jacket under my left armpit. What had really happened was the bullet had taken his canteen off his hip, then gone on through my sleeve. Definitely too close for comfort.

I told him he'd have to make a lost or missing report form along with a requisition for a new canteen and submit it to the company property sergeant. Don't think that went over too well with him. He didn't respond to my wise crack and we both instinctively left that hot spot immediately.

He took cover in a nearby shell hole with one or two others, and I found the very small one with the radio man already there (not wounded at this time). During the next 20 to 30 minutes was one of the hottest fire fights one could imagine. In my opinion we inflicted heavy losses on the opposition, but in the long run we were the losers. I believe we had 67 men along in this attack, and only 19 survived, with most of them being pulled through the escape hatch into your tanks.

As the noise and utter chaos subsided, and in between digging the shell hole deeper and larger to accommodate the two of us, and hastily taking a peek to keep abreast of the situation; I noticed a lone Marine lying out in the open and certainly exposed to gunfire. I saw that he would occasionally move his head very slowly once in a while to see if he might find cover some place. I finally got his attention and signaled for him to get up and run to our shell hole.

Well, he started crawling. I knew that he was going to be discovered, so I yelled and signaled again. By this time he had covered about 15 yards of the total 25 yards he had to go. He got to his feet, started to run, and made a dive for our hole.

As his head and shoulders became visible to me, a shot rang out, and right in front of my eyes—not four feet from my face—I saw dust fly out of his bed roll mounted on his back pack and blood gush out of his lower jaw and jugular area. Of course, death was almost instantaneous. He had fallen draped halfway in our hole where he remained 'til we ran out of there 36 hours later. To this day I don't know his name or where he was from.

Ten minutes later I heard a voice calling, "Lieutenant, where are you?" Also, sounds pounding the ground, some one running. Then a sniper shot rings out. There is a deep grunt ... a moan ... then the sound of a body striking the ground just a few feet outside our position. I heard one further sound from that direction, indicating there was life left. There was "Ski," a big sergeant, 81mm mortar forward observer.

Just as you guys arrived, he jumped in the hole with me and the wounded radio man. Hardly room for the two of us, let alone adding a 220 pound man with a pack and map case on. Anyway Ski's head protruded up too high, and just as I gave the signal for individuals to charge over and pick up the radio man (I had placed him on a shelter half earlier) and run with him, a sniper shot rang out. It struck a corporal in the arm just as he stood up to come and assist in carrying the wounded man. The bullet zipped past my head, then struck big Ski about one inch below his left eye. On being hit he slumped into my lap, bleeding on my feet and clothes, dying in less than two minutes.

During all this time you guys were givin' them hell with machine gun fire to cover our getting out of there. Sure would like to hear the story of the tankers involved in the dramatic rescue of our group. **W.A. O'Bannon (Obie).**

Crossing The Runway

The most peculiar—and certainly terrifying—event I experienced on Iwo Jima came about as a result of a message Major Robert H. Houser, my commander, gave me to deliver to our executive officer on the beach. The message was a written note, not suitable for radio transmission.

I left the company area on the north edge of Airfield No. 1 and made my way to the north-south taxiway (which was adjacent to aircraft revetments on the near side) and skirted Beach Yellow, where we had landed on February 21. I did not anticipate a problem in crossing the taxiway to the far side as enemy fire had not been very heavy there for a day or two.

Just as I prepared to cross the runway, I observed two men running across from the other side, then sprawling on the tarmac when several explosions erupted near them. I judged the enemy fire to be 77mm artillery. Unhurt, they finally made it across to my side.

This was no great thing, I thought. surely the Japs were not taking pot shots at individuals with that big gun! I paused a little before starting across, to get a better feel.

Then, a single Marine started his trip across the strip at a gallop, and was flattened by another near miss that was nearly on target. Amazingly, he got up and made it safely across.

What to do? I reasoned that it was getting late in the day and didn't have time to re-route my crossing further south, toward Suribachi. I waited a little longer, and other individuals tried the crossing with similar results. Luckily the Jap was not a very good shot ... and each Marine made it across ... but I reasoned that for a Jap gunner to shoot at individual Marines—instead of troop concentrations—was getting a little too personal.

I laid on the lee side of a revetment a bit longer; then, holding my breath, took off for the far side. No surprises ... then __!!! Halfway across, an explosion right in front of me knocked me backwards to the tarmac. In an instant another shell struck a little further away. I was bewildered that I was still mobile enough to resume the trip—not knowing that I'd not been hit by shrapnel.

It must have been a queer sight to see this long legged Marine scrambling the rest of the way to the protective side of a sand dune, where two men who had also just made the crossing lay on the sand.

The three of us looked at each other (rather wild eyed, I suspect) and then literally rolled over with unbridled laughter.

The rest of the day went much better ... it had to. **Maury T. Williams Jr., Dayton, OH, Weapons Company, 21st Marines**

Marine Engineers Provide Potable Water On Beach

Remember, when we went to Iwo and there was no water—only ocean water and surface water. So, after a few days, our distillers were in place on the beach, producing potable water from sea water. So, aside from the water on board Navy ships at the start of the operation, that was the source of water for the fighting men on the island.

General Erskine gave us quite a compliment on the achievement.

We also found that if we dug a hole in the ground, warm surface water collected. So we took a three inch pipe and made a shower arrangement ... warm shower ... for whomever came down. We found as we used more water from across the sulphur, that it got hotter and hotter. So, we devised a system to pump ocean water up into a big tank and blend it in with the hot water to provide warm showers.

A lot of the fighting troops didn't know the showers were down there, but the Air Corps people sure did. When they came back off their Tokyo run, they would come down and take their showers, and thought it was just fantastic. **Art Cassaretto, Delhi, CA**

Steam Heat On The Island

Steaming hot spots caused by sulphur springs were a common sight on the island of Iwo Jima, and helped to warm more than one can of C-rations. Here two Marines warm their coffee at a sulphur pit where the pipe was used in a Japanese steam bath. (Defense Department, Marine Corps photo)

My First Dead Marine
February 25, D+6

A/1/9 received orders to move North and pass through the lines of the 21st Regiment. It was here that I saw my first dead Marine.

He was lying on his back, the right side of his face was missing. The remains of a cigarette were locked between two fingers of his right hand. It appeared that during the night this Marine had been standing up smoking and the Japanese had fired at the glow of the cigarette, snuffing out the last puff and the life of the young man.

The Marine apparently fell straight back with his right arm frozen in a bent position indicating he was lowering the cigarette from his mouth when he met his death.

I remember looking at his leggings thinking that when he last laced them he didn't believe he could be killed. I feel certain that he acknowledged the fact that there would be deaths, but he, like the rest of us, felt it would happen to the other guy.

Windage ... Too Far Right
February 26, D+7

On this morning, we went into the attack in an attempt to take the high ground which circled the northern end of Airfield No. 2. As our machine gun squad moved

One Must Never Underestimate the American's Fighting Ability.

Lt. Gen. Tadamichi Kuribayashi

forward, using shell holes and whatever other cover we could find, we received heavy small arms fire, coming from the high ground. Ahead of me, I saw Fred Westwood get hit and go down with ammo boxes flying. It was just after this that I got up and ran for a shell hole a few yards ahead. I had just about made it when my left foot felt like it had been hit by a baseball bat. This caused me to fall into the hole for which I was heading.

I used the sulfa powder and dressing on the wound. I then decided that the hole I was in was too shallow. Using my trenching tool, I started to dig; however, every time I threw out dirt more rifle fire came my way. The digging ended and the frustration began.

When a Japanese mortar barrage ceased, I started making my way back, under small arms fire, to an aid station which I finally reached with the help of

A Hero, Too

On Iwo near the end of the battle, my platoon was dug in on line below a cliff and close to the ocean at the far end of the island. It was a dark night, no moon or stars and very quiet.

Suddenly we could hear a clanking metallic sound in front of our lines. We had no idea what caused the sound nor could we see anything out there. Then a mortar flare popped high in front of us. Caught in the white glare was a lone Jap soldier. The flare obviously had startled him for he froze standing straight up. Immediately my platoon fired at him. What a target, about 50 yards and not moving. Just before the flare burned out, he fell in full view.

The next morning we moved out and found his body. He must have been carrying at least 12 empty water canteens—about all of them had bullet holes. Apparently he volunteered to go through our lines for water for his unit who were hiding in some nearby cave.

To change Admiral Nimitz's words some, "Courage was a common virtue on both sides of the lines." **William I. Pierce, Minnivale, IN 46410**

two walking wounded. I was told later that they never saw anyone move so fast on one leg as I did.

I was moved from the aid station to the beach and then taken out to the USS *Newberry*. While being taken down a companionway of the *Newberry*, someone called my name and when I looked up I saw Westwood waiting for medical treatment. He had also been hit in the left foot. Fortunately, the Japanese who had shot both of us was off "off to the right" on his windage. **Karl Appel, Baywood Park, CA**

Help ... But No Help

H.J. Klimp of St. Simons Island, Georgia won't forget (and hasn't) how landing on Iwo, with flares and small arms all around, he was able to get help from a bulldozer for this truck, stuck in that soft beach sand. Trouble was, the dozer pulled the bumper off his truck—leaving the Marines (E-2-12) to fend the rest of the way inland by themselves ... in the face of little Japanese hospitality.

Gibson Waited For Others But The Shells Didn't Care

When we were at the second airfield on Iwo Jima, some distance to the rear of our 60mm mortar section, stretcher bearers were bringing the wounded for evacuation to the beach. At this position behind a small hill, down a slight ravine, a weasel (one of the vehicles with tracks instead of wheels) and a trailer would take the wounded to the aid station on the beach ... from there the wounded went to the hospital ships.

Among the wounded was a good friend, Norman L. Gibson from Kentucky. He was in very good spirits in spite of being wounded. He told me he really had it made. This was his second wound and having gone through the campaigns of Bougainville and Guam, he was sure to be sent home.

I said goodbye to him and went back to my position. Some time later I went back to the area of the wounded and Gibson was still there. I asked him how come, and he said he thought it best that some of the men with more serious wounds go first.

Later when I went back to the area, I noticed the wounded were really piling up and Gibson was gone. I asked why the weasel wasn't back making more trips. I was told the weasel and trailer had received a direct hit down near the beach.

Gibson, listed as Killed in Action later, must have been part of that load.

There was really no place on Iwo where a man was safe. Men were being killed and wounded in rear areas, as well as up front. But there were certainly heavy casualties among the officers of rifle companies.

Our company (A-1-9) was a typical example. We landed with eight officers, and as the fighting progressed we received two replacement officers. Of these 10 officers, seven were killed and three were wounded.

It was near the second airfield when stretcher bearers came upon our gun position carrying Captain Fowler, our company commander. In spite of being wounded, and heavy action going on around us, he asked to be put down so he could say goodbye to some of his men. Though he was not to be with us anymore, he certainly boosted our morale. **Edward J. Duch, Chicago, IL, A Company, 9th Marines**

Hospital Not Safe From Jap Attack

On Guam, I had been wounded in the leg by a Jap sniper during an attack against Jap positions on "Banzai Hill" while with 1-B-21. On Iwo, I was hit in the foot of the same leg.

My company had been advancing towards an airfield when we halted for a brief rest. Two of my buddies and I sat down on what appeared to be an ordinary pile of rocks. Before we realized we were sitting on a Jap pillbox, the lone Jap occupant hurled a hand grenade out of the opening. My two buddies and I scattered and were not hit by the grenade but fire from the Jap rifle killed one of my friends and badly wounded the other.

A Jap who had the look of a crazy person, charged me, and I pumped eight shots into him before he finally fell, only a couple of feet in front of me. It was at this time I was hit in the right foot but didn't realize it until the close range fight ended.

I was taken to the division hospital located in a tent on the beach. There were about 25-30 Marine casualties here, all on plain cots. It was here I had another narrow escape at the hands of Japs.

Along about 10 that evening, I heard loud voices coming from an enclosed area located directly in back of our hospital tent; this was where some captured Japs were being guarded and they were giving the sentry a hard time.

Suddenly three of these prisoners escaped, and screaming, charged into our hospital tent which was totally dark. The sentry couldn't fire at them for fear of hitting a patient. One of the Japs leaped on my cot; I knocked him off, but he came back and I was able to get a stranglehold around his neck. A Marine in the next cot passed me a knife and I stabbed the Jap once and passed the knife back. This Marine then stabbed the Jap several more times and killed him. The other two Japs were also killed by Marine patients in this hospital tent. **Earl M. Eckert and Glen Allen**

When We Filled The Line Of The 4th Division

John J. Wlach with F/2/21 was one of the first of the 3rd Division to land on Iwo. His battalion relieved the 4th Division in the center of the line.

On our second day on Iwo, February 22, when we relieved the 4th Division, the Japs had manned a defensive line with minefields on either side of the road to the second airfield.

Examining that line disclosed countless thoroughly mutilated bodies ... intestines showing, halves of bodies missing. A very small immaculately khaki clad Jap officer lay on his back. There were no visible wounds, however, a purple countenance and ensuing rigor mortis indicated death was from severe concussion.

To the far right in a stone above ground emplacement appeared a heavy machine gun with broken bodies strewn about. I asked Sergeant McEwen, our company armorer, to remove the bolt (which I still possess) with a serial number 7777.

Crawling about, Corporal H. Ridings and I reached the far left of the line and there, before our eyes, in the above ground emplacement, stood three Jap infantrymen, shoulder to shoulder, with their heads resting back on the close confines of the stone enclosure.

They had chosen to die by holding their rifles to their heads and depressing the triggers with their bare large toes.

The Solution Was Get In Close

Sometimes on Iwo, the rifle companies had to do what could not be accomplished with the heavy stuff— the artillery and the blasting away by tanks.

One example was five Marines of C Company, 21st Marines who destroyed 55 enemy positions in six hours. With Corporal Hershel Williams of West Virginia blasting his way through with flame thrower and TNT ... and the others covering him with their rifles ... the group was able to accomplish close-in what heavier weapons could not reach.

Corporal Williams was awarded the Medal of Honor for his action, and the other four are recognized as heroes as well. **Thomas F. Murphy, Andover, MA**

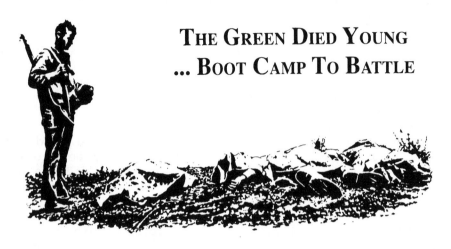

THE GREEN DIED YOUNG
... BOOT CAMP TO BATTLE

Walter R. Gustafson, who grew up in Olympia, Washington, had hardly received his diploma from North Central High School before he was in the Marine Corps. He offers reflections of common experiences in boot camp, and in training in Pendleton when a rattlesnake coiled around his rifle butt. A small guy, he got the BAR, and was thrown into combat early in the fiercest of all ... Iwo Jima.

We got underway with people and relatives waving and a band playing ... then life became dull as we proceeded toward Guam. People were seasick ... weather was hot ... hold was intolerable partly from the guy in the top getting seasick (those stacked below were the beneficiaries of his illness) ... much time spent in the chow lines, three times a day ... But, washing clothes became no problem—you tied them on the end of a line and trolled them behind the ship.

At Eniwetok Atoll troops were allowed to enjoy warm beer and coca colas ... then after getting heat rash and jungle rot, we did arrive at Guam—some six or seven weeks after leaving San Diego.

We ... the troops of the 28th Replacement Draft were anxious to be assigned to one of the rifle regiments of the 3rd Marine Division ... In January 1945 info was passed which substantiated rumors that the 3rd Marine Division would go to Iwo Jima.

The division arrived near Iwo on D-Day, February 19, 1945 and all 28th Replacement Draft would soon be landed to fill the ranks of the 21st and 9th Marines, both with severe losses ... some companies having no more than 15-20 men out of the original 200.

The impression of Iwo Jima from two miles at sea (aboard our troop ship, the Holloway) was of some kind of factory—either building or destroying something. A haze of dust over the island, planes bombing and strafing, Navy ships shelling, artillery from three regiments and supportive artillery—all adding their power to the factory.

The sky was blue, but the island was bare and brown. You could hear the island, the battle noise carried over the sea for miles. The product of this factory was hundreds of DUKW's bringing wounded to ships that had medical facilities ... the hospital ship couldn't do it all.

The landing of our contingent from the Holloway was uneventful as the beach was secure by this time and the MLR was a couple of miles north of Mt. Suribachi.

Ernie (Ernest Gerlach, Spokane, Washington) and I were told to dig in for the night because we were receiving artillery fire ... no place was safe yet. Things were confused and we were ordered to grab a stretcher and carry two cases of C-rations to the front and bring back a wounded Marine. We did this for a few days ... saw dead Japs frequently ... the whole ones were bloated and covered with flies ... they really looked like big dolls ... sometimes we saw only feet with rubber shoes on them—shoes that had a place for the big toe.

The sulphur smell of the island, the steam coming from holes in the fumaroles in the ground, along with the smell of everything else, made the place a brown—yellow asteroid—it had no human or familiar characteristics.

Several men in our unit were wounded or killed carrying stretchers ... you're in the open and fair game for enemy snipers behind the lines ... yet, there were really no lines. I learned the Marine Corps brass were reluctant to use the replacements in front line companies because they were so green—but had little choice ... the survivors learned fast, had luck, or took fewer stupid chances.

What can you do when you're in a deep hole and a mortar round drops in on you? Nothing. You could tell when this happened because it looked like the hole was painted red.

When assigned to a 21st Marine unit, our first night on the line was uneventful. The night was constantly lighted by flares thrown up by company mortars and destroyers lying close to shore ... the popping of flares went on all night, creating a moonscape movie set.

The second night I was doing a bulkhead stare at steam coming from a fumarole a short distance from my hole. It was 3 or 4 in the morning, and very dark.

A figure came toward me through the smoke and I said, "Start talking, Mac" then fired two rounds at it with my M-1. It jammed because I was pulling the trigger too fast. Fortunately we had found a BAR and loaded it with tracers just for the hell of it. So I grabbed it and fired off the whole 20-round clip. The effect was pretty and there was an explosion toward the figure. The sergeant hollered ... Did I know what I was doing? And, did I throw a grenade? I said, "No."

At daylight we went over and found an enemy soldier. I had hit him in each kneecap, and he fell and held a grenade to his chest, committing suicide.

One of the old salts, field stripped him and took his flag which was wrapped around his body ... I was given his wallet and some postcards.

On two successive mornings, I saw something I'll never forget—there was a road of sorts running parallel in back of the front line, and I saw a Recon truck (like a 3/4-ton pickup) with its bed stacked crosswise with stretchers containing Marine bodies. I guess they had been killed the night before, and the truck was taking them to the rear.

Fear came in three packages for me, usually under heavy artillery or mortar fire: 1. Hope that you're killed instantly; 2. Maybe you will get just a small wound; 3. You just want to make it without getting hit. Compounding the problem was that sometimes our own artillery landed almost on top of us.

When I wasn't petrified with fear or otherwise occupied with the campaign, eating took on an important meaning. We had C-rations mostly, with some K's thrown in. C's came in two flavors then—meat and vegetable hash, and meat and beans. I loved the hash and had plenty because some of the troops threw them away. When we received a new C-ration—spaghetti and meatballs—everyone wanted them. If one was issued to me, I'd trade it for two hash. We heated our canned rations in fumaroles after learning to puncture the tops to avoid explosions.

Foxholes were heated by volcanic radiations and felt good on cool nights. One couldn't lie in one place too long though as the bottom of the hole became too warm for comfort ... one small luxury, at least.

Toward the end of March, we heard that some elements of the 4th and 5th Divisions were leaving. Our unit stood down for a day and visited the 3rd Marine Division Cemetery. It dumbfounded me—rows and rows of crosses and Star of David markers. Helmets were removed and men who found buddies buried there cried.

One day we found a United States Army Infantry unit alongside of us and knew we'd leave soon. The only subsequent action before leaving nearly finished us though.

We were exhausted. Three or four of us were in a rather large hole and that night we all fell asleep with no guard on duty. An enemy soldier came up to our hole but, fortunately, an alert Marine a few feet away shot at him. He got away, but 10 minutes later Romano's LMG section saw him on a trail and killed him. They made a lot of noise over this, screaming "Gung Ho" and "Semper Fi" so as to impress the Army troops nearby.

The next morning we left the island. As we walked by the Army, we saw a soldier lying dead alongside a foxhole. We found he had stood up during the night to relieve himself, and that was the wrong thing to do.
Walter R. Gustafson

Effects Of The Dead

The 3rd Division was again given the mission of reserve force. This time for the campaign to capture the Japanese island of Iwo Jima and it turned out that the division played a large part in the campaign. I regret to say that I was ordered to remain at our base camp on Guam in the Regiment's (12th Marines) rear echelon.

Among my duties at base camp, I was to go through the seabags of any of our men killed in action, and even those most heavily wounded and returned to the States. This turned out to be an interesting job, however uncomfortable. I was conscious of prying too far into another's life, of seeing more than it was my business to see. The job was considered essential, though, in order to somewhat relieve potentially greater heartaches for bereaved loved ones.

While actually going through the personal effects, there had to be two officers working together. My partner was Lieutenant Jake Segal, the H&S Battery S-1 personnel officer.

There were always many letters the Marine had received and these had to be carefully studied to determine which were from the wife and which from the girl friend on the side. If the latter, the letters from the wife's rival were simply destroyed. When in doubt as to who was who, all the letters were destroyed. Similarly, pictures were studied and, when in question, destroyed. Resulting small packages were forwarded to "next of kin." Austin Gattis, Washington, DC

"A Big Bullet"

"I suppose you know Bob Glass was killed," said the editor to Paul Brown, legendary coach of the Cleveland Browns and Massilon Ohio football teams? The coach was silent a moment.

"Yes," he finally replied, "And I bet it took a big bullet to stop him."

Bob Glass, a 1st lieutenant with A Company, 9th Marines, was fatally wounded February 27, 1945 during Iwo Jima fighting. Before joining the Marines, he was a running-back for three years with Paul Brown's powerful Massilon Ohio High School football team. He earned third team All-Ohio as a sophomore, and the next two years was first team All-Ohio.

Glass then was recruited by Tulane University where he starred for the Green Wave football team, and earned All-American mentions for his powerful running.

> *More than 76,000 marines and 58 chaplains had taken part in the bloodiest battle of the Pacific War. None would ever be the same. The struggle left its mark on their bodies and minds, and it had scarred their memories with the most violent 36 days of fighting in American history. None of them would ever forget Iwo Jima.*
>
> Father Donald F. Crosby, SJ

Praying Needs No One Religion

One of the early nights on Iwo we had moved further up on the airstrip and set up a defensive position. Three of us were in a hole - Fitzgerald, Isaacs (a Jewish boy from New York City), and myself.

I always carried a rosary of plastic beads on a string that would not rust in the salty atmosphere. Fitzgerald and I took turns saying the rosary... and we taught Isaacs how to say it too.

We told him to count 10 "Hail Marys", then say "Our Father." But Isaacs would say only eight "Hail Marys", or do it wrong some way, and we would caution him, "That's another 'Hail Mary', dammit."

But he learned and we survived the night.

Steve Vajda was in Regimental Weapons Company, 21st Marines, Westwood, NJ.

Nice Doggie

Iwo was an interesting place all right.

The story of the dog (the war dog assigned to the company) that nearly took off my hand was only half told. Damn dog bit me.

That dog could do two things— he could tell if there was someone in front of you. (I asked the trainer if he could tell whether it's a Jap or not ... that's all I care about ...)

"No," was the answer. The trainer said, "He does deliver messages." "Can you send a message for us?" "No, we aren't equipped to do that. It takes two men to work with one dog to send a message. Got to have one man to send the dog out."

That damned dog. I picked up his blanket ... just tossed it over ... I was going to take him to dinner with me. That dog went "Awaahrr" and ran his claws the length of my arm. But I forgave him. "I shouldn't have taken your blanket. That was a silly thing for me to do." I've always been a conciliatory fellow.

I went out the next day (there was no grass alive on Iwo—everything had been killed—grass, trees, and there were a lot of holes). We were looking in an enclosed area, looking for anything that would show up as an open cave or emplacement. I went first because I could go with the dog, I'd be taken care of. As I was walking along, I got off balance and I stepped back a little bit ... and damn if I didn't step on that dog's foot.

Same thing again "Awaahrr, rrr, rrr."

I survived, not because of lack of force on the dog's part ... but because of no teeth. Fortunately, not too far behind was the trainer. I told him, "Get that dog out of here, or I'll kill him." I think I would have, if he had bit me again. What I didn't tell—until this minute—I think I would have shot the trainer too. **James E. Harper, Atlanta, GA**

151

More than one group of Marines were thankful for the presence of chaplains so that at the end of the Iwo Jima hostilities they were able to receive comfort in giving thanks to God. Here Catholic Father Lonergan holds divine services for members of the 1st Battalion, 9th Marines. (Defense Department photo, Marine Corps. Photo by Sergeant R.R. Robbins.)

The Voice Of God

The most memorable experience I had in the service happened not many days into our assault and subsequent capture of the island of Iwo Jima in 1945.

One day while we were pinned down on the island, and it seemed that all hell had broken loose, Marines were being killed or wounded all around me, I heard this powerful voice—which sounded audible say to me—"Son, unless you call upon me, in a matter of moments you are going to be dead."

Well, when you are already scared nearly to death (and who wasn't) and you hear a thunderous voice speak the above words to you, anyone with good sense would conclude ... this must be the voice of God; and I called upon God.

My prayer was, "Oh, Dear God, if you will get me off this island alive, I will serve you and live for you," and that's all I remember praying.

The next thing I knew, I woke up aboard ship and a Navy Corpsman was saying to me, "Well, fellow, I am sure glad to see you coming around. You have been knocked out for 48 hours."

To which I replied, "48 hours, what happened to me?" He said, "I don"t know, they found you unconscious on the island and you still had a pulse, so they brought you aboard ship."

Dear Marine friends, it was about seven months after that before the war ended with Japan, at which time I committed my life to God and was called into the ministry. Now, after 45 years of ministry in God's great army, I am looking forward to one day soon being a participant in making a "beachhead" in heaven. May I invite other ex-Marines, who I know had a similar experience while in action, join me in God's great army. **The Reverend John A. Jackson, Marion, IL, C Company, 9th Marines**

Hey, There's A Jap In The CP

Somewhere in the Pacific (Delayed)—One of the favorite stories of Father Vincent J. Lonergan, USNR, a chaplain of the 3rd Marine Division's 9th Regiment, has to do with the strict military courtesy observed by a Marine during a hot moment on Iwo Jima.

"He saw a Jap in the CP (command post) one night," the chaplain said, "and yelled out 'Hey, there's a Jap in the CP!'"

"Well, the major spoke up and barked 'who said that?' So, Carvino (Marine Private First Class James J. Carvino of Brooklyn, New York) answered with complete dignity, "It is I, Carvino, Sir!"

"The major said, 'I'll see you tomorrow, Carvino!'"

But Carvino, who had seen the Japs by the light of trip flares, accounted for two foes who had been attempting to make their way into the command post.

"I saw the two Japs and opened fire," Carvino said, "One fell but the other rushed past me, in toward the command post, and that was when I said 'hey, there's a Jap in the command post.' I had a BAR, but I was trying to fire so fast, I was only getting one shot out at a time.

"He ran behind a rock where I couldn't see him and then the Jap I had shot got up again, but another Marine finished him off and we went to find his pal. We tossed a couple of flares and saw him, still behind the rock, with a grenade in his hand. I killed him with the BAR."

Did the major "see" Carvino the next day?

"He sure did," Carvino said, "but he wasn't mad anymore. The mistake I had made was in telling the Japs, as well as our own men, where our command post was. But getting the two of them helped my case with the major, I guess."

(This story was copied from a news release that had been prepared by Private George Liapes, a combat correspondent from Berkeley, California and formerly of the San Francisco Chronicle. The copy of the release had been sent to the parents of James Carvino by the United States Marine Corps many years ago.)

Who Says Japs Can't Shoot?

Toward the end of the Iwo Jima Campaign stiff Japanese resistance remained near "Cushman's Pocket." K Company, 21st Marines attacked this area by sending a tank into a draw adjacent to the pocket. When the tank entered, the Japs smoked it. As I moved into the draw a Marine who had been hit in the foot by a grenade came running through the smoke. Another Marine and I removed his legging and shoe, applied pressure near his knee and stopped the blood flow. A Corpsman bandaged the wound and our sergeant told us to take him back to the command post.

After this task was completed, I started to return to the action, when a major ordered me to remain in a foxhole nearby. Later our runner returned and I asked him where the guys were. He replied, "In the same location." I asked the Major if I could return to my squad. Reluctantly he gave me permission, so I started across the open ground toward the draw. When I arrived at the draw, I ran into what was left of my platoon. We were immediately pinned down by sniper fire. I was lying flat on the grounds and my corporal was next to me in a slight depression. All of a sudden

he raised up on his knees and was killed by a sniper.

Our sergeant told us to withdraw about 300 yards to the rear. Most of the men made it in one dash. I was the last man out and hit the deck about half-way back when I felt a shock against my back, but placed no significance to it.

That night when I took my pack off I noticed a bullet hole through the canvas cover for my shovel. Removing the shovel, I found that it had deflected the bullet, thus saving me from being

wounded or killed. I found our later that our runner had been killed apparently by the same sniper, and that our tank had been knocked out. The next day the 9th Marines passed through our lines. One of their tanks was destroyed by a Jap operating the gun in the K Company tank which had been knocked out the previous day.

Forty years later while doing research on the 3rd Marine Division, I came across an account describing this action. **Eric S. Ruark, Baltimore, MD**

REMEMBRANCES

Samoa … observation post on a mountain top … living in villages with natives … hunting wild boar … diving for octopus … wedding ceremony (not his) that lasted three days and nights.

Guam .. was one of 15 "volunteers" chosen for mission that captured a Japanese general along with 25 of his personal troops, women and children. I carried a small Japanese boy on my shoulders with him clinging to my helmet.

Iwo Jima … First night in rain filled foxhole. I dug a hole big enough for two of us stretched out … took off our outer clothes to dry out … Japs attacked us in our skivvies … never took clothes off again.

Inspiring to see that flag go up on Suribachi …

On the hospital ship it was about 3 a.m. before they got to me. The next thing I knew it was daylight and the doctor was standing over me with a glass of bourbon. "Drink this," he said, "It will make a new man out of you …" **John McKinnon, Oakland, MS**

It Was Just A Leg

We got word that one of the officers in H&S Battery had lost a leg and had been evacuated to the hospital on Guam. He was Major Minetree Folkes from Richmond, Virginia. We went to see him right away.

What do you say to someone who's just lost a leg? Do you ignore it? Do you look the other way and talk of anything other than his permanent injury? Should we act as if we were attending a wake or a funeral? The closer we got to his bedside, the more uncertain we were as to how to greet our friend and talk with him.

No sooner had we exchanged "Hellos" than all our questions were answered, however. The atmosphere in that tent ward was actually jovial. We were amazed and blurted out, "How can you be so happy when you've just lost a big part of your body, Minetree?"

"Hey, anyone would sing with joy to get off that hell hole of Iwo alive and only leave a leg there! Besides, look at that guy. He keeps all of us in stitches all the the time! And, he lost both his feet from a land mine." The young fellow Minetree was referring to was an ensign in the Navy Underwater Demolition Team.

Minetree Folkes return to the States, was fitted with a prosthesis, resumed his law practice and, subsequently, ran for political office and won the election. **Austin Gattis, Washington, DC**

Delivering The Goods

The postoffice or any of the package firms could learn a bit from VMR-952 based at the NAS, Agana, Guam. Guy Wirick, one of the flying Marines with them, can vouch how his people would deliver anything: food, ammo, mail—virtually anywhere. This included delivery to the Iwo invasion. As crew chief, Wirick made a mail drop and picked up some hostile small arms fire from the Japanese. **Guy (and Doris) live in Crown Point, Indiana.**

They shall grow not old,
as we that are left grow old.
Age shall not wear them,
nor the years condemn.
At the going down of the sun
and in the morning
We will remember them...

"For The Fallen"
Lawrence Binyon

Relieved By The "Doggies"

After Iwo was secured, we were ready to go back to Guam. An Army unit was due to come on the island to relieve us.

They landed, heard a few shots being fired in anger, and someone said, "Wait a minute. We're not combat— we're occupation troops!" So we had to spend another two or three more weeks ... sealing caves and making sure no Japanese were hiding on the island.

The Army Air Corps fighter plane squadron, newly stationed on the island, had come and set up tents with decks ... even though we were still in foxholes. Fighter planes based on the island could now take off and accompany the B-29 bombers flying over from Guam and Saipan. (This was one of the objectives for capturing Iwo, of course.)

About the second night the pilots were there, some Japanese broke out, went down and threw grenades in all the tents. So, a bunch of Marines were gathered up to corral and beat back the Japanese. A number of the Air Corps people were wounded.

When we were sealing caves, we were assigned a war dog to help us. He stayed right in our company area. This was a mean dog ... I think it was because he ate with us, and we fed him the C-ration stew. No wonder he stayed mean!

Captain Jim Harper liked dogs, and was gradually making friends with this one. Each day he would get a little closer, a little closer ... to the point where he thought the dog might really be a pet. The next day, the dog almost took his arm off.

It finally was near time to go back to Guam ... and we were pretty much sleeping three to a foxhole. Weather was cold at night, and we had managed to get an extra allotment of blankets. The quartermaster came up one day, and said, "You're going to have to turn in your extra blankets." "When are we due to leave?" "In four days." "Then we"ll give you the blankets in four days."

We thought quartermasters didn't always worry about people, as much as they worried about blankets and 782 gear. That isn't true of course, but rifle company people, by nature, are a salty bunch.

On the day of our departure, we were down at the beach with whatever gear we were carrying with us. when word came, "You're not going to go on board ship until tomorrow." Here we were, all on the beach, with no food, and starting to rain.

A Seabee outfit, a rather large one, was nearby. They took us all in. They had their own bunks, but had covered king-size foxholes as well. We stayed in those that night, and kept dry and warm. The Seabees fed us, both that night and in the morning. **Robert Cudworth, Camillus, NY**

I Need More Training

I would like to recount an event that took place on board ship when we were heading toward the Iwo Campaign. Captain Fowler, First Sergeant Samuels and I were talking when a man from A Company approached Captain Fowler and said, "I don't think I have enough training to land." And Captain Fowler told him, "Oh you"ll make out all right." First Sergeant Samuels wanted to throw him overboard.

Anyway, later on in the campaign, I heard my name being called one day, and I looked up and saw this same man standing on the hillside with a box of grenades, hollering "Horton, Horton." About that time the Japs opened up on him and one of the shots hit the box of grenades. Whereupon he dropped the box of grenades, raised his hands to heaven and began to curse the Japanese.

I don't know why they never fired again at him. But he reached down and picked up the box grenades and came down to where I was. This man, by the way survived the whole campaign without a scratch and he made a contribution ... did his best ... and that's about all any of us can do. **Carl Horton, Pioneer, CA**

Could Have Been A China Marine

Canio J. DiGerardo, who was a wireman with the 21st Marines on Iwo, found that Jap snipers don't like Marine wiremen.

He had been sent to repair a broken telephone line from HQ Company to A Company. After making the repair, and on his way back to the command post, a Jap sniper was zeroing in on him. DiGerardo jumped in a shell hole ... and each time he tried to move from there, the sniper reminded him that he was still on duty. This went on for two to three hours, and finally Canio decided to make a break for it, with dusk coming.

On arrival back at the command post, Mike the wire chief, said sympathetically, "And where the hell did you go ... to China?

Canio and his wife Jean live in Brooklyn. After his discharge in 1946, he went to work for New York Telephone Company, joined the organized Marine Reserve, and was called back to active duty during the Korean War.

A Haven For Crippled Planes

Scarcely had the flag been planted atop Mt. Suribachi on Iwo when crews started work on Motoyama Airfield No. 1 to get it in condition for planes to land.

The first plane—a returning B-29 ... crippled during raid on the Islands of Japan ... made a forced landing on March 4, 1945. It was the first of 852 such forced landings that were to take place over the next three months. The Air Force estimated that over 9,000 lives were saved, and the value of the planes recovered was something like 510 million dollars.

By March 16 (D + 26), all organized resistance was declared at an end; but for the next two weeks, mopping up would be taking place and that's dangerous work too.

For the size of the battlefield, and the troops involved—both ours and theirs—there has probably been no other battle in our history which surpasses that of Iwo Jima in magnitude and fury.

Statistics show that on the eight square miles of HELL, there were 85,000 men—61,000 Marines and 24,000 Japanese. Statistics show too that if all the men who participated in the operation—including aboard ships, flying bombers, manning landing crafts and barges— had also been placed on the island at one time, there would not have been standing room left. **George B. Lyon, Florence, AL, 3rd Motor Transport Battalion**

Fortunate Transfer

We were in Regimental Weapons Company, but the 9th Marines rifle companies had so many casualties on Iwo that we were given the word to supply some replacements. My fox-hole up to this time had been a shell hole, caused by a mortar round explosion, and partially covered by a piece of corrugated tin to protect against flying debris.

Myself and two of my men, Private Frank Knoderer and Private Steve Michallichick, were ordered to report to I Company of the 3rd Battalion. We served with I Company until around March 16 when organized resistance fell apart and fighting slowed down.

Private Knoderer and I headed toward the cliff area overlooking the ocean where Weapons Company, 9th Regiment was located. (Private Michallichick, our other crew member, had been killed while serving on this detached duty).

When I reported back to First Sergeant Adams, I got news that the shell hole I used for a foxhole had received a direct hit only hours after I left. I was lucky.

We spent a few more weeks on Iwo Jima blowing up caves with the 75mm gun mounted on our half-tracks, and then back to Guam. **Joseph R. Baricko, Silver Hill, MD**

The Gold Pocket Watch ...

Toward the end of the Iwo Campaign our unit, the 2nd Battalion Ninth, was in reserve for a few days. Consequently, a couple of buddies and I decided to go souvenir hunting.

We started to look to our right which had recently been secured by the 4th Marine Division.

It was rugged terrain, sand stone cliffs which enclosed a former Jap troop concentration. There were at least 100 Japs there, splattered all over the ground in a variety of grotesque bits and pieces—probably caught by artillery ...

We thought we had hit the mother lode of souvenirs when a Jap sniper's bullet slammed into a sandstone wall over my head. It didn't take a consensus for us to immediately vacate, since we had no idea from were the round was fired.

On the way back, we found three dead Japs and proceeded to rifle their uniforms and packs. Shortly after, one buddy let out a whoop and holler since he found the most exquisitely designed hugh gold pocket watch ...

Continuing back we ran into some tankers to whom my buddy showed his recently acquired trophy. One of the tankers bought the watch on the spot for $800 ... At the time that was equal to a year's pay for a corporal. **Walt "Jibo" Wittman now lives in Bradenton, FL with his wife Jackie.**

Iwo Marine Plays Possum While Japs Pick Pockets

Stripped of part of his equipment by Jap souvenir hunters, a wounded Marine played possum inside Japanese lines for one night and lived to tell about it.

The story of the amazing deceptiveness was told by 2nd Lieutenant John H. Leims of Chicago, Illinois, commanding B Company of the 9th Regiment, 3rd Marine Division. The wounded Marine was Lieutenant Leims' messenger.

In taking a hill, the Marine, whose name was withheld, was seriously wounded. Later, the company was forced to withdraw and the Marine could not be evacuated.

Next morning, the hill was recap-tured. Leims searched for his messenger and found him being carried on a stretcher to an aid station. "He was weak, shaken up, but alive," reported Lieutenant Leims.

In between gulps of steaming hot coffee, the wounded messenger pieced together this story of his fantastic experience.

"When the Japs came, I tried to act as dead as I could. My heart sounded like a trip-hammer. I was sure they could hear it pounding. I tried to hold my breath as the Nips stripped me of my canteens and cartridge belt. But I must have looked pretty dead. They kicked me into a shell hole after they had taken what they wanted."

The Devoted Horseman

On what was probably the last night of the Iwo Jima Campaign, Corporal John B. Nelson, 2nd Battalion, 9th, was posted to prevent escape of Japanese stragglers from a draw on Iwo's northwest coast where the enemy had conducted the last organized resistance (area of Cushman's Pocket).

As Nelson was setting up, the light of a flare showed four Japanese soldiers on the trail. When they froze in surprise, Nelson shot and killed three. The fourth ran away, but Nelson says he was caught in machine gun fire.

Among the souvenirs Nelson took from the enemy was an oilcloth packet containing neatly tied loops of a horse's mane. He kept it for years but an appeal in *Leatherneck Magazine*, Quantico, Virginia by General Erskine, prompted Nelson (retired telephone employee of Marion, Indiana) to give up the souvenir.

The mane had been carried to Iwo Jima by Lieutenant Colonel (Baron) Takeichi Nishi, Japan's most famous equestrian and gold medalist of the 1932 Los Angeles Olympic. It was from his prized and loyal mount Uranus.

The mane now rests in the Monument of Repose of Souls at the Remount Section of the Japanese Army at Hombetsu in Hokkaido. Colonel Nishi had once been assigned there.

The Overlooked

It happened a lot; and why their commanding officers, even colonels on up, did nothing about it still puzzles people like Chris Doumis, Scout and Sniper and instructor in the same harmful trade.

Doumis was in on the forming of the 3rd Marine Division, coming over from the Second Division and assigned to 9th Marines. Doumis went to Scout and Sniper School and was in on the invasions of Bouganinville and Guam—where they were "hit very hard."

Here were real pros in combat, skilled at reconnaissance, rubber boats, infiltration; and by the time they got to Iwo were hard, seasoned, and lucky to be alive.

Yet ... as Doumis writes ... "The new Marines from the States, slick and clean, had better ranks (when they all met on Iwo) than the "war beaten, rugged Marine scouts.

"Most of the scouts were still private first class. I was a sergeant but every scout should have been a sergeant, at least.

"During the war it was hard to make rank overseas" ... which might or might not have been an excuse.

However, lieutenants became captains, and colonels still became generals. **Chris Doumis, Hollywood Beach, FL**

It's Official ...
JAPAN
SURRENDERS

Admiral C.W. Nimitz signs the surrender document for the United States aboard the USS Missouri (BB-63). Principals at left are: General Douglas MacArthur, Admiral "Bull" Halsey, and Rear Admiral Forrest P. Sherman, Deputy Chief of Staff for Admiral Nimitz.

Home

I came home via San Francisco on a baby flattop. It was a pleasant trip with a short layover in Honolulu. When I left the ship to go ashore there, the first salute I received was from a BAM (Beautiful American Marine).

We'd heard about the women now wearing our proud uniforms, but hadn't seen one before this. I didn't know whether to kiss her or salute back!

From Hawaii to San Francisco it seemed so strange to have lights on the ship ... I mean on the open decks. The ship even sailed a straight course, no zigzag, no submarines, no general quarters, no danger.

I'll not try to put into words the sensation that went through me when I saw the Golden Gate. I just couldn't. Tears of emotion are even now blurring my eyes so I can hardly see to write, after all these years.

The day we landed, headlines on an "Extra" announced the bombing of Hiroshima. I caught a train east. In Chicago, another "Extra" heralded the destruction of Nagasaki. The day after I arrived in DC, the Japanese surrendered. **Austin Gattis, Washington, DC**

A MATTER OF PINS

The short intense combat, common to "island hopping" battles can do strange things to the men. On Iwo we fought this intense battle continuously for about a month—26 days. Near the end of the battle, I had almost a full platoon of raw recruits fresh from "boot camp." They performed well, but some were nervous.

In A Company, 1st Battalion, 21st Marines, I had one man who got a powder burn on his throat from firing his own carbine. No bullet wound, just the burn. Unbelievable! Then several nights later I could see him throwing grenades, but heard no explosions. At day light I approached him and found out that he had been so frightened that he had forgotten to pull the pins. I had a hard time getting a foxhole partner for him after that! **William I. Pierce, Merrillville, IN**

From Iwo To Wappinger

There's still a soreness for the heart in the Japanese family picture which Jim Gabriel, Wappinger Falls, found in the pocket of a dying Japanese soldier on Iwo Jima. But there's just as much pride for the once youngster who grew up in Hell's Kitchen in New York, and lived through the campaigns of Bougainville, Guam and Iwo. A chunk of hot Japanese metal hit him in the chest on Iwo, but he was there long enough to see the flag go up over Suribachi.

Jim's a restaurant man and once had his own place in Queens. Like most of the salts, he's retired now ... **James Gabriel, Wappinger Falls, NY**

Good Fortune Near Motoyama

On or about March 13, 1945, K Company, 21st Marines were situated in a reserve area west of Motoyama Village on Iwo Jima. I dug in for the night next to a small concrete pill box that was still intact. I was disappointed because two other Marines had staked claim to the pill box.

After an uneventful night, I entered the pill box the next day and was engaged in cleaning my rifle when a mortar barrage landed in the area. I heard later that there were nine casualties.

When I returned to my foxhole outside of the pill box I found that a direct hit had blown my combat pack to bits.

How lucky can one get? **Eric S. Ruark, Baltimore, MD**

Memories ... Nearly All Bad

On the evening of March 6, the sergeant said we (F Company, 9th Marines) would wake up early the next morning and not make any noise. We moved out around 4 a.m. in a night attack, walked to the right on the hill, and started forward. Terrain very rough.

Flares lighted the way ... and we had to "freeze" in position when flares lighted the sky. Certain flares, extra bright, made the terrain look as though it were glazed with ice.

By daybreak, we advanced about 70 yards with Corporal Harry Woods leading our squad. He came to a two and one-half foot ridge, took a left, the men followed. I came to the ridge and started to drop down when Japanese machine guns in front opened up. I could feel the bullets grazing my back. A split second slower, I would have been shot.

These were the first shots fired by the Japanese, as they raked the top of the ridge with machine gun-fire. I could not set up my machine gun ... shooting was coming from different directions ... we were pinned down with many wounded and killed.

I threw hand grenades in the direction of fire. Tanks came in for support and later picked up the wounded. After the tanks left with the wounded, the Japanese again opened up with all weapons. Then before noon all firing ceased. I hear a Japanese saying, "Banzai, Marines, you die!" a weird feeling came over me.

Japanese started to infiltrate, throwing grenades into foxholes. I moved back about 10 steps to a large mound ... looked like a good position to stay in. Men to the left were lying in prone positions ... no one was moving ... "Oh, my God, I'm the only one left." Furiously I started picking up chunks of rock to build a parapet for protection. A Marine raised up, I motioned him to come over and help me. Four more came over. I said, "Let's make it large enough for all of us, we're pinned down and trapped. There's no way to get out right now."

My number two gunner walked up and said, "Help me to the aid station, I'm shot in the hip." I asked him to stay ... we would be shot if we tried to move out ... he walked upwards a few feet and was cut down ... I felt bad, if only he had listened.

Private Donell Demoises tapped me on the back. He had been badly wounded in the face. I told him to lie down, gave him first aid, told him we were pinned down and couldn't get out right then.

Maybe it was afternoon of the next day I saw a radio antenna in a shell hole about 25 yards to my right. Later, a Marine raised up shoulder high, his hands asking me how many were with me. I held up six fingers. He motioned with his hand downward indicating for us to stay—that help was on the way. (I later learned that Marine was Amos Jones of Hattiesburg, MS).

No more shooting that afternoon ... I wondered why ... but about 4:30 p.m. on March 8 there was a lot of shooting with all Hell breaking loose. Two tanks were coming toward us ... shooting their way in ... the same way they had left the previous morning.

When they got close, I got up and stepped forward waving until one tank stopped in front of me. The hatch popped open and the driver yelled, "Get the wounded and get out. We will give you fire cover." One of the men helped me put Donell Demoises on my poncho, and we ran as fast as we could to the aid station. The men in the hole with radio also go out.

At the aid station the doctor immediately gave Demoises plasma as I watched. When he got through, he took care of me and told me not to go back up on the front lines that night.

I don't remember when I rejoined the group, but I had another machine gun and two ammunition carriers. Advancing, we got near a ridge with Japs inside giving us problems. Tanks were called in for support, and one flame-throwing tank shot flame all over the ridge. Some Japs were running with their clothes on fire. I watched in awe.

With the ridge secured, we advanced toward the seashore, handling resistance that had to be taken care of. When our company assembled, I got to see who was left of the original men. Six

of us left out of 52 men from the machine gun platoon and about 26 men total for Fox Company. When the day came to leave for Guam, I was more than ready to leave that place.

Reunions: Joy and Sorrow

At reunions from 1988 to 1991, I met various Marines who talked about the pre-dawn attack ... Amos Jones of Hattiesburg who was in the hole with the radioman ... William Mandac one of those in the trap ... W.A. O'Bannon in command of Fox Company at the time ... Roger H. Radabaugh one of the tank crew and his driver, C.C. McCabe ... and I hope to meet others like Warren McPherson, the tank driver who opened the hatch and told us to get the wounded out.

Before the reunions I never talked to anyone about my experiences on Iwo Jima. I closed my mind to it. In July of 1985, I was in a gasoline fire with burns to my body, and it caused me to have flashbacks to Iwo Jima. Renewed my war memories ... and had nightmares about it. I want to thank everyone who helped us get out of the trap in Cushman's Pocket. **Louis R. Machala now lives in Dallas, Texas**

Pearl Harbor Thru Iwo ...

The Odyssey Of "Tanker" Joe Garza Jr.

The disbelief on December 7, 1941 soon turned to a patriotic fever. That week, I was out recruiting all my friends into the Marine Corps. I had four enlist. On January 16, 1942, I told the recruiting sergeant to sign me up ... I was 17 and ready. So, began an odyssey with a corps that created an espirit and dedication bridging 48 years.

"Forget Pearl Harbor ... remember Wake Island ... that's where the Marines were fighting against all odds," D.I. Corporal Dill bellowed out at the civilians in the San Diego Recruit Depot Quonset hut. "Before we get done with you people, you'll be Marines."

A short tour of duty with D Company, 2nd Tank Battalion, 2nd Marine Division followed boot camp. We learned to drive all the Army discard tanks, Mormon Harrington, M2A4 Liberty radial engine and diesels that we started with shotgun shells. In late summer they selected some of us to form the nucleus of D Company, 3rd Tank Battalion in the 3rd Marine Division.

USS *Bloemfontein* (January 1943) was reserved for us to Auckland. D Company would not survive long—it was deactivated soon after arrival in New Zealand, and the Marines were scattered throughout the 3rd. Seven of us found ourselves with the 3rd Tank Battalion Scouts and Snipers (Recon). A lot of training, but also good liberty, was on the menu in New Zealand. I will always remember the grassy knolls of the area and the camaraderie.

Guadalcanal was an adventure in June 1943. Washing Machine Charley, the Mitsubishi on night harassment visits from Rabaul, kept us awake at the beginning. It soon became a nuisance to jump into the foxholes by our tents.

We had some of the fringes we left in the states ... made our own apple jack—particularly tasty with assorted bugs. But, we craved fried chicken. We'd heard of a native village some miles East, but we had to cross a crocodile infested river (or so it was reported). For trading we took a couple of obsolete green pad protectors and my old lemon wood bow I'd brought from Auckland.

Getting there was no problem. We traded all our goods and left with a big red rooster that had passed the Social Security bench mark. We stayed too long and had to cross the waist-high river in the twilight hours. Our side arms were ever ready as we crossed and saw many a pair of beady eyes shining on the water. Someone yelled, "Whatever happens, save the rooster."—we knew our priorities well.

The river was some 30 feet wide, and we set a record for that crossing. We fried "old Red" the next night out in a jungle clearing, and realized how really tough Guadalcanal roosters can get to be.

But then, we were here for a war, and so off to Bougainville ... a perimeter surrounded by slimy muck. Absolute quiet except for nice and restful jungle noise ... that was to be the 3rd's opening game. We set up a 50 calibre machine gun on Cape Torokina, and were part of the firing power against the Nip planes that zoomed in from Rabaul to work over our transports.

The first and only dog fight we experienced there, or ever more, had Marine fighters, Navy fighters, New Zealand fighters, and Nip Mitsubishis, Zeros, and Zekes filling the sky. Our tracers sought out the Mitsubishis as they came in for close drop on the transports. It was a good feeling—we were finally getting back.

We moved up the Piva Trail to Hill 500 where we built our bunkers, and waited for them. The first night on the line was unforgettable. We'd set up our concertinas with the usual empty C-rations cans with rocks for first alert. Three of us in the bunker gave us a 2-on, 4-off watch.

We strained to see through that blackness of the Bougainville jungle nights. Our ears were set for the slightest step of a foot coming down on the dark jungle ground. After 15 minutes of tenseness, our reaction time was an automatic split second. A burst of firing to the right was followed by a demanding reserve of eyesight and hearing. You knew you were seeing movement by large masses of men; but, you waited for a first alert—C-ration cans were our best sentinels. This was our greatest threat—the waiting and uncertainty.

Christmas found us back on Guadalcanal, enjoying a great turkey dinner—cold storage of many years notwithstanding. The first campaign can make a difference. We looked at replacements joining us as Marines who were the links in the chain of legacy. They, too, would soon be tested and we knew the Corps would prevail.

Training gave way to the usual guessing of where-to next. Boarding the transport (*Fuller*), we began the long journey to the Marianas as floating reserves for the 2nd and 4th Marine Divisions on Saipan.

Guam, the first American possession, was to be re-taken. The landing—a long walk across the exposed shallow reef—added another dimension to our history. The breakthrough found us as tank ordnance on the high ground perimeter, dug-in to block off an approach. The first thing that went into our foxholes were cases of crabmeat cans and bottles of Asahi beer. We guarded all this with our friends—the land crabs.

As we established a high ground tank farm, we spent some time on combat patrols looking for strays. We had finally moved to the high rent district ... our tank farm was on a high plateau surrounded by coconut trees and a banana plantation. It quickly reverted to training, and fighting off the scourges of dengue fever and impetigo.

Most of us had been in the Pacific for more than 18 months with two campaigns behind us. We knew we needed at least one more rotating ... and we had no expectations of anything less, so we were not disappointed. The next campaign was soon building, as training intensified.

Boarding our transport in early February 1945 had, as I remembered, the same excitement. However, we felt this would be a different engagement. The weather turned cold, with a brisk wind ... nothing we had experienced for a long time ... and the seas got rough as we made our way up to Motoyama No. 1 terrace. The infantry didn't like that as they were standing close to our tanks. The Japs laid in a heavy mortar barrage. We had about 10 casualties in that first encounter. We knew it was going to be a long stay.

Special troops, born out of necessity, created the 3rd Tank Battalion, who had been training with Norman Herrington scout tanks and light tanks using Gueberson diesel and aircraft radial engines—all hand-me-downs from the Army. Such was the beginning of "tanks" that proved themselves worthy on Guam and Iwo Jima.

We said goodbye to Iwo, and all those we left behind, shortly after the first of April. I traded a .45 sidearm that I carried through all the campaigns to an Army lieutenant for a fifth of bourbon. It was a good trade. I was going and he was staying.

Following a quick return to Guam, we had a reunion ... rationing the bourbon I had carefully packed. We were soon rotated back to the States in later April. I finished my tour at the Tillamook NAS, Marine Barracks ... it was the best of duty stations. Separation came on October 16.

I used the G.I. Bill to earn my degree in engineering at the University of Southern California. Then, got my master's degree in business. Economics from the Claremont Graduate School. I worked my way up from chief engineer to vice president of Aircraft Systems with the Conrac/Bulton Mark IV Industries. **Joe Garza Jr.**

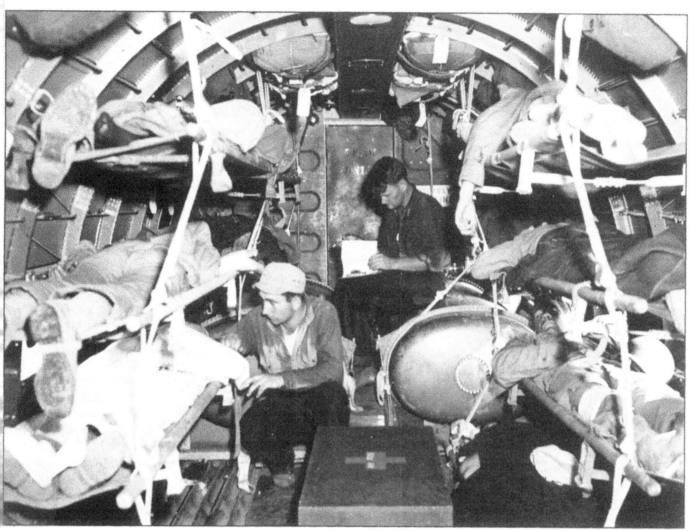

These wounded Marines, stretched out in a C-47 transport plane, were the first to be evacuated from bloody Iwo Jima. The Leathernecks were flown to Guam for hospitalization. Previously ships had removed casualties to advanced bases for further flight to Guam and Pearl Harbor, but this was the first flight to land on Motoyama airfield number one and take off with patients. The field was shelled at the time the plane was loading. Pharmacist Mate First Class John Drust Jr. (left) and Chief Pharmacists Mate Silas V. Sturtevant are the corpsmen attending the wounded. (Official U.S. Marine Corps photo)

Lots Of Friends For Rosplock

Eugene J. Rosplock has never had trouble meeting people or even finding them. On the 'Canal—in those days as far away as the moon—at Tetere he bumped into George Huyler, Bill Hughes, and Joe Colucci ... and on Iwo, Marty Hughes with the tanks ... and all from home town Elmira ... On the way home who steps up but the ship's sailor Tom Devlin ... Meeting people is still his luck bumping into old World War II buddies Charles Cramer, Joseph Leniart, Ken Cowley, Thurman R. Batchelor, Howard Layton, Richard Eber, Paul Stender, Jack Steward, Stanley Okdakowski, Con DiGerardo, Bob Gunderson, Jim Garner, Marlowe Williams from 1st Battalion, 21st ...

Rosplock remembers more than people: the dog-fights over Henderson Field, the creek full of blood and Japs after the Guam banzai, the Elmira buddy killed on Iwo (didn't mention name), the close shaves on Iwo and the big barrage balloon on Guadalcanal that got loose and soared like a balloon from the circus.

Anti-Freeze

When discharged from the staff of "The Boot" at Parris Island in October 1945, long standing orders were followed: A visit with General Allen H. Turnage, Assistant Commandant, United States Marine Corps.

Attractive sergeant receptionist informed me that General Turnage would be in his office sometime after noon chow.

Chowed with 1st Lieutenant T.O. Kelly (my bunkie on Guadalcanal when Kelly was Division Sergeant Major) suggested that the two of us visit with the General. "Too busy." remarked Kelly.

Returned to the office of the assistant commandant. Attractive receptionist sergeant accompanied me through the office of an attractive 1st lieutenant secretary, through the office of a colonel administrative assistant.

"How the hell did you get out?" asked General Turnage. I replied, "A Sergeant Major at Parris Island." "Pappy" (a regular) also was discharged.

What was not told the General: Pappy Bunch phoned Hancock, informed him that the dispatch from Headquarters Marine Corps freezing certain personnel for six months had been buried. Buried, that is, until, "Pappy" Bunch made sure that Bunch and Hancock's names were on discharge orders.

Bunch returned to his native Arkansas. Hancock returned to Hershey, Pennsylvania. **Brent I. Hancock, Hershey, PA**

Oh Those Seabees, How We Loved 'em

For as long there has been a 3rd Marine Division, the Seabees and the Marines have lived and worked, fought and died together. They achieved a camaraderie unknown to other inter-service unity. This is demonstrated in a famous slogan of the Pacific War:

So when we reach the isle of Japan,
with our caps at a jaunty tilt,
We'll enter the city of Tokyo
on the roads the Seebees built.

The contributions made by the Seabees in World War II are aptly described in an article written by (then) Lieutenant (jg) W.B. Huie, USN, published in the October 9, 1944 issue of *Life* magazine. The following passages from that article especially highlight the bond forged between the Seabees and the Marines.

Construction Expertise With Fighting Spirit

It was during the bitter months at Guadalcanal that Marine and Seabee first went under fire together. The professional soldier and the essentially civilian machinist eyed each other skeptically until the Seabee saw the Marine operate on a Jap "rat nest" and the Marine saw the Seabee operate under bombs and shells on Henderson Field. Then each decided that the other was good enough to play on his team. They shook hands and immediately started the joshing, never-ending argument as to which is the toughest and which one is "winning the war."

The Marines chided the Seabees about their higher average age. Because the accent is on mechanical skill, the average age in many Seabee battalions is as high at 31. "Never strike a Seabee," the Marines jived, "for his son may be a Marine." The Seabees countered by stamping out "Junior Seabee buttons and announcing that after any Marine had killed 10 Japanese in hand-to-hand combat he might be adjudged "tough enough" for one of the awards.

The sharpest banter concerns the jealously guarded Marine prerogative for always getting everywhere first. For years Marine poets have been proclaiming that when the Army and Navy get to Heaven they will find Marines guarding the streets. Now the Seabee poets insist that when the Marines get to Heaven they will find that Seabees have built the streets.

The Seabees seized upon an incident on New Georgia Island and rubbed it mercilessly into Marine hide. The Marines landed on New Georgia on June 30, 1943. As they ran out on the beach looking for Japs, a party of white men stepped out from behind some trees and waved to them. Marine jaws dropped as the party approached.

"Major, the Seabees are always happy to welcome the Marines," said Lieutenant Robert L. Ryan, CEC, USNR, of Ventura, California to the Marine major.

There was much loud cussing and ground stamping before the Marines could appreciate the Seabee sense of humor.

Actually and seriously, the Seabees do not pretend to be assault troops. Specialized units like the demolition and pontoon squads may lead the way but the Seabees are primarily concerned with landing the equipment, constructing the war "plant." Guadalcanal is the book illustration of how the Marine-Seabee team functions.

The 6th Seabee Battalion, led by Commander Joseph P. Blundon of Keyser, West Virginia, arrived on Guadalcanal three weeks after the initial Marine landing. The Seabees pitched their tents around Henderson Field, which was still under Japanese artillery fire, and began a four month battle in which they distinguished themselves as much for raw courage as for construction know-how.

Construction VS Destruction

The battle was to turn on whether or not the Seabees could keep an airstrip operating under the almost constant Japanese attack. It was Japanese destruction versus American construction, with the Japs trying to blast holes faster than the Seabees could fill them. To meet the challenge Seabee Commander Blundon developed "the world's fastest technique for filling craters." He ordered foxholes dug right alongside the airstrip. He computed the exact amount of gravel necessary to fill craters caused by various-sized bombs and shells; then he loaded this measured material on trucks which stood ready in revetments. He found that one 500-pound bomb would tear up 1,600 square feet of Marston mat, so he had packages of this amount of matting distributed along the strip like extra rails along a railroad.

When the attack came the Seabees were ready. Squatting in their foxholes at the very edge of the strip, they would wait until the bomb hit. Then, with the attack still in progress, they would leap out and race for the crater. Trucks, with motors already running, would roar out of revetments to dump their measured loads. Like jitterbug dentists filling a tooth, the Seabee compressor men and pneumatic hammer operators would jump in the crater and start packing. Bucket brigade lines would start tearing out the crumpled Marston mat and passing up the new.

"We found that 100 Seabees could completely repair a 500 pound bomb hit in 40 minutes, including the replacing of the Marston mat," Commander Blundon reported. "The Japanese required from three to four hours for the same job, and then they only filled the hole with dirt. They had no compressors, no pneumatic hammers and no mats."

The supreme test for the Seabees came on October 13-14, 1942, when 53 500-pound bombs hit the runway in 48 hours. Even the Seabees cooks had to leave the galleys and join the endless battle to fill the craters. Often, during this man killing stretch, our fighter planes soared over the field with near empty fuel tanks while the Seabees fought to clear enough strip for them to land.

"We won that battle," Commander Blundon concluded, "not only because of our crater filling technique, but also because we had battled constantly to enlarge the landing surface. By October 13 we had enlarged the Henderson strip to 5,600 x 150 feet, yet in emergency we could operate fighters off a clear surface of 2,500 x 75 feet. Thus we really had four fighter strips at Henderson and, in addition, we had roughed out a sage grass covered strip on which we could land a fighter in extreme emergency.

Two other Seabee performances must be fitted into the Guadalcanal record. In the early days there were no lights on Henderson Field, yet fighter craft would have to be landed at night in emergencies. At such times Seabees would hold flashlights and form a human boundary around the landing strip. Death would literally hover over these men since the planes were often partially out of control, and even the brush of a wing tip would sever the head of any man holding a light.

At one time the aviation fuel shortage became terribly acute. Cargo planes, extremely vulnerable to Jap attack, were flying in some of the fuel. Destroyers were dashing in at night to bring more. Still there was not enough. The Seabees, using their self-propelled pontoon barges, undertook to help relieve the shortage by making the 20-mile trip across open water to Tulagi. Part of this trip had to be made by daylight. Stoically, while the Jap dive bombers attacked, the men rode these slow, chugging barges, knowing that the slightest hit would send them and their gas up in a puff.

"That's the easiest job in the Navy," the Seabee bargeriders claimed. "If the Jap don't hit you, you got nothing to worry about. If he does hit you, you never know it".

Wrong Move Means Death

Coldest, bloodiest of all Seabees are the members of the demolition units. They are "hard rock men," as nerveless as clams, who carry TNT and bangalore torpedoes onto an enemy held coast and blast out obstructions to a landing. They are the real advance agents of amphibious warfare. Since a single wrong move by any one of the five members of a demolition team can wipe out the entire demolition unit, much confidence within a unit is im-

perative. So the units are formed with great care. The men live together, train together, work together, and only in rare circumstances is the personnel of a unit changed.

The Seabees perform such necessary functions that civilians may wonder why they are a new organization. Who filled their role in previous wars?

Answers to this are found in the nature of the present conflict. Prior to December 7, 1941, it had always been the policy of the Navy to have the majority of naval construction done by civilian workers. The Navy's small, highly trained civil engineer corps, regular commissioned staff officers with special technical qualifications, planned, designed and supervised the installations, but all major construction was done through civilian contract. This system was being followed, with growing difficulty, until bombs began to fall on Pearl Harbor. Fourteen hundred American civilian workers at Wake, Guam, and Cavite became Jap prisoners.

The realities demanded a new military organization. The bombing plane had changed the nature of war to make naval shore construction the center of attack. And since the Pacific War was to be an air-sea war, fought over a limitless,

unprepared battlefield, it would involve more construction than any nation had ever before contemplated. As the mobility of our amphibious forces increased, the demand for more mobile and advanced bases would increase.

The Navy decided to build the new organization on the nucleus of its Civil Engineer Corps. The CEC then turned to the engineering societies, the construction industry and the labor unions and asked them to provide the personnel for the Seabees. Engineers and superintendents became officers in the CEC and skilled union operators poured into the ranks. Eighty percent of the men in the Seabees carry union cards.

There are now 8,500 officers in the CEC, and the Seabees have grown until there are almost a quarter of a million of them. Their chief is Admiral Ben (not Benjamin) Moreell, who joined the Navy's CEC during the first World War. Admiral Moreell is as tough and jut-jawed as the boss of a 235,000 man road gang should be. And he is proud of the toughness of his Seabees. Of them he has said, "The Seabees are the men who built America's cities, dammed her rivers, strung her wires and dug her sewers. They are the kind of men you can depend on to finish a job."

They Called
Them The
Base-Builders

Those Seabees—Trained Like Marines, Thought Like Marines ... But They Ate Better

What follows is a stirring and nostalgic account of the experiences of a World War II member of the 25th Seabees. Roy D. Exline, ex MM2C (machinist mate second class) is now of High Point, NC. He was attached to the 3rd Marine Division from Guadalcanal through Bouganville and Guam.

We Started Right

The 25th Naval Construction Battalion was activated August 15, 1942 at Bradford, Virginia ... and the Marines trained us from day one.

We saw the 18th and 19th Seabees march by in their Marine khakis and they went to the 1st and 2nd Marine Divisions. At the time we didn't know we were going to the 3rd Marine Division.

We went to Camp Elliott where we fired on the rifle range, then went on a seaborne maneuver out of Diego in the Coast Guard ship *Arthur Middleton*.(A real stinker of a ship, it later sank in the Aleutians.) Then the day we landed from the maneuver we walked overland to Camp Pendleton. It was here in November 1942 that we drew our Marine uniforms. While drawing my clothing I saw Tyrone Power getting his boot haircut in Diego.

My Seabee contingent shipped overseas to Auckland, New Zealand aboard the USS *Mt. Vernon* with the 9th Marine Regiment. The big thing in New Zealand with the Marines was hikes (including a 60 mile survival hike where we carried all our food the first day—two strips of bacon, handful of raisins, and two pounds of rice for supper and breakfast).

Everyone had to finish the hike, so the doctors and corpsmen were kept busy. Our average age was 33, while the 3rd Division Marines averaged 20 years.

One day we landed from the President Hayes and were bivouacked on the beach. One old boy of ours was about 40, a steel worker all his life, had a real harsh voice, and could really cuss. We had got in our foxholes when we heard him cussing—the air was blue—because someone had stepped on him in his foxhole. There were a few seconds of silence and then a calm voice said, "You're talking to Captain ___, United States Marines." I'll put you in the brig so far they'll have to pump sunshine in to you."

Then we heard our old boy say, "I didn't know who the hell you are." We were all snickering, and the captain probably was too—we poked a lot of fun at our guy for days after this ...

To The Canal

We arrived at Guadalcanal July 6, 1943 to disembark and move into our new area, 20 miles south of Henderson Field in a coconut grove, one mile in from the beach. We had made three false stops, and nearly everyone had dug foxholes, except at this location. We had dug slit trenches on the side of the grove next to the beach.

About 8:30 or 9:00 p.m., we heard some ack-ack fire up toward Henderson ... and soon we could hear Washing Machine Charlie coming straight for us.

A friend and I were the nearest to the street, a new road with good ditches on either side. I heard nine clicks and the darndest sound I'd ever heard seemed to be going down the back of my shirt. I went for the ditch with a good 20 inch vertical wall to protect me ... hit the bottom of the ditch about the time the first bomb went off. It flounced me about a bit, and I was trying to keep my tummy off the ground, my mouth open, and my ears covered like I had been trained to do. I saw the flashes and coconut fronds coming down, but knew I wasn't hurt.

The other fellow had jumped in behind and grabbed both my legs in a death grip so I couldn't get up, he thought I was wounded. He finally released me and we got out of the ditch. Had barely stood up when here comes that SOB Jap again, and we hit the ditch again. He didn't release any bombs the second time, but he scared us good.

The adjutant decided to call the roll. His voice got stronger as he progressed down the muster roll and had reached the name of a little Italian replacement, shouting his name for the third time before we heard, "Heeere" from way over on the edge of the grove. Our guy was still down in the slit trench, and we only had one quart of water each to clean up with.

September 20 on Guadalcanal, the Japs had near missed us again, but did hit the 19th Engineers sawmill, and wounded some Marines in HQ and A Companies. We saw the searchlights come on up toward Henderson and there was one plane in the lights, but no ack-ack. Suddenly from the right came a streak of tracers and that Jap plane was afire and fell in a long arc into the mountains.

Suddenly another burst of tracers reached another Jap plane we hadn't seen and it was arcing almost exactly like the first ... only about 30 seconds apart. Talk about cheering. Everyone on the Canal was cheering ... yelling like a bunch of kids.

I always wondered who that P-38 pilot was. I didn't learn his name until years later. I read *Forked-Tailed Devil: the P-38* by Martin Caidin (a wonderful book) and there documented on page 245, the pilot was Lieutenant Harry Meigs II.

Sometime in late summer, 1943 there was to be a division parade and everything on wheels had to go by the reviewing stand—even if it had to be towed. We had a roller we were unable to get parts for before the parade, and I remember thinking how ridiculous it was towing that thing past the stand. We had prepared the parade ground beforehand, using our rollers to roll the six foot high grass down ... and it made a very good parade ground.

(It made us think back to our first 9th Regiment parade at Camp Pendleton where we stood over four hours in the rain, and Colonel Lemuel C. Shepherd straightened the shovel on my 1918 Army pack on his way past me. He was my hero ... a Marine's Marine ... I'm proud I served under him.)

When we landed at Bougainville, an 18 year old friend said the first thing he saw was a Marine running after a Jap and reaching for him with his Ka-Bar knife. Everyone in the 3rd Division seemed to have "Ka-Bar Fever" ... honing or playing with his new issue knife. I still have mine and every once in a while get it out and feel the edge with my thumb.

Our area of Bougainville was beside the Laruma River where it emptied into the ocean, and our work consisted mostly of building roads to the lines. We got our share of bombings, because the Jap bombers could find us easily between the Laruma and Torokina Rivers.

Our Marine buddies would come back from the lines to get better chow, but they would never stay after the first night, saying, "Hell, I can sleep more in the lines." Charlie seemed to stay over us all night, diving and pulling out. We never knew when he was going to "lay his eggs" and go home.

The Marines wanted BARs and Thompsons (sub-machine guns), so we traded all of ours for their carbines. After all, they were between us and the Japanese.

I went through my first earthquake here and at first thought it was a Jap artillery piece firing from up in the hills. Reminds me that there was a Jap gun up in the hills shelling the point ... not many rounds at any time so they wouldn't give away their position.

One important thing I witnessed here was a daylight air raid in the forenoon when I wasn't on duty. I heard a machine gun firing, looked up, and right over my head was a Zero. The air battle went north and I saw a plane come down, and I was sure it was a Jap. Then another was falling and I felt sad because I thought it was the American I had seen get the Jap. A parachute did open, the pilot landed in the water, and a PT boat just split the water getting out to him. I learned years later (in 1952) that the Jap flight sergeant picked up by the PT boat "talked" and told the United States Navy more about the Japanese Air Arm than they had known up until that time. I had "witnessed history."

By Christmas, most everyone was back on the Canal and we were drawing replacements, new equipment and receiving new training in street fighting ... We knew we would be "going in" nearer Japan and any thinking person knew things were going to get rougher.

We See Naval Power

We boarded our old friends—the Ugly Seven—including President Adams, President Hayes, President Jackson, Hunter-Liggett, American Legion, George Clymer, and Crescent City. When we stopped in the Marshalls ... while Saipan fighting was going on ... we knew we were in on a "biggie." At times we could see three other convoys.

A most vivid memory on D-Day at Guam was the power we had and the barrages we laid down. The count was something like ... 14 aircraft carriers, I've forgotten how many wagons, six heavy cruisers, and numbers of destroyers on the horizon. The 4,800 rockets fired during the last 10 minutes before our troops hit the beach was very impressive.

We took numerous casualties on Guam and I had several misses. (I still have a bullet that hit the fender of my truck. It landed in a camouflage net I was carrying on the fender, while I was standing on the running board). Several of our men were decorated for their actions on D-Day.

I took the first load of cascajo (coral) across the island. The quarry was on the hill outside of Agana, and I was told to go

as far as I could, fill the last bomb crater, and return. It was scary. An equipment operator is sure at a disadvantage. I wasn't able to shoot back.

The road was narrow, dead Japanese were in the road alongside some of their trucks, and it was here I noticed the Japs had on heel-less socks. (We probably are producing more of those than anyone in the world today.) I had to pass some caves on an upgrade when I was moving slowly. Don't believe anyone when they tell you the hair doesn't rise on the back of your neck like a dog's ... I know better.

Next day I pulled out of the quarry, and was starting down the grade toward Pago Bay and a swarthy looking chap came out of the jungle and asked what time it was. I called the time to him and he yelled back, "American or Japanese?" I realized what he asked and I've always pondered ... did I give a Jap the time of day?

One of the last jobs we did while attached to the Marines was to help the 19th Engineers build a water point beside the Pago River. The cascajo quarry was on top of the hill toward Talefofo. As I was bringing a load of coral down the hill to the new water point I chanced to look back on a little trail to my right and there stood a Jap watching the road. I knew if I stopped and tried for him that he had every advantage, so I kept going. Every time I went by the rest of the day I had my truck in a gear I could speed up with.

Next afternoon I had just picked up two Marines, and when I got to the little trail a man jumped out and flagged me. He scared me, but I noticed him to be a soldier ... soaking wet. He said five soldiers had been below a cliff walking in the edge of the water when they took Japanese rifle fire from up on the cliff. One soldier was wounded badly in the upper arm, and another, a boy from Michigan, had leaped behind a rock and killed two Japs with his M-1. The man, who had flagged me, said, "Will you go in and get the wounded man?"

I said, "Yes, if I can turn around." He said I would be able to, so I gave my carbine to one of the Marines (the other had a .45) and in we went. The soldier was in a little clearing, lying down and bleeding badly. I was still carrying two canteens and he drank both of them.

I knew where the Army hospital was located, so we got him into the back of the dump truck, his buddies and the Marines cradled him, and I ran 53 miles per hour to the hospital. I went back to the quarry, never told my com-

manding officer or anyone, but always hoped they were able to save the boy's arm.

While we were repairing the road over to the Pago River we were under sniper fire many times.

One time with sniper fire, everyone took cover. We kept a BAR man with us all the time, but he was unable to locate the Japs. A young truck driver, who had a .45 jumped behind the bulldozer blade, stuck the pistol out around the blade ... only the pistol and his hand showing ... and was yelling, "Where are the SOB's?" We really kidded him.

We had a battalion "fall in" one day and a captain from the 9th Regiment addressed us, telling us the enemy attack had been partially successful and that they needed us for reinforcements in the lines. We were to draw extra ammo, hand grenades, have two canteens of water, and standby. We were so seasoned by this time all we needed was more hand grenades.

We "stood by" for 24 hours, then the crisis was over, and we weren't needed. The 25th Battalion had been Marine combat troops for 24 hours. I think with the Marine training we had that we were mentally prepared to give a good account of ourselves.

Guam was secured August 19, 1944 and the 25th was returned to the Navy. We missed the 3rd Division, but we could still visit with our buddies there and they could visit us. When the 3rd Division was pulling out for Iwo Jima, I felt a touch of sadness ... as though I didn't belong any more.

When I was attached to the 3rd Division one thing happened to me that is still affecting my life.

A Lifetime Promise

I learned a 97 lb. school teacher, who had come home with my sister from college, had joined the WRs. I thought our paths might cross, so I sent home for her address. She was stationed at El Toro, and came back through San Francisco so we didn't see each other in service. However we were married October 7, 1947 and are still together.

We attended the 1980 3rd Division reunion in New Orleans and it was wonderful. One other Seabee from the 25th was there and we had a great time.

I wouldn't exchange my experiences with the 3rd Division for anything; and when my tour of duty is over in this existence, I would like a Marine firing squad and a Marine bugler to sound "Taps" over my carcass.

The Road Was Not Easy, But They Proved They Were True Marines

by Tom Bartlett*

The Marine Corps obviously did not want to enlist "Coloreds." Major General Commandant Thomas Holcomb testified before the General Board of the Navy on January 23, 1942, that "there would be a definite loss of efficiency in the Marine Corps if we have to take Negroes."

This was in response to Executive Order Number 8802, issued by President Franklin Delano Roosevelt on June 25, 1941. The Presidential decree established fair employment practices and stated, "all departments of the government, including the Armed Forces, shall lead the way in erasing discrimination over color or race."

Of the military services, the Marine Corps was the last to accept black troops. First to enlist were Alfred Masters, George Thompson, George James, John Tillman, Leonard Burns and Edward Culp. Then came James Brown, George Glover and Davis Sheppard. They were first to enlist, but Howard Perry was first to report to what would become known as Montford Point (adjacent to Camp Lejeune), North Carolina. Jerome Alcorn, Willie Cameron, Otto Cherry, Lawrence Cooper, Harold Ector, Eddie Lee, Ulysses Lucas, Robert Parks Jr., Edward Polin Jr., Emerson Roberts, Gilbert Rousan and James Stallworth checked in.

Recruit training began in September 1942. Obie Hall from Boston, became first man in the first squad in the first regular recruit platoon.

More than 20,000 black recruits enlisted during the early 1940s. Montford Point consisted of 120 prefabricated huts, including 20 which housed 16 men each. There were also two warehouses, a theater, chapel, dispensary, and headquarters. Colonel Samuel Woods' staff consisted of 23 white officers and 90 white enlisted Marines.

Marine recruit training has never been "easy;" but when blacks reported for instruction, more often than not, they were trained by southern drill instructors. Many of the officers had received military schooling at southern institutions, such as VMI or the Citadel, reflecting deep roots of prejudice.

Edgar R. Huff, one of the first to arrive at Montford Point recalled, "The white officers and staff NCOs didn't want us. We formed the 9th Platoon and our drill instructors gave us a good talking to. He says, 'The best thing you people can do is sneak out after dark. Go on home. Why try to play ball on a team that doesn't want you. But, if you decide to stay, I'll make damned sure that you'll be sorry.'"

Following graduation from recruit training, Huff was retained at Montford Point, becoming a drill instructor. Gilbert "Hashmark" Johnson was also made a drill instructor after successfully graduating from boot camp. (Montford Point was renamed Camp "Hashmark" Johnson in honor of the legendary Marine. Huff retired from the Marine Corps as a sergeant major, following more than 30 years in Marine green.)

During World War II, black Marines complained they were used as mules. "We carried supplies and ammo on our backs. We were stevedores; two legged mules. We toted, but few fought."

They referred to themselves as "The Chosen Few." They were making history, landing, fighting, and bleeding at Guam, Peleliu, Iwo Jima and Okinawa. Blood flows red, whether the wounded is black, white, red, or yellow.

No longer would the training be segregated. They would upon graduation become, "Marine Green."

On November 10th (Marine Corps birthday of the Corps) in 1945, Frederick Branch was commissioned a second lieutenant of U.S. Marines. He became the Corps' first black commissioned officer. On September 8, 1949, Annie E. Graham enlisted from Detroit, MI, becoming the Corps' first black female Marine.

During the Korean War, units were integrated, although blacks were certainly in the minority. But during the Vietnam War, blacks held more than their own. Private First Class James Anderson Jr. was killed in action; he threw himself on an enemy grenade to save his buddies. He received a posthumous Medal of Honor. Sergeant Rodney Davis, same reaction. Same Medal. Same fate, as was that of Private First Class Ralph Johnson, Private First Class Oscar Auston, and Private First Class Robert Jenkins Jr.

But there were problems in Vietnam during the late 1960s. There were race riots at Camp Pendleton and Camp Lejeune as white militants sought out black power enthusiasts; KKK material and meetings were raided. Both sides suffered casualties. It was a "no-win" situation and soon, representatives from both sides met to talk.

Vietnam was the proving ground for blacks, who had nothing actually to prove to anyone. The fact that they enlisted was proof of their love of country. It has been written that there are no atheists in foxholes; there's damned few bigots there, either, where every man depends on his buddy, especially on a two-man listening post or ambush site.

On September 30, 1991, Headquarters Marine Corps reported that there were 194,040 Marines serving on active duty. Of that number 120,095 were white enlisted; 35,261 or slightly more than 20 percent were black.

Lieutenant General Frank Petersen was the first black Marine aviator; a combat veteran of Korea and Vietnam, he became the Corps' first black general.

In September 1949 Montford Point was deactivated as the recruit depot for black Marines. More than 12,735 blacks served their Corps and country during World War II.

Today the camp is known as Camp Gilbert "Hashmark" Johnson, and it is home for many Marine Corps schools for enlisted personnel.

Blacks desiring to enlist in the Marine Corps would receive recruit training at either Parris Island, SC on the East Coast, or at San Diego, CA on the West Coast. No longer would the training be segregated. They would, upon graduation become, "Marine Green."

*Tom Bartlett
Managing Editor, Leatherneck Magazine
Master Sergeant, USMC Retired

Guam And Iwo Early Tests For Black Marines In Combat

On July 21, 1944 on Guam, the 2nd Ammunition Company was in direct support of the 3rd Marine Division landing north of Agana, and the 4th Ammunition Company was in direct support of the 1st Marine Provisional Brigade landing near Agat. That night a platoon of the 4th intercepted a party of explosive-laden Japanese headed for the brigade ammo dump and killed 14 of them.

Four black Marines companies, as part of the 8th Field Depot, fought at Iwo Jima.

Okinawa would see 11 black Marine companies employed.

At Guam ...

The heavy naval bombardment had levelled most of the beach defenses of Guam, but there were still some antiboat guns operative, and Japanese mortars and machine guns were active. The fire was particularly devastating on the 1st Brigade's beaches and in the waters offshore—and the black Marines were in the thick of it, unloading cargo from LSTs standing off the reef.

The 3rd Division had landed in a natural amphitheater with the Japanese holding the high ground overlooking the beaches.

On the night of D Day one of the platoons of the reinforced 4th Ammunition Company, which was guarding the brigade ammo dump, intercepted and killed 14 Japanese soldiers laden with explosives.

> "We Montford Point Marines served in the Marine Corps in World War II, we are proud of that experience, and we do not want to be forgotten."

Black Marines had the job of unloading landing craft and amphibious vehicles while under almost constant enemy shellfire ... it and the clinging volcanic sand made life on the beaches a living hell.

Early on March 26, ten days after Iwo Jima was officially declared secure, a well-armed column of 200-300 Japanese, including many officer and senior NCOs, slipped past the Marine infantrymen who had them holed up near the northernmost airfield and launched a full-scale attack on the Army and Marine troops camped near the western beaches.

The units struck included elements of the Corps Shore Party, the 5th Pioneer Battalion, Army Air Forces Squadron, and an Army anti-aircraft artillery battalion. The action was wild and furious in the dark; it was hard to tell friend from foe since many Japanese were armed with American weapons. The black Marines were in the thick of the fighting and took part in the mop-up of the enemy remnants at daylight.

Two members of the 36th Marine Depot Company—Privates James M. Whitlock and James Davis, both received Bronze Star Medals for "heroic achievement in connection with operations against the enemy."

Colonel Leland S. Swindler, commander of the Corps Shore Party and also commander of the 8th Field Depot, was particularly pleased with the action of the black Marines in this battle and in his report for Iwo Jima stated that he was: "... highly gratified with the performance of these colored troops, whose normal function is that of labor troops, while in direct contact with the enemy for the first time. Proper security prevented their being taken unaware, and they conducted themselves with marked coolness and courage. Careful investigation shows that they displayed modesty in reporting their own part in the action."

The largest number of black Marines to serve in combat took part in the seizure of Okinawa in the Ryukyu Islands, the last Japanese bastion to fall before the atomic bomb and the threat of invasion of the home islands combined to bring the war to an end.

Some black Marines also served in Japan as part of occupation units, but these units were rather quickly combined, then deactivated, and the troops sent home for reassignment or discharge.

(Excerpted from *Blacks in the Marine Corps* by Henry I. Shaw Jr. and Ralph W. Donnelly; History and Museums Division, Headquarters U.S. Marine Corps.)

Two black Marines take cover on the beach at Iwo Jima on D-Day, while the shattered hulk of a DUKW smokes behind them. (USMC Photo)

First Lieutenant Eleanore Little in an-off duty moment at Jack Dempsey's Restaurant, New York City in 1943

The Women Marines
The Way We Were ...
And Are Today

When the guns stopped booming over the Pacific in the summer of 1945, there were 23,000 women in the United States Marine Corps.

They had come largely in a multitude, young women as courageous and red-blooded as their brothers, most just after January 29, 1943 when Secretary of the Navy Frank Knox gave the oath of office to Jerseyan Ruth Cheney Streeter of Morristown who would lead them as director of the Marine Corps Women's Reserve.

There was already a proud and dormant Marine Corps tradition of Women Marines. Three hundred and five donned Marine Corps greens in 1918.

Major Ruth Cheney Streeter (the first woman major in the corps) said the women actually released enough men to form the 6th Marine Division ... helping the Corps to pave the way (God forbid) for the invasion of Japan ..." It (the 6th) could not have been formed had not women taken over the home jobs of 18,000 men and 1,000 officers.

For them it was not a world of doilies but it certainly was a military world of respect ... You can take that now from Eleanore Little of North Freedom, Wisconsin, 1st lieutenant. She had WWII service with 3rd Marine Air Wing, and was the control tower officer at Parris Island, and Cherry Point ... plus much more.

"I was never treated better in my life," she recalls now after a successful life of raising two sons, operating motels, resorts, apartments ... " and with no greater respect than when I was a Marine. I never heard bad language, or even four letter words, at my stations ... none like I hear now from junior and high school boys as I walk by ...

Like the men, holds Eleanore, they are "one of the few and one of the finest.: Real strength of character is demanded of men and women alike. How well you know? Watch how she carries herself ... and watch her at work ... There is no better proof."

There were in January 1991, 8,679 enlisted Woman Marines, and 677 officers. Eleanore is proud of them all— and herself, too.

Colonel Hazel Benn Recalls How It Was
On Her Second Hitch In The Corps

The Women Marines organization—one of the most pleasant formations of the United States Marine Corps—was organized during WWII. Hazel Benn, who retired as a colonel, tells some of her experiences during her second hitch in the Marines.

In 1951, I was recalled to active duty at HQ Marine Corps to be head of education and internal information. I remained in the billet for over 23 years, and one of these 23 years (1967-1968), I was also head of Special Services at Headquarters Marine Corps.

When I first returned to active duty, women constituted about ten percent of the corps. I was particularly privileged, however, to write policy for the whole Marine Corps in my areas of responsibility, and also was assigned temporary duty as a member of the inspector general's team. In such duty I was attached to the divisions and wings to inspect education offices, enlisted messes, and special service areas. In the latter two areas I obtained appropriate answers and information from the heads of these sections at HQMC. Then, when I arrived in the field it looked as though I had considerable knowledge and experience of messes and Special Services.

However, it did not prepare me for some of the situations I encountered while inspecting these all-male, Fleet Marine Force commands. Some examples follow of what could have been "near disaster."

Support For The Troops

On my very first trip with the Inspector General, I talked to the division Special Services officer before going down to a regimental office. Education was no problem ... that was my primary area. Special Services became a different situation.

When I went to one office to see what gear they had on hand, there was nary a jockey strap in sight. I looked at other items, all of which had been purchased by Headquarters Marine Corps in the branch in which I served. Then I asked the Sergeant to open cabinets.

He did this willingly and readily, except for one—which was locked. I insisted that I wanted to see its contents, but this sergeant practically got down on his knees begging me not to make him open it.

I left it then and reported the incident back to the division Special Services officer. The next day the division officer told me he had been informed that when the sergeant found out a "woman" inspector was coming, he had taken all the male pinup pictures from all the cabinets in the stock room ... along with the jockey straps ... and locked them in that cabinet. The colonel said that the particular sergeant was so proper that had I made him open the cabinet he probably would have fainted from embarrassment.

I suggested that, since I had sometimes helped coach a boys' high school basketball team ... and since the purchase orders for the jockey straps and all Special Services gear he left out, were prepared at a desk about two away from mine at HQMC ... I was thoroughly familiar with them. We had all sorts of Special Services gear, including jockey straps, hanging on the walls.

(Little did I know at that time, or until I became head of Special Services, that Headquarters had been ordering jockey straps in "pairs." I was informed by the woman who prepared the purchase orders that my male predecessors had signed the orders without looking at details. She didn't know what they were used for, and she had concluded that it would be cheaper for the Marine Corps to buy them in pairs!)

His Embarrassing Memory

After the jockey strap experience, I went to another regimental Special Services office. A handsome lieutenant had all his Special Services gear out for inspection. However, while he was responding to my questions, I realized he was holding a football player's crotch guard in his hand. My reaction was to ignore this and keep talking.

As time wore on, he obviously didn't know how to get rid of it. His hand seemed stuck, and he was obviously uncomfortable. I turned from him and went to the next room; and said nothing more to him. But at the end of the day, I told the division officer because I thought it was laughable.

Years later at Headquarters Marine Corps, as I was walking down a corridor in the Navy Annex (I was in uniform), I was suddenly stopped by a young major. He blurted out in a high voice, while turning red, "You don't remember me, do you?" The former handsome lieutenant, now a major, had been assigned to HQMC. I never did understand why he stopped me for I hadn't noticed him. My suspicion had always been that his former division Special Services officer had given him a bad time over that item of football equipment.

Women Marines assemble at Mt. Holyoke College, Massachusetts in 1943. First Lieutenant Eleanore Little is at far left.

> *Although the next battalion was just around the corner, he drove and drove, back and forth, among the streets.*

Streaking To Inspection

The Regimental Special Services Officer (SSO) went with me to a barracks where pool and ping-pong tables were installed in one end of the floor, and Marines were billeted in double bunks at the other end. A partition with a door separated the two areas. The door had a large pane of glass approximately 36" x 36" in the center so one could see activity on either side. On the SS side I was inspecting the condition of the tables. It was obvious as I went down the room that troops were having junk on the bunk inspection, since they were standing rigidly at the foot of the bunks.

Suddenly I noticed a naked man run from the shower, located just beyond the door on the right side. He finally got a towel around him as he raced down the corridor between the men and the bunks. To my amazement another stark naked Marine charged out of the same shower without a towel, and followed the other naked one.

Keeping on with my duties, I then walked back out the door at the SS end of the building. When we got outside, the SSO said, "Didn't you see those naked Marines?"

I replied, "Yes."

He said, "But you never let on—your expression never changed." My answer, "I didn't have to let on. They're the ones to worry. They weren't dressed for an imminent inspection, and they'll never live down the fact that they were observed by a female Marine inspector."

I wondered later if the situation was "planned" or if it just did occur by accident.

Driven To Distraction

On one of my earlier IG trips, I was due out to the area where the supporting battalions were housed. An obese driver with a station wagon picked me up at the BOQ. I was somewhat startled to see him dressed in dungarees, at least during

an I.G. All commands usually put their best foot forward to make a good impression in order to pass the inspection. The temperature was expected to reach the middle 90's, so it was warm early in the morning. I had been in the area to be inspected before so I knew the ease with which one could get around between battalions.

When we reached the first organization, I suggested that my driver might want to come in and sit in one of the offices and read while I inspected. He came to the office with me but ... instead of sitting ... followed me from room to room and building to building. I became quite firm against his continually looking over my shoulder. I suggested that he bring a book to read the next day.

No Time For Books

However, further talk revealed that he was illiterate and couldn't read—even if he had a book. So I directed that he stay out of my way but that he not get into anyone else's way either, because the troop inspectors were also in the area.

When I was ready to go to the next battalion, I deliberately got into the car to make him go with me, so I'd know where he was when I wanted to return to the main camp. Although the next battalion was just around the corner, he drove and drove—back and forth—among the streets. I got angry and we came to the desired building in no time at all.

He pulled the same driving stunt going to the 3rd Battalion, which was commanded by one of my friends. The commanding officer came out to meet me at the car. This battalion had the mess school for the division. The commanding officer suggested that after I inspected, I eat with him at the school mess. He also asked my driver if he would like to eat as well.

The driver's reply bordered on insubordination, but my friend did nothing. In essence the driver said that he didn't want to eat in any division school-operated mess. He didn't even offer, "Thanks for asking."

The commanding officer and I ate our lunch. When we returned to the car, the driver was sitting in it, so we assumed he had stayed in the car in the heat.

I asked him to take me to the next battalion—again around the corner. He asked me when I planned to return to main camp. I replied that I planned to go

at 1700, but if he kept driving all over the area, when both he and I knew better, we would leave at 1800 or later.

He Had A Problem

He said, "Oh gosh, I'll have to sit in the car in this 95 degree heat while you inspect, or stand up against a wall, for hours." After considerable exchange of words about his remarks, for he'd never stayed in the car before, he explained his problem to me.

"You know that lieutenant colonel asked me if I wanted to eat. I told him I didn't want to eat in his old mess hall. But the car got so hot, I decided that I might as well eat. I went in and stood in the chow line. I was leaning up against the wall and the line moved. I stood up straight to move. There was a great big hook behind me, and it ripped a big hunk out of the seat of my pants. So I either have to sit in the car in the heat, or stand up against a wall."

> *There's nothing like a good mustang officer and a sergeant major.*

I reacted quite angrily and told him to get me around the corner to where I wanted to go, with no foolishness. This he did. When he stopped the car, I ordered him to stay in it until he was given permission by an officer to get out. I went into the building and found the sergeant major and the adjutant, who was always a mustang officer. I said to the adjutant, "I asked to see you because I need help. I'm going to inspect the SS and education areas of your command, but I need you personally to do an inspection for me." I told him what the driver told me about the seat of his pants, and asked him to determine what the situation really was.

I saw him make the driver get out of the car, and he looked at the seat of the pants. From the window where the sergeant major and I were watching, it was obvious that the driver was in deep trouble. The driver rushed back into the car and took off at good speed.

The adjutant came back into the office. He said that the first problem was that the driver was out of uniform. No driver of the division or base was authorized to wear other than the uniform of the day when driving either a sedan or a station wagon. And yes, there was a huge hole in the seat of his pants—about 4" x 6". Furthermore, he had informed the driver that, while I was inspecting, he go back to main camp ... get into the proper uniform ... and he would inspect him upon return. He knew that the driver would have to rush, but he considered it "good for persons who couldn't obey well known orders."

The driver returned. He was not very talkative on our return trip for his own command had been called, by the command being inspected, to put him on report.

(There's nothing like a good mustang officer and a sergeant major)

Smarter Than His Kids ...
And Quicker

The most rewarding gift I received during IG inspections was in the area of education. In one of the regiments I found a tech sergeant (gunny) who was the court recorder for the regiment and the education NCO.

Early in our meeting I learned that although he was the court recorder, he had completed only six years of school. The fact that he had to coordinate the high school and post secondary education programs was a real problem. Education officers (commissioned at the regimental level) had 10 or 12 duties, including education, so spent little time on education programs.

In spite of his lack of formal education, the gunny was doing an excellent to outstanding job.

He had an obvious inferiority complex, however, because of his lack of education in performing his Marine Corps education duties. Worse than the lack of formal education to help Marines was the fact that he had two sons graduating from high school in a few months. When this took place, they would be above him academically, and he seemed not sure how he would cope with the latter situation.

I stopped inspecting and sat down with him to talk about his own education. I explained the General Education Tests (GED) to him and how he could possibly have his high school diploma before his sons. His particular state of residence had a policy where an adult could get a real high school diploma with adequate scores on the GED's. I explained that senior staff NCO's usually came out with scores in the 90th percentile; much higher than required. I

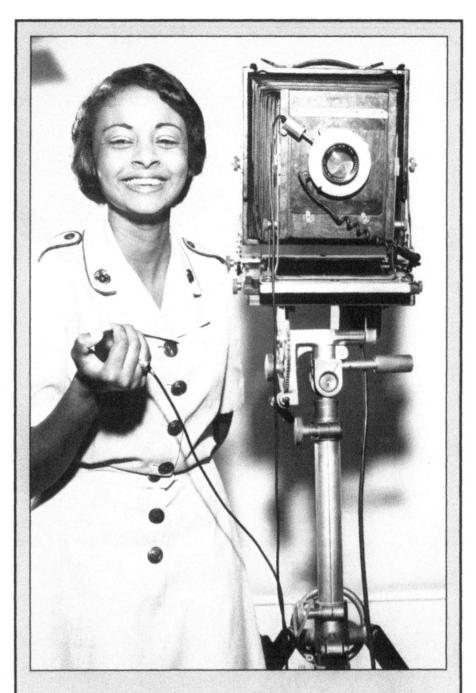

Give A Big Smile

This woman Marine photographer shows her subject what she wants in the way of facial expression. A private first class in 1966, she received on-the-job training for her assignment at Marine Corps Recruit Depot, San Diego.

also explained how his particular primary duty of court recording would give him a bonus few others would have. He promised he'd take the tests.

Before I left the command I talked to the sergeant's officer and asked him to encourage the gunny. In very short order, after I got back to HQMC, I got a phone call from the sergeant. His lowest score on five tests was a 94 plus percentile. The other four scores were 96 or above. He had the scores sent to his state, and, through his high school principal, received a real diploma. His call was to say, "Thank you." He felt that he could work better with civilian school officials, the Marines, and his sons were prouder of him.

This was a "Thank you" I'll cherish all my life, and will never forget.

THE MARINE FIGHTING KNIFE, UTILITY

In the early days of World War II, when a fighting knife was issued to every Marine going overseas, they were called Ka-Bars ... because many of the early ones were made by Ka-Bar Company, Olean, New York.

Twelve inches long overall with a 7-inch blade, a good many were used in hand-to-hand combat. But these sturdy, handy knives had a multitude of other uses—opening cans, cutting tent pegs, knife throwing at coconut trees, digging holes, cutting leather, opening coconuts, and so on ...

The military demand was so heavy for utility "fighting" knives that other companies were also called on to produce these all-purpose tools.

The Camillus Cutlery Company of Camillus, New York ... who had been making knives since 1876 ... produced fighting knives for various branches of the service, and to this day manufactures survival knives and fighting knives for our troops and foreign units as well.

Other companies—such as Pal and Robeson Shuredge—made the 12-inch knives for the Marine Corps also. Some of the 7-inch blades were blued, some parkerized, and others plated. All met the Marine Corps "specs" for overall and blade lengths, but each manufacturer had slightly different marking on the handle and blade.

The markets for knives were extensive during World War II. The Camillus firm alone supplied almost 12 million pocket knives and two million fighting and other sheath knives.

Some firms, like Camillus, and Ka-Bar, continued to make fighting knives for Marines in Korea, Vietnam, and the Persian Gulf ... and are making them today. It's sort of a 50th anniversary for some of them too.

THE WAR DOGS

This is one helluva story ...

It's about the stuff of which dogs are made: puppy dogs tails, courage, loyalty, and the last full measure of devotion.

There are hundreds—maybe thousands—of guys reaching their seventies today who might never have reached 25 had not one day a dog, somewhere in the islands, been a part in their lives.

You can take that from Dr. William W. Putney (then captain) today a doctor of veterinary medicine in Woodland Hills, California. He headed the 3rd Platoon. Or you can take it from William T. Taylor (then a captain) now of Baton Rouge, LA, who headed the Second Dog Platoon and unofficially commanded the unofficial dog company.

The two platoons slated for the 3rd Division had 110 men and 60 dogs when they left San Diego in February 1944. They joined the 3rd Division in the coconut grove at Tetere Beach, Guadalcanal.

Not only were the handlers trained dog specialists, but were skilled scouts as well. Man and dog they searched out the enemy, waited his coming, caught him by surprise or when creeping into our perimeter. Together they found snipers, routed bushwhacking stragglers, searched our caves and pillboxes, ran messages and protected foxholes like private homes.

The dogs ate, slept, walked, talked, lived with their handlers. If either died, the other cried. Three handlers and 15 dogs were killed with the division on Guam—particularly because the Japanese knew the threat of the canines and fired on them first.

Dogs brought cheers from Marines on the line. Their presence meant a good night's sleep. The dogs would be sure to call if the enemy were near.

Dr. Putney recalls how the dogs, including the random non-working stragglers who joined, were always as close as blood relatives. One such was Bobby, a pick-up mutt, found crushed in the tread rut of a tank. There wasn't a dry eye at the burial as he was nestled into a makeshift coffin. However, at mid-ceremonies Bobby showed up out of the bush, and peered down into the grave. Bobby survived the war.

Handlers found Japanese war dogs also. One was trained to hunt down his old masters. Another was ever-loyal after Dr. Putney removed a .45 slug from under his ear.

Private First Class Marvin M. Corff, now a retired DVM in Oregon, recalls how "Rocky"—a Doberman—spotted an ambush and saved innumerable lives. And, all recall "Missy," who was found dead ... punctured by half a dozen Japanese rounds.

Captain Taylor says the dogs carried messages through shot and shell and personnel (Japanese or American). Particularly on Bougainville. pointed out 1st Lieutenant Clyde Henderson of the 1st Platoon where radio communication was a mite better than yelling, but not much. Captain Taylor later took the 3rd Platoon to Iwo Jima, while the 2nd Platoon eventually joined the 2nd Division.

Eighty-five percent of the Marine Corps war dogs were Doberman Pinschers, and the rest were German Shepherds. Handlers who requested to keep their dogs after the war were generally allowed to do so by the owners who had enlisted them.

For the record, here were assignments on embarkation for the invasion of Guam: 1st Lieutenant William Taylor with 12 dogs and handlers and three NCO's—9th Marines; Platoon Sergeant Raymond L. Barnowsky with 12 dogs and handlers and three NCO's—21st Marines; and Sergeant Alfred E. Edwards, 12 dogs and handlers and three NCO's—3rd Marines.

The remaining 24 war dog handlers and other personnel, plus their War Dog Hospital (just above Fonte Ridge on Nimitz Hill) with supporting

First Lieutenant William W. Putney in his war dog hospital, or sick bay, near Talefofo Bay on Guam. He later became chief veterinarian of the United States Marine Corps.

supplies and equipment, were under direction of 1st Lieutenant William W. Putney.

Later promoted to Captain, Dr. Putney became chief veterinarian of the United States Marine Corps. He is today in private practice.

Home At Last ... Dogs Are Discharged

These discharges from Marine Barracks, NAD handsomely pose for their final photograph on Mare Island. All Doberman Pinschers, they are some of the 23 dogs that were soon to be discharged from their sergeants' ratings and returned to civilian life. Marine trainers with the dogs are, from the left—PFC Michael R. Ryan, Pvt. Walter J. Reagan, PFC Raymond W. Harris, Pvt. James E. Burden, PFC James Gutierrez Jr., PFC Donald E. Elmer, PFC Melville J. Garrett, PFC Eben F. Scherber Jr., PFC Paul E. Huey, and PFC John A. Ciotti

From Trunk To Display

What did you do with your dress greens?

Fred H. McCrory of Vestavia Hills, AL was in the Marine Corps for 37 months ... 30 months of that being overseas in the Pacific. The greens didn't get much wear, dungarees being more the uniform of the day on Bougainville and other islands.

Fred has donated his greens to the state of Alabama, Department of Archives and History, and they will be on display at the military artifacts collection in Montgomery.

Before relinquishing it, Fred tried on the size 40 uniform and found that it fit perfectly.

"We are delighted to have the uniform in our collection. It is rare to find a Marine uniform from an Alabama enlisted man that is in such good condition," said Bob Bradley, the archives' chief curator.

McCrory, who is serving his second term as a Vestavia Hills councilman, made the donation to the state archives because, "World War II is being forgotten, although it was a difficult part of the history of our country. Our young people need to have reminders of the price paid for our freedom. Freedom doesn't come easy."

Senator Howell Heflin and "A" Company Boys

RECALL HOW IT WAS ...

It was a "family" reunion in Washington just a few years ago (1986) when enough of Able Company (1/9) were there to skirmish up a hill ... gathered and heard an old 1st Lieutenant share with them what happened and why they were blessed and lucky enough to be there at all.

Howell T. Heflin was the speaker, big as ever, distinctive as ever, as he talked about things we knew so well—atabrine, jungle, and pride. He was introduced by Lieutenant William S. Pfeifle, Woodside, California, former New Zealand hut-mate, and they all talked about Captain (then) Conrad Fowler, Lanett, Alabama, who had been company commanding officer, and about everybody else as well.

Memories And Emotions

I suppose as we get nostalgic ... and realize that we live a relatively short period of time ... a lot of things are memorable. Probably the emotion that remains with us the most is the emotion of pride. A Marine has great reason to have that pride.

For example, when you came out of boot camp you felt you could whop any swabbie, any dog face you might see. You had been through boot camp, and you could take anything that anybody might try to dish out to you.

You found out later there was a lot more "dishing out" that would occur, and you would go through many hard-

ships. Then you joined a very special outfit, the 3rd Division, an outfit in which you could take a lot of pride. We probably had as much pride in being a part of "A" Company, as any measure of pride we had ... a little something extra ... icing on the cake ... the pride of being a real Marine's Marine.

We had a special company, a special group of men, a special company commander in Conrad Fowler. While he was described as spit and polish, he created a special espirit-de-corps among his men. One felt confidence in him. He was a great leader, and whatever he did you felt confident that he had thought it through well ... and that he would be there with you. I think that was true of all our officers and NCO's, as we look back at them.

Looking back ... I joined the company and division at Camp Pendleton. After about two weeks we had to go down to San Diego to help in the loading of a ship.

Then on our way to New Zealand, as we crossed the equator, most of us were initiated into the Royal Order of Neptune, and graduated from pollywogs to shellbacks.

You remember how hot it used to get in the holds of that Mount Vernon ship as we got into the warm climates ... and perhaps you don't want to recall— and certainly not relive—those particular days of cramped quarters and hot holds.

We arrived in New Zealand, and found that being on the other side of the equator in late December or early January, made for some extremely hot weather. Remarkably everyone down there was singing "I'm Dreaming of a White Christmas" even though there was not a snowflake in sight.

In the little village of Papatura, some of us enjoyed the fare of steak and related food ... we remember the strange dialect of the people who lived there ... vehicles driving on the wrong side of the road ... and mutton. I tell you I ate more mutton in New Zealand and on Guadalcanal; so much, that it was about 15 years later before I could think of putting a piece of mutton in my mouth. But nowadays they talk lamb chops ... and I like lamb chops.

We had some hikes on New Zealand, as you remember. I remember one month, we had about 15 hikes of about 20-30 miles each. I wrote to my folks; who, in turn wrote to Captain Fowler ... saying we had been on a 300 mile hike. On being quizzed, I said, well I was just adding it all together. But my feet did get awful blisters on some of those hikes.

At Guadalcanal "Washing Machine Charley" would fly over periodically to harass us.

Also at Guadalcanal, we took enough atabrine (to prevent malaria) to turn our complexions yellow. And while there, we got more mutton and the Army got the beef. Sometimes Marines were called on to unload ships that came up from Australia and New Zealand. I remember as goods were being transported from the ship to an Army unit on the island of Guadalcanal that sometimes a side of beef was tossed from the truck into tall grass along the side of the road. That was a way to have a respite from the diet of mutton; and when we went to pick up these provisions from the tall grass, it was referred to as "Midnight Requisitions."

Then we went to Bougainville on November 1, 1943. There we had several casualties ... including some caused by bombs dropped on us by our own planes.

A lot of us can remember the Thanksgiving Day dinner that we started to eat, when we were told we would have to immediately go up to what was known as the Battle of Piva Forks, as well as Hand Grenade Hill. Obviously, Thanksgiving meal was interrupted.

Back at Guadalcanal, the raisin jack was an interesting memory. Also, shortly after we got back to Guadalcanal the rains came. I remember waking up one morning and finding a regular river of water going right through our tent.

We started training and heard we were going to Rabaul. I'm glad we didn't, because that was probably one of the most heavily fortified positions that the Japanese had. Rabaul and Truk were two very well fortified Japanese locations ... and the Japs would really have been waiting for us.

Do you remember when we went from Guadalcanal to Guam on a very long hot ride? We were kept in reserve for several weeks ... partly in case we were needed as reinforcement for the invasion going on at Saipan and Tinian.

We did land on Guam, pretty much as planned ... although I didn't last very long. I had gone over the first hill, but got hit on the second. A couple of stretcher bearers carried me out.

Many of you went onto Iwo Jima where the organization was awarded the Presidential Unit Citation. I don't remember the exact number, but A Company suffered more than 100% casualties on Iwo Jima ... when you figured replacements added to the company during the campaign. No question: taking this island meant a great deal to the United States. From there, we were able to fly planes to accompany bombers flying to Japan, and eventually help end the war.

We often get asked the question of whether Harry Truman made the right decision in dropping an atomic bomb? I have no problem ... I think he used by far the best judgment. There would have

been a million casualties if we had gone in and invaded Japan. Atomic weapons produced many casualties at Nagasaki and Hiroshima, but with the number of American lives that were saved, President Truman made the right decision ... and brought an end to World War II.

Speaking again of emotions. We look back at casualties and friends whose lives were lost. Another emotion ... when you are talking with friends about someone who has been hit by a land mine ... this certainly leaves a sad feeling in the depth of your heart ... uneasiness about it.

An amphibious landing, as you head toward the beach, and the front of the boat goes down so you can go ashore ... the mortar and artillery shells landing about you ... certainly are emotional experiences.

I'd say if any one thing came out from our experiences in the Pacific is that "there truly are no atheists in foxholes." I know that we all prayed a lot ... for our friends, for ourselves, and for our families. I'm convinced that somehow Almighty God gave divine guidance to each of us, and that we're here as a result of his mercy.

We all have great pride in our organization and as the Marine Hymn says, "If the Army and the Navy ever gaze on heaven's scenes, they will find the streets are guarded by "A" Company Marines."

Son of a Methodist preacher, graduate of Birmingham Southern, Howell T. Heflin, commissioned in October 1941, was still in training when the Japanese hit Pearl Harbor. Heflin commanded the 3rd Platoon, A-1-9 and received the Silver Star on Bougainville for action on "Hand Grenade Hill," and was wounded twice on Guam. Graduate of University of Alabama Law School, Howell Heflin was to become the chief justice of the state and is credited with revamping the state's judicial system. He was elected to the United States Senate in 1978, where he has served since.

JASCO, Joint Assault Signal Company, is another innovation of the Marine Corps. Third Jasco attached to the 3rd Marine Division, was commanded by (then) Major John Ellis from its inception, throughout Guam and Iwo Jima, to its return to the United States in 1945. He is the author of its brief history that follows. The function of this unit was to support assault forces by calling in fire from the Navy's big guns, from the air and from artillery. Every surviving assault Marine owes his life in no small part to the efforts of JASCO.

3RD JASCO
A MARINE CORPS ORIGINAL

JASCO was born and cut its teeth during World War II, and parented the modern day ANGLICO, the Marine Corps Air Naval Gunfire Liaison Company that participated with distinction during the Desert Storm Operation.

The Marines learned early in World War II, at Guadalcanal, the importance of supporting arms to the infantry battalion assault. Close air support and the employment of naval gunfire "came of age." JASCO provided the spotter, the liaison, and the communications for the commander to utilize this additional firepower.

1st JASCO, activated at Camp Pendleton in October 1943, was sent overseas to join the Marine Corps Fifth Amphibious Corps and was attached to the 4th Marine Division in Hawaii.

The 2nd JASCO completed training at Camp Pendleton and was sent overseas in February 1944. One officer (Lieutenant Jenkins) from 2nd JASCO was detached and assigned duty with 3rd JASCO for continuity and assistance in training the latter company.

In March 1944 the officers and men of 3rd JASCO reported to the Communications School, Amphibious Training Command at Camp Pendleton "for further instructions."

The Marine air liaison officers, the Marine forward observer spotters, and the naval gunfire liaison officers were given special training at the Naval Air Station, North Island and at the Naval Amphibious Base where the air support and naval gunfire schools had expanded to meet the demands of the war.

The JASCO was composed of three major sections: Shore Party Communications ... Shore Fire Control ... and Air Liaison—plus the company headquarters. During an amphibious landing, a team from each section was assigned to each of the nine infantry battalions of the Marine Division as well as to the three infantry regiments and to the division headquarters.

The Shore Fire Control personnel, at the battalion level, were divided into a spotting team often employed at the infantry company level, and a Navy gunfire liaison officer team. The Marine air liaison officer not only called in the air strikes but also directed the pilot onto target from his front line position.

The shore party teams provided the needed logistical communications laterally among landing beaches, but also inland to the supported infantry battalion or regiment.

The first three JASCOs also included a Navy enlisted component of 131 signalmen and radiomen that were intended to provide ship-to-shore communications for the Navy beachmaster on each of the battalion landing beaches. (3rd JASCOs Navy enlisted were transferred back to Navy control when the company arrived at Guadalcanal in May 1944).

The group had traveled from San Diego to the canal on the "Mormacwren" via Espiritu Santo, in an "unescorted," but uneventful, Pacific crossing. The unit included 43 officers and 456 enlisted.

At Guadalcanal the 3rd Marine Division was well into the final preparations for the Guam (Stevedore) landing in the Marianas. Air, Naval gunfire, and shore party teams were assigned to the infantry battalions and regiments and participated in division rehearsal exercises in May on the beaches of Guadalcanal. The division special staff officers—as well as concerned air, naval gunfire and shore party—closely monitored the assignment and employment of the teams.

In mid-June the entire division awaited orders off Saipan where that operation was going slower than expected, and the Naval Task Force Commander sent the division to Eniwetok for exercise and recreation ashore. One report notes that the war dogs were least affected by the 50 days aboard ship in hot, overcrowded conditions where the supply of cigarettes was running low.

Then in July—with a typhoon threatening—the task force carrying the 3rd Division arrived off Guam in preparation for the 0830 H-Hour landing next

morning between Adelupe Point and Asan Point. The first operation of the war to recapture United States Territory from the Japanese was ready.

On the landing the three JASCO components were committed in the manner intended by their christeners at headquarters Marine Corps when "birthing" the communications company. A 3rd Marine Division directive assigned an air liaison (AL) team, naval gunfire (NGF) liaison and forward observer teams and a shore and beach party team to each infantry battalion, a NGF and an AL team to each infantry regiment, plus the AL and NGF teams at division headquarters.

(Some commanders questioned the need to use embarkation space for these unknown communicators; but, for the Iwo Jima assault, commanders asked for larger teams.)

JASCO did its job well on Guam and proved its worthiness as a number of action reports showed:

"The NGF officers of the JASCO kept the star shells placed so that they silhouetted the advancing masses of the Japanese and kept infiltration down. From this (D-day, July 21) night on, the JASCO detachments played an increasingly important role in the operation."

"A JASCO officer took command of the company (A Company, 1/3 on July 22) when the company commander was evacuated ..."

"... the JASCO team kept naval guns dropping high explosives on the reverse slope (2/9, July 26)."

"For the operation the 3rd JASCO remained only an administrative unit ..." (The infantry commanders at Guam, through their 3rd JASCO liaison teams, were able to make effective use of available close air support and naval gunfire.)

"Scout, torpedo and fighter bombers were most effective against flat targets not reached by flat trajectory naval gunfire."

"Teams of 3rd JASCO landed less than 20 minutes after H-hour, much sooner than necessary. They had their ship-to-shore and lateral beach communications set up by 1100 hours, but there was no traffic until more than two hours later. Four JASCO teams waited on the beach for more than four hours before their services were required."

Following the Guam landing, the 3rd JASCO Air Liaison teams were detached to General MacArthur's command and worked with Army units in MacArthur's return to the Philippines—the Morotai and Leyte campaigns.

After their baptism ... and successful growing up ... on Guam, JASCO went on to play an even more significant role in action at Iwo Jima. Their need had been proven.

"Liberty"—At Any Price

Upon our return to the States after the war, our contingent of about 1700 men was taken to Camp Mathews at LaJolla, California. Most of us had been overseas for about two years and had not been to a liberty port or even slept in a house.

We were denied liberty because many of us did not have a dress uniform. Some of us did have a uniform; my dress greens having been in the bottom of my seabag for two years.

Knowing that we were about to be processed for discharge; that Camp Mathews was enclosed by a 3-strand barbed wire fence; and that the posted guards did not have ammunition—some of us decided to go on liberty anyway.

Corsairs making rocket strikes against Japanese positions. (Defense Dept. Photo-Marine Corps)

Upon my return a couple of days later I learned there had been no formal assemblies and no roll calls. Rumor had it that some of the guards reported that we were leaving in substantial numbers and were told by the officer of the day, "You had better let those people alone or one of them may take your rifle away and stick it where you wouldn't like it." **E.T. Passons, Sulphur Springs, TX, Passons served with the Combat Correspondents in WWII.**

Deactivation Of 3rd Marine Division After WWII
By Robert B. Van Atta

Just over four months after the climactic conclusion of WWII and the major accomplishments of the 3rd Marine Division which sped that end, the fighting unit was deactivated.

Brigadier General William E. Riley, division commander, on December 28, 1945, announced the end in a dispatch to the commandant:

"I regret to inform you that the 3rd Marine Division passed out of this world—for the time being at least—at 2400, December 28, 1945 ... The same high morale that characterized this division during the Pacific Campaign was fully evident in all units until the final disbandment. As a result, the passing was painless and the spirit for which this division was justly famous was maintained until the end."

The deactivation process was a continuous one, involving occupation forces, discharge point system changes, and some rotation. It was slower getting started from the occupation standpoint because the division's assignment at war's end was to remain as a force in readiness in the Marianas in event of unexpected difficulty or treachery in the occupation process.

When these fears had not materialized by October, the 1st Battalion, 3rd Marines, was ordered to Chichi Jima in the Bonin Islands, and the 2nd Battalion, 21st Marines, to Truk, bypassed Japanese strongholds. The division was authorized to reduce its strength to a force of two reinforced regiments, and the 3rd Battalion, 9th Marines, became the first to be disbanded.

Coincident with this, a continual reduction in numbers resulted from periodic drops in the point score for reserves and regulars with expired enlistments. The flow of men reassigned within the division or sent to the Transient Center of Guam for other assignment or return to the United States.

A draft from the 9th Marines was sent to North China in November, and the remainder of that regiment sailed for the United States and disbandment in December. At Christmas time, the division furnished two more drafts for occupation duty in China.

One last gasp of combat action came after an Island Command patrol was ambushed by hidden Japanese remnants, December 10. The 3rd Battalion, 3rd Marines, took part with other Guam-based units in a sweep of jungle areas.

Typical of the deactivation process was that of the 12th Marines, in which the writer of this segment was involved as acting regimental sergeant major. Regimental H&S-12 and other elements were folded into the surviving 2nd Battalion, 12th Marines, November 17.

Then, after continued depletion, the surviving battalion's last 12 men were transferred by staff returns to Division Headquarters Battalion, December 15, along with a like number from the 21st Marines. By Christmas, only six men remained in the 12th area, chowing at Headquarters Battalion. They had gathered in one tent for the final nights' sea stories and sleep before moving to the Transient Center, December 27, as one of the final elements of the division.

The next day, the 3rd Marine Division ceased to exist until it was formally reactivated January 7, 1952, during the Korean War.

REFERENCES: The 3rd Marine Division, by Aurthur and Cohlmia, 1948.
The 3rd Marine Division News-Bulletin, 1945 issues.
Intra-Division Transfer Order #35-45, December 14, 1945.
12th Marine Regimental Change Sheet #321-45, November 17, 1945.
2nd Battalion, 12th Marine, Change Sheet #349-45, December 15, 1945.
Personal recollections of Robert B. Van Atta.

GENERAL EDWARD CRAIG

A LEGEND IN HIS OWN TIME

Retired General Edward Craig—beloved and respected—is a cornerstone of the 3rd Marine Division. During World War II, he commanded the 9th Regiment in the Bougainville and Guam campaigns, and was operations officer for the V Amphibious Corps during the Iwo Jima Campaign.

In later years he was president of the 3rd Marine Division Association and has attended numerous chapter and regional reunions.

The 3rd Division was deactivated after World War II and thus was not active during the Korean conflict. But Edward Craig was active—he was commanding general of the advance elements of the 1st Marine Division ... a "reinforced regimental combat team" numbering about 5,000 men which landed at Pusan in South Korea in August 1950.

Time Magazine devoted its August 14, 1950 front cover to a handsome Brigadier General Craig; and part of its story makes some nostalgic reading for World War II veterans of the 3rd Marine Division—especially those in the 9th Regiment.

Excerpts from the story ... Last week advance elements of the 1st Marine Division—a "reinforced regimental combat team" numbering about 5,000 men—landed at Pusan in South Korea. The Marines, who carried, along with their shiny new equipment, a large ration of glory in their packs, arrived at the critical moment in the Korean fighting—which was also a critical moment in United States history.*

*Marines have landed in Korea five times in the last 80 years to protect United States interests, but only once before did they have to fight. In June 1871 as a result of Korean attacks on United States ships and sailors, a United States assault force including 105 Marines stormed several forts on the Han River. Marine casualties: one dead, one wounded.

Korea would not turn out to be much less tough than Guadalcanal; it might turn out to be tougher. The commander of the Marines in Korea, Brigadier General Edward Craig, has recently expressed an opinion on how his men would handle themselves in this situation. Said he: "They are ready for anything that can be thrown at them."

"You Gotta Be Mad." The Marines are supposed to be the assault troops of the United States Navy, and that is the role they like best. They have always considered themselves an elite corps. One Marine historian has described them as "a number of diverse people who ran curiously to type, with drilled shoulders and a bone-deep sunburn, and a tolerant scorn of nearly everything on earth. They were the leathernecks, the old breed of American regular, regarding the service as home and war an occupation ..."

Conversation in a Foxhole. Peace, of a sort, came to the 1st Marine Division. Its first postwar station was Tientsin, China, where it helped Chinese forces manage Japan's surrender. One regiment returned to the United States and was disbanded; most of the career fighters were sent to Guam where they lived in wretched, ramshackle huts. On Guam, they came to know better the tall, quiet, professional general whom they had "taken aboard" in China as assistant division commander.

Edward Arthur Craig had also fought his way across the Pacific, in battles other than theirs. He had been a combat commander (9th Regiment, 3rd Marine Division) at Bougainville and Guam, a crack operations officer for the V Amphibious Corps at Iwo Jima. He won the Navy Cross, Bronze Star, Legion of Merit.

An officer who fought with Craig on Guam relates: "After I ran ashore, the bullets were raining in from several pillboxes, so I dived into the nearest foxhole. Who in hell was in there but Eddie Craig. He was lying there with a phone and a notebook, talking to a runner. He was so quiet and collected he could have been at a desk in the Pentagon. 'We got to get those damn pillboxes!' I yelled at him. 'Now sit down there a minute,' Craig says, 'we'll get to 'em.' He just looked at me and smiled. In a few minutes we had the pillboxes. There's one thing about go-

ing to war with that man; there's no need to worry about who's running the show."

Craig used to drive around the front lines in a mud spattered jeep, toting a carbine (he is an even better shot than most Marines). Some Marines claim that Eddie Craig has steel wool instead of hair on his chest and a 40mm gun barrel for a backbone. But he is no military tyrant. Like many another Marine Corps officer, Craig believes that the welfare of enlisted men comes first. On Bougainville (which rhymes, in Marine parlance, with Hoganville), officers slept in foxholes if the men slept in foxholes, ate whatever rations the men ate. On postwar Guam, although the roof leaked in Craig's hut, he refused to detail carpenters to repair it until they had finished work on the enlisted men's recreation club (with six bowling alleys.)

Also Edgar Guest. Eddie Craig was born (1896) in Danbury, Connecticut. His father (who is 78 and to whom Craig still writes; a letter once a week) is a one time Army medical officer. Eddie went to St. John's Military Academy at Delafield, Wisconsin, jumped at a chance for a second lieutenancy in the Marines, sulked because he saw no action in World War I. His first overseas duty was in the Dominican Republic.

Craig is now regarded as a model of decorum, but there is evidence that in his youth he was something of a gay blade. On weekends he used to ride at breakneck speed into the town of San Pedro de Macoris on a noisy, dust-spurting motorcycle, seriously disturbing a Marine Captain attached to Santo Domingo's Guardia Nacional, who rode into town at the same time on a mule named Josephine. The mule rider, Gregon Williams, is now chief of staff of the 1st Marine Division and he and Craig are close friends.

Craig climbed up the ladder of routine peacetime duty from Haiti to China, acquired the reputation of a steady, thoroughly professional soldier. His first wife was an invalid for a dozen years; during those years, Craig spent all his spare time at her bedside. He rarely appeared at officers clubs. She died in 1943; he got the news just as he was about to go into battle at Bougainville. He married again in 1947 and has become a contented homebody. Of an evening, he likes his wife to read to him from the poems of homespun versifier Edgar A. Guest.

What Makes a Marine. In postwar Japan, Craig spent several months teaching amphibious tactics to Douglas MacArthur's 1st Cavalry and 24th Infantry Divisions, now in Korea. From April

1949 to his departure for Korea, Craig was the 1st's assistant division commander at Camp Pendleton, California, in charge of training under bull roaring Graves B. (the big E) Erskine, a stickler for perfection who "turned over" (i.e., relieved) 15 colonels in one year. To Marines, the fact that Craig survived under Erskine is the proof that he is good.

At Pendleton, the 1st's postwar training was the most rugged and exacting that any peacetime United States outfit got. Explained one Marine officer, "A kid reports for boot camp and we challenge the s.o.b., we dare him to try and be a Marine. We give him so much of that in boot camp—and even flunk some of them out—that when he gets out, he's the proudest damn guy in the world, because he can call himself a United States Marine. He's nothing but a damn private but you'd think he's just made colonel."

The Marines continued training under live ammunition, a practice which the Army discarded but recently resumed. This year they rehearsed an amphibious demonstration, "Operation Demon III" for the Army's Command and General Staff College. One of the division's companies ran off a cold weather landing exercise in Alaska; a regiment put on an airlift assault on cactus covered San Nicolas Island off the California coast. If and when the time comes for the United States units to break out of the beachhead in Korea, Craig's great store of amphibious know how will come in handy for assault landings behind the North Korean lines.

I'll Do My Best. Craig drives his men unflaggingly. They grumble about it, but they worship Craig. At Pendleton's Navy relief carnival last month, when Craig had already been ordered to sail for Korea with his combat team, a Marine corporal approached Mrs. Craig and said; "Excuse me, ma'am, but I'd like to talk to you, if you don't mind." As Mrs. Craig continues the story, "The man said he was a little drunk and he was, but he wanted to say that he had been posted away from the brigade to another outfit. Then he said: "But you see ma'am, I want to serve with General Craig. I want to go with him.'"

Mrs. Craig was moved to tears, and the corporal gallantly offered her his handkerchief. He went with Craig to Korea.

In Korea last week, General Craig had quite a reputation to uphold—for himself and for the Corps. Said he: "My boys will do their best. I'll do my best. Let's hope we'll be good enough."

Copyright 1950, Time Warner Inc. Reprinted by permission.

The Division In Deactivation

The December 1945 deactivation of the 3rd Marine Division was accomplished on Guam, Marianas Islands when the division colors were subsequently retired to museum status at Quantico, Virginia.

This did not remove the identity of the division from the military/public eye, as returning members who had time to serve, or those who opted for continued regular service, proudly exhibited the CALTRAP shoulder patch design. This colorful emblem would be officially authorized to be worn for yet a few short years, along with all other combat unit-designating insignia, both ground and aviation.

Rebirth ... A Brigade: The second genesis or rebirth of the division was fore-ordained with the formation and activation of the 3rd Marine Brigade, commanded by Lewis B. "Chesty" Puller, now a brigadier general at Camp Pendleton, Oceanside, California.

In early 1951 General Puller was enthusiastically dedicated to putting his new command on a war time footing and returning to combat. This was central to offsetting alleged statements to the press, where in he had been "quoted" regarding the virility of personnel entering the military service in that era. Although the validity of the press interpretations remains questionable, there is ample evidence that the contents brought forth the national ire of the membership of the WCTU, PTA's, and mothers in general.

(Continued on page 179)

Operation Blueheart

Sandwiched between ever-increasing interesting and responsible assignments to the 12th Marines, Captain Vincent J. Robinson, USMC encountered an unusual tell-to-to-your-grandchildren experience.

Sunday, June 25, 1950 found him in the Tokyo area of Japan assigned to the 8th Army, breakfasting with the COMNAVFE duty officer, a Marine Corps friend (Lieutenant Colonel Joseph P. Sayers) from Coronado days. The meal was interrupted and the duty officer returned to his office after receiving an official call to receipt for important messages. He left with the assurance: "Keep the food hot ... I'll be back shortly."

Some two hours later he called the duty officer and advised that there was no point in waiting longer. He suggested that I return to my base camp at Camp McGill, headquarters of a 1st Cavalry Division unit. North Korea had just attacked South Korea, and United States personnel were soon to be involved.

Members of the training team were immediately attached to the United States Army's 1st Cavalry Division at Camp Drake to assist in their deployment to the combat zone. This was enhanced by a directive issued shortly stating that "based on CINCFE's instructions, Naval Forces Far East will seize, occupy, and defend by amphibious assault a beachhead in Korea on or about July 18, 1950 in the vicinity of Pohang." The scene shifting to Pohang entailed more than a decision. A lot of preliminary work had to be done.

The first and worst shortage was intelligence. Except for day-by-day photographs of Pohang taken and supplied by the United States Far East Air Force, little was available. For instance, it was not known if an LST could negotiate Yongil Bay and make a high and dry landing on the proposed beaches. In the limited time that remained, there was only one way to find out ... first-hand.

A two man reconnaissance team (Marine Captain Robinson and Navy Lieutenant (j.g.) George Atcheson III ... son of our former ambassador to Russia) flew from Tachikawa, Japan up to Pohang, Korea on July 9. Atcheson was an underwater demolition expert who would survey the harbor. Robinson, acting as a representative of the 1st Cavalry Division, would be responsible for an analysis of terrain features from the high water mark inland ... as well as intelligence matters.

Pohang's military status was doubtful. Red troops were reported only a few

miles north. To defend the city, only two uninspired ROK Navy battalions were available ... led by ROK Lieutenant Commander Lee. They knew that the Communists were only 10 miles from the city, and the intervening defense positions were abandoned.

The previous night, ROK troops had skirmished with the Reds only seven miles up the coastal highway. If the invaders would make a determined assault on the city before July 18, erstwhile sailors would be of little impediment.

Landing on the Pohang strip, six miles south of the city, the recon team politely commandeered a truck early on that July morning and rode into town, naturally anxious to gather information without divulging the secret of an imminent landing.

However, at Pohang's outskirts, townspeople were clustered along the dusty road, waving arms, hands and flags, at the Americans ... a sea of smiles. Noting the unusual number of men and boys of military age in the celebration, a disconcerting thought occurred to the Marine captain, wondering why the cheerleaders weren't out defending their country.

The recon team arrived at Pohang's naval headquarters shortly after 1100. After the usual interpreter problems, a ROK soldier was located who spoke rudimentary English—Corporal Lee Hong Chok had worked at Pusan with the American Military Group.

The party inspected the gray stone quays and the channel leading to the harbor. Atcheson, at his first opportunity, slipped away and measured the channel's depth. With a small lead line, he ascertained this to be six feet of water at the wall, nine feet in mid-channel. At its widest, the canal was only 36 feet—but could accommodate the smaller landing craft.

Meanwhile, Robinson and the ROK escort walked up the sandy beach, past innumerable fishing boats spreading their salty nets to the sun. An old rusted LST was spotted out in the harbor. With the assistance of ROK Corporal Lee, questions were asked as to the ability of that particular ship to come into the beach. Supported by nods of the curious fishermen nearby, the information was confirmed that if that ship tried to land, its ramp would "go under water." Not good.

Robinson then reconnoitered the area immediately behind the water's edge. Sand dunes, atop which the South Koreans had dug trenches, paralleled the beach. Behind them, 75 yards from the sea, ran the coastal road—not a good one, but solid enough for movement of artillery and armored vehicles. Amazingly, all prepared defensive positions were below the military crest, on the

Amazingly, all prepared defensive positions were below the military crest, on the reverse slope facing inboard away from the sea.

reverse slope facing inboard away from the sea.

Atcheson had changed to his frogman's uniform and commenced offshore measurements. The beach gradient was shallow—a loaded LST would touch bottom 150 yards off the beach, its blunt nose still in five feet of water. The smaller LCM's could get within 75 yards of the shore, the ramps would still be in four feet of water. Even the LCVPs could get no closer than 35 yards, leaving the average grunt to wade ashore, hip deep in water. Not ideal, but not impossible.

The two Americans nosed around for the rest of the afternoon, gathering facts, making notes, comparing data, drawing rough sketches. The local police chief became a part of the scene when he volunteered, through the interpreter, to show how justice was meted out to "enemies."

A visit to the jail proved that the "enemy" consisted of seven teenage girls, who were accused of being "too friendly" with suspected collaborators of North Koreans. Their cells, so called, were purposely constructed to impose physical punishment. Vertically, the occupant could not stand to full height, nor stretch out to full-length horizontally. A human kennel.

While gathering information, came forth the knowledge that the earlier troops to land would be coming from bases in Japan, and one relevant bit of information stood out as a "health and safety" factor. Korean road traffic conformed to American style on the right hand side. This was in opposition to the Japanese system, to which all occupying troops had become accustomed.

Upon request, Commander Lee presented, with much ceremony, a silk Republic of Korea flag to Captain Robinson, when he learned that the South Korean standard was unfamiliar to the two Americans. (Subsequently, copies of same were hurriedly manufactured and distributed to lower echelons to distinguish friend from foe.)

"The information gained and disseminated in personal briefings to Rear Admiral James Doyle, the task force commander, and to Major General Hobart Gay of the 1st Cavalry Division was invaluable in the success of the later landings." - *Battle Report* by Captain Walter Karig, USN 1951.

King Neptune ... And The 180th Meridian

The first sea voyage I ever took was my trip to Asia in 1956 as a United States Marine. The trip was very routine, but since I worked in the troop office, I did not have many chances to go topside during the daylight hours.

Rumors were running rampant through the ship that we were entering the "Domain of the Golden Dragon."

The night before we reached the 180th Meridian—also known as the International Date Line—the PA system cracked and burst into vocal activity.

"Now hear this, now hear this. This is King Neptune speaking. All landlubbers who trespass in my domain will be brought up on charges in the morning and will be prosecuted by a jury of sea dogs appointed by me."

Because I worked in the troop office, I had many charges against me ... ranging from hindering communication from the ship to King Neptune to plotting a landlubbers' mutiny.

In the morning when I went to trial, I was found guilty and had to kiss King Neptune's navel. King Neptune was the fattest sailor on the ship, dressed in a grass skirt, with swim fins on his feet and a rag mop for his hair. His court of salty sea dogs were also attired in strange garments.

Because I was found guilty, I had to walk the plant. The plank was put over the rail of the ship by the assistant executioners, and it must have been a 30 foot drop to the sea. The chief executioner blindfolded me, led me to the end of the plank, and gave me a shove. I felt very helpless falling through the black void. When I hit the water I tried to surface for air and take off my blindfold.

When I got the blindfold removed, I saw that the plank was parallel with the ship's railing and that I had only fallen three feet to land on the deck of the ship, where I was hosed down and had buckets of salt water thrown at me.

It was a good baptism. I will never forget it. **C.P Storm (Bud), New Milford, CT, C Company, 9th Marines**

Passover Seder—Chaplain Lieutenant (j.g.) Reuven Siegel, USN, with then Lieutenant Colonel Sidney J. Altman (now Colonel, USMC, Ret.) and Catholic Chaplain E.A. Slattery celebrate Passover with the troops in a service in Korea. The service commemorates freedom of Jews from Egyptian slavery.
Colonel Altman arranged the logistics for a similar Passover Seder in 1944 on Guadalcanal before the Guam Campaign, with food from Australia, New Zealand, and Stateside. Colonel Altman received the Silver Star for action on Bougainville where as commanding officer of E/2/21, he survived an ambush on the Numa-Numa Trail then initiated an attack against the defending Japanese. He now lives in northern Virginia.

Training Of A Brigade

The newly forming brigade was founded upon its sole regiment, the 3rd Marines, with attached combat and combat support units. The growth of tenant units at Camp Pendleton shrunk existing limited training areas necessary for the professional sharpening of the burgeoning brigade. As an example the repeated attacks on the all-too-familiar geographical high rise landmark, known as Horno Ridge, ultimately became the major sterotyped tactical unit exercise, albeit non-firing. Safety factors reduced the proposed impact areas to near zero.

Training thus became an imperative priority, as the brigade was being fleshed out rapidly with called-up reserve personnel and two year enlistees.

Of necessity, in the fall of 1951, Lieutenant Colonel Earl J. Rowse (interim C.O. of the 12th Marines pending the reporting-in of Colonel Leonard F. Chapman) double-hatted as commanding officer of 1st Battalion and made a mission oriented request through channels for that unit to spend three fruitful weeks at Camp Irwin, California. This was roughly 50 dusty desert miles north of Barstow, California. An Army site, it had been a tank training area during World War II but was now in caretaker status.

Shortly after 3rd Marine Brigade's lucrative field artillery training—accomplished with night and day firing exercises, displacements and thorough MOS indoctrination of new personnel—this desert site, due to the exigencies of the Korean War, was upgraded to Fort Irwin, a permanent Armored Training Center.

Public Law 416, 82nd Congress, signed by President Truman, provided for three active Marine divisions and three Marine air wings, and coequal status for the Commandant with the joint chiefs of staff when matters of direct concern to the Marine Corps came under consideration.

Activation At Camp Pendleton

Next step in the evolution process was the January 1952 activation of the 3rd Marine Division at Camp Pendleton. Former brigade units were absorbed as command passed to Major General R.H.Pepper. In due process, the 9th Marine Regiment returned to its parent organization, followed subsequently by the activation within the division of the

The Smallest Grunt

During the three years that I was a peace time grunt, including a tour of the Orient with C-1-9-3 in 1957 and 1958, I was always the second smallest man in every outfit I served in. This caused problems.

For example, in consideration of my size, USMC always insisted on honoring me with carrying the heaviest, most awkward and most miserable items that could be found in a Marine assault battalion,, such at the AN/PRC 10 radio and extra battery, the inner base-plate of an 81mm mortar, or that World War II weight lifter's dream, the Browning Automatic Rifle (BAR).

I wasn't really a foul-up, but I usually didn't get anything right until the second or third try, so it sure seemed like I was a deliberate foul-up.

The first time that I hiked with our company up and down, and round and round, the foot-hills of Fujiyama, my sleeping bag got untied. It unraveled down my back and trailed out behind me, so that I resembled something highly un-Marine-like, perhaps like a strange, new species of jungle bird. But exhausted as I was, with sweat stinging my eyes, I never noticed what had happened. I just felt kind of silly and embarrassed because everyone was looking at me and howling with laughter, but no one gave me the dope.

I bravely marched on, kind of smiling back but puzzled, until our company reached our bivouac area. Then the skipper, Captain Plenty Mad, called me into his command post. "Private! What in the goddam hell are you trying to do to my Marine Corps?," he inquired of me with booming concern, and then informed me that in his years in USMC he had never seen such a spectacle ... and then confessed that he did not intend to ever see such a spectacle again.

I strongly agree with him that he should never again have to see such a thing. Then his good friends, my squad sergeant, platoon leader, further impressed on me the importance of eliminating all traces of jungle bird images amongst Marines. They convinced me that it would not be good for me to repeat the day's activities.

But the next morning it happened again. This time no one was laughing or wise-cracking. Now I was worried. I was scared. I knew a military punishment was awaiting me. I asked my 18-year old self, should I confess or should I try to escape? But confess to what and escape to where? Was my crime mutiny? Was I going to have to swim home across the Pacific?

Before I could figure out any answers to these questions, we were back from the field. Our platoon leader, Lieutenant Young Mad, hollered me into his office. I was standing at attention, face-to-face with my military future.

He provided me with an excellent example of a tirade against fouling up by an officer of Marines. He was nicely warming up to telling me what kinds of juicy punishments he had in store for me.

But while he was thinking he was dealing with another John Dillinger in the making who wasn't even trying to be a good Marine, in fact I was feeling great frustration because even though nobody knew it, at least I was trying to be a good Marine. But for my best efforts, I was just getting constantly chewed out. Tears welled up in my eyes. A Marine crying! Add feeling ashamed to my frustration, humiliation and anger.

Just then Lieutenant Young Mad looked through his own anger to see my tears. He was flabbergasted. Instantly, he realized that what he thought was my incompetence was not. He saw correctly my determination to be a good Marine. And I knew he saw this.

"PFC Colaizzi," he bawled at me. "Yes, Sir," I roared back. "Carry on, he ordered. "Aye, aye, Sir!," I most loudly responded.

As I turned a snappy about-face and marched out of his office, I was hotly flooded by the sounds inside me of my own private Marine Corps band: Da da da da da da da dum, da da da da da dum. Not long after, I was promoted to corporal. **Paul F. Colaizzi, Pittsburgh, PA**

Dr. Paul F. Colaizzi, now a psychologist in Pittsburgh, PA, is founder and director of the Pittsburgh Institute of Existential Therapy.

Wrong Sub (In Many Ways)

An officer, several NCOs, and I were to take a group of new Recon Marines aboard a submarine for rubber boat training about 1960. The procedure was that after inflating the boats, the sub would crash dive, and the Marines would float free in their boats in darkness to land near Camp Hansen, Okinawa.

We were to embark on the sub at 0800, and were in formation at the pier when a sub arrived, right on time. The officer and I went aboard to make arrangements with the sub's commanding officer on how we were to carry out the three day exercise. He was unhappy, because he was planning some R&R for his crew.

After getting others of the sub's staff involved in planning the exercise, the port commander joined us and wanted to know what we were doing aboard that sub. It was the *Blackfin*, and we were supposed to be using the Tiru. It (*Blackfin*) had arrived at the designated place at the right time, and the other had not. A sub is a sub!

Several hours later, the Tiru arrived, and being aboard the sub was a harrowing experience in several respects, even before our launch. During our first attempt to launch, my job was to have the Marines get the rubber boats from storage, inflate them, then give a "thumbs up" to the periscope, after which I would re-enter the sub through a forward hatch.

The sub was rigged for diving as I sauntered nonchalantly over to the hatch. It was tightly secured! The sub started to slip between the waves with me in starched utilities and spit shined boots. I gave a flying leap into the first boat (some less complimentary descriptions were given by others).

The sub submerged just a bit, then caused a foul-up with the first boat washing up on the second and second onto the third. There wasn't enough water for clearance, but we fought the boats free and slithered backward into the deeper water—later learning that the sub had malfunctioned and couldn't complete the dive.

We had some other mishaps with that sub on the exercise, and were happy when it was finally over. **Sergeant Major (then 1st Sergeant) J.R. Skinner, USMC (Ret.), C-3rd Recon, Jacksonville, NC**

The Day The 9th Marines Took Fuji

We shipped into Yokosuka and disembarked late afternoon. It was already dark, so we never did get a first look at Japan. By the time the trucks brought us to our new home, it was pitch black out and we were ready for our first drydock sleep in over two weeks, so we really didn't much care about seeing anything.

I was an 18 year old private first class. Before I hit the sack that night, I learned I had been assigned as a BAR man to "Chargin'" Charlie Company, 1st Battalion, "Striking" 9th Marines, 3rd Marine Division, FMF. But as to exactly where I was, I had no idea.

The time was 1957, and the occupation of Japan by United States ground troops was nearing its official end, and USMC was looking for a grand finale for its happy times in Japan.

Next morning, I ran outside for roll call. Then I saw it. Spectacular, breath-taking, enormous, majestic, holy, graceful, powerful, and big. It was splattered across the whole sky, rising clear up to heaven. Of course, I was looking—gaping—at the perfect yet asymmetrical form of snow capped, sky reaching Mount Fuji. Fujiyama filled my eyes and my consciousness. It was a case of sudden beauty and awesomeness, one of the experiences that has lasted in vivid memory throughout my life.

But these feelings of wonder and majesty were not pure. At the very moment of my awesome awareness, I was feeling miserable—because I knew instinctively that sooner or later, we Marine infantry were going to execute that finale by climbing that big thing. At the time, I didn't know that we would be the first major military unit to do so. I already felt the weight of my BAR grow to a ton in anticipating the climb up Fuji, and I already hated Japan's national symbol, which had nothing at all to do with World War II.

Then I saw where we were: actually on the slopes of Fuji. Gotemba was the name of the nearby town, named after the Gotemba Trail that ascends Fuji. Scuttlebutt had our barracks on the site of a POW camp which held Gregory Boyington after his capture by the Japanese at Truk.

Part of 1st Bn., 9th Marines as they climbed Mt. Fujiyama in Japan.

Training in the foothills of Fuji was strenuous, but we all thought it was well worth it just to be able to spend a single liberty evening in Gotemba town. Liberty was great. Life was great. But all the while, under the surface, we brooded about the inevitable announcement. It came two months later, in June. It was bad news. But it was good news, too.

Yep, we had to climb Fuji. That was bad; but Fuji is a national and sacred shrine in Japan, which did not permit its being mounted by anyone carrying weapons of any sort. That was definitely good. We wouldn't have to lug the 10 pounds or 20 pounds of our M-1's or BAR's, bayonets or helmets. Without that weight, we figured we could damn near run all the way up that mountain.

We were so fired up to conquer the beast that either our eyes deceived us or somehow we managed to not notice what even every photograph clearly reveals, namely, that there is always or nearly always snow on top of Fuji. It was June, and we were dressed for June, and just didn't believe in snow, I guess.

No one climbs straight up Fuji but follows a widely winding trail that has hardly any noticeable grade to it, so that the climb was long but not at all steep. We climbed till nightfall, then camped right below the weather line.

Next day we quickly reached the snow, and quickly cursed our "snow blindness." Through snow and cold, we finally scaled the summit, admired the view and congratulated each other on our feat. Then, on the first day of summer, we froze on the top of Fuji.

Our misery was prolonged mainly by the photographers who want to snap us poised on the lip of this extinct volcano. When we were finally released from the cold ordeal of summit sitting, we ran down that mountain in two hours flat even though it had taken good parts of two days for the whole regiment to climb it.

As a result of our lack of warm clothing, many of us caught the flu, which then quickly spread around the whole world, and was called the Asiatic Flu of 1957. We started it.

We didn't carry our weapons up Fuji, but carried Fuji sticks—squared off wooden poles with Fuji bells and flag attached. I still have my flag and bells flying in my study to remind one and all of the day the eagle, globe, and anchor, and me, rose 12,395 feet into the air.

I've been itching to do it again for 33 years. I'm ready. **Paul Colaizzi, Pittsburgh, PA.**

PFC Paul Colaizzi is shown with his Fuji Stick as 1st Bn., 9th Marines were atop Mt. Fuji.

historic 4th Marine Regiment, which replaced the original 21st Marines.

With these infantry units intact, the natural expansion of all the skeletonized supporting units to bring the new 3rd Marine Division up to speed became a positive fact.

Two notable events occurred in the first six months of 1952. The first of these was the successful conclusion for purchase of a vast amount of desert real estate at Twenty-nine Palms, California, initially designated as Marine Corps Training Center.

Then, under the aegis of Marine Corps Base, Camp Pendleton, it allowed several units from the Oceanside base to take advantage of the open fire and maneuver area for advanced training. These units, moving through an as yet unpopulated government installation, made motor marches under independent control via the Hemet-Beaumont-Banning-Morongo Pass route ... then to the high desert areas of Yucca Valley and Joshua Tree.

Among these pioneering units was Lieutenant Colonel Rowse's 1st Battalion, 12th Marines which set up the original Camp Wilson tent city area, just west of the Dead Man Lake dry bed. (This camp was so named, on a permanent basis, in honor of Colonel John Bushrod Wilson, first commanding officer of the 12th Marines, who commanded the regiment throughout World War II.)

Other units which took advantage of this remote, isolated—yet ideally suited for rugged terrain and training—were engineering, motor transport, and reconnaissance units.

Dead Man Lake was the location of the first intramural softball game in March of 1952. Contrary to fantasy that the desert remains in a warm-to-hot status, all players in that first game were heavily adorned with mittens on both hands, utility caps, sweaters and scarves, combat jackets, and hoods. The beer ration—which was the winner-take-all award—had to be stored in the gas Kelvinator refrigerator of battalion sick bay in order to keep the precious liquid from freezing (and going flat) in the external temperatures.

General Shepherd Visits

The second landmark situation in April 1952 was the arrival of the Commandant of the Marine Corps, General Lemuel C. Shepherd, who had been a

(Continued on page 183)

"Green Side Out" At Lake Yamanaka

When 3rd Marine Division units moved to Japan after World War II a warm relationship was established by our Marines and the administrators and children of the Lake Yamanaka Orphanage.

Members of the 12th Marine Regiment were the principal patrons in providing regular financial assistance to the orphanage, but members of all units at South Cuddle or North Camp helped to construct and maintain facilities as well as to entertain and look after the young orphans.

The orphanage was also visited by Commandants of the Marine Corps during their Far East inspections and other senior officers—all warmly received by the director, Sister Mafalda Morando, an Italian nun affectionately known to her staff and Marines as "The Sergeant Major."

During a 1962 visit by Major General John Condon, commanding general 1st Marine Air Wing and CG TF 79—of which the 3rd Marine Division was the principal ground component—Sister Mafalda proudly escorted the general through the spotless buildings and Quonset huts that made up the complex. General Condon noted the uniformity in color and asked Sister, "Why is everything here painted green?"

Without any hesitation the director proudly replied, "General, green is the only color paint the Marine Corps had."

Thousands of Marines who passed through Camp Fuji—as the training camp was known after North, Miandla, and South camps were deactivated—remember the Yamanaka Orphanage well. And, the scrubbed cleaned, well-disciplined children remember the green camouflaged Marines who gave them financial support, love and affection as well.

Christmas Away From Home

The Christmas holidays were particularly happy ones because of the generosity of Marines who had gifts sent from home, purchased in town or at the PX, or who made generous contributions to the Christmas fund.

At the well remembered Christmas of 1961 when I was serving as commanding officer, Camp Fuji, Sister Mafalda was escorted to Yokohama to buy clothing and new uniforms for her children. The truck came back well filled with a happy and grinning nun sitting next to the Motor Transport driver. Even Santa Claus, in the person of Lieutenant Colonel Danny Regan, commanding officer, 3rd Tank Battalion, made his appearance and passed out clothing and gifts to each child and staff member while the more that 100 Marines present added their personal gifts to the orphans perched on knees or laps. Marine Corps Christmas dinner, which had been prepared at Camp Fuji and transported

For The Children...

At Christmastime in 1961 Sister Mafalda Morando and some of the children at Lake Yamanaka Orphanage are shown with the new clothes the Marines had bought them. Colonel Ed Danowitz at left is holding his godson - Edward Katsumi- sponsored in baptism.

In April, 1983 when he was then Commandant of the U. S. Marine Corps, General Robert Barrow paid a return visit to Lake Yamanaka Orphanage and presented a gift to Sister Mafalda.

over the mountain to the orphanage in Marine green trucks, was served with Captain Russ Adamczuk, Camp Fuji maintenance officer, supervising

A Time To Pray

Sister Mafalda in later years told how 3rd Marine Division Marines, alerted for deployment into Vietnam and combat, came quickly to the orphanage to pray a last time in the quiet chapel and to ask Sister Mafalda's blessing. For many it was the final farewell to "The Sergeant Major" and her children.

The years of the Vietnam War were quiet and lonely ones for the orphanage, but in August 1969, when the 9th Marines returned to Okinawa, laughter and smiles returned to

the sisters and orphans at the Lake Yamanaka Orphanage with each visit of a new Marine in green.

When I visited the orphanage for the last time in November 1983, Sister Mafaldo Morando was still in charge, having recently returned from a furlough in her native Italy—her first departure from the orphanage in 10 years. The old green Quonsets had been replaced by sturdy cinderblock buildings which housed the 54 new orphans, ages two to six.

Today, the orphanage contains plaques and mementoes from almost every unit in the 3rd Marine Division. It is a fitting memorial to every Marine who at one time served in the division and left part of himself at the Lake Yamanaka Orphanage. **Colonel Edward F. Danowitz, USMC (Ret.)**

When the 3d Bn., 9th Marines, and supporting BLT units were in Thailand in the summer of 1962, distinguished visitors at camp were not unusual. (l. to r.) Lt. Col. H.W. Adams, Battalion CO; Thai military aid to the king (behind); Brig Gen O.R. Simpson, USMC; Col. V.J. Croizat, USMC (behind); Thai Prime Minister Sarit; Her Majesty the Queen of Thailand; and His Majesty the King.

In The Media Spotlight

In 1962 BLT 3-9 was training in the Philippines when we were suddenly ordered to land in Thailand to aid that country amid Communist activity threatened from the Laos border.

While establishing and improving the camp there (Camp Rama I), the team conducted conditioning hikes in extreme heat, helicopter patrols and operations, ambush exercises, and training with and for the Thailand Border Patrol Police.

To strengthen Thai-United States relations, our BLT patrols scoured the countryside and physically marched through every village. Motorized patrols debarked troops on the outskirts, then embarked again on the other side.

Many distinguished visitors came to our camp, including top military officials of both the United States and Thai, civilian leaders of both countries, and even the king, queen, and prime minister of Thai.

After we left, the team, including supporting units, was awarded the Marine Corps Expeditionary Medal, said to be the first award since 1941. During the time we were there, we received international publicity and made the major news magazines.

While deployed there, I was told in a rather joking vein just how lucky I was by several contemporary senior Marine officers. The claim was that this was the first time in Corps history that a battalion commander had the unchallenged and uncompetitive support of a squadron of choppers and a squadron of close air support A4Ds. **Excerpted from material provided by Lieutenant Colonel H.W. Adams, USMC (Ret.) 3-9**

wartime commander of the Striking Ninth Marines. General Shepherd's visit was timed for the presentation of colors to the newly activated 3rd Marine Division and its organic units.

The divisional units were stretched geographically at Camp Pendleton from the San Luis Rey gate northerly to the far reaches of the San Onofre-San Clemente tent areas, plus those units based across Highway 101 near the amphibian vehicle areas.

The entire division was assembled on the appointed day with all personnel and mechanized equipment for the proud presentation of unit colors. CMC General Shepherd took the review along with the division commander, General Pepper. This was a massive and lengthy "Pass in Review" as troops of all units marched on foot, followed by their rolling stock with token vehicular crews.

The 3rd Marine Division had graduated from a paper tiger into a once again identifiable combat command.

After small scale training throughout the year, a combined major 3rd Marine Division exercise was executed in December 1952. This coincided with the arrival of the first permanent 70 man detachment sent from Camp Pendleton to organize and maintain security of the desert training site. This new Marine Corps Training Center was growing apace with the appearance of roads, barracks, utility plants, quarters, chapel, airfield, water facilities, and infamous and pungent Lake Bandini.

Training and amphibious exercises progressed on schedule, geared to the continuing involvement of the 1st Marine Division in Korea, and need for possible support there.

Location In Hawaii

In March/April 1953, strategies dictated employment of an available airground task force to the mid-Pacific. Consequently, the 3rd Marines, along with all attached and supporting units, contributed the nucleus for a new 1st Marine Brigade, along with aviation units supplied by the 1st Marine Air Wing. This new brigade was relocated to MCAS, Kaneohe Bay, Hawaii.

Time To Move

Shortly thereafter, in mid-1953, with cessation of hostilities, the peace/truce at Panmunjon, all forces were withdrawn to both sides of the demarcation line in Korea. Thus the return of the 1st Marine Division to its home base at Camp Pendleton became an established fact.

(Continued on page 185)

In Thailand in the summer of 1962, "march throughs" of Thai villages were a key part of BLT 3/9 efforts to improve Thai-U.S. relations and show support for anti-communist activities there.

Gobble, Gobble
... And A Long Way From Home

It was around April Fool's Day, when 3rd Battalion, 3rd Marines climbed aboard ship in Naha. Operation PHIBLINK was afloat to the Philippines, an amphibious landing to run around the boonies a few days, and float back.

This was an annual affair and the duty experts—i.e., taxi drivers, tailors, and bar girls—would all tell you the same thing: "Marines comes back Okinawa in three weeks." But the year was 1957.

We went over the side at Dingalan Bay and made a mad dash inland aboard Charlie Company tanks. The first night we bivouacked alongside the road in absolutely nowhere. The only bodies there were some friendly, ringed-black and white snakes. Next morning, trucks took us to a hill beyond Clark Field where we would play war and be entertained by our company gunny, Tech Sergeant Victor L. Martinez of Colebrook, Connecticut.

As far as gunnies go, he didn't cuss all that much except when he would sing. Like: "You are my friggin' sunshine ..." And as far as chaplains go, our battalion had a good one. He made daily rounds, talking to individual Marines, not just jawin' with other officers. And he was quick to let the company commander know when something displeased him ... like the day he happened by when the gunny was singing. You could always tell when the chaplain was coming after that ... some smart ass would start singing gunny style.

There's A Drop On My Mashed Potatoes ...

Battalion got the field mess set up, which is big happy time for grunts; and damn if some dumb nut doesn't complain about his mashed potatoes getting rained on. The chaplain wanted to know what us headquarters guys were going to do about it, as we sat on the grass in the command post (a tent fly stretched between some bushes and our jeep trailer ... and was the only damn shelter on the hill).

... We could have told him we ate only after serving the rest of the company, a squad at a time, and things like fruit cocktail, ice cream, and mashed potatoes usually ran out before we got to eat.

... We could have told him we hoard, trade, and steal "C" rations so there are extras for the platoon sergeants to keep their chow hounds from grumbling.

But, like they say in Brooklyn: "didn't nobody say nuttin'".

Some of us got to visit the MEU tent. That's a Medical Evaluation Unit operated by Section 8 sailors and Marines who run around the boonies picking up things like spitting cobras, bamboo vipers, and 15-foot pythons that bite like alligators. They told us there were no king snakes around, but there were kraits—the highly poisonous, many banded krait. But what did "they" know?

About the time we got 'hot' groceries squared away so that sometimes even the guy that eats last (the company commander) gets a goodie, we get orders to move out. We climb aboard trucks, but not before everyone gets to dump out his pack on a poncho, and repack his gear. There's no sense in carrying around extra weight ... like a stowaway king snake.

The trucks take us to Subic Bay Naval Station and drive around the bay to the fuel pier on the far side. That's the Cubi Point side where four Seabee battalions removed a mountain so they'd have enough flat real estate to build an air station. There's a nice new air station on the hill, but the docks are a restricted area of fuel tanks, jungle, and our APA. So what, as long as we're going home to Okinawa.

But we don't go home, we just sit there.

Everyday the ship's PA system tells us we'll sail as soon as they receive orders. When a convoy of trucks brings provisions, they also bring ammunition and field rations which may, or may not, have significance.

About that time they stop using the PA system and our officers pass the word: We are waiting to sail to a destination not named for security reasons, and the location and movement of our battalion was now classified. (Of course, being there's no war going on, scuttlebutt has it some admiral wants to protect his pretty base from the Green Machine.)

Every damn thing on the base was uniform-of-the-day after 1630, which translates to no dungarees, and no slack for deployed troops who have been living out of a haversack. But we did get liberty and, with no help from the luxury liner we lived on, there were 3rd Battalion Marines all over the campus in fresh starched khakis.

During the day we could take platoon size conditioning hikes off base. Normally, a conditioning hike is something the seasoned grunt tries to avoid; especially when they are headed by a young lieutenant who wants to break world record for speed, time, and dis-

tance. But, damn, if they weren't some of the biggest platoons you ever did see. Had to turn back some volunteers.

It turned awful hot and humid, and the two canteens each man carried were finished long before the hike was. We stopped in a village where the lieutenant let us buy what cokes they had.

An old English-speaking gentleman offered us water. It took a little bit of talking, but the lieutenant let us form a relay line into the man's yard and up the front porch where canteens were handed up to little kids who filled them and passed them back—a bucket brigade of canteens.

The water resupply was going great until a six foot python slithered out of the house. That gosh darn snake came out like he lived there—which he did—right between the bare feet of them little kids. And damn! if them big bad Marines didn't go ape.

They dropped those canteens, fell all over themselves trying to get out of there. That started a stampede and the canteen brigade in the yard crashed through that man's beautifully woven fence.

The little kids screamed and laughed so hard they couldn't stand, and just about every Filipino on that end of the island came over to help them laugh.

We were pretty much red-faced. The Marines tried to mend the fence, but, hell, they weren't Filipinos and didn't know diddle about weaving a tree branch fence. We passed the hat, collected a few bucks which were given to the old man for his trouble. Know darn well those kids laughed long after we slunk down the road, out of there.

Suddenly one day ... our APA was ordered to anchor out in the bay, with all troops aboard, and no unnecessary ship-to-shore movement of personnel. Spring in the Philippines can be a real steamer ... the bay is surrounded by mountains ... and our APA was not air-conditioned. Every day the ship's captain would weigh anchor and reposition the APA so we could be on the side of the bay that had a breeze.

After a few days, some heavies—Navy captains and Marine colonels—started to raise hell. We piled into Navy busses one day and went to Driftwood Beach—a beautiful white sandy beach with new concrete snack bar and beach house—for a party. Soon we moved into tents in that area with mess hall and club set up in some old frame buildings, and a fresh water pipeline from the base. A shuttle bus started regular runs to Subic and liberty call!

One event may have caused us to be moved to our new quarters. A platoon leader, a favorite lieutenant affectionately called "Iron Mike" Powers of New

Jersey, had the 'hots' for a Navy nurse he was to meet later at the officers club. The officers club at Subic Bay was only a few feet high ... a foundation dug down into the ground with a roof on top made it a great way to keep cool in the days before air conditioning. Iron Mike came down the street in front of the club with his platoon and—to let the nurse know he was passing by—gave the command to "stomp cadence." That's just like "count cadence," but you stomp your feet instead of counting. Then the platoon did a "to the rear" and stomped again, another "to the rear" and a stomp.

After that third stomp cadence there were people running out of the club waving their arms and carrying on. The club must have been build on a swamp, for when those Marines did that first stomp the whole place shook and the

Over the side at Dingalan Bay

folks inside thought there had been an explosion. Drinks started dancing off the bar, the Navy officers were wearing their good humor suits, and must have been pretty busy jumping around to keep from getting splashed.

Early one morning, we boarded ship and headed toward Indonesia where we would be the landing force, if needed, to protect American lives and property. Steaming around the South China Sea in midsummer was hot ... one of the water distillation units broke down ... we were allowed on deck in skivvy shorts and shower shoes and could jump under a saltwater shower at any time.

Back to Subic bay, where we snapped back into a regular routine.

Marine Corps birthday and Thanksgiving were coming ... but no turkeys in sight. So the colonel commanding Marine barracks at Subic Bay purchased enough frozen butterballs with his Special Services fund for our battalion to have a turkey dinner.

Meanwhile, the Army quartermaster back on Okinawa was having problems. It was their job to provision the Marines and, with units deployed and not returning on schedule, they had a glut of turkeys. So, to solve their frozen food storage problem, they located the Marines and shipped us our share of holiday birds.

There is nothing immoral about serving turkey three times a day but it should be illegal. Chow like hamburgers and beef stew became exciting. Up until then, I always thought a box lunch with

sandwiches of turkey to be a pleasant change over from salami or baloney. Not so, now.

We sailed south again, through the Strait of Malacca and across the Indian Ocean, to be in the Persian Gulf for the Lebanon crisis ... passing several possible liberty ports. After a couple weeks in Singapore, we returned to Okinawa to give up our permanent barracks at Sukiran (Camp Foster) to the 9th Marines who had just migrated from Japan.

I must have developed a lapse of memory because I cannot remember exactly where we were that Christmas. But, wherever it was, I know I did not eat any turkey. **Al Beveridge, Old Route 22, Copake, NY**

To Japan And Iwo Jima

High level strategy triggered the requirement for combat-ready Marine presence in the Far East. So it came to pass that the 3rd Marine Division deployed in July from Oceanside, California to dispersed tactical field bivouacs on the Island of Honshu in Japan proper.

During the training missions in the WesPac area, a deja-vu opportunity was presented to 3rd Marine Division units as one of the amphibious exercises was carried out on the Island of Iwo Jima.

This island had been militarily significant in February 1945 to the 3rd Marine Division and the rest of the Marine Corps. It had created a "snug harbor" for B-29 crews limping back to Marianas bases following bombing raids on Japanese homeland targets, but it made a grim reminder of multitudinous casualties. This geological marvel was composed of black lava sand beaches, dominated on the left flank by the imposing Mt. Suribachi, renowned in the famous Joe Rosenthal flag-raising photo, and a constant wind. Truly this island had to be seen to understand the inherent hardships that the individual Marine faced when securing this landmark against very hostile forces.

Permanent Camp In Okinawa

After a two year stint in the home islands, the 3rd Marine Division began moving elements to Okinawa in 1955 and completed the troop and equipment transfer in 1957. Permanent camps and training sites were created.

During the next eight years, because of training commitments with forces of other countries (Japan, Korea, Nationalist China, Philippines, among others) and the requirement to "show the flag," the division became fragmented throughout the Far East. Added to the administrative woes was the introduction of transplacement battalions, for training on a rotating basis.

An amphibious landing was made at DaNang, Republic of Vietnam, in March 1955 with the result that the battle colors of the 3rd Marine Division now flew once again in combat.

On The Move

The dash inland aboard Alpha Company tanks.

Tales From "The Rock"

After a few weeks at Tent Camp #14 in Camp Pendleton, and 10 days aboard ship, it was great to finally reach "the Rock" (Okinawa) in August 1955.

I had extended one year in order to "get to Japan," but when the orders were handed out at Pendleton, a sergeant merely went down the line, pointing to each man in turn, and hollering out—"Japan, Okinawa, Hawaii." But he just shrugged me off, saying, "don't tell me my job. It's Okinawa for you."

Life with the 3rd Shore Party battalion on Okinawa was rugged in those days. I was assigned to White Beach which my outfit shared with the 3rd Amtrac guys. Now there was a love-hate relationship! We sometimes punched it out with them at the local slop chute. And, we didn't think much of them off base either. But we stuck together when the swabbies came in on liberty.

One night, we joined forces with them to clean up on the sailors who racked up one of the Marines at a cabaret in a nearby (allegedly communist) village of Heshikiya—"whipping every sailor they could find between there and Yonashiro," according to newspaper reports.

After a free-for-all involving some 200 Marines and sailors, police said that Marines even went down to the beach where the Navy vessel was anchored, but turned back after an unsuccessful attempt to board the ship. (I heard the

officer of the day aimed a machine gun at the Marines and they backed off.) Some 300 leathernecks surrounded a sailor's club at White Beach but left there around 1 a.m.

At White Beach there was no water for showers when the "fleet was in" because the destroyers had to replenish their tanks, and it somehow interfered with our supply. We had frequent dysentery because the bubbling GI cans we dipped our mess gear in were never perfect; muggy heat and plenty of mosquitoes.

But hey, we were the vanguard of the helicopter renaissance, the first gyrenes to load them with equipment and troops for attacks. We thought they were a novelty and could never replace the amphibious landings with their APAs, LSTs, and other craft. Little did we know they would form a backbone in Vietnam (Indo-China as we knew it.)

We went on maneuvers one day and had to load troops into the choppers for an attack. As the infantrymen clamored aboard, our job was to give them, from the ground, a heave-ho that would get their equipment-laden bodies through the hatchways. I had all but one trooper loaded. He was short and wiry, but not moving fast enough to my liking. When he finally arrived, I grabbed him by the seat of the trousers and threw him through the hatchway, chiding him with, "Move your butt, Marine."

As I was checking the safety belts of the team prior to take-off, I heard him say softly, "Jeezus, Gilbo, I didn't know you had it in you." He turned out to be Staff

Sergeant Dunlap, my junior DI when I was but a timid 17-year old boot at Parris Island.

In early 1956, this young corporal found escape from White Beach when the 3rd Marine Division Drum and Bugle Corps showed up for an honor guard ceremony. My cousin with the Division band on Japan had tipped off the drum major that I could play the trumpet. Sergant Ziska, the drum major, had the bugle master take me out into the boonies for an audition. On our return to Top's office, "Pappy" Seibert, an older corporal, told the drum major, "The poor guy's lip is in terrible shape, but he knows the trumpet. I think I can whip him into shape."

Despite the first sergeant's protestations, the drum major got me orders on the spot, telling the Top, "General's orders, Top, I can confiscate musicians where ever I find them." Before I knew it, I had my seabag packed and was aboard a bus with the corps headed for Camp Napunja, and a new life with H&S 9th Marine Regiment. That's where I made buck sergeant.

I left behind some good buddies, and fond memories of my introduction to the "Nissans" that graced the little dirt-floored bars where tired Marines could buy a quart sized bottle of beer for Y100. The girls were not very sophisticated, but they were a cheery bunch who played black jack with a vengeance. I remember a favorite haunt somewhere between White Beach and Tairagawa. The place had kerosene lamps and a mama-san who treated us like kings to fried Spam and rice, and a wind-up phonograph that played a 78 rpm record, entitled "It's Only A Paper Moon." It was the only record they had. We thought we were in paradise.

Then, there was "Doc," who during morning roll call warned us not to frequent the house of ill-repute outside the gate, the "house with the yellow curtains" because somebody had gotten syphilis there. It worked for awhile until somebody discovered that the corpsman and his buddies went there every night, and of course, had the place to themselves.

Life with the "Striking" 9th Regiment was quite different. Spit and polish, morning guard mounts, and rehearsals. During maneuvers, we manned .50 caliber machine guns and traveled with the infantry. Shortly after I joined the unit, we saddled up and moved our quarters to Sukiran, the Army bastion.

Our first morning, the drum major marched us down to the Army band's headquarters where, before the first rays of sunshine shone through, we serenaded Colonel Sanches and his musicians with the Marine Corps Hymn. Bottles, peanut cans, newspapers and other trash greeted us as the dogfaces poked their angry faces out the windows and shook their fists.

Life was good at Sukiran. Fewer bugs, a mess hall with Okinawan help and trays, and plenty of great places to pull liberty. Sukiran was also the place where I experienced my first typhoon and earthquake (the worst to hit the island in 30 years we were told.)

Then, there were 10 days leave in Japan where I went with $50 in my pocket and came back with $2 left, having eaten well, traveled extensively, and stayed in good hotels. I also spent some time at Itami Airbase and caught hops to Tokyo.

I had been greatly impressed by James Michener's novel *Sayonara* and had followed the hero's travels around Japan, seeing an all-girl show at the famous Takarazuka School, and even dancing with local girls at a nearby dance hall.

My drunken attempt to visit the showgirls that evening came to an abrupt end when I heard the unmistakable locking and loading of a carbine, and a Japanese security guard call out from the darkness on the school grounds, "No stay here, GI. You go home." Oh well, it was still the best liberty I ever had.

There were never ending maneuvers with the Ninth. On one occasion, I had been told to find a defensive spot overlooking the road leading into the old Japanese airfield at Yonabaru. It was getting dark, but I had my crew set up the "fifty" on a cliff with a good view of the road. We bedded down for the night in the foxhole and enjoyed the cool breezes from the ocean.

At dawn, while dining on C-rations, we found a trench right next to ours that had been used in World War II by the Japanese. At the bottom, hidden by weeds, we discovered bone fragments and decayed pieces of unfamiliar canvas. They no doubt belonged to the Japanese defenders 10 years before. They must have thought the place would be a great defensive position. A grisly bit of irony. We were glad when we pulled out of there, feeling the presence of old ghosts.

Qualifying on the M-1 was quite a feat on Okinawa. Our outfit was sent to fire as a typhoon was gathering. I fired 178 in the heavy wind. High man on the range just barely qualified. But we had to keep to the tight schedule, and got out just before the typhoon moved into the area a day later with 140 mile-an-hour winds. Maggie's drawers were waving all day long.

Leaving "Okie" was no picnic, even though we counted the days and cut short-timer sticks to mark our future departure.

Some of us had girlfriends, "honeys," and had been breaking curfew to see them for most of the year. There we were aboard the USS General Walker on a dreary November morning, listening to the Navy band play "Anchors Aweigh," and watching our girlfriends waving from the dock, their eyes covered with sunglasses to keep us from seeing the tears.

That was the quietest departure for home I had ever seen. The railings were lined with Marines fighting to hold back the tears. No cheers, nothing, just lots of waving and handkerchiefs being raised to the eyes.

The ship pulled out and headed for one last liberty in Yokohama where it would pick up more men and set course for San Francisco and Treasure Island. Some of the Marines would have married their girlfriends, but such marriages were out of the question for Marines back then. The girlfriends, the hardships, and the friendly people are now just part of the "the good old days." **Pat Gilbo, Rockville, MD**

A Third Division Marine uses the M-60 machine gun, which was new at the time, during a Far East maneuver. (U.S. Marine Corps Photo)

Saga Of BLT 3/3
IN ITS TRIP ROUND THE WORLD

In 1956-1957, the men and officers of BLT 3/3 take a trip in which they uphold the United States Marine Corps worldwide reputation for smartness and neatness.

A Birthday To Remember

Towering Mt. Fuji, gleaming under its first mantle of snow looked down on the peaceful scene November 8, 1956 as the 3rd Battalion, 3rd Marines bivouacked in the Juliett area, resting from a day of field work. The coming morning would be the fourth day of a regimental field exercise.

The field telephone rang in battalion headquarters in the early morning hours and 1st Lieutenant Ian Campbell shortly learned that the message he was receiving was not part of the field problem. The battalion was being ordered to break camp and immediately prepare for return to Middle Camp Fuji.

Battalion Commander Major Al Sollom had awakened during the brief conversation, and learning the content of the message, he and his operations officer set out for regimental field headquarters, while Lieutenant Campbell began alerting the companies.

Soon there were blinking lights, the clash of equipment, the bark of sergeants, the roar of jeep and truck motors. By 0400 the battalion had cleared the Juliett area and shortly before dawn was back in Middle Camp. While the battalion was still en route to Middle Camp, the regimental executive officer, Lieutenant Colonel John A. White relieved Major Sollom, who took over as battalion executive officer.

The transfers that were to change the whole appearance of the 3rd Battalion—that were to beef it up to T/O and reinforce it into a self-sustaining battalion landing team—had begun.

From South Camp and North Camp, from McGill and Okinawa, they came— the riflemen, communicators, machine gunners, chaplains, photographers, demolitionists, tankers, shore party men, engineers, 4.2 mortar men, the 75mm recoilless rifle platoon, the amtrackers.

They came, and the 3rd Battalion became BLT 3/3.

Shortly after reaching Middle Camp November 8, the battalion officers were summoned to a briefing. First Lieutenant William G. Soden was late for the conference. A leg injury had prevented the commanding officer of Hotel Company from being with his company on the field problem. Entering the conference room, he glanced briefly around and with set features walked to an empty chair. The hip to ankle cast Lieutenant Soden had worn from October 31, until just a few minutes before, was gone. If the battalion was mounting out, Hotel Company's skipper wasn't going to be left behind.

The briefing was short and many questions were left unanswered. Time was too precious to go into full details or indulge in idle speculation. The battalion had to be prepared to move out within two days.

At 0930 Friday, November 9, Lieutenant Colonel White met with Brigadier General Victor H. Krulak at 3rd Marine Division (Rear) headquarters in Camp McGill. Shortly before noon, the Colonel flew back to Fuji, carrying with him verbal orders to move the battalion's supplies and equipment to the Yokosuka docks.

Camp closing activity continued through Friday night and into Saturday. Finally the battalion was ready to move, and a break was called to attend to some unfinished business. It was November 10, the 181st anniversary of the founding of the United States Marine Corps, and the men gathered in the mess hall for the traditional birthday dinner.

In the middle of that birthday dinner, the battalion received "Execute orders." Desserts were hurriedly eaten, final cups of coffee gulped. By 1330 on that memorable birthday, the trucks were rolling again. Later that night, the battalion had joined with its reinforcing elements and BLT 3/3 was aboard the attack transport USS *Telfair* and the loading ship dock USS *Oak Hill.*

And They're Off - But To Where?

The hurriedly formed task group sailed from Japan November 12 with an announced destination of Okinawa and a scheduled landing exercise. Care of weapons reached a new high as the men silently wondered what the future held. At all hours of the day, the decks were packed with men cleaning and oiling rifles and pistols. Machine gun crews,

rocket teams, and mortarmen cleared small corners to hold gun drill. No one meant to be caught short.

News broadcasts aboard the ships instantly halted all conversation. Alert and serious, the men crowded near to catch each word. They listened to reports of fighting in the Middle East and Russian brutality in Hungary. Then they returned to their weapons.

The battalion staff sections labored to complete plans for the Okinawa exercise. Intelligence reports were prepared, briefings held. On the company level, boat teams were assigned and revised. Everyone looked forward to the practice landing and raid. Almost one third of the BLT were new men, and the officers and staff NCO's wanted to see them in action. If something bigger was in store, this would be an excellent shakedown.

Just five minutes before the ships were to drop anchor off Okinawa to begin the exercise, orders were received to proceed to Subic Bay in the Philippines. The voyage was getting bigger ... and the scuttlebutt was running fast.

Daily training aboard ship was stepped up, despite space limitations ... physical exercise was given a high priority ... top level briefings held for officers ... platoons and squads held daily schooling.

Approaching the Philippines, the seas began to mount. Two storms were raging in the area and both seemed determined to test the embarked Marines' sea legs. Marines and sailors alike took to their bunks as the task group came about to ride out the fringe effect of the typhoons ... and later reports indicated more than 100 persons died in sea disasters during the storms.

The towering waves finally subsided and the ships hurried towards Subic. But once again the radio crackled, bringing new orders. Now there could be little mistaking the direction. The ships were turning west, towards the troubled Middle East.

A Pause In British North Borneo

Westward, the weather improved. The sun was brighter in the clear, blue skies ... men cultivated out-of-season suntans ... The South China Sea was calmer ... only the leaps of the tropical flying fish seemed to disturb the surface. It was almost a pleasure cruise ... but the infrequent news reports and the daily training schedule were constant reminders that the voyage may have more serious meaning.

188

Refueling was to have been at Subic, but now took place at sea, east and north of the Malacca Straits. Morale on ship climbed when a small quantity of mail was taken aboard.

By this time the Middle East situation had eased ... fighting had ceased in all but scattered sections ... a United Nations police force was on its way into the troubled area.

The task group received orders to change course and proceed to Brunei Bay, British North Borneo. Anchoring off Victoria Island, the small town of Lebuan offered only limited recreation facilities, but still a welcome chance for the Marines to stretch their legs after 15 days aboard ship.

Ashore, swimming areas were available, the two British clubs were opened to the United States Force, and on organized liberty parties, Marines became acquainted with coconuts, breadfruit, and bananas—and like their comrades of 15-20 years previously found that a bayonet was a handy tool for opening a coconut. But, in addition to play there were conditioning hikes, and wet net training conducted from ships to lowered landing craft.

Throughout the world, Marines have become almost as famous for their peacetime acts of charity as for their wartime combative prowess. Borneo received first hand knowledge of this when BLT 3/3 donated over $200 to the St. Anthony Boys School there. Another $800 was mailed to the Yamanaka Orphanage, the 3rd Battalion's favorite charity in Japan.

The task group stayed in Borneo for 10 days, and then was ordered to proceed into the Indian Ocean.

The Bazaars Of Pakistan

On December 1, two days after leaving British North Borneo, BLT 3/3 passed the point of no return. Ahead lay southern Asia and the Middle East.

While rounding the island country of Ceylon in the Bay of Bengal, the task group was ordered to pay a good will visit to Karachi, Pakistan. As the ships hurried toward Karachi, liberty uniforms began to appear from seabags and the Marines put away their rifles and began to spit shine dress shoes. This was the first large United States Marine unit to appear in that part of the world.

When the ships tied up in Karachi on December 12, the appearance was totally unfamiliar. Gangling camels stalked through the streets, pulling carts and wagons. Brightly colored bazaars clamored for the Marines' attention, and tall, heavily-veiled women silently hurried through the city.

Many friendships arose between the Marines and their counterparts in the Pakistan military. In the open street bazaars, Pakistan sailors aided the United States visitors in the friendly bargaining for leather goods, carved chests, fabrics, chess sets, and knick-knacks. The Marines also learned something of the Moslem religion, symbolized by the sacred cows which walked through the city.

A hurriedly picked Marine rifle and pistol team competed with marksmen from the Pakistan Navy during the stay. Led by the sharp shooting of Technical Sergeant Steve Ohina, the Marines gained a two point aggregate victory. Other rifle shooters were Captain Fred MacLean Jr. and staff sergeants Robert Waltman and Herbert G. Hasse, and Sergeant Gordon S. Rhodes Jr. The pistol team included Captain F.R Hittinger Jr., Lieutenant Richard D. Ross, Sergeant Waltman, Sergeant J.R. Bown and Staff Sergeant C.B. Hasse.

On December 15, the Marines again put to sea and retraced their steps across the Arabian Sea, heading for their next port of call.

Planned displays and exhibitions played a large part in maintaining the United States Marine Corps worldwide reputation for smartness and neatness. Entering every port, sharply dressed Marines were drawn up on deck to render honor to ships of foreign countries. While in port, a Marine honor guard was held in readiness to greet the many dignitaries visiting the ships.

In Karachi and Bombay, a Marine drill team performed for the local military forces. Led by 1st Lieutenant Richard J. Walters, the platoon from India Company drew both applause and laughter from their audiences as they went through both standard formations and trick maneuvers.

Many Friends Made As BLT 3/3 Visits India

Bombay—gateway to India and second largest city in the teeming country— became BLT 3/3's third port of call on December 17. The Marines visit coincided with Indian Prime Minister Pandit Nehru's trip to the United States for talks with President Eisenhower, and also with Indian Navy Week.

The visiting fighting men were immediately struck by the modernness and cleanliness of the city. Wide, heavily trafficked boulevards moved through the canyons of tall, modern buildings. Luxurious hotels, pleasant restaurants and large stores vied for the men's attention.

The city presented many paradoxes in architecture. The harbor fronting buildings presented a coast line like that of Miami or Rio de Janiero in Brazil. Moving into the city, ornate ironwork and overhanging balconies reminded the Marines of New Orleans. And further inland, was the old city where on long narrow, twisting streets, merchants displayed the native handicrafts in fabrics, metalware, and carvings of wood and ivory.

A gigantic Christmas mail delivery— making home seem a little closer—was a highlight of the stay in Bombay.

The Indian Navy opened its officer and enlisted men's clubs to the visitors. In a move to aid friendship and understanding, Indian officers were invited to dine aboard the United States ships, and United States officers attended dinners and receptions both ashore and aboard Indian ships in the harbor.

Christmas And New Years Celebrated In Ceylon

On Christmas Eve the Marines arrived in Colombo, the capitol city of Ceylon. The ship was strung with giant strings of colored lights from stem to stern, and Christmas trees were set up below decks. Christmas services were held both Christmas Eve and Christmas Day, with a Marine and Navy choir rendering seasonal carols.

Christmas dinner featured both steak and turkey, while many of the men were invited into private homes. Another group met Hollywood movie stars William Holden, Alec Guiness, and Jack Hawkins, who were on location in Ceylon and celebrated the holiday with them. Official receptions were held by United States Embassy officials and Ceylon Navy officers. Ceylon is noted for its star sapphires and the gems were a big attraction in the local markets. Feeling like wealthy merchants, the Marines would fit a jewelers glass in their eye and minutely compare stones.

Leaving Colombo on December 29, the task group sailed around the southern tip of the country to Trinhcomalee, large British Naval base on the northeast

189

coast, arriving there New Year's Eve. Despite the holiday atmosphere, work went on. Marine working parties functioned round-the-clock to help the Navy replenish the ships from two supply ships which rendezvoused there, and a harbor island was used to brush up on field work.

The Britishers took the measure of the Americans in all but one of the sporting events held during the stay. Volleyball teams from Hotel and H&S Companies lost in straight games, while the British entries swept the first four places in a 12-boat regatta to prevail by 15 points. The Marines' wounded pride was slightly salved when a combined Navy-Marine officers team employed brawn and power to counteract finesse and experience and held the British wives field hockey team to 1-1- tie.

The Cruise Enters The Third Month In Singapore

Moving ever closer to their home bases in Japan, the far traveling Marines made their sixth stop in Singapore, Malaya January 11, Singapore seemed to be a meeting place for all races and cultures, and the Marines had no trouble fitting into the varied scene.

The shops offered a wealth of bargains—native silverwork and rice straw pictures on black satin to the latest stateside-made shirts and suits. The Marines found it difficult to completely see the city in the three days they stayed there.

King Neptune Holds Open House

Leaving Singapore January 14, the task group sailed south towards the equator, mythical realm of Neptunus Rex ... and for those making their first equator crossing January 15 became a day to remember.

Ceremonies were complete with a wake up call at 4 a.m. ... typical briny deep breakfast of boiled hard rocks, burnt toast, fresh wiggly eels, boiled seawood and salty coffee for land lubbers and beachcombers ... facing the royal court of the Royal Judge, the Royal Baby whose grease covered belly had to be kissed, the Royal Prince with his electric shocking wand, and the Royal Surgeon who painted the patient with all matter of lotions and ointments ... the gauntlet of enthusiastic shellbacks armed with belts and boards ... the Royal Barber and his electric clippers ... the garbage pit ... and the baptismal tank.

Shopping In Hong Kong

Immediately after crossing the equator, the task group came about and steamed northeast across the South China Sea to Hong Kong, final port of call before Japan.

After seven days in Hong Kong, it was again time to move on and the ships sailed into the rough waters of the China Sea. As the nights became colder, the Marines knew they were getting close to Japan and the end of the history-making voyage.

On the morning of February 4, 1957, Mt. Fuji again loomed above as the ships entered Yokosuka harbor. It had been 88 days aboard ship, with some discomforts ... but even more pleasant memories. Training had been worthwhile, even though not as severe as many Marines were used to undergoing.

But on one point there was no speculation. Throughout the cruise the men of BLT 3/3 not only upheld, but increased, the United States Marine Corps' worldwide reputation for smartness and neatness.

This material has been excerpted from a booklet prepared by 1st Lt. W.M. Morris, Public Information Officer, 3rd Marine Division Headquarters.

Training - Training - Training
WHEN 3RD MARINE DIVISION WAS STATIONED IN JAPAN

In 1954

Members of the 3rd Battalion, 4th Marines helped make history when they participated in "Operation Comeback"—the job of transporting 14,000 Chinese ex-prisoners of war to Formosa, seat of the Chinese nationalist government.

Moving to Kobe the troops left the trucks on the dock and filed aboard the USNS *Marine Serpent* which sailed as soon as they were embarked.

Operation Flag Hoist

Of all the Marine training exercises in the Far East, the one most widely known in America was the full scale division landing exercise that the 3rd Marine Division carried out against Iwo Jima in March 1954. The "second" landing on Iwo was far more pleasant than the "first" one in 1945.

The operation actually began a month before the division hit the beaches when the 2nd Battalion, 4th Marines embarked for Iwo to prepare for their role as the enemy "Aggressor" forces, assigned to the defense of Iwo against the coming division landing.

In the early dawn of March 21, elements of the 3rd and 9th Regiments assaulted the beaches.

Carrier based fighters lent close-in support to the assaulting leathernecks.

Tank infantry teams moved inland from the beaches. Memorial services at the end honored the dead of World War II.

Trans-lex 1

Dubbed "Operation Merry-Go-Round" this was a complex, two-part serial feat that involved the 1st Battalion, 9th Regiment and 1st Battalion, 4th Regiment.

Late in May the 9th Marines having completed month long maneuvers at "tent city," Fuji, were loaded aboard flying boxcars at Atsugi and flown to Itami Air Base. There the disembarking members of the 9th met the 4th Marines, which had travelled to Itami by train from the 4th's home base at Nara. The 4th climbed aboard and flew to Atsugi, entraining from there to replace the 9th on maneuvers at "tent city."

An involved logistical feat, Trans-lex 1 went off without a hitch and wrote a new chapter into the growing story of the Marines' mobility.

"Move Out" shouts the NCO's of 3rd Marine Division personnel who are taking the beaches of Iwo Jima for the second time in one generation. (Defense Dept. Photo-Marine Corps, taken by TSgt. C. Adelman)

Unforgettable Memories Even On 2nd Iwo Landing

I was a jeep driver for Colonel Nihart and was assigned to a team to secure camp before boarding ship. Anyway I took some officers to San Diego to board ship, and when I got there, my jeep was taken from me and I was put aboard ship by mistake. I had to hitch-hike back to camp while in dungarees and had one hell of a time explaining myself at the main gate.

Then when we made maneuvers at Iwo Jima in 1954, I was with H&S Company, 1st Battalion, 4th Marines as a driver. Even though a driver, I went over the side of the ship into the LCPs and onto the lava beaches of Iwo. Memories of this landing will be with me the rest of my life, and I'm glad it was one experience I had the opportunity to take part in.

The thoughts that went through my mind ... the hardships the men had to endure in securing the island, while under the hostile fire of the Japanese ... the struggle the men had fighting their way up Mount Suribachi.

I walked up a road to the summit, which was very steep, and looked out over the ocean for miles and miles. I heard the Japanese could see a hundred miles or better from the summit.

On the beaches were LCPs rusting away and rocking back and forth. There were also bayonets rusting away.

There could be no better symbol than the statue of the flag raising on Mount Suribachi to symbolize the United States Marine Corps. **Al Simmons, Morgan Hill, CA**

Iwo Jima Revisited

Jose J. Gallegos of Santa Fe, NM says he'll never forget the very close buddies who were with him in the 6th Wave on a maneuver landing February 19, 1954 on Iwo Jima.

Jose was with the 9th Marines, which he joined in 1953, to be assigned to Camp Otsu, Japan in that twilight period between the Korean War and Vietnam.

Though the 3rd Division did not participate on Korean soil, the 3rd was indeed a reserve division for the 1st.

Pictured L to R and seated are: PFC Hubby, Indiana; PFC Hill, Tennessee; PFC Jose Gallegos, Jose is a barber at Paul's in Santa Fe; PFC Newberry, OH; PFC Ramos, Puerto Rico; and PFC Shellhammer, Pennsylvania

191

Paddling On A Treadmill

During Operation Blue Star on Taiwan in 1960, C-3rd Recon was assigned as "rubber boat company," to land several hours ahead of the main landing. We trained in a disembarkation technique that worked well in a calm sea.

The day before D-Day, rough weather came up. At scheduled debarkation time from the destroyer, it was taking over two feet of water across the aft weather deck with each roll, on top of other problems. The ship's captain had to come to a full stop to put the boats over the side, in hard rain, oncoming darkness, and offshore wind gusts over 40 knots. We faced a 1,500 yard paddle.

After about 45 minutes of furious paddling, I was startled to hear a voice from above and behind me, "Having problems down there?"

It was the bow watch on the destroyer, and we obviously were making less progress than we thought. However, finally our platoon's four boats labored over a coral shelf and through a natural coral bulkhead opening which turned out to be a Benjo ditch emptying from a Chinese village above.

We moved up the ditch and through the village above, waking every dog. But we found that we had landed about a mile south of where we should be. To get to our planned command post, we had to climb a mountain between, after the rowing and Benjo ditch experience. Our mood? Should be obvious.

Came dawn ... and with it an empty landing beach. No Recon Marines. No ships. We sent patrols to locate the other Recon platoons, without much success. We found one, but could not break radio silence to call the others.

When the division landed, it turned out that some Marines were lost at the sea, and the command problem was to count noses. Two boats from the lead destroyer had blown to sea and were adrift in the Formosan Straits.

There were some hairy experiences involved before all were rescued, but a potentially dangerous situation was averted by Marine discipline and common sense. **Sergeant Major (then 1st Sergeant) J.R. Skinner, USMC (Ret.), C-3rd Recon, Jacksonville, NC**

Cold Weather Landing

In 1962, 1500 Marines from the 3rd Marine Division in Okinawa took part in a cold weather amphibious landing exercise on the east coast of Korea. Major General Robert E. Cushman, Jr., center, Commanding General of the 3rd Division explains the nine day exercise to Lt. Col. Ki Hwa Kim, ROK, Marine Corps at right, and Nam Ho Kim at left, a journalist for Pacific Stars & Stripes. Although the cold weather and choppy water made for realism, the Marines knew the exercise was really a practice because they were met by smiling Koreans waving flags.

I Remember Uncle Dave

The Kadena Air Force Base football team was on their way to the Far East championship. The Kadena team had tromped over all the Army, Navy, Marine, and other Air Force teams ... They were unbeatable and no one wanted to play against them.

Then the 9th Marine Regiment arrived on Okinawa, being relocated from Japan. It was a lengthy migration, interrupted by field deployments, and they had missed much of the football season. The 9th Marines entered the play off and, after a series of some of the hardest played football you ever did see, eliminated the "champs."

After the final game, following a brief talk by an Air Force lieutenant general, the 3rd Marine Division commander was called to the podium for the trophy presentation.

The three star general went on to say: "Following last Saturday's game an empty whiskey bottle was found under the Air Force bleachers, and there were at least 50 bottles found on the Marine side of the field." After much animated applause the Air Force general asked the Marine brigadier, "Can you explain that, general?"

"No sir, I can't," answered the Marine. "But I'd sure as hell like to know what that Marine was doing on your side of the field. **Albert J. Beveridge, Copake, NY**

This view of Camp McNair, Japan with a definite cool, snowy atmosphere is far different from the jungle-like weather that most Marines were used to. (Photo submitted by James Sprungle, Jr. who was with K-4-12)

Sowing The Seeds Introducing
Large-Scale Helicopter Carriers

There was no way that the antenna of the AN/PRC 10 on my back was gonna' get through those goddam jungle mountains. Then the cobra struck at my boonies. Its fangs didn't penetrate to my skin but damn near did and its force knocked me further into the flesh-cutting jungle, scratching my hand and drawing a thin line of blood. I always figured that the cobra was an enemy of the United States and that I deserved a decoration, especially since the rest of my squad bayoneted and butt-plated that sonovabitch to death before it could get away.

That squad was the first squad of the first platoon of C-1-9, and we were in the Philippines. It was 1958 and we were the lead elements of Exercise Strong Back. This was the first large-scale helicopter training exercise for USMC, and the largest exercise in the Pacific since World War II. We were under the overall command of Major General David M. Shoup. Major General Francis M. McAllister was 3rd Division CG, and Brigadier General August F. Larson was CG, Helicopter Assault Force.

We were operating from the USS *Thetis* Bay, the world's first helicopter carrier, which brought us from Okie. We boarded at Naha, and a week later, on March 1, 1958, touched down in the mountains near Dingalan, Luzon.

VIPs from all over SEATO observed the exercise as guests of the 7th Fleet, commanded by Vice Admiral Wallace M. Beakley. "It displayed to them the speed and power with which the United States could come to the aid of any threatened allies." (Triad, March 14, 1958).

Even John Wayne

But much more important for us was the fact that John Wayne was there to watch us! He wanted to see the "new Marines in action, and we wanted to show him. For indeed we were pioneering helicopter assault strategy in our training on Okie, and this was our "public" debut. He was impressed. Everyone was.

Even back then we knew where we were headed. Our NCO's who had been in Korea had us all singing their song—"... So put back your packs on, the next stop is Saigon. Cheer up, my lads, bless 'em all." And we knew it was true, even if a sukoshi early. But we were Marines, playing our game with a gun that wasn't yet loaded.

By the time it was loaded again, college and graduate school had ruined my vision to the extent that by 1965, I was unacceptable for re-enlistment in USMC and USAF. So I only knew the life of a peacetime training 3rd Division Marine, which is why everyone ignored my views about my cobra; for I viewed my fang-marked boondocker as an authentic combat trophy.

Later that year, back with the 6th Marines, I lost those boonies and the seabag. They were in on a real operation in a place called Beirut, Lebanon.

Of course, it was not all helicopters with the 1st Battalion, 9th Marines (commanding officer, Lieutenant Colonel William L. Dick). There was the time we took a little walk—all the way around Okinawa, 141 miles! It took almost a week. We had more blisters than the Marines in the movie "Battlecry." And I was still carrying the company radio (commanding officer, Captain James B. Vanairsdale).

Always Faithful

During Strong Back, Private First Class Cotton in our platoon fell down a mountain and broke his leg. He was lying at the bottom of a narrow valley. The nearest helicopter was called in. It wasn't Marines. It circled us above the valley, declined and flew away.

A less near helicopter was called in. It also wasn't Marines. It circled us above the valley, declined and flew away.

Finally, the least near helicopter was called in. It was Marines. It circled us above the valley, dropped down with inches to spare, loaded up our casualty, and flew him safely away.

Exercise Strong Back served as an outstanding demonstration of the readiness, mobility, and power of the United States 7th Fleet with the Marine Air Ground Team.

Operation Wetback

This operation involved a three day maneuver by the 3rd RCT, supported by Panther Jets from VMF 224 and Seafuries from the HMS *Warrior* and artillery from the 12th Marines 1st Battalion. The landing, opposed by a battalion from the Army's 29th RCT, was successful in seizing three objectives: Kin Airfield, its dominating ridge, and Hog Hill.

The Marines And The Kids

The men of the 3rd Marine Division reached deep into their hearts and their pocketbooks and came up with something that meant more than just international friendship for Japanese orphans; it meant warm clothes, a firm roof, and plenty of "gohun."

The story of the 12th's service to the Yamanaka Orphanage actually began in October 1953 when a chaplain noted the destitute condition of the orphanage located near the 12th's base at Camp McNair.

Beginning in the 2nd Battalion, members of the 12th began fixing things up with hammers and saws and with donations to a "Milk and Rice Fund." Soon the whole regiment had joined in, and through a series of carnivals, the 12th raised thousands of dollars toward the construction of a huge new building. In June 1954 a final carnival sent the building fund over the top and in July, the completed structure—long a dream—became a reality.

The Vietnam War was unlike any other in United States history, and Marines battled inherent Asian tradition, stacked political decks, weather extremes, and hostility at home to try to stop the spread of Communism in southeast Asia.

Marines were the first combat troops in that embattled area, the last to leave, and cared for a greater share of United States fighting than in any previous war.

Recollections of this longest war in United States history by Marines of the 3rd Division span a kaleidoscope of weather, insects, and animals ... jungle privation and hardship ... hidden hazards ... the brotherhood of combat ... human concern ... and personal distress.

On these pages are their personal remembrances, the agony and the humor, the frustration and the accomplishment, most as viewed in retrospect a quarter of a century later.

They are a tribute to those Marines who have sorted them from the ordeal of experience, the events of difficult days. We honor them in printing their memories.

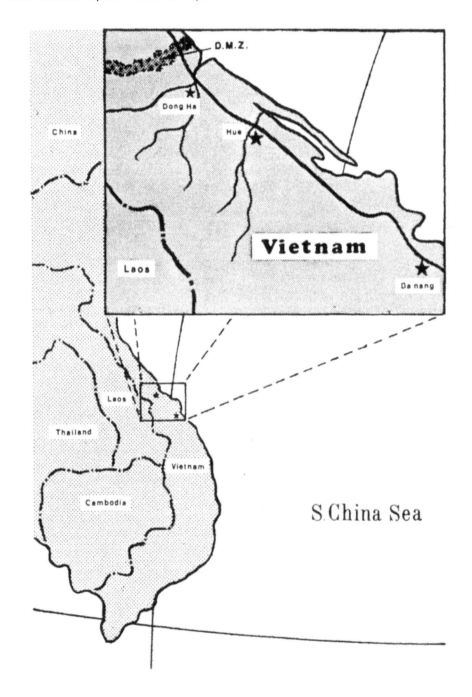

How To Meet A Next Door Neighbor

This story has an unusual ending. The sequence started with a patrol one morning, with a lot of noise and action in the hills, in an area near Tam Boi. Early the next morning, as it continued, dog teams were brought in and I was right up front with them, passing the previous day's ambush area.

Our formation was a wedge, and I was behind the dog team with a short timer to my right. Going up a gradual slope, I looked just to the dog's right and saw this bush get up and move. I turned back to the short timer and asked if any Marines were in front of us.

Before he answered, I opened up. I sent about 14 rounds at the bush and it went down. I yelled that there were gooks out there. No reply. I yelled "frag out" and tossed one toward the bush. It went off and a split second later I heard curdling screams. Unfortunately, they came from me.

A gook saw me throw the frag and threw something at me. It went off real close. I got a piece across the lower back next to the spine, fragments in the lower right leg, and a large one snapped my left femur clean about an inch from the hip joint. The short timer got it also.

I don't remember with total clarity the next few hours—morphine and shock make things blurry. No medevac was possible, so all night long I lay immobile in the dark with other WIAs.

Eventually, I got to Quang Tri and a MASH-type situation. For two days, I waited my turn as more seriously wounded kept coming in.

That's how I met my next door neighbor for the first time. He was a corpsman doing his thing there after a bush tour. He and his wife had moved next door in Wisconsin while I was in boot camp. I had met his wife and kids, and he knew my parents. But I had to go 12,000 feet from Me! James (J.D.) Manville, H-3-9, Stevens Point, WI.

An Overview ...
The 3rd Marine Division And The Vietnam War

Colonialism and Communism were the two major forces which created the conflict in southeastern Asia that occupied the 3rd Marine Division from 1965 through 1969. Marines, the first combat troops in Vietnam and the last to leave, shouldered a greater share of United States fighting there than in any of the nation's previous wars.

Marine traditions of readiness, adaptability, tenacity, and combat efficiency were again enhanced. But it wasn't easy. Political and governmental restraints, and the unpopularity of the war with the American public, inflicted a tremendous burden on the fighting forces.

The officers and men of the 3rd Marine Division, in the process, added more names to that historic list of gallant distinction in the nation's history—names such as Dewey Canyon, Khe Sanh, Con Thien, Quang Tri, Rockpile, and others.

The Background

The colonial period in Vietnam began in 1859, when the French, there since the 16th century, captured Saigon and organized the colony of Cochin China. In 1884 the French declared protectorates over Tonkin (North Vietnam) and Annam (Central Vietnam). By 1900 other French protectorates in the locale included Laos and Cambodia.

Throughout Vietnam, a nationalist movement arose after World War I and gained momentum during the Japanese occupation of World War II. In 1946 prolonged and bloody guerilla warfare broke out between the French and the Vietnam government.

Complicating the situation by that time was the pervasive and militarily strong Communist influence spreading throughout much of Asia.

The long standing French colonial presence ended in 1954, when the French and Vietnamese garrison at Dien Bien Phu surrendered to Viet Minh Communist forces.

The country was divided at the 17th Parallel, north latitude. The northern zone, including Hanoi and Haiphong, went to the Viet Minh as the Democratic Republic of Vietnam (North Vietnam). The Republic of Vietnam, known as South Vietnam, embraced the area south of the dividing line.

A campaign of terrorism began in 1957, aided by the Communist government at Hanoi and including opponents of South Vietnam President Ngo Dinh Diem. By 1958, the Diem government was involved in a guerilla war.

In late 1960 the Communist "National Front for the Liberation of South Vietnam" was formed, with the Viet Cong (also known as the VC or Victor Charlie by Americans) as its military arm. Large numbers of South Vietnamese who had gone north when the country was divided began to return south to join the resistance against the Diem government.

Small groups of United States Marine Corps officers and NCOs began a training program which included a two-week period observing the counter-guerilla tactics employed in Vietnam in 1961. That year also saw the beginning of greatly increased American aid to Vietnam as a part of the widespread effort to help those fighting Communist takeovers.

The first Army helicopters with operating and maintenance crews arrived at the South Vietnam capitol of Saigon in December 1961, a prelude to Marine aviation involvement in transporting South Vietnamese forces in "search and destroy" operations the next year.

Helicopter operations by the Marine Corps greatly accelerated in 1962, and on October 6 a search-and-rescue copter, supporting the Second Army of the Republic of Vietnam (ARVN) Division, crashed near Tam Ky, killing five Marines and two Navy personnel.

The Action Is Stepped Up

The South Vietnamese government of President Diem was overthrown by an Army coup, November 1, 1963, and Diem

(Continued on page 197)

Never Forget The Radio

I was a radio operator assigned to the POL (Petroleum, Oil, and Lubricants) platoon at Cau Viet, at the mouth of the Dong Ha River. We lived in our own hootch with a bunker attached to it by a slide.

Our job was to stand six-hour radio watches, handling the fuel from Navy ships, plus piling sandbags from NVA cave and track artillery in the DMZ. Things were very hot during the TET offensive and Khe Sanh siege.

You could only dig a two and one-half foot hole until you hit water, so protection from incoming fire was sandbags on top of our hole. We needed lots because we were a good target. The fuel was kept in 10,000 gallon rubber bladders laid out in neat rows on berms of sand.

A lively evening erupted August 23, 1968. About 9:30 p.m., NVA artillery rained about 90 rounds of 122mm and 151mm on us, and 15 fuel pods were transformed into a roaring conflagration. When the incoming stopped, an amtrac came over to drive us to the fire.

I was the radio man, but the driver yelled down, "You won't need your radio; I've got one." Like a fool, I left my radio behind. We reached the fire, jumped off, and the amtrac drove away with its radio.

Sergeant Weeks yelled to me, "Hippie! Call the Navy and have them send over a bulldozer. Maybe we can save the last 20,000 gallons of fuel." I told him I would if I had a radio, but it was back at the hootch.

This brought some choice verbiage and the promise of a court martial if I didn't get a bulldozer. Running in the direction of the bunkers to call up on the land line, I shouted at the two guards not to shoot as I approached from the rear.

They told me their land line was dead, but that the next bunker 300 yards over had a radio, and to "watch out for concertina wire from the old line about 100 yards ahead."

I thanked them and started at a dead run, but after about 20 yards, I hit squarely a triple strand of concertina wire, fell right in, and was stuck! Luckily, I had my flak jacket on. I could hear the two grunts loudly laughing. Was I pissed! But I still had to get the bulldozer.

"Get me the hell off," I yelled. They did. From head to toe, except for chest and back, I was covered with little gashes. I ran the rest of the way to the amtrac and burst into light all covered with blood.

I finally located the bulldozer and walked back to the fuel fire with its 20-foot high flames. What a perfect target, I thought, and just then two rounds of artillery whooshed over. I was in the sand before they hit, but now had sand in all those cuts.

After hitting the sand three more times, I reached my destination and the sergeant, by that time directing the bulldozer. Sergeant Weeks smiled and asked, "What the hell happened to you, hippie?"

"It's a long story. How about if I go to the aid station."

Twice more I hit the dirt before I reached it, where an old Navy medic cleaned my cuts and told me to consider myself lucky, since my mistake had caused no dead or wounded.

From then on, at all times, I ate with, rode with, and walked with my radio. **John A. Lundquist, ForLogSup, Milwaukee, WI**

A Letter That Wasn't Mailed

We had quite an adventure on what I remember as Alamo Hill. We were to hump over to this hill being held by a sister company (Lima, I think), and they left a squad there with a gun team until we arrived.

By late morning, we were urged to hurry. When we got there, they were being shot at by unknown numbers of well-camouflaged gooks. Our fire team was right behind the point team, and when we got to the hill perimeter, I was told who I'm supposed to cover and where he is ... naturally, directly across the open area of the hill.

Off I go crouched low and as fast as I can run with my gear. Near where he is supposed to be is a medium sized fallen tree. I yell his name. He answers. Then he tells me the gooks have a damned .50 caliber, sniping with it, and that I'm not in a good spot to move out of. So I stay by my new tree fort.

One of his buddies yelled for grenade support, the gun team just down the hill with gooks real close. I can't see over the hill or the gun team, but following his aiming directions, I half lob, half throw, a frag. It's close to target. They ask for one closer so they can dash right after the blast. I let one go, a bit too close.

They came up the hill half mad and half glad. More of our company arrived and was filling in. However, I have to take a piss. My options: Get up and dash for cover, or stay prone and let go. John Wayne would have dashed. I pissed.

We secured the hill and dug in. Then, a crack nearby. A gook had seen the squad we replaced as it left, and assumed the hill was empty. He came walking near my hole. The next morning was a dandy, and I started to write my mother a letter, the "I'm fine, don't worry about me" type.

My turn came for hole watch, and I'm sitting on the back side of a three or four man hole when AK goes off, real close. I dropped straight down and in milliseconds four other Marines are on top of me after beautiful head first dives and slides. I'm literally being crushed and suffocated at the bottom on a jungle hole in late morning direct sunlight with a flak jacket on, a third of a ton of weight on top of me, and a 16 with a chambered round next to me.

The gook fired one round to make us hole up, then came the Chi Coms, so close I could hear them burning. First one was short, second one long. In response to my screams to get off me, a couple guys dove into the next hole and we emptied our clips at him.

I got out of the hole, pissed that I hadn't seen him coming up the hill ... I had been damn near shot at close range ... and nearly crushed to death. To make matters worse, his second chi com had landed quite near my gear.

My poncho looked like a green piece of Swiss cheese wrapped around a bush, and three of my canteens had been flung around like polo balls. Another looked fine until I drank—cordite flavored warm water from small frag holes.

My letter to mom was burned and holed. I never mailed it. My LSA was all over everything, like a herd of bull elephants had creamed all over my gear.
James (J.D.) Manville, H-3-9, Stevens Point, WI

Starlite/Researcher

It is unlikely that there is anyone who knows more about the details of Operation Starlite and the personnel involved than former Staff Sergeant Ed Nicholls, Kendalia, TX.

Nicholls, "one of Colonel J. Muir's riflemen" of the 3rd Marines, has spent more than 11 years gathering information on personnel and other involvement in Starlite, in August 1965 the first regimental size United States combat action since Korea.

It has taken that long "for a mailman to get that far," Nicholls explained. He served two tours in Vietnam in 1965-1966 and 1969-1970, and continues to display great pride in his Marine service. He also served as a CWO in the Army.

His extensive compilation of Starlite data includes command lists, press coverages, ship lists, unit rosters, and much individual detail including casualty information.

---

assassinated. The new provisional government was promptly recognized by the United States. By the end of 1963, the total of United States troops in Vietnam had reached 16,500 from all services.

A major step up came in August 1964 when North Vietnamese patrol boats clashed with United States Navy ships in international waters in the Gulf of Tonkin. The United States retaliated by bombing North Vietnam naval stations.

The United States Congress, at the behest of President Lyndon Johnson, quickly passed a joint resolution to 'take all necessary measures to repel any armed attack" against United States forces. The House unanimously passed it; the Senate vote was 82-2.

That Tonkin Gulf resolution gave President Johnson broad power to escalate American military presence in Vietnam.

South Vietnam, the country in which the United States was becoming more of a presence, covered 66,263 square miles in the shape of a slender north-south crescent bordered on the east by the South China Sea. A population of over 15 million people was of divided ideology because of past problems with its government and the VC infiltration and support. The zone of primary concern militarily included its five northern provinces, divided by a quirk of geography into three compartments.

The northernmost provinces of Quang Tri and Thua Thien were set off from Quang Nam to the south by the Hai Van mountains. In turn, Quang Nam was similarly separated from its southern neighbors, Quang Tin and Quang Ngai, by the Que Sons mountain spur. Than geography greatly influenced coming military operations that involved the 1st and 3rd Marine Divisions.

1965
The Third Arrives

At the request of the South Vietnamese government, the United States military advisory forces were soon augmented by the first combat troops. And the first to answer the bugle's strident summons was the 3rd Marine Division, in early 1965.

In early March 1965, forward elements of the division made an amphibious landing at Da Nang as the 9th Marine Expeditionary Brigade. Battalion

(Continued on page 199)

First Mission - Da Nang Defense

The first mission of 3rd Marine Division combat troops in Vietnam on arrival in early 1965 was defense of the Da Nang airbase.

Da Nang was an anomaly. In a few years, it had grown from a provincial airport to a major airbase where both civilian and military activities of both Vietnamese and Americans were clustered around a single 10,000-foot concrete runway.

At the same time, just beyond it and within mortar range were some 250,000 Vietnamese residents of various political inclinations. Just before the Marines arrived, it was decreed that a warren of Vietnamese dwellings must be cleared out of a 400 meter zone around the base, but it took some time to relocate the 7,000 persons involved.

There was also, on the west side, the inevitable "Dogpatch" of bars, laundries, tailors, photographers, and souvenir shops. The perimeter which enclosed the base consisted of a ring of dilapidated concrete blockhouses, steel watchtowers, an unreliable lighting system, and belts of rusting barbed wire.

Marine helicopters had operated from Da Nang since 1962, three years earlier. When the 9th MEB arrived in March 1965, it was the contention of United States military leadership that defense of the Da Nang area should be Vietnamese.

When the 9th arrived, it shared in close-in security for the airbase plus defending about eight thinly populated square miles of high ground west of the field. One battalion moved there, while another remained at the field for security against the customary Viet Cong guerilla action.

Who Is In Command?

In the early summer of 1966, the battalion area of 2/9 was just south of Marble Mountain near Da Nang. I was a fire team leader and lance corporal in Echo Company's 2nd Platoon. One night, Echo Company was called back to the battalion area and our platoon was ordered to set up in the hamlet just outside. My fire team had position directly on the road.

As morning came, a jeep approached our position. The driver was a corporal, and I could see a lieutenant colonel in the driver's seat and a sergeant major in the back seat. The jeep slowed at our position, we let it go, and it continued on. A few yards down the road, it stopped and the colonel got out. "Come here, Marine," he commanded. I did.

He asked, "Don't you salute?" I told him we were instructed not to salute in a combat zone. He said, "Salute, Marine." I did. Then he asked who was in command here. I said I was. The sergeant major became convulsed with laughter.

The colonel asked then if I knew who my battalion commander was. I identified him by name, and the colonel replied that was he, and stormed away.

It was quite a joke in the battalion area that a lance corporal had first refused to salute the colonel, then had the audacity to tell him the lance corporal was in command.

I still felt I was in command of the fire team at that spot. **Vincent Rios, E-2-9, South San Francisco, CA**

The Day Of The Rooster

After 3rd Recon joined the rest of the battalion at Quang Tri, the unit moved to a new site next to the shore party area. Relations were not too harmonious after a few tear gas grenades were tossed back and forth, and one thing led to another.

One night, a Recon member lobbed an illumination round from an M-79 into the shore party area. This might not have been a problem had the round been high enough, but the prankster's aim was low and it started a fire. Shore party people, upset, formed a line to assault Recon.

Officers and the battalion chaplain got wind of the situation and intervened before things went too far. The chaplain proposed a cookout for both units to improve relations. The day of the cookout was great, with steaks, beer and soda. The chaplain also decided to raise some money for the Dong Ha children's hospital.

He brought a wild NVA rooster into an open area, charged a dollar each for a chance to catch the rooster, and promised a case of cold beer to the Marine who caught it and returned the rooster alive to the chaplain.

About 300 Marines surrounded the chaplain and one frightened NVA rooster. The chaplain threw the rooster into the air and the chase was on. The chaplain apparently failed to establish fair play rules and legal pursuit, however.

The rooster ran through two rolls of concertina wire with the 300 Marines in hot pursuit. Pursuers in front never had a chance to stop at the wire, and about 175 "buddies" unfortunately used them as a human bridge over the wire.

A Marine finally snared the rooster. Rules required him to hand it to the chaplain to complete the game. But fair play disappeared in the quest for that beer. After the fray, it looked as if a 155 round had landed. Marines were untangling themselves from the wire, some limped toward sick bay, some were trying to get up off the ground, and most were nursing fresh cuts and bruises.

The chaplain was trying to figure what went wrong, and no one remembers who, if anyone, got the beer. Whether relations improved between the two units was doubtful, but all remember "the day of the rooster." **PFC Dave Doehrman, B Co., 3rd Recon Bn., Fort Wayne, IN, maintenance manager for Ryder Truck Rental there.**

Landing Team 3/9 spearheaded the arrival of Marine ground combat forces, including the first elements of the 12th Marines. An airlift brought 1/3. An immediate mission of 3/9 was the security of Da Nang air base.

Da Nang was the northernmost of three South Vietnam jet airfields. Others were Ton Son Nhut and Bien Hoa, close to Saigon.

Other units were helilifted to Hue and Phu Bai to assume defensive responsibilities in that area. During April and May, more units of the 3rd, 9th, and 12th arrived. The 4th arrived at Chu Lai in May to rejoin the division. The 3rd set up on the Da Nang perimeter with the mission of eliminating the Viet Cong from that sector.

The advance party of the III Marine Expeditionary Force, commanded by Major General William R. Collins, also 3rd Marine Division CG, arrived at Da Nang, May 3, and was redesignated III Marine Amphibious Force (III MAF). Forward 3rd Marine Division headquarters was also established at that time at Da Nang. A few days afterward, the Chu Lai area was added to 3rd Marine Division control.

At the time of the division's arrival, it was estimated that 12 battalions of Viet Cong were within striking distance of Da Nang.

Earlier, activity involving Marine helicopter units already on the scene was stepped up in early 1965, with Viet Cong attacks on United States bases and installations, and other guerilla activity. The first extended Marine combat clash with enemy guerillas was a patrol action, April 22 near Da Nang airbase.

Shook John Wayne's Hand

Some of the famous people that came to Vietnam to boost troop morale came to see us, although we were never able to see one of the USO shows. One of the first was Frank Sutton, who posed in his "Sergeant Carter" fashion as photos were taken with us. I also saw Archie Moore and Martha Raye, who later sent me a picture.

One day we came off bridge security to the 9th Marines area somewhere south of Da Nang. Three or four of us went to look for John Wayne, who we heard was in the area. We located him with a small group of Navy and Marine officers dressed in khakis, as was Wayne.

Grungy from two weeks of patrolling, we nonetheless went up to him. John Wayne asked to see the M-79 we carried, inspected it, brought it to his shoulder and sighted in on distant targets. He observed that it "was an awesome weapon."

We were thanked for coming, and started back as I moved to a position where I was able to shake John Wayne's hand as they passed. It meant a lot to me to see John Wayne and to have shaken his hand. **David Torrel, F-9th Marines, Eveleth, MN**

Starting The Nation's Longest War

The stage was rapidly developing for what became the nation's longest war. When the United States first went into Vietnam, the "black pajama guerillas" (VC) were the principal enemy, and were losing according to military observers. But the North Vietnamese began an infiltration of from 4,500 to 5,000 men monthly, troops with a trademark of "aggressive fanaticism."

The Marines were a force trained for offense and attack, and had to adapt to counter-insurgency warfare. There was a complicated command structure

(Continued on page 201)

One Hairy Morning In 'Bodeland'

As a member of WestPac reaction forces at Okinawa, I volunteered for a combat zone in late 1963 and was sent to Jungle Survival School forthwith, perhaps because of my half-Cherokee Indian heritage. Ten of us arrived in-country at Vietnam for what was later named "Recondo" training at Nha Trang.

The four of us who graduated met Generals Harkins and Krulak and several "spooks," then were incorporated into Oplan 34A, billeted at Tan Son Nhut, and assigned to the southernmost trail outlets in Cambodia as part of "other theater" activities. On February 20, 1964, I took my first steps into the bush of "Bodeland," charged with locating enemy base camps, rest areas, supply dumps and caches, vehicle depots, and troop movements (if any). Also POW and KIA friendlies' locations, map construction and correction, jungle density recordation, topography and new roads and trails, downed aircraft locations, identifications of plants and trees and other flora and fauna.

The most important (and "hairy") aspect of my activities was mapping the trail and its intersections, widths, outlets, bifurcations, canopy coverings, and its pedestrians (number, type, units, armament, etc.)—especially the VC 9th Division, reported by a contact as operating in the area.

With some natural ability for stealth and survival, plus considerable formal and personal training, I was able to make it through three tours without ever being seen!

I was heard once, though. In 1966, at a time when I changed position in a tree (most of my time was spent in trees), I was detected by a "hawk" team for the "Cu Chi National Guard" (25th), quite alert to jungle sounds.

I heard "Dung lai," and saw at least five floppy hats. I yelled "Wetsu (followed by an obscenity)," which in turn was followed by silence. After a few seconds, one of the team said, "Identify yourself."

My response was: "Studies and observation group. I'm studying plant life right now, but you better didi your asses out of here right mau, 'cause I'm expecting a whole shitload of gooners through here rickety tick."

Somebody said "Roger," and they left, making a lot more noise than when they came in.

I patted myself on the butt, lucky not to have gotten it shot off because the team was looking to take prisoners rather than shooting first.

This was one incident in about 36 missions into Cambodia, and even up to Ban Tasseng in Laos, locating and mapping the trail. **John S. O'Neal, 1/3, Upland, CA**

Reflections After The Fact

Our intentions in Vietnam were honorable, and many of the results we achieved were commendable, but the end was deplorable. In 1966-1967, I participated in more than 300 combat patrols with F-2-9, ranging in size from two-man LPs to battalion sweeps. I still have some gook metal in my knee from one of them.

On two-thirds of the patrols I was on, nothing happened. In retrospect, I believe defoliation of the jungle restricted the movements of the VC/NVA. This lessened their opportunities to ambush us, which gave me a greater chance of survival.

After leaving Nam, I was met at Minneapolis Airport by three brothers and a cousin. Their handshakes and those of others were enough. I didn't need a parade. It was good enough to be back. **David Torrel, F-2-9/F-2-26, Eveleth, MN**

Canteen Water, Warm Type

On my arrival at Da Nang from Okinawa in January 1969, there is an ammo dump exploding about a click away with then unfamiliar shock waves. Finally, I catch a C130 to Quang Tri and assignment to L-3-9 at their rear area at Dong Ha. In short order, I learned to sleep through the solid "Ka-toonks" and to roll outboard from the tent side on all the screamed "short rounds, hangfires, and oh shits."

After a few days, we received convoy shotgun duties. Six to nine of us would climb in the back of a sand-bagged "six by" and assume a 360 degree observation/fire arc amidst the flatbeds, tankers, and other six bys. There were guard trucks every 50 vehicles or so, with quad 50 dusters spaced throughout the convoy, and gunships overhead.

Our orders were to advance rapidly toward hit and run attacks, or dismount and assault if it was a full gook assault. All the drivers had 16's slung in their cabs. It was very dusty as the convoy would move along at a pretty good clip, with about 50-yard intervals between vehicles.

We rolled westward through one or two villages, past the Rockpile and into Stud. It took until mid-morning to sweep for mines. There was asphalt for a little ways out of Dong Ha; otherwise, it was just dirt and dust.

Some of the guys would throw C ration cans out of the back of the trucks to see if the little gook kids could retrieve them before the next truck came along. Some of the guys would just throw the cans near the kids on the sides of the roads.

We would spend a couple of hours at Stud while the trucks were unloaded, and then head back. I always remember that the 5th Mech guys were well stocked with Coca-Cola. We had canteen water, warm type. **James (J.D.) Manville, H-3-9, Stevens Point, WI**

because of the variety of United States and other military forces, including the South Vietnamese, and governments involved. And the rules of engagement were that the United States forces had to receive enemy fire before firing back. In later action around the Demilitarized Zone, the North Vietnamese had sanctuary against ground action until they emerged from it.

There were four phases of operations established: (1) securing coastal areas; (2) operations against the enemy from coastal bases; (3) movement inland to secure additional bases; and (4) operations from inland bases.

Three-stage operations were planned: (1) security of base areas with limited patrolling; (2) deep patrolling and offensive operations outside assigned areas of tactical responsibility; and (3) long range search and destroy and reserve reaction operations.

After arrival of command echelons in early May, the 9th Marine Expeditionary Brigade was disestablished, and the III Marine Expeditionary Force became senior Marine command.

Activity accelerated rapidly, especially since in mid-1965 the South Vietnamese army was believed near collapse.

Major General Lewis W. Walt succeeded Major General Collins as commander of III MAF and 3rd Marine Division, June 4. Four days later, the United States Military Command in South Vietnam was authorized to send American troops into combat alongside South Vietnam forces at the request of that country's government.

As additional 3rd Marine Division units arrived in July, by month's end there were three infantry regiments on hand. The 3rd Marines were west and north of Da Nang, the 9th Marines south, and the 4th at Chu Lai, along with the 12th Marines and other supporting units.

The first regimental size United States Combat action since Korea took place August 18-23, Operation Starlite, an effort by division units to trap the Viet Cong on Van Tuong peninsula. The operation's code name was meant to be Satellite, but a harried clerk typed it as Starlite.

In September, what became an annual institution was inaugurated with Operation Golden Fleece, to prevent the rice harvest from falling into enemy hands.

The 3rd Recon Battalion on October 18 started Trail Blazer, a six-day deep patrol operation to determine enemy strength and intentions toward the Da Nang area. Other fall efforts included Dagger Thrust on the peninsula south of

Qui Nhon, and Harvest Moon between Da Nang and Chu Lai.

By November 1965, the monsoon season had brought an average of an inch of rain a day in parts of South Vietnam. Canvas turned green and moldy ... rough seas made unloading difficult ... roads dissolved into greasy mud ... and rain, drizzle, and clouds filled the valleys, all of which frayed Marine tempers.

Another disruption to Marine grunts throughout the war were the top-ranking United States military command officers who coptered from "civilization" out into the field, but who were hardly persuasive in starched and ironed uniforms. This was especially true for Marines for whom the realities of war were heat, wetness, malaria, leeches, repetitive patrols, ambushes, sapper attacks, and anti-personnel mines, as they listened in filthy, sweat-soaked utilities.

As 1965 ended, the United States was at war in force with over 39,000 Marines in Vietnam.

1966
War In Earnest

Late January and much of February 1966 saw Double Eagle I, biggest amphibious landing since the Korean War, and the largest sustained operation in Vietnam to that point. In southern Quang Tri, it involved Task Force Delta with 4th Marine units. In late February, Double Eagle II resulted in seizure of a large cache of weapons north of Chu Lai.

Utah in early March produced a fierce four-day battle and defeat of a North Vietnamese regiment, southwest of Chu Lai.

The combined commands of III Marine Amphibious Force and 3rd Marine Division were separated, March 18, when Major General Wood B. Kyle became division CG while General Walt continued in force command.

When the 1st Marine Division established headquarters at Chu Lai, March 29, it was the first time since World War II that the Marine Corps had two divisions committed to combat in a war zone.

Complicating the time from March through June 1966 was a South Vietnam political crisis from an ARVIN command change, what was called the "Buddhist Revolt." Created at the time was a major threat to Hue, particularly after the fall of A Shau to the North Vietnamese. The revolt ended when the removed general went into exile.

The division, particularly the 3rd and 9th Regiments with support units, con-

(Continued on page 203)

A "Grunt" Gets Used To "Humping"

World War II vets had all kinds of names for the often excessive loads they were required to carry ... but one of the words that Vietnam veterans used was "humping." This story by Gunnery Sergeant John E. Conick, written in 1965 for general reading, describes it very well.

Vandegrift Combat Base, Vietnam

He calles it "humping." If you haven't tried "humping," ask a Marine rifleman—a "grunt." He'll tell you about it.

"Humping is an everyday occurrence south of the Demilitarized Zone within the 3rd Marine Division. After assault landings from helicopters, it is up to the leatherneck infantryman to stalk the hills and search for "Charlie"—the North Vietnamese and Viet Cong soldier.

The origin of the words "grunt" and "hump" becomes obvious as a Marine "saddles up," or straps on his gear. The equipment is loaded around his waist and on his back.

Back packs—carrying all the things needed to sustain life and a few luxuries—contain sundry items from a toothbrush to an inflatable air mattress.

In the pack will be up to five days rations. Although designed to give nourishment, an experienced Marine will supplement the rations with hot sauce, onions, or whatever else may enhance the taste of pre-cooked foods. Also included with the pack are his "snoopy blanket," (a camouflaged nylon cover), socks, poncho, rain gear, and shaving equipment.

Strung around his chest may be bandoleers of ammunition, while his cartridge belt contains more pouches of ammo. Also hanging from his belt are canteens containing almost a quart of water each. Regulations normally require two canteens, but the field Marine usually carries at least four.

Other items include rifle cleaning equipment, gas mask, first aid packet, and entrenching tools to dig a foxhole.

Then there is still more equipment placed on the Marine's body. His rifle, smoke or fragmentation grenades, flak jacket, and helmet. Radios for instant communications, parts of mortars and ammunition for the mortars are also carried.

The weight? No one has stood still long enough to determine how much this assortment of gear weighs. The average marine would gladly lighten the load if he could.

Experts have made estimates of 60 to 80 pounds, about one-third of the man's body weight.

No two Marines carry the same equipment, but each leatherneck would testify that his load must weight at least 200 pounds. He would also tell you that his is carrying the heaviest load in his company.

Day in and day out carrying this burden, obstacles are placed before him. The long slanting slopes of ridge lines, frequently wet and slippery; jagged mountain cliffs; dense jungle growth; and trails that twist and turn every few feet. To these natural obstacles the enemy has added mines, poison punji stakes, and other headaches. But the leatherneck continues to march, troop, stomp, pace, or hike. He calls it "humping." He frequently "grunts" as he does it.

Take Five

Marine Lance Corporal Dennis L. Butts, 19, of Carlyle, IL takes a break after humping his heavy pack and radio during an operation against enemy forces in Quang Tri Province. Butts, a member of 1st Bn., 9th Marines, is one of thousands of grunts who must carry heavy packs, radios, and other equipment as they search for the enemy in the jungle terrain just south of the Demilitarized Zone. (USMC Photo)

Patroling Without A Bolt

One day I got tagged for a patrol, toward Laos. I wound up walking backward as "tail-end Charlie" all afternoon. We were definitely on to some main roads. The ground was pulverized in places, and there were craters everywhere.

We saw the drive train and axles of a truck that must have been blown off the road, way down on the side of a hill. It sort of looked like a fish skeleton you see washed up on shore.

About halfway through the patrol, the point man waved us to the deck. It was very quiet. The bombing had driven Mother Nature's creatures away. As I crouched there, the guy further up from me loudly whispers, "Oh shit!" I look at him and he just points.

We are staring at the ass-end end fins of one big mother _____ bomb that is half buried right next to the road. After a few minutes of sphincter silence, I asked him if he saw any cords or strings attached. If this baby goes off, it won't be as terrifying as the normal booby trap that just mangles. They won't even find one of our atoms.

It didn't, and we finally got back to our home hill. We took some sniper rounds, and one of the guys thought he saw the source. We were all given permission to fire. I did, but nothing happened. I looked down and my bolt was playing peek-a-boo with me—it wouldn't lock in.

I had just walked rear guard on a hazardous patrol with a very useless M16. **James (J.D.) Manville, H-3-9, Stevens Point, WI**

---

ducted an extensive pacification sweep and clear operation in the Da Nang area. In fact, a number of 3rd Marine Division units were extensively involved in such operations throughout the year.

Pacification and search and cordon operations were anti-guerilla efforts intended to break down Viet Cong village infrastructures. Hamlet residents were given free food and entertainment to occupy them, and rudimentary health care. Communications, supply, and fire support help went to local militia responsible for security. South Vietnamese forces searched for Viet Cong, propaganda, and arms. Civic action programs were carried out to provide better relations with the Vietnamese people and free them from the threat of Viet Cong terrorism.

From Hastings To Chinook

Much of July was involved in Hastings, the largest Marine multi-battalion effort to that point. Six Marine battalions, along with five from the ARVN, were committed to action 55 miles northwest of Hue against an enemy force of regulars, estimated at 10,000.

Objective was defense of the DMZ, and the famous Rockpile became a key observation and communication point for that and later operations. Rockpile was a 700-foot toothpick of a mountain in the middle of an open area near the DMZ, with sheer cliffs. That open area was in an area dominated by rough terrain and heavy vegetation.

The first two battalions of the 26th Marines, activated early in the year at Camp Pendleton, landed at Da Nang in August. That regiment became part of the 3rd Marine Division.

Hastings was followed by Prairie I, which began August 3 and continued into 1967. In mid-September, an amphibious search and destroy operation, Deckhouse IV, was carried out just north of Dong Ha as elements of the 26th moved into action. The Prairie series, which continued until May of 1967, involved some of the war's heaviest fighting, with the 4th Marines heavily involved.

Some other operations in mid-1966 included Texas, north of Quang Ngai in the An Hoa industrial area with 2/4 ... Georgia by 3/9 ... Jay with 2/4 ... and Macon with the 9th in the An Hoa area.

(Continued on page 205)

Operations Hastings Sees Every Fanaticism

Operation Hastings marked a transition for the 4th Marines as the level of fighting increased from minor Viet Cong skirmishes to confronting well-armed, well-trained troops.

In mid-March 1966, the 4th Marines pulled stakes in Chu Lai and sailed north to Phu Bai, just north of the old Imperial capitol of Hue.

My MOS was the usual for any Marine: grunt plus .. in my case, radio operator. Though assigned to HQ Company, several of us "radio ops" were routinely TDY to the line companies during combat operations. This provided us the opportunity to experience "the bush."

For Operation Hastings, we had set up a forward command post in Dong Ha, which had been secured in April. From there, G-2-4 landed in Khe Sanh, while K-3-4 dug in at ConThien. I was assigned Company Radio Op with K Company. There were six of us from HQ Company and four from K Company to make up the radio team. We manned an eight hour radio watch, backed by eight hours on perimeter duty.

We were continuously shelled at ConThien, barely managing two hours of uninterrupted rack time, even though there was very little in the way of armed conflict. We had fortified our position with concertina wire, booby trapped with claymores and such. We also had an artillery battery of four 105mm guns that periodically pumped out rounds at suspected enemy mortar positions. As a reinforced rifle company, we had about 270 Marines.

One dark, starless night, shortly after being relieved of radio watch at midnight, I hustled out to the bunker I shared with a couple of grunts whose names I don't remember.

I leaned into the sandbags and gazed at the concertina. It was strangely quiet ... eerie. So dark I could barely distinguish the twists and curves of the wire.

Then, in the distance, came noises like shouting and laughter ... approaching quickly. My partners had heard it too and were up and alert. We knew this would not be a welcoming committee.

Our line sergeant instructed us to hold fire until he commanded as the artillery cranked out some "candlelight." Though the sounds had grown much louder, we could not see any movement until, suddenly, there they were! Hundreds of Viet Cong seemed to surround us, bearing down at an outright run. I was petrified! My palms felt slippery around my M14. Then all hell broke loose!

As Charlie crashed into our concertina, we opened up on them. It was weird though because they weren't shooting back. They were throwing rocks, grenades, and junk at us. It was as if they were sacrificing themselves to set off our booby traps. There were two waves of assault like this, but the third was a whole different ballgame.

We sustained slightly moderate casualties at this point. Then the fourth wave came on with a fury. I flashed on a scene from "Pork Chop Hill" and cringed a little. So did the other two guys, and probably everyone on the line.

Our line sergeant ordered us to beat feet to our secondary perimeter. The NVA was storming over our concertina using the dead Viet Cong as stepping stones. Having expended a great deal of ammo, we had fixed bayonets, though I wasn't too sure they would be of much help.

Air strikes had been called for; but because of the swarm of Charlies, the payloads were dropping on us. It's like a bizarre dream F-4's strafing the whole area, Charlie all over us, it was hell.

Third Division Marines load wounded men abroad a HUIE MedEvac copter during Operation Hastings. The operation, one of the largest at that point in 1966, took place a few miles south of the demilitarized zone between North Vietnam and the Republic of Vietnam. (Official USMC Photo)

We held Con Thien, though we suffered terrible casualties, but it fractured the 324B Division of the North Vietnamese Army.

There would be other close encounters throughout 1966 until May 1967—including two stints on the Rockpile, but Operation Hastings will forever be etched in my memory. I salute all the guys of K/3/4. They showed what courage is all about.

In addition, I would like to pay homage to G/2/4 at Khe Sanh. Theirs is a truly sad story ... the Khe Sanh siege gets too little attention.

After Operation Hastings, the 4th Marines distinguished itself as one of the premier grunt units in the 3rd Marine Division. **Christopher J. Crawley, Etiwanda, CA**

Welcome To Vietnam

When I arrived in Vietnam to assume command of the Maintenance Battalion, Force Logistic Command, the officer I was replacing was getting ready to leave the next day.

An extra cot was prepared for me in the commanding officer's hootch. I needed rest but couldn't sleep. The artillery was firing off in the distance, up into Elephants Valley, which was the major terrain compartment northwest of Da Nang. To the uninitiated, it sounded like the guns were outside my door.

In the morning as we walked to the mess hall, our eyes cried. The quick explanation was that either animals or people had walked on the gas pellets outside our perimeter during the early morning hours, a routine occurrence.

The pancakes were terrific, tears and all. As we left the mess, Marine jets were attacking targets in the foothills leading into Elephants Valley. We watched from miles away as the aircraft dove. We could see the flash, and counted the seconds for the thunder to reach us.

"What the hell is this? Is the whole war on my camp's doorstep?", I asked. The reply was, "No sweat, you'll get used to it."

And I did. **Col. W.F. Sheehan, USMC (Ret.), later Division Supply Officer, Fayette, MO**

An Hoa, a budding hydroelectric and industrial complex southeast of Da Nang, developed into a major Marine combat base.

A major tactical change came on October 10 as the 3rd Marine Division was displaced northward to Thua Thien and Quang Tri provinces to conduct offensive operations designed to prevent a possible North Vietnamese push across the DMZ. The 1st Marine Division was assigned to the three provinces to the south.

The final days of 1966 featured Harvest Moon and the beginning of Chinook. Harvest Moon was aimed at relieving the threat to two district headquarters in the Que Son Valley, and Chinook was an attempt to deny Viet Cong access to rich coastal areas. A number of division units were involved in both.

As 1966 came to an end, Marine Corps strength in Vietnam reached 65,789, nearly one-fourth of the Corps.

1967
The Beat Goes On

By 1967 3rd Marine Division units in Vietnam were participating increasingly in larger scale combat efforts, after the heavy focus on battalions as the war's major unit operational level.

Elements of the division began Prairie II, February 1, a continuation of earlier multi-battalion action. It lasted 46 days, followed by Prairie III from March 18 until April 20, in which the 3rd Marines saw some of the heaviest fighting, mostly in rolling grasslands.

In January, 1/9 went ashore in the Mekong Delta as a part of Deckhouse V. Chinook, which had begun in December, ended northwest of Hue. The 4th was primarily involved, with some elements of other regiments. Another in the Deckhouse series, VI, in early March also involved the 4th among others.

Command changes had Major General Bruno A. Hochmuth succeed Major General Kyle as division commanding general, March 20 and Lieutenant General Robert E. Cushman replaced Lieutenant General Walt as force commanding general two months later. In April, regimental headquarters of the 26th Marines completed that regiment's move from Okinawa.

Amidst these command changes, the first battle of Khe Sanh began when a patrol from 1/9 made contact with an enemy unit northwest of the city. Subsequently, a number of division units were engaged in bitter fighting which continued until May 12.

The 26th Marines relieved the 3rd Marines at Khe Sanh, May 13, and began Operation Crockett in that area. It lasted until July 16. Khe Sanh became relatively quiet in early summer before an uneasy period later on.

Death Of A Commander

Five days later, other division units initiated Operation Hickory in the southern portion of the Demilitarizeed zone with a landing by 1/3 (Special Landing Force Alpha). Operation Buffalo, July 2-14, involved 3rd Marine Division elements supported by Marine aviation and Army artillery.

Early in August, 1/3 (Alpha) conducted Beacon Gate southeast of Hoi An along the coastal boundary of Quang Nam and Quang Tin provinces. Alpha then shifted to 1st Marine Division operational control with Task Force X-Ray for Operation Cochise at Quang Tin.

Tragedy struck November 14 when Major General Bruno A. Hochmuth was killed in a helicopter crash five miles southwest of Hue on the way to Dong Ha. Assistant division commander Brigadier General Louis Metzger assumed command until Major General Rathvon McC. Tompkins arrived. The Hochmuth tragedy spawned the 3rd Marine Division Association Memorial College Scholarship Fund as a memorial to the education minded general.

Meanwhile, Vice President Hubert H. Humphrey arrived at Da Nang, November 1, and presented the Presidential Unit Citation streamer to the 3rd Marine Division.

During 1967, division units also participated in operations Ardmore, Scotland I, Cimarron, and Lancaster in the areas of Camp Carroll, Con Thien, and Gio Linh.

By the close of 1967, the 12th Marines had become the largest de facto regiment in Marine Corps history, with 11 battalions, including three from the Army, under regimental control.

At year's end, III Marine Force strength in Vietnam was at 81,115, including 77,679 Marines with 3,436 Navy personnel. This was more than 27% of the entire war-enlarged Marine Corps.

(Continued on page 207)

A CAMOUFLAGED ARTILLERY POSITION

Tale Of An NVA Lizard

Well into the monsoon season, my 3rd Recon Team Broad Weave II was on patrol northeast of Khe Sanh. In 24-hour-a-day rain, it was impossible to stay dry when we were out in the bush. Ponchos would cover up your gear, and they're noisy moving through heavy vegetation. The temperature never got much below 50, but felt a lot colder when we were soaked to the skin all the time.

Due to difficult terrain and weather, I was alternating point duties between the regular point man, the secondary point, and myself. I had just taken over the lead in approaching dusk, and began looking for a good harbor site. Moving around a large bush, I suddenly came nose to nose with an NVA, an NVA lizard, that is! Right out at the end of a tree branch and about six inches away at eye level, it almost caused me to jump out of my skin!

Despite the scare, it didn't run away. In the cold, and being a reptile, it remained there. Some kind of iguana, about 18 inches long, I carefully pried its feet away from the branch and pondered on what to do with it.

A good spot for a harbor site was nearby, so we settled in for the night. To keep my new pet safe, I carefully curled it into a ball and tucked it into my binocular case. After setting out claymores and giving nighttime artillery registrations to Rainbelt Bravo, I made a sort of closed basket cage out of sticks and vines and put the lizard inside.

After an uneventful night, we prepared for being extracted from LZ about 300 meters away. We were actually within site of the base, but it would have taken hours to get back so we happily sat in the rain and mud to wait for choppers. When the UH-34s came, we got the smoke grenades and air panel ready and as they landed, we lowered our heads into the prop wash and began boarding.

As patrol leader and last to board, I anticipated difficulty with all the normal gear and weapons in getting in the door, plus my lizard. I handed the cage to the door gunner, but could not see his reaction behind his goggles and helmet to possibly the most interesting POW that had ever boarded his chopper.

After debriefing and turning in our gear at Khe Sanh, I went to the sergeants hootch, took my "prize" out of its cage, straightened it out, and set it on the cot next to mine. The hootch was comfortably warm and didn't seem to affect the lizard, or so I thought.

After about 20 minutes while I cleaned my rifle, everything became a blur. Someone came in the door which made a sudden noise. It startled the now thawed-out iguana which bolted the length of the cot, banked off the wall, and screamed past the bug-eyed Marine as it went out the door, never to be seen again.

Back in its own world, the thought of having to eat C-rations with Marines was perhaps too much for its reptilian mind. **Sergeant Steve Johnson, B and E Companies, 3rd Recon, Donelson Road, Jamestown, NY, recalled to active USMCR duty in Desert Storm, Association officer and board member.**

A Tough War

After a month or so in Vietnam, I became point man and went out on a company-sized patrol with the lead platoon as point ,man. We moved out through a large clearing that was once a rice paddy, with dikes still there and used as footpaths. Since the VC often mined the dikes, we would walk on their sides. When the sides were mined, it was back on top.

We started on top, then got off and onto a side. After moving no more than ten meters, there was an explosion behind me. I hit the deck. Another explosion...they had mined the sides and we had walked past them when we were on top. We moved to a tree line and formed a perimeter immediately.

There were more explosions, then sniper fire as we shot toward various spots in front in an attempt to kill or at least discourage him. Even the gunships accompanying the medevacs managed only to slow him down during the evacuation of casualties. After the gunships left, an A-4 made three strafing runs and the sniper fired at the plane.

When we were able to form up again, my squad had six men, the platoon less than 20, and company strength was less than 50% with the casualties.

We learned the hard way it was to be a tough war against a formidable opponent. **David Torrel, F-9th Mar, Eveleth, MN.**

Non-Partisan Bees

In late 1967, I was a patrol leader. Just south of Khe Sanh, we had been followed for several days without contact. In the area where the river makes a U-turn, at the bottom between two very steep hills getting water, the rear point heard what he thought was a rifle butt hitting a tree.

Tactically in a very bad position, we began climbing the hill to our front northward. The hill was so steep we were grasping for anything and climbing with hands and feet when, suddenly the "whisper" of bees was heard. In amazement and wonder, the stings began.

We got to the top in what must have been record time, and the thousands of bees suddenly disappeared. The troops were positioned defensively to await the company when a squad count revealed a man missing. I went to the edge and quietly called his name. Surprisingly, he answered, "Sergeant Hutton, I can't see."

Thinking he had been stung in the eye area, I asked what was wrong. He said his eyes were closed, so I said, "Open them." He replied, "They're open." Much relieved, I told him to come up. The reply came, "Sergeant Hutton, I lost my rifle." I told him to look around, and he indicated that he found it. Then I heard, "Sergeant Hutton, I lost my LAAW."

Told to take a quick look and to get up top with the rest of us, he found it and shortly appeared and joined us. I was never so happy to see anyone in my life. Not only that, but the kid Marine involved turned out to prove himself many times and I was glad and better off for knowing him.

We were in bad shape due to the stings, but apparently Charlie found the combined prospect of the bees and our hilltop position too much and we never saw him. **SSgt. J.L. Hutton, B Co., 3rd Recon, Jacksonville, NC; battlefield commissioned, did second tour in Vietnam and retired after 30 years as a major.**

He is completing his degree so he can teach learning-disabled children.

A Rat For A Bed Partner

In late 1966 when our Recon outfit arrived at Dong Ha and the DMZ area, there were only a few scattered old French mission buildings used for a mess hall and command post. Tents set up in rain and mud were eventually improved with pallets and plywood. But the rats ...!

The size of a small to medium dog, they had red eyes and long spiral pink tails. One night after lights out, we heard a slurping noise. Squad members sleepily yelled to their buddies to knock it off. When this failed, flashlights came on, and in the middle of the empty cot across from me was the biggest rat I ever saw, licking a metal mess tray.

The lights were hurriedly turned on, and an irate band of Marines started throwing everything available at the massive rodent. In the chaos, the rat leaped across the floor and jumped on my cot. One lance corporal told me to hold my foot still while he blew the rat away.

The combination of fright and chaos combined to make me unable to remember later what happened, but I was told the rat was shot.

A few years ago, in a contact with a former Recon member, I was trying to describe circumstances that would help him remember who I was. Finally, he asked, "Are you the skinny guy that had the rat in bed with you?"

We slept with lighted candles placed around our cots until we got used to our persistent nocturnal visitors seeking food scraps. (Editor's note: The rat may have been a descendant of those on Pavuvu in the Solomon Islands back in 1944.) **LCpl. G. Edward Merrihew, C-3rd Recon, Santa Rosa, CA. After discharge in 1967, he came back to the Corps as an officer in 1973 and was a 1st lieutenant when he came out in 1975. He now issues environmental permits for Calpine Corporation in California.**

Operation Mixmaster

There were a variety of colorful names for the hundreds of military operations involved in 3rd Marine Division participation in Vietnam. One, however, that was administrative in nature was known as Operation Mixmaster.

It involved transfer of thousands of Marines between units to "smooth out" the rotation dates of personnel within battalions, primarily in the last few months of 1965.

This avoided large losses of personnel at one time from rotation, and stabilized battalions to minimize peaks and valleys of experience.

1968
A Busy Year And The Tet

The year of 1968 turned out to be another eventful and busy one as the war persisted.

The second battle of Khe Sanh began when 3/26 attacked a North Vietnamese unit between Hills 881 north and south, January 20. There was heavy fighting the balance of the month along the Demilitarized Zone, and the 26th completed a move to Khe Sanh, also in January.

During the last three days of the month, North Vietnamese and Viet Cong forces launched the Tet offensive with attacks against military installations and 105 South Vietnamese cities and towns. The objective was to freeze American troops in base areas while the towns were attacked. This included ground assaults on five provincial capitols, involving ultimately over 80,000 North Vietnamese regulars and Viet Cong guerillas. Hue control was gained, but only for a short time.

The 3rd Marines took part in Napoleon/Saline along the Cua Viet River in February to insure free shipping and deny enemy access to rocket positions which could threaten Dong Ha and Quang Tri.

The city of Hue was regained and declared secure, February 24, after a 25-day battle. Combined Marine, Army, and ARVN forces accounted for over 4,500 Communist troops killed and the capture of large amounts of supplies.

Statistics released in early March dramatically confirmed that Marines had borne the heaviest burden of Vietnam fighting, with about 20% of the troops in country and 40% of the casualties. And at the end of March, force commander Lieutenant General Cushman commanded 163,000 combat troops (including Army forces), more than any Marine general in history.

Early March saw sharp fighting in several areas along the DMZ ... enemy entrenching operations near the 26th Marine perimeter near Khe Sanh ... and an outbreak of fighting northeast of Con Thien and near Cua Viet on the South China Sea. As the weather improved at that time, an exodus of major North Vietnamese units began and pressure on the combat base at Khe Sanh, which had been the focal point of enemy activity, began to ease.

(Continued on page 209)

A Long Night ...
All In A Day's Work For A Platoon Commander

For several days prior to an incident of May 28-29, 1968, we of F-2-3 (I was a platoon leader) skirmished with various NVA elements, giving ground, taking hilltops, moving from ridge to ridge, and keeping the enemy on the run in Operation Pegasus. We had been in the area since April 15, straddling Highway 9 about two miles east of Khe Sanh combat base.

By May 1, we were engaging the enemy on virtually a daily basis. Fox Company set up its position at the end of a finger ridge pointed toward KS combat base, three miles northwest. An excellent OP, it provided good observation west into Laos and south into the higher mountains.

Sharp-eyed lookouts May 26 spotted NVA files snaking over hills to the west, and the FO called in strikes that were right on target. The next day, we saw a convoy coming from the combat base ambushed about a half-mile outside the gate by brush-concealed NVA ... and we saw support come to reinforce the convoy, in broad daylight on a clear, beautiful day that we could see the blue of the China Sea over 20 miles away.

Mistakes ... Mistakes

The following day, we planned to move but were told by the battalion commanding officer to stay because we were having such good results. This was a mistake! And since we were low in supplies, everything from ammo to C-rations, we called for helo resupply, another mistake.

The helo brought surprises, including large tins of canned shrimp (probably purloined from the Army) and a case of fresh head lettuce. It also, with its late afternoon arrival, announced to all our exact position. As dusk descended, well-fed Marines (from exotic shrimp dishes) repaired to night watch positions.

My position was just back of the tip, near the crest of a ridge under a small bush where I hung my poncho liner during the day. We had a Starlight scope, and our platoon manned a listening post about 75 yards west of the platoon position, down a gentle slope that was a likely avenue of approach.

Others were placed nearby, and the FO had pre-registered artillery fire when we first arrived. I turned in early, wrapped in my poncho to ward off the chill. My corpsman was next to me, then radio operator Lance Corporal Chapin.

About 1 a.m., Lance Corporal Randy Huber woke me with news he had observed about a dozen NVA at the base of the hill through the scope. I ordered an alert ... radioed the commanding officer ... checked with the LP .. and requested illumination rounds as what was to be a long night began. I told the LP to quickly return to the lines, and they reported seeing figures between them and us in the star-shell light.

"Friendlies!"

Suddenly, three NVA jumped in an already cramped hole. They said "Friendlies, friendlies," but were covered with explosives prepared to die. We shot them, and kept firing our mortar.

The radio operator joined me for the move to the opposite side of the ridge top, but in the dark forgot the steepness of the hill and tumbled down into a ravine. The handset was ripped from the radio, so we couldn't report. We crawled up the steep slope, identified ourselves, and made it back to the security of the lines.

Soon, I hard Viets talking a few meters away. I told the operator to get as many grenades as he could from the command post and report the situation to the lieutenant there, asking for artillery support (with no radio of our own to call it in). Returning to the platoon position to direct defensive efforts, several wounded were reported by corpsman Frank (Doc) Saraniecki.

Doc went to the command post, scrounged a load of grenades, and exposed himself to fire to deliver them to the men. NVA had extensively penetrated our position, and a request for a mortar concentration didn't work out. Casualties mounted with heavy fire coming over the perimeter, and the fire team that sounded the alert was cut off. They couldn't return to our perimeter, but were able without compass or maps to get to the battalion command post and report our desperate straits.

As dawn broke and casualties increased, the Fox CO and I called for napalm. In minutes, the Marine F4s were washing the hillside. Some carried upward, though, and ignited the elephant grass, forcing us to scamper off the hill.

Waiting for the flames to die, we saw Echo Company (Captain William Russell) hurriedly making way up the ravine to our position. They linked with us, and our lines then overran the few NVA left from the F4s work.

By noon, our ordeal was over. Of 13 dead Marines 11 were from my platoon. Doc Saraniecki was alive, severely wounded, a hero for his resupply effort. I walked off the hill with a nasty hole in my back, then the next day caught a six by to Med D at Dong Ha.

After an operation, I returned to the company in June. It was different. There were many replacements which gave it a new look, and the veterans were subdued and apprehensive. Not too long afterward, the battalion was airlifted to Quang Tri for a three-day stand-down, first since April 1, almost three months earlier. **Ray Dito, F-2-3, San Rafael, CA**

In The Marines ...

"I Matured Very Quickly And ...
I Learned The Measure Of A Man"

In January 1965, I went to Vietnam, assigned to the 3rd Battalion, 3rd Marines at Chu Lai . That assignment had us performing perimeter security along with sweep and clear operations while an airstrip was being built. When our assignment ended, I moved to Marble Mountain with the 9th Marines.

After college I joined the Kansas City, Missouri Police Department and have worn that uniform for the past 21 years. In June 1990 I was appointed the 38th police chief for Kansas City ... serving 430,000 people within Kansas City's 322 square miles.

My life's work in policing has certainly been influenced by my Marine Corps experiences. Like almost everyone who joined the Corps, I matured very quickly. I also learned life-long lessons about teamwork and the value of steady leadership ... that ethnic heritage and social position are not the measure-of-a-man. These lessons have served me well. **Steven C. Bishop, Kansas City, MO**

A New Commanding General, Raymond G. Davis

The Khe Sanh battle (Operation Scotland) came to an official end March 31 after five months, under 26th Marine control with elements of the 9th. Also on March 31, back in "the world," President Lyndon Johnson announced a de-escalation of the war and initiated peace talks.

As April evolved into May, the 320th North Vietnam Division attacked 3rd Marine Division elements in a bitter six-day action at Dai Do hamlet, northeast of Dong Ha. Heavy fighting continued for most of May.

A new commanding general for the 3rd Marine Division arrived on the scene, May 22, when Major General Raymond G. Davis, a winner of the Medal of Honor in Korea, replaced Major General Tompkins, who became deputy commander of III MAF.

During much of 1968, the 12th Marines fired a number of artillery missions into the DMZ and the southern panhandle of North Vietnam in support of other division operations.

The 9th Marines captured Dong Tien mountain, northwest of the Rockpile artillery base, from enemy forces, September 9, after the 3rd Marines took Mutter's Ridge, north of the Rockpile. These actions prevented the enemy 320th Division from crossing the DMZ.

A week later, the 26th Marines suffered heavy casualties when North Vietnamese troops fired 200 mortar rounds into Marine positions near the Rockpile. By late October, Mameluke Thrust, in which parts of the 26th Marines participated, ended southwest of Da Nang.

November action included Garrard Bay, a search and clear southeast of Da Nang by elements of the 26th Marines, and Dawson River, also a search and clear in Quang Tri by 9th Marines units. In December the year long Napoleon/Saline search and clear under division control was concluded.

Some other operations in which division elements participated during the year included Kentucky, Houston, and Meade River, along with extensive pacification efforts.

As 1968 came to an end, Camp Carroll, a major Marine combat base in northern Quang Tri, was deactivated

(Continued on page 211)

"Nothing Heroic"
Busy Life Of Combat Engineer

Andrew W. Sabol, now of Shavertown, Pennsylvania, wrote the following piece. Every Marine who served in Vietnam can identify with the many vignettes he describes during his tour of duty. When he submitted this piece, his letter included this sentence: "Nothing heroic, just a year in the life of a combat engineer, May 1965 to June 1966." I guess it takes a Marine to define all that he endured as "nothing heroic."

Sick of barracks duty at Camp Lejeune in 1965, I put in for duty on Okinawa (The Rock). When we reported for staging in California, we were told we would be going to Vietnam. Given a beer party before we shipped out, it took us about 18 days to reach Okinawa with stops in Hawaii and Japan.

When we landed in Okinawa, Marines waiting to return home made fun of us because we had a whole year to do. (In contrast, when we left Vietnam later we didn't say a word to the new troops. They were facing a different situation ... and we knew a lot of them wouldn't make it home alive.)

Okinawa was a shock to us because at New River we had women in the barracks doing laundry, working in the mess hall, even showering in our barracks ... but there were no women at Okinawa.

We were on "The Rock" for about six weeks doing a lot of running to keep in shape ... went to camo and demolition school ... and found great liberty was just a five minute walk outside the front gate.

We were all just waiting our turn to go to Vietnam—"Going South" was the term. While loading onto a landing craft, a sailor was giving one of our Marines a hard time. The Marine never said a word but, before you could blink, gave the sailor a butt stroke to the gut. I then realized I was with a different type of Marine.

Before we landed in Vietnam a colonel told us he was sorry that a change in our destination meant we would not be landing with a bayonet in our teeth.

Firing In The Mountains

The USS *Pickaway* landed us at DaNang around July 4, 1965, and on our way into the beach we heard firing in the mountains. "Hear that?" said an officer, "That's your brother up there."

After landing we were taken to positions south of DaNang. We were told to dig in, but the ground was like cement. Finally we moved right to DaNang Airbase to support 2/9 ... and much of our work consisted of walking parties working mostly on culverts and outhouses.

In August we went on a lot of patrols and sweeps. At first we carried a mine detector, but this was a waste because we never really used it, and it slowed the patrol too much.

One time I got shot at when we were in the middle of a rice paddy, acting as a blocking force. We weren't supposed to fire back, but most of us did anyway. It was like a game at first ... but we wouldn't think that way a year later.

Our engineering lieutenant was wounded by a booby trap and was sent home. It seemed like lieutenants were not going to last very long in Vietnam.

I was attached to 1/9 for Operation Golden Fleece, and one of their officers asked me to find one of the platoons while we were under fire. I didn't care for this, but I did it. Later in the day we were pinned down by heavy fire. A Marine next to me raised his rifle to fire, and had his rear sight shot off.

Later, we headed back to Base Camp as 105mm rounds were fired into targets. As engineers we had to pick up quite a few 105 rounds that didn't go off, take them to a safe area and destroy them. I hated carrying these rounds, because you always wondered if they were going to explode.

Right before the rainy season a woman reporter, Dickie Chappel, stayed with our outfit for a while. She was later killed after leaving our outfit.

During the rainy season we worked on the main supply route ... including a couple miles of railroad track. During that rainy season we couldn't even take our boots off because our feet smelled bad.

One of our new men, who had been in Vietnam for only eight days, was killed by a land mine. He had told me the day he was killed that he enjoyed it over there.

Training New Men

During November 1965, three of us were attached to 3/9 to do demolition work ... mostly getting rid of duds. Instead of promoting us, sergeants came into the outfit from the States. We would train them, and then they would take

A Marine Engineer finishes setting a charge in Viet Cong cave. (Defense Dept. Photo - Marine Corps)

over. It wasn't fair, but we didn't mind because most of them turned out to be good men, and in the field, rank didn't mean too much. We all shared the work.

The two Marines with me often drank beer all night and shot marbles they had made from melted candle wax. They were both from the south and good men.

The first couple of days in a new location, we slept in a tent but it was really wet inside. (I slept on a mine detector box to try to keep dry.) We asked if we could move into one room of a house nearby. It was small and the three of us barely got into it, but it was dry.

Thanksgiving Day came and special dinners were flown into our camp. We engineers didn't eat any, and so it ended up that we were the only ones who didn't get the runs.

One of the replacements who joined us was a Sergeant Dodson. Later in the war, he was captured by the enemy, but managed to escape.

We usually went out on three-man demolition teams. One time a new man was put in charge of us. He was told to watch out for us, but he was the only one wounded. He put a charge in a cave that had water in it and the charge floated back out. He pushed it in again and was waiting to see if it would float out again when it blew up in his face.

One day we worked on a supply road all day in the rain. After that, I drank a bottle of Japanese whiskey ... and my bunkmates said I just passed out. I had guard duty that night, but one of our sergeants took my place. A great guy, he never even said a word about it. One time we were pinned down by automatic gunfire and this same sergeant stood up to make sure that we were all okay.

Free Time Was Great

In one location we used to hop a Viet bus and go to PX back at our base camp whenever we had free time. We were able to buy watermelons, pineapples, peanuts, sodas, just about anything off the natives. The natives did our laundry, developed our film, cut our hair, just about any service for a cheap price.

I remember when we were first in Vietnam we used to patrol with some Green Berets ... but that didn't work out well. We never seemed to make contact with the enemy, and thought maybe the Green Berets were overly cautious. But looking back on it now, that might have been due to their experience.

after more than two years. By that time, many Marine camps were fortified garrison towns inside barbed wire, but still occasionally subject to rocket and mortar attacks. However, huts had in many cases replaced tentage, and macadam covered red mud.

On the unhappy side, officers and NCOs particularly were coming back for second tours in Vietnam, after being harassed by demonstrations and anti-war sentiment back in the United States. It was hardly an appropriate tribute for the extremely difficult service they had rendered for their country.

1969 "Vietnamization" Arrives

A gradual "Vietnamization" of the war in 1969 slowly reduced the Marine Corps combat role, helped by the January 1 suggestion of South Vietnam President Nguyen Van Thieu that the ARVN was "ready to replace part of the allied forces."

Dewey Canyon, one of the best remembered operations when 3rd Marine Division Vietnam vets gather at the annual reunion, began January 22 in the Da Krong valley of Quang Tri, with the 9th Marines and supporting artillery (2/12).

Marine Corps and military historians have labeled the high mobility action as the most successful regimental size action of the war. It was completely dependent on helicopters which lifted troops into the enemy area for the nearly two-month operation.

Dewey Canyon, along the Laotian border 35 miles west of Hue, was a large bowl in mountainous, jungle covered terrain. The NVA started reusing the route after a large period of disuse, and had extensive facilities such as underground garages, command posts, and supply storage in addition to artillery sites and billeting areas.

Purpose of Dewey Canyon was to deny enemy access into critical populated areas of coastal lowlands, including the A Shau Valley. The operation blocked enemy resupply and infiltration, despite continuing cloud cover which complicated Marine helicopter supply and evacuation procedures.

Among the operation's accomplishments was discovery of the largest single

haul of arms and ammunition, so vast extra Marines had to be called in to help destroy it.

As the Viet Cong-proclaimed truce ended, February 23, Communist forces launched a major offensive throughout Vietnam. As February faded into history, so too did 3rd Marine Division operations Scotland II (after ten months) and Kentucky (after 16 months).

The March-April period also saw command changes for both III MAF and the division. Lieutenant General Herman Nickerson replaced Lieutenant General Cushman, March 22, at the force. Twenty days later, Major General William K. Jones succeeded Major General Davis as division CG.

The 3rd Marines carried out Maine Crag from March 15 until May 2, a sweep and clear south of the old Khe Sanh combat base.

After a relatively quiet first three months of 1969 for most of the division, north of Da Nang along the DMZ and Laotian border, the North Vietnamese renewed action in April. The 9th Marines encountered heavy resistance between Cam Lo and the Rockpile as Operation Virginia Ridge began, April 30, with the 3rd Marines also involved.

Other operations over the next few months included Purple Martin ... Apache Snow, a sort of Dewey Canyon replay (9th) ... Herkimer Mountain, north of Cam Lo (4th) ... Russell Beach (26th) ... Utah Mesa near Khe Sanh (9th) ... Idaho Canyon (3rd) ... and Cameron Falls, southwest of Quang Tri (9th).

By mid-July, Phase One of President Nixon's "Vietnamization" or troop withdrawal plan was initiated, with the departure of 1/9. On August 14, 3/9 left to complete the 9th Marines redeployment to Camp Schwab, Okinawa. Over the period from August to November, the 12th Marines moved to Camp Hansen, Okinawa. The 3rd Marines returned to Camp Pendleton in October-November.

The 4th Marines were also redeployed to Camp Hansen in November.

The 26th Marines remained in Vietnam, reattached to the 1st Marine Division.

Third Marine Division headquarters redeployed to Camp Courtney, Okinawa, in November, completing the division's withdrawal after more than four years in South Vietnam. During that period of time, the division was credited with 28,616 enemy killed, 499 prisoners of war captured, and 9,626 weapons taken.

(Continued on page 213)

When trying to go through a rice paddy at night to set up a blocking force or a sweep in the morning, there would always be one Marine to fall off the dike ... and let loose with a string of curses.

We used to volunteer for patrols and sweeps because it seemed better duty than your typical engineer construction work. In the field we were often our own bosses ... and since we didn't wear rank, the grunts didn't know if we were sergeants or privates first class. As long as we did our jobs, no one bothered us ... and this was good duty compared to barracks duty in the States.

Sometimes right before we swept into a ville, the infantry would get the order to "fix bayonets." This sent a chill down my spine ... after all, I was there too.

I lost my glasses in a rice paddy one night and had them replaced. I guess being lucky was more important than being able to see!

We supported various grunt outfits for different operations; and sometimes when the operation was over, the grunts didn't send us back to our outfit. I suppose the infantry just wanted some extra fire power.

Field Ingenuity

... It seemed like we never had enough containers for blasting caps, so we used the package that smokes from our C-rations came in. They were just the right size. Never had enough crimpers either, so we would crimp caps with our teeth or the bolt on our rifle. If we ran out of fuse lighters, we would split the fuse, insert a match, and then light it. Little tricks get the job done.

Like living in the Old West, we carried loaded rifles, no matter where we went, even to the showers. We were 19-year olds but were treated like 15-20 years in the Corps. What a thrill it was to hitch a ride with a grunt going for supplies on a mule atop those railroad tracks. He would "open it up" and you'd expect to fly off the tracks at any moment.

Once while working with 3/9, we found a lot of rice in a column that supported a building. We figured that something was hidden in there; so, instead of taking all the rice out by hand, I decided to put a small charge of C-4 at the bottom of the column. I expected just a small blast, so I stood right outside the building. When it blew, the wall I was standing by fell in on me. The grunts, including their commanding officer, were

laughing like crazy. The enemy must have had explosives in there.

The grunts usually liked us engineers because we kept them supplied with C-4 which was great for heating up C-rations.

When we were at the rear, it seemed like the other troops would look down on the infantry and attached units. But when you put on that combat gear and mount up it was a different story. It was like we knew something special ... and "they" would never know it.

Celebrities of Celebration

While in Vietnam I saw Hugh O'Brian, Ann Margret, and Bobby Rydell. It wasn't planned, but on New Year's Eve at 12, everyone in 2/9's base camp opened up with their weapon or flares. Battalion headquarters thought we were under attack.

It always amazed me how the grunts could pick their way around in the dark ... even when you couldn't see the guy in front of you—only two feet away.

There was one grunt captain in our group who really cared for his men. Even with just a couple of enemy sniper rounds, he called in an airstrike or any other firepower he could get.

While on Eagle II Operation, south of DaNang, I'll always remember the generosity of an old Mama-san. We were sweeping through the valley and she comes with her teapot, offering all of us a sip of tea.

Then one cool night we were sleeping in a garden between the crop rows. Suddenly this old lady comes out of her hooch with some blankets to cover us up ... another act of kindness.

It used to get us mad that we couldn't get jungle boots or utilities, when everyone in the rear seemed to have them. Another thing that bothered ... the grunts were rotated to the airbase every once in a while—but we stayed in the field, rotated to different outfits.

One time we went out on an ambush; but before we could get set up, we were ambushed. The ambushers were using a lot of M-79 rounds, but they weren't doing any damage. When the fracas was all over it turned out we had been ambushed by our Viet allies. They were laughing like crazy, but we were really ticked off. Lucky they didn't wound any of us because the grunts wanted to blow them away.

One new guy in the outfit just refused to go out in the field when it was his turn. He was finally forced to go, and the first day out, he was wounded. Maybe he had a feeling.

Once in a while I would smell flowers, and within a couple days one of the guys would be wounded. It was getting to that point in my tour where I felt somewhat shaky. The fun of it was long gone. In the day you'd only hope that you'd live to see the night, and at night you'd pray that you would see the new day. I hated to see a buddy get hurt, but sometimes you'd think—better him than me. We had 34 men in our outfit, and 24 of them had Purple Hearts. I considered myself lucky not to get one. When I left, infantry squads in 2/9 were down to six men.

Going Home

Before going home, we went to the rear area of the engineer battalion, and that was good duty. Most of our work was cleaning machine guns. If we went on working parties, we had Vietnamese to do the work, and we supervised.

A few days later when we boarded ship to go back to the States, a Marine colonel told us no one would bother us, and we would have a relaxing trip home. I'll always remember those 18 days on the warm Pacific ... sunbathed all day long. When we pulled into San Diego there were recruits to handle our seabags ... we didn't have to carry a thing.

The winter greens and other gear we had left on Okinawa were damaged by a storm while we were in Vietnam, so we were given vouchers for everything we had lost.

While being processed at San Diego, we were on the parade field late at night. Suddenly, we heard a woman's voice way across the parade ground, like a voice in the wilderness, "Welcome Home, Marines" ... and we all let out a cheer. I'll always remember how good that made me feel.

My one regret is that I didn't keep in touch with the fine men ... good buddies ... I served with. Sergeant Brydges, Sergeant Scalf, Sergeant Traxler, Lieutenant Shumaker, Lance Corporal Lynch, Lance Corporal Warren, Lance Corporal Jincks, Lieutenant Regan, Private First Class Walls, Lance Corporal Nedric, Private First Class Hall, Private First Class Mitchell, Sergeant Promsberger ... and many more forgotten names, but not forgotten Marines. **Andrew W. Sabol, Shavertown, PA**

Surprise Reception

When the first ground combat units arrived in Vietnam for the initially limited mission of strengthening security at Da Nang Airbase, the reception was somewhat unexpected.

Battalion Landing Team 3-9 was the first to land, from ships, while BLT 1-3 was airlifted in and began to set down two hours later on the morning of March 8, 1965.

The skies that morning were gray and sullen, with stiff northeast wind causing the surf to run five feet or more, delaying the landing briefly. It was just after 0900 that the first wave of the 3rd Battalion, Ninth crossed the beach just north of Da Nang.

The area had been thoroughly swept by two Vietnamese battalions, and there was air cover. BLT 3-9 did not expect the beach to be defended, but 9th Marine Expeditionary Brigade commander Brigadier General Frederick J. Karch and the combat equipped troops were in for a surprise.

There was an elaborate official welcome, including a group of giggling Vietnamese girls who proceeded to decorate the leading edge of the landing force, including General Karch, with garlands of red and yellow flowers.

Necessary Roughness

In August 1966, I was a lance corporal and the supply man for Bravo Company, 3rd Tank Battalion, situated at Hill 55.

The C-rations were stored in back of the company office tent. From this tent, next to the perimeter, a field stretched for 800-1,000 yards and then met a treeline.

One afternoon a tanker and myself were loading C-rations from the company office tent into a jeep trailer. Suddenly a bullet from a sniper's rifle struck the stones close to the jeep trailer. I immediately told the office personnel that a sniper was out there.

The company commander didn't waste any time. He immediately had five tanks come up on line and they blasted that rice paddy for a solid 2-3 minutes, firing at anything and everything because we didn't know exactly where the shot came from. A patrol from 1/26 could not confirm that we hit anyone. **Alan L. Bachman.**

Kicked Some Butt

It was August 1966, and my outfit was assigned to protect the 105mm and 155mm guns at Cam Lo. I was with A Company, 4th Marines, and my MOS was 0311-Grunt.

We were being attacked by the 324 B Division of NVA which had overrun the perimeter, causing considerable damage.

Their suicide companies hit and came through the first strand of concertina wire, but the second strand stopped them. Some of the enemy laid across the wire so others could walk across over the top of them. The enemy was all over the place.

We had been instructed to stay in our foxholes and shoot anyone outside of a foxhole. And, of course, it was difficult to see at night. Flares were shot off constantly, but the illumination was intermittent, not steady—and the poor visibility added to the confusion.

This action continued for several hours, but was subsiding by daybreak. When the sun rose, we went out to get a body count. Some of the dead had been pulled away from the compound with communication wire.

One of our key people was a machine gunner in a bunker at the point of the perimeter. With his .50 caliber machine gun, he really kicked some butt when the enemy came through the wire. He never stopped to change his gun barrel, and it was warped from his steady, effective firing.

Many of the bodies were mutilated ... and the smell from the heat was becoming awesome. I mean BAD. **Dennis Mansour, Iselin, NJ**

Epilogue

In 1970, the American troops remaining in Vietnam went into Cambodia for two months to destroy Communist military sanctuaries, and in early 1971, South Vietnamese troops crossed into Laos to try to cripple Communist supply lines from Hanoi.

After North Vietnamese and Viet Cong troops attacked the South Vietnam area across the DMZ, Quang Tri was lost on May 2. From May until the end of the year, there were heavy American bombing raids of Hanoi, Haiphong, and other North Vietnam military bases.

At the end of 1972, President Nixon ended the American air offensive and resumed peace negotiations with the Hanoi government. A truce agreement was signed by many of the parties at Paris to end the war January 27, 1973, and American fighting ended.

By the end of March 1973, United States troops were withdrawn and American prisoners of war released. But the internal and ideological conflict of long standing continued in the war-torn country. The situation disintegrated further, however, on a relatively quiet basis until December 1974 when NVA divisions moved into several key areas.

By the end of March 1975, the South Vietnamese abandoned Quang Tri city and province, and the NVA had taken Quang Ngai, Tam Ky, Hue, the former Marine base at Chu Lai, and Da Nang airbase.

The 3rd Marine Division was not done yet in Vietnam, as elements were called back into the country during the period from March until May 1975 to participate in evacuation operations in both Vietnam and Cambodia.

With the South Vietnamese army in complete disarray, Saigon surrendered April 30, and was promptly renamed Ho Chi Minh city by the Communists.

An emergency helicopter evacuation removed the few remaining Americans from the South Vietnamese capitol. Eleven Marines taken from the embassy roof were the last military personnel to leave.

On May 7, 1975, President Ford proclaimed the final end to the Vietnam War, the longest in the nation's history.

All-Weather Classic
Dewey Canyon

Operation Dewey Canyon differed from most contemporary 3rd Marine Division operations. While it involved mobile USMC forces against sizeable enemy resistance and other customary tactics, it emerged as a conventional regiment in the attack with all three infantry battalions on line.

Dewey Canyon took place along the Laotian border, 35 miles west of Hue and 50 miles south of its principal support facility at Vandegrift combat base. The monsoon season was in its final month, with no significant rainfall but cloud cover for periods of as along as 11 days. That made normal helicopter resupply and evacuation impossible.

The area of operations was generally mountainous and jungle covered with complex terrain compartments, forming a large bowl. A few kilometers south was A Shau Valley. The Da Krong River began its northward flow where A Shau opened to the south.

Hill 1228 (Tiger Mountain) dominated the area between the two valleys. Route 922 entered the A Shau area from Laos. To the west, on the Laos-Vietnam border, Co Ka Leuye, a razorback ridge dominated.

Several developments created interest in January 1969. The NVA opened Route 922 into the A Shau after many months of disuse. Anti-aircraft activity increased, with high performance helicopter, and reconnaissance aircraft receiving fire. Road traffic in Laos shortly doubled to more than 1,000 trucks a day sighted at times.

Enemy forces displayed more presence. Sophisticated wire communication nets were sighted. Reports indicated enemy movements back into the Da Krong River area, perhaps for commitments into other critical area.

Among these, the enemy could attack populated areas as far south as Da Nang with speed and surprise. Primary purpose of Dewey Canyon was to deny enemy access to critical populated coastal lowland areas, and also to interdict his access route.

From late January until mid-March, enemy resupply and and filtration was blocked by allied pressure, and many facilities destroyed.

In his book, the *Illustrated history of the Vietnam War (Marines)*, Brigadier General Edwin H. Simmons summarized Dewey Canyon:

"On January 22 General Davis sent the 9th Marines under Colonel Barrow into the valley in Operation Dewey Canyon, a high-mobility operation completely dependent on helicopters. Three fire support bases, Shiloh, Razor, and Cunningham, with others opening as the regiment moved forward.

"The heaviest fighting involved the 1st Battalion, 9th Marines, in the center of the line. By March 19 Da Krong Valley seemed cleaned out, 1,617 enemy dead had been counted and a huge amount of material taken, including 1,461 weapons and hundreds of tons of ammunition and other supplies. Rated the most successful regimental operation of the war, the cost to the Marines was 121 dead and 611 wounded."

Unit integrity was a major factor in operational success. Battalions often had to fight uphill for high ground objectives, and as battalions moved in line, strong contact was maintained.

An innovation in the bad weather was the successful handling of casualties in the field when medevac was not possible. Unfortunately, the bad weather also gave the enemy a chance to prepare for the latter phases of the operation.

Among the materials captured dur-

This Leatherneck gun crew from the 12th Marine Regiment, and their 105mm howitzer engaged a North Vietnamese Army 122mm artillery piece in an hour long artillery duel during Operation Dewey Canyon. Nearly 100 rounds were exchanged before the Marine cannoneers at Fire Support Base Cunningham silenced the enemy artillery piece. (Official USMC Photo)

Col. Robert H. Barrow, then commanding the 9th Marines drops a line from his sandbagged CP in Vietnam. This was just before the battle of Dewey Canyon and when the Third Division was under command of General Raymond G. Davis. Col. Warren H. Wiedhahn, the "4" (logistics) of the 9th was later also to command the regiment. Col. Wiedhahn, now retired, became president of the 3rd Marine Division Association. General Barrow became Commandant of the U. S. Marine Corps. (Official USMC Photo)

ing the operation was a well-equipped field hospital, and 1/9 captured two 122mm field guns, accurate to a 13-mile range. They were the largest field guns captured to that point in Vietnam.

Many large caches of enemy supplies were uncovered in bomb craters. During the operations, resupply and casualty replacement were continuing problems. By the end of February, the three battalions had advanced south to the Laos border, and were mopping up both the enemy and supplies, but suffering from those continuing resupply problems.

By March 18, with the extraction of 1/9, Operation Dewey Canyon was officially terminated. The 9th Marines success against a determined enemy was the result of long teamwork and close communication within the regiment.

This summary except for the small insert from General Simmons' book, was adapted and excerpted from a Marine Corps Gazette article under the same title by 1st Lieutenant Gordon Davis, son of the division commanding general.

Make A Gunny Sergeant Blush?

Most of the young Marines in my battalion were remarkably unselfish. They gave time and dollars to the Wakahn children's hospital, which was built by Seabees inside the FLC compound. The World Council of Churches had sponsored two doctors and their families from the United States to come to Vietnam to treat children who were victims of war.

It was a beautiful, practical, Christian effort that did a lot of good. It touched me to see the Marines playing with the children, reading to them, trying to help; many were just kids themselves. The children were truly pitiful, many with lost limbs, burns, and all kinds of injuries.

In their impish way of having some fun, the Marines also taught these kids some expressions that would make an old gunny sergeant blush. From the mouths of children, the effect of these litanies was sometimes quite startling.

But the concern and genuineness of the help these Marines provided far overshadowed their impious English lessons! **Colonel W.F. Sheehan, USMC (Ret.), later Division Supply Officer, Fayette, MO**

"We are going to Marine. When we can't Marine anymore, it's time to retire and go sit on the porch."

Maj. Steve Shivers
(Reprinted from Marine Corps Gazette)

Lieutenant Barnum And The Medal Of Honor

In 1965 Lieutenant General Victor H. (Brute) Krulak designed a program in which selected individuals from security forces at Marine Barracks, Pearl Harbor, were sent to Vietnam for observation duty, after which they would return and teach young Marines.

Lieutenant H.C. (Barney) Barnum and 1st Sergeant Natchez were the first selected for what Barnum thought would be a short-term observation tour with some artillery fire support duty. Soon after he got there, he found himself commanding a rifle company, organizing a counter attack against 15 to 1 odds, and standing on a knoll with arms outstretched to guide helicopters toward enemy targets.

Three months later, he was awarded the Medal of Honor.

When Lieutenant Barnum arrived at Vietnam in mid-December of 1965, he was assigned to E-12 south of Da Nang. From there, he went to join F-7 as an artillery forward observer. The personnel of F-7 and H-9 had been combined because of heavy casualties of both. Barnum recalls:

Fox Company took off on a battalion mission, Harvest Moon, between Da Nang and Chu Lai. Terrain was hilly terraced with rice paddies, a lot of bamboo, walking muddy and difficult. Weather was wet, cold at night because we were constantly soaked to the skin. Miserable.

After a few skirmishes along the way, on December 18 about noon, we heard shooting up forward, and a lot more shooting. Rice paddies on both sides had narrowed the company (H) front, and it was ambushed. Captain Paul Gormley, company commander, was one of the first casualties, probably hit by a rocket, but not killed outright, and the initial rounds killed his radio operator.

Everyone hit the deck as we experienced heavy fire—small arms, mortars, recoilless rifles. I could see where fire was coming from, so contacted the 12th Marines as a forward observer for artillery support, although we were at maximum range from where the 155s were located.

Others were hit trying to help the wounded. Private First Class Savoy and I moved the captain back since enemy rounds were dropping in our position, but he died in my arms a few minutes later. Under heavy enemy fire, I tried to organize our headquarters so fire could be directed toward our right flank. We were able to rescue the radio.

I was not in the Fox Company chain of command, having been there only four days. Most of the company didn't know my name, but with the commander killed, the troops looked for a leader. I called the battalion commanding officer, told him of the situation, that I had assumed command, and outlined a counter-attack plan.

He asked about the exec and platoon leaders, but I told him we were pinned down with no knowledge of where the other officers were. He said, "You are in command ... pass the word and make sure the platoon leaders and platoon sergeants know."

In organizing a counter attack, we found we had been ambushed by a regiment and odds were 12 or 15 to 1. A saving grace was our TRC 25 radio, one of the first to be assigned line troops. With it, all units were on the same frequency and everyone in the coordinated team could hear what was going on. By listening, we knew that everyone was under fire.

Artillery support was limited because it was firing at maximum range. Weather was heavy, and fixed wing aircraft could not be called in. We did have air support with four armed helicopters which did a magnificent job.

We fired white phosphorous with a rocket launcher to mark enemy targets. I went up a small knoll to point out targets when a helicopter called, "point your arm at the target and we'll fly in right down the axis of your arm." They made run after run, firing at the enemy, and after running out of ammunition, kept making passes to keep the enemy down.

They Called Our Bluff

After the copters had bluffed for nearly an hour, the enemy realized they were out of ammo and started firing back, forcing the copters to leave. About then, the battalion commander radioed us, "It's getting late. You have to join us."

We were about 200 yards away from the rest of the battalion. I had the engineers blow down a couple of trees for a landing pad for copters to take out the wounded.

There were many heroes that day. One was Corpsman Wes Barard from Chicago. Wounded earlier, once when treating the captain, he wouldn't take morphine so that he could tell others what to do to treat the wounded. He insisted that everyone else be put on the helicopter first, during which time he was wounded a fourth time.

Since we had to make a dash to join the rest of the battalion, we put all extra gear in a pile so we could make that effort unencumbered, then the engineers blew it up to prevent any value to the enemy.

With the ceiling closing in, which may have helped, we made squad rushes across some 200 meters of open ground to Kee Foo village to join the rest of the battalion. The wounded were helped by others as we made it across in good shape.

I then met the battalion commander for the first time, accounted for everyone, counted casualties, redistributed food and ammo, and reported back. A day or two later, we proceeded to Da Nang where a new company commander awaited and I was relieved to join my unit.

During the next few weeks, I had some trouble and went to the hospital where I learned that General Walt had recommended me for the Medal of Honor. I returned to Pearl Harbor to guard officer duty in February 1966, and received word there that the award was to be made. **Condensed from a tape on which Colonel Barnum told of the action in Vietnam.**

Medal Of Honor Citation

Captain Harvey C. Barnum, then Lieutenant, H-2-9, was awarded the Medal of Honor for his actions, December 18, 1965, at Ky Phu in Quang Tin province. The citation read in part:

"For conspicuous gallantry and intrepidity at the risk of his life above and beyond the call of duty. When the company was suddenly pinned down by a hail of extremely accurate enemy fire and was quickly separated from the remainder of the battalion by over 500 meters of open and fire-swept ground, and casualties mounted rapidly, Lieutenant Barnum quickly made a hazardous reconnaissance of the area, seeking targets for his artillery. Finding the rifle company commander mortally wounded and the radio operator killed, he, with complete disregard for his safety, gave aid to the dying commander, then removed the radio from the dead operator and strapped it to himself. He immediately assumed command of the rifle company, and moving at once into the midst of the heavy fire, rallying and giving encouragement to all units, reorganized them to replace key personnel and led their attack on enemy positions from which deadly fire continued to come. His sound and swift decisions and his obvious calm served to stabilize the badly decimated units, and his gallant example—as he stood exposed repeatedly to point out targets—served as an inspiration to all ..."

Church Services held in the field by Kilo Company, Third Battalion of the Fourth Regiment. (Defense Dept.. Photo - Marine Corps)

"I Remember ..."

I remember painting our faces ... carrying the M-14 with grenade adapter tube (some grunts thinking it was a silencer) ... seeing the bombed out moonscape from the B-52 Arclight strikes, water in the holes reflecting the moon ... walking off Hills 881 North and South and 861 ... being in slow motion after contact, time standing still ... watching rockets walk into the base ... a birthday cake being delivered by hand.

Robert Stack, the actor, came to the base. I remember he had clean utilities and new boots ... Captain Reynolds ... contact, being tracked and surrounded, blowing the claymore at night ... crossing the Rao Quan River on a rope ... one of our sergeants messing with us about haircuts after a patrol ... radioman Costello, and waking up and finding him in the cot next to me in a stateside hospital a month or more after I was hit, both of us in transit.

I don't remember firing a weapon. I remember watching the face of the NVA point man come into the sights of my M-60 ... the bodies ... waking up on patrol when we were all asleep and Alpha Relay was calling us ... Sergeant Johnson with his bush hat and compass asking for Arty to find where we were; we followed the azimuth, not the trail, so we lived.

I remember Lieutenant Pfeltz's revenge patrol ... the scout dogs drinking all our water ... a new guy medevaced for heat ... coming out of the jungle over that prehistoric valley and looking across to Co Roc in Laos ... the leeches inching their way to us ... the jets napalming the hill across the river and to our backs ... dragging in supplies dropped by parachute.

I remember running back from the command post bunker and diving into a slit trench as a rocket hit above our heads ... the straps from our packs digging into our shoulders "'til they were numb ... blood in our boots ... and the Reverend Stubbe pronouncing over the dead. **Lance Corporal Ken Cope, B Company, 3rd Recon, Attica, NY; seriously wounded in 1968 Tet offensive and medically retired, doctors told Ken Cope he would never walk again, but he set out to prove them wrong—and did.**

"With My Pants Down ..."

My platoon from 1st Recon was sent TAD to Phu Bai and attached to Bravo Company, 3rd Recon, for Operation Hastings in the summer of 1966. My first platoon leader was Lieutenant Green from Texas, and my comrades were "Hillbilly," "Canuck," "Snake," "Water Buffalo," "Brig Rat," and others—whom I came to love as brothers.

While on patrol in the mountains west of the DMZ, we ran into an NVA staging area and were surrounded for two days before a company of grunts—the most beautiful sight I ever saw—fought their way into the area and routed the enemy to save their fellow Marines and brought to real life the pride and traditions of the Corps.

While that patrol could have been my last, their action saved me for others. On one, while in a "360" perimeter, a sudden case of the "quick step" hit me. Immediate relief was urgent, and I was granted permission to exit the perimeter. After notifying my buddies, I moved about 25 yards in front of the position, found a large boulder and thick foliage, dug a hole, piled my equipment, and disrobed sufficiently to accomplish my urgent mission.

While I was crouched bare-bottomed with my M-14 over my knees, a single gunshot sounded nearby, followed by a lively exchange in Vietnamese which sounded like it was just on the opposite side of the boulder. I feared that not only was an attack on the platoon imminent with the enemy grouping right before me while I relieved myself directly in the line of assault!

The image of being killed, wounded, or captured with my pants down was so frightening that I immediately became constipated!

I recovered, rearmed, and regrouped and returned to the platoon with such speed that my buddies who witnessed the mad dash nearly burst with suppressed laughter.

The anticipated attack never came, no more shots were fired, no more voices were heard (they had also heard the shot and voices), and the platoon finally moved out of the area without further incident. What exactly happened is still a mystery.

I never again had a call of nature bad enough to leave the perimeter alone! **Sergeant James H. Rowe II, B Company, 3rd Recon, McSherrystown, PA**

For Puerto Rico The Cost Was High

Lance Corporal Antonio (MAO) Ramos Y Ramirez de Arellano of Cabo Rojo, Puerto Rico lost his right hand when he tripped a booby trap in Vietnam, and it all helps make clear just how much more Puerto Rico gave of her sons in that Asian War.

"Ten per cent of the Vietnam deaths were Puerto Ricans. We surely paid our price in blood." For that reason, Puerto Rico Marines and soldiers deserve their place in 3rd Marine Division history.

Patrolling ... Serious Business

During the early summer of 1968, B/1/3 was a sea of new faces, replacements for our combat losses suffered at Dai Do in May. None of us were upset at leaving the Gio Lind-Cua Viet area. Routine patrol duty at the Rockpile was a boring relief. As a scout I didn't see my job as many did. I walked point on every patrol, even three times a day. My S-2 officer, Lieutenant Sheridan Watson Bell III, was continuously ordering me off the point.

I knew though that my training made me the best person for that job, not some Marine who was walking point because he was a new guy, or a ten percenter. I paid meticulous attention to detail.

One evening while on OP Wynn with a squad, I observed the night sky blossom and felt an earth tremor. I ran an azimuth and oriented my map. There was an awesome explosion from Dong Ha. I guessed that it was fuel or ammunition.

My understanding is that a V.C. mortar scored a direct hit on our ammo dump at Dong Ha where a 30-day advance supply of ammunition for the 3rd Marine Division was stored. A 10-day supply was stored at a new facility at Quang Tri combat base. Headquarters advised that it was a likely target. B/1/3 was assigned to secure the ammo dump.

This was somewhat of a break for us because Quang Tri was our rear base camp area. It meant a trip to the showers every couple of days, cold sodas and hot meals, and an opportunity to see a dentist or a doctor about those types of things that may bother you—but aren't serious enough to get you medevaced.

On the other hand, we were assigned a specific and important mission with a large number of untested personnel in our ranks. Although they were untested, they were still Marines and we therefore expected them to perform admirably ... which they did.

Nine squads rotated the patrol duties with the company sending out three patrols a day on a rotating basis. I walked point on each patrol.

The patrols were routine but much more than armed strolls through the dusty fields. They were potentially life-threatening events. As we progressed from checkpoint to checkpoint, I took in everything. I had studied the maps before every patrol ... studied the surrounding terrain with binoculars from our base camp. I knew where depressions were that could conceal enemy ambushes. I carefully scrutinized tree lines, hedgerows and rubble of abandoned buildings.

If I saw something, or knew of an area that was a potential danger to our patrol, I was usually allowed to go with a fire team to check it out—either ahead or off to one flank—while the patrol progressed.

There were a few complaints from people whose attitude was, "There's nothing out there, let's walk this route and get it over with." They did stroll along on patrol with their rifle resting behind their head across their shoulders, their arms draped round it, hands dangling loose ... until one patrol was ending and we were coming in on the back road.

Twenty feet on the road from the gate and guard shack entrance, I was still searching when I spotted a sand-colored trip wire across the road. I called out, "I've got a booby trap, don't anybody move."

Rifles snapped to outboard position and the men began a visual all-around before moving off to the side of the road.

I knew this wasn't a command-detonated mine ... or I would already have been dead. I graduated from the Land Mine Warfare and Demolitions School at Camp Pendleton, and felt confident in my ability to handle a simple booby trap or grenade.

On hands and knees, I followed the trip wire to a 105mm round, and backed away immediately, calling for a Klondike. My heart was racing, I couldn't get enough breath, and my pucker string was really yanked while we waited on the engineers.

I had never tripped a wire or led a patrol into an ambush ... my job was to get my buddies safely from place to place ... but I almost blew it on this patrol.

Sadly, I can't even remember the names of the guys who were on that patrol with me, but they live in my mind, and my heart is one fantastic entity—the United States Marines. **Submitted by Michael R. Conroy.**

Heaven Can Wait ...

Location: Vietnam, Quang Tri Province

In mid-August of 1968, I was a 19 year old rifleman in Bravo Company, 1/3. The company had taken part in a battalion size operation against the North Vietnamese in the mountainous jungle north of Camp Carroll, sending them retreating into the jungle. Naturally, the effects of combat along with the nature of the terrain played havoc with our bodies and clothing. Our bodies were dirty from humping the jungle trails, digging foxholes and crawling on the ground. We were tired and our muscles ached from overuse and the weight of equipment. Our jungle utilities were dirty, mud-caked and torn by the terrain.

We patrolled the area for days and could not locate the elusive NVA for their desire to avoid contact, the operation terminated and we looked forward to well deserved rest.

Bravo Company was helicoptered to Camp Carroll in the late afternoon and we were led off the roadside to await orders. As we waited, we remembered the previous assignment of perimeter duty at Camp Carroll back in mid-June. That was an easy two week duty for Bravo Company and we loved it. Our minds return to the present and were astonished at the disarray of the base. It seemed like the base was being dismantled with the existing structures being torn up. The company is tired from the long wait and we are finally moved across the roadway and are seated on the ground in front of a row of canvass tents. Oh brother, we are finally going to get a break, we thought. We could sleep all night without interruption? However, still no word and we still waited.

Several members of the company used the idle time to gather empty canteens from the men, walk to the mess hall and fill them. My own canteens are empty so I decided to gather up the squad's and fill them at the Camp Carroll mess hall. I wait my turn in line until I fill up the canteens using the 'water buffalo' located outside the mess.

Once finished, I decided to enter the mess hall to see if I could manage to get a drink of cold milk which my taste buds would find more refreshing than drinking warm water. As I drank a glass of cold milk from the milk dispenser, I am confronted by a mess hall staff NCO ser-

A Search Mission

geant. I was a sight to behold: My body caked with dirt, jungle fatigues soiled with dried mud, torn and tattered. At first I thought that my appearance made the staff NCO angry because of the way I wandered into his clean mess hall. He asked my outfit and I knew that he was going to report me to the company commanding officer. I replied that I was with Bravo Company, 1/3, and that the company had just come off an operation in the mountains. He looked me over again and saw the string of canteens on a web belt that I was holding in my other hand. Seeing him glancing at the canteens, I informed him that I had just finished filling the canteens with water from the 'water buffalo' outside until my passion for a drink of cold milk overpowered me to enter his mess hall.

Evidently, the staff NCO felt sorry for my appearance and allowed me to drink my fill of cold milk. He asked if I could use anything else. I replied that I could use anything that he could spare.

I was astounded as he filled my arms with loaves of bread, canned juices, canned fruit, canned meat and other commodities. I couldn't hold half of the stuff that he was giving me, let alone carry it back to my squad. It was Christmas in August as he knew that I couldn't possibly carry back the food stuff in my arms. He retrieved them and placed all the goodies into several cardboard boxes and stuffed the boxes into a jeep and personally drove me to the company assembly point where I directed him to stop at my squad's location.

I was a sight to behold again as I was expected to bring back only filled canteens of water, but a jeep load of various food commodities! I was the hero of the

hour as I instructed that the food be distributed among the men. I thanked the staff NCO for his kindness and generosity towards us as he left.

Not everything comes easy in the Corps as we didn't even get a chance to open a can of food nor to chomp on a piece of bread when our orders finally arrived and we learned that we wouldn't be spending the night at Camp Carroll. Bravo Company was ordered to saddle up and be trucked out to our next destination, the Rockpile. Our 'Christmas feast' was temporarily delayed. War is hell. **Michael Rodriguez**

Proud To Serve 3rd Marine Division

Our 9th Motor Transport Battalion was attached to the 3rd Marine Division at Dong Ha in 1967-1968, a "gung ho" outfit faithful in its duties to the division and proud to serve with it.

Convoys over the DMZ were a daily routine, and very hazardous duty for our drivers and vehicles. We did most of the hauling from Dong Ha Airbase to Quang Tri to help build the new airbase there.

In addition, the 9thMTBn participated in pacification with a battalion medical team which visited Dai Do village to help Vietnamese families.

Many vehicles were damaged by rocket attacks and mines, and repaired at Da Nang or Okinawa. **Gunnery Sergeant Charles W. Hall, USMC (Ret.), 9thMTBn w/26th Marines, Lawrenceville, GA**

"No Firing Tonight, No Celebrations ..."

By mid-afternoon, July 4, 1968, the captain called all the platoon commanders together. The battalion commanding officer said, "There will be no firing of weapons tonight, no celebrations." The captain made it clear we were to see that orders were followed! We passed the orders on and got ready for the night routine at A-3, the northern most outpost in South Vietnam.

A-3 had been built for a battalion plus sized unit and was now manned by a company plus. It was a sprawling complex of bunkers, wire and trenches, with more of the same protecting the various interior sections of the compound. There was also a section of Army tanks, dusters and quad 50's for air defense. Yes, air defense, because of the almost nightly sightings of Russian helicopters to the north. The evening sky was unusually clear and stars provided the only light.

I was out making the rounds of my platoon positions when we first noticed the sky to the southeast. Long streams of tracers going almost straight up. Then more and more from the same base. As if answering smoke signals, another outpost started their own spontaneous celebration and it was not just Army units, for soon the whole horizon to the south made "mad moments" at infantry training school look real small time.

Battalion headquarters was at Con Thien to the southwest, and they soon joined the fun. That was the last straw. First the quad 50's in a long burst (that's impressive if you haven't been up close), then the 40mm dusters and tankers also joined the display. As the captain started screaming, "Cease fire!" over the radio, every grunt on the lines opened up with his personal weapon. What a roar, but only about equal to the captain's screaming that he was going to court-martial all his platoon commanders.

The captain decided direct control was the answer and he left his bunker yelling to "cease fire." I ran the other way trying to regain control. Soon it was over and I was back in my bunker when the call came, "Report to command post." We gathered outside consoling each other about what went wrong and gathering up our nerve to face him.

I don't remember a thing he said except that all of us platoon commanders were less than competent. But, I do remember what the captain looked like. His clothes were torn from head to toe and he was bleeding profusely. He had evidently made it full speed to the first row of concertina. His radio operator told us the next day how long it took him to help extricate the captain. Never again did he exhibit "Stateside discipline." He should have earned a Purple Heart but I don't think anyone dared explain in writing how he was wounded.

A Fourth of July to remember. **Kent Wonders, Encantada, Tucson, AZ**

The Colonel Ate Ham

At Dong Ha for Thanksgiving of 1966, we celebrated with turkey and stuffing and ham, good chow and lots of it. Assigned to mess duty, I helped the cook do everything but cook, such as peeling spuds and taking the garbage out.

ARVN kids from a camp next to ours customarily congregated at the fence. My old man back in Brooklyn sent balloons to give to the Vietnamese kids, so I naturally became a "favorite Marine." I'd show them the one magic trick that I knew, and they would laugh and giggle. We had a good time together, despite orders not to go near the fence or deal with them.

Almost everyone at Dong Ha had eaten that Thanksgiving dinner except the colonel (word was that he was not coming to the mess hall), I kept bugging the good-natured cook for two leftover turkeys. He finally gave in. I took the turkeys to the back fence, ripped them apart, and distributed them to the Vietnamese kids.

All of a sudden, hundreds of kids, old women, dogs, and whatever came "out of the woodwork," yelling and running at the fence. Throwing the remains over the fence, I fled back into the mess hall.

Shortly afterward, the colonel appeared at the front door of the mess hall for his Thanksgiving turkey dinner, along with an onslaught of majors and captains asking the mess sergeant where the turkeys had disappeared. The affable mess sergeant "saved my butt" and said he ran out, but had an extra ham in the oven (which I missed).

Medical Attention In the Field ..did not always rate a private room for consultation

The colonel had ham for this Thanksgiving dinner! **Lance Corporal G. Edward Merrihew, C-3rd Recon, (43 Occidental Circle, Santa Rosa, CA) later worked with issuance of environmental permits for Calpine Corp.**

A Christmas Eve In Vietnam

The irony of blending Christmas and warfare crackles throughout the following story describing Christmas eve in Vietnam 1967. This was written by E. Michael Helms, now residing in Youngstown, Florida. It first appeared in the October/November 1989 issue of *The American Veteran* and then, slightly altered, in his book, *The Proud Bastards.*

As we neared the village, I began to pick out scattered but distinctive notes during lulls in the blustery wind. It sounded vaguely familiar somehow ... that was it ... Christmas carols ... Christmas carols ringing out over the Vietnamese countryside.

December 24, 1967. It had been a grueling patrol for the men of first platoon, "E" Company, 2nd Battalion, 4th Marines. Although we had encountered no action all day, tension remained high and nerves stayed on edge. After all, no one wanted to die on Christmas Eve. Even if the enemy cooperated by honoring the holiday cease-fire, how were the mines and booby traps supposed to know hostilities had been suspended?

Now, as the sun showed signs of retiring for the day and we approached a small hamlet near the outskirts of Quang Tri City, I began to breathe a little easier. Less than a mile to go and we'd be back within the relative safety of our perimeter where our company was providing security for the naval construction battalion. They were building a new airstrip in the northernmost province of South Vietnam.

Visions of a promising hot turkey dinner from the Seabees' mess hall, a rare good night's sleep, and no war for a day began to fill my thoughts. I had been "in country" for less than two months, but had already seen enough combat to relish the thought of the yuletide truce.

There was that sound again, like faraway chimes dancing on the gusting wind. It reminded me of the bell tower of the Baptist Church back home, and a wave of homesickness swept over me like the clouds rolling in from the east off the South China Sea, pulling their gray blanket cross the retreating blue sky.

As we neared the village I began to pick out scattered but distinctive notes during lulls in the blustery wind. It sounded vaguely familiar somehow. That was it! ... Christmas carols ... Christmas carols ringing out over the Vietnamese countryside!

Up ahead the weathered stuccoed walls of a Catholic Church stood nestled among neatly-trimmed hedgerows, the result of dedicated French missionaries, no doubt. I searched in vain for a belfry, and wondered at the source of the melodic chiming. Then as our platoon passed within 75 yards of the front of the sanctuary, I noticed loudspeakers mounted on either side of the wide double-door entrance.

We angled obliquely to the left up a gradual slope as the recorded strains of 'Silent Night' filled the air and warmed lonely hearts with remembrances of family and home. I turned to Private First Class Morton who was helping me guard the rear of the column: "How about that, Chuck? Don't that just beat anything you ever saw, hearing Christmas music in a place that th..."

A sudden volley of rifle fire erupted behind us. A cold fist slamming my insides, and chilling fear raced up and down my spine. The dreaded hissing and snapping of hot invisible messengers of death began filling my world again.

"Oh God, oh God!" I screamed silently as I sprinted madly for the crest of the slope ... "Not gonna make it, Chuck, we're not gonna make it!"

The rest of the platoon had gained the crest and were firing back furiously at the Viet Cong who were set-in among the hedgerows which paralleled the church. My legs were turning to mush and my heart had jumped up and knotted in my throat, almost choking me. The crest and safety seemed so far away— too far away. Behind us the ground exploded in showers of clods as automatic weapons sought us out. "Can't make it," I thought. "Go down, go down, gotta go down ... No choice, it's too hot, too hot, gotta go down."

I sprawled to the ground, my rifle and helmet tumbling away. "Oh Jesus, oh God" came the pitiful muted prayer as a lone tree midway between Morton and me was being denuded of bark and turned into a million toothpick-size splinters. "Stay down, Chuck, don't move!" There was no chance except to lay there and play dead.

> "We cowered helplessly as the platoon poured a fusillade of fire back at the enemy and hostile rounds continued popping and cracking and laughing and taunting and kicking us with dirt in our terror and humiliation."

The ground churned around us and the air above was wickedly thick with sickening whines. My fingers clawed the earth in fear and desperation, and every muscle and nerve was tensed against the bullet I knew that any second would tear through my body. We cowered helplessly as the platoon poured a fusillade of fire back at the enemy and hostile rounds continued popping and cracking, laughing and taunting, kicking us with dirt in our terror and humiliation.

It finally stopped ... I glanced over at Morton. A sick, tormented gaze haunted his eyes, and I knew I was looking into a mirror. I felt sick and wanted to flee from this nightmare, to just run and run until I could turn the corner and be safe at home.

Morton and I struggled to our feet, gathered our rifles and helmets and what was left of our sanity, and scrambled up the slope to the waiting platoon. We hurried on into the deepening shadows toward our lines, as 'Silent Night' faded in the distance.

Reminders ...

"Holy, Holy, Holy," ... I hear the church choir sing today ... "Lord God Almighty," ... and my mind goes back to that day on the gently rolling deck of the USS *Princeton*.

Marines are standing at parade rest. Three symbolic M-16's with bayonets are stuck in sand bags. We're singing ... "early in the morning" ... the hymn's words seem so distant then as we remember 43 friends who had been beside us just a few days before—most killed in a single mortar attack.

It was called LZ Margo. Our BLT was part of a four battalion operation northwest of Camp Carol. 2/26 landed at LZ Margo and soon broke down into company-sized patrols headed north along steep ridges into the DMZ. Contact was initially light, but the lead companies confirmed what intelligence had warned—our area was defended by an NVA Weapon and Mortar Company. As we neared what later turned out to be base camps and supply points, resistance became stronger, but we were moving and had the advantage until order changed. "Return to LZ Margo ASAP and wait." Unknown to us B-52 strikes were being planned. The protests began immediately at the company level and continued all the way to General Davis, 3rd Marine Division Commander. He seemed to understand the tactical

situation but the order stood, so hundreds of Marines returned to the hill known as LZ Margo.

The last of the units crossed a river, climbed the hill, and dropped off packs while others were just moving to their perimeter positions when the first rounds started landing in the battalion command post area. All too soon the results of too many men in too small of an area was a reality. I don't remember which was initially worse, the anger or the fear.

Hundreds of mortar rounds later, the barrage let up. As assistant S-3, I got permission to work the medevacs. It was hours before all the wounded were taken

out—thanks to many Marine pilots and crew men who took the risk to land. I remember the face of one young pilot as I pleaded with hand signals for one more minute so we could get one more wounded Marine aboard—his thumb came up as he looked for the mortar round he was sure was on the way.

In the intensity of the caring for the wounded and organizing the loading into the helicopter, time came to a standstill. At some point, up walked my platoon sergeant from the platoon I had commanded. He was gently carrying Corporal Bradford, the 2nd squad leader from F-2. "Lieutenant, you have to help, please," as he gently put his limp body into my arms and walked away.

Just two more events about that 24 hours are always fresh in my mind. Fighting sleep while trying to check and recheck the accounting of KIA's (about 35 as I recall) and WIA's (about 180). And the memory of standing near General Davis the following morning when he left his helicopter and walked among the body bags. He stopped and came to attention. After a long pause he asked, "Lieutenant, where is the colonel?"

September 17, 1968 ... "Holy, Holy, Holy, Merciful and Mighty, God in Three Persons, Blessed Trinity." Reginald Heber, 1826. **Kent Wonders, formerly platoon commander of 2nd Platoon in F Company, 26th Marines, and later assistant S-3 of 2/26, lives in Tucson, AZ**

Untraditional Christmas

It didn't look much like Christmas on that day in 1968 when these Officers posed on Fire Support Base Dong Ha. Left to right, Col. Stormy Sexton, Admiral McCain, Lt. Gen. Cushman, Maj. Gen. Raymond G. Davis, and directly behind Gen. Davis is Lt. Gen. Stillwell, U.S. Army

The Story Of Bravo's Revenge

Khe Sanh stays in the minds of those who guarded that clay. Forty thousand NVA regulars were reported massing around Khe Sanh. General Westmoreland felt the main enemy thrust was to be a major assault against Khe Sanh's 6,000 Marines, but that superior air and artillery support would give the United States a decisive military victory.

At Khe Sanh, we lived in fear that in the heavy night fog, the enemy led by tanks could penetrate the wire and and mines and overrun the base. Hill fighting had been intense at times, and mortars hit every minute or so throughout the day and much of the night for 77 days.

For a 20-year-old grunt just three months from boot camp, the Khe Sanh experience burned deep in the psyche. Seven replacements flew into Khe Sanh, and the chopper barely touched down as NVA gunners dusted the landing zone. Two of us were assigned to Bravo Company weapons platoon, 1-26.

It didn't take long to understand the job of a 60 mortar ammo humper—open crates, open crates, open cannisters, stack ammo, carry crates, and clean gun. With little to do but drill and keep low, we became quite proficient. We knew the yardage chart by heart and were good at estimating angles of fire.

On March 30, 1968, our section leader volunteered the 60s as part of a sweep to recover 25 Bravo Marine bodies KIA in an earlier NVA ambush. When we were fully equipped and loaded, our squad leaders checked us. When he opened my grenade pouch, where I had just put six brand new grenades, he told me to stand still.

I had never taken new grenades from cannisters and never thought about the pins being straight. In the anxious dark, we worked the pins back into place. We moved out in the early morning dark, keeping touch contact with the person in front, with hanging fog after it started to get light.

After leaving the base perimeter, we spread out and were near the front when small arms fire suddenly erupted, and continued with grenade explosions all around. We had caught an NVA company by surprise in their trenches. I grabbed a grenade and tossed it in a spider hole I saw, and it went down and exploded.

We moved forward from trench to trench, hole to hole, and bomb crater to bomb crater. Khe Sanh-based artillery blocked NVA reinforcements from aiding their surrounded comrades, but we were exposed to intense enemy mortar and small arms fire as the sun burned the fog away.

We moved to one crater to help the wounded from an 81mm mortar blast, and I had used all my bandages and went to help one seriously hit Marine who looked at me and asked, "Don't you know who I am?" I was stunned. It was John DeBok, with whom I had gone overseas. I had to go forward, but yelled for a corpsman to care for him.

I moved ahead to join fellow squad member Doug Green, and hit the ground when I looked up and saw a round drifting toward us. It landed close and I felt a sharp, burning pain in my side and felt some sticky blood. Green looked and told me that it was just a couple "little pieces."

We climbed down a little crater, set up our 60, and fired the 15 or 20 rounds we were carrying. It seemed that the incoming stopped, but our base continued to fire on the tree line and our far right flank. We looked for our wounded, then started to fall back to our base.

When we got to our trench line, we realized we had made it safely and I went for a cigarette. But my friend DeBok's blood had soaked them. I went to sick bay and a corpsman plucked three pieces of shrapnel from my side.

I couldn't find DeBok. Later, after the Vietnam memorial was dedicated, several of us went to Washington and were indescribably relieved not to find his name there.

For ten Marines, life ended that March 30. Eighty were wounded. But for each of our dead, 16 NVA died. Two days later, we finished our mission with recovery of our Bravo comrades. The NVA were gone, and the siege was over. **Robert W. Hanna, B-1-26, Beavercreek, OH**

During a base camp lull in the early summer of 1968, these Bravo Company of 1-26 Marines included, left to right, Bob Hanna, John Kuhn, and Rueban Nace. Kuhn was the alert leader who put Hanna's grenade pins securely in place.

The Eccentric Bird Man

I was a Navy corpsman attached to D Company, 3rd Recon, in 1967-1968, located at Phu Bai, Quang Tri, Khe Sanh (with the 26th Marines), and Dong Ha.

I remember a strange guy, name unknown, who went by the handle "Bird Man." He would dress up in an LZ marker, jeep driver's goggles, and helmet ... and sometimes little else. He would then go dashing in and out of the hootches (not caring whose) acting rather strange. I suppose he was our "Klinger" from TV's MASH. **HM3 Lee P. Weber, Co. D, 3rd Recon (president and publisher of the Pacific Daily News on Guam)**

Artillery Helps Save The Day

I had been in Vietnam a full year already when Operation Buffalo commenced on July 2, 1967 with an ambush by a full NVA company on Bravo and Alpha companies of 1/9 about 1000. L-3-9 was choppered in from the field about 1500. For the next several days and nights, we engaged the NVA in continuous combat, hand-to-hand at times.

On the Fourth of July, we reached the bodies of 1/9 on the field. The assault continued for the next several days before a temporary lull in the fighting. As an artillery forward observer, I was searching the terrain for targets of opportunity.

Suddenly, north of us, across the Ben Hai River in the DMZ on the North Vietnam side, I spied several troop trucks in a convoy obviously speeding toward the Ben Hai to cross and attack us. Plotting the grid coordinates and shooting the azimuth, I called the skipper (Captain T.T. Shirley of L-3-9) and a fire mission.

As FDC plotted back at Con Thien combat base, the air FO ran up and offered the big naval guns on this target plus eight-inch or 175mm artillery from other combat bases. Together, we called in the artillery using my coordinates and spotting, and it destroyed many NVA troop trucks and KIA about 75 NVA soldiers. We felt quite good about our role in preventing the NVA from reaching the battle, and helping turn things in favor of the 9th Marines.

In eight days of battle, over 1300 NVA were killed while we sustained 167 KIA and 953 WIA. This effectively stopped the NVA attempt to overrun the Con Thien combat base and sent the NVA running back to North Vietnam in defeat. **Lance Corporal David P. Martin, l-3-9, Trenton, NJ**

PFC Heifner Always a Leader

Private First Class Heifner, the all-American boy, a 19-year-old Marine, athletic, strong, smart, aggressive, a leader, and desiring to do his job right. We were both new to the platoon. I first noticed him when we took a daytime break from our nightly patroling and ambushing north of Phu Bai. While some slept inside the CAP Compound (Combined Action Platoon), some played football. I finally stopped the game for fear of being court-martialed because of the growing list of WIA's from the game. but Private First Class Heifner's leadership showed up in football and patrols.

After moving to Hill 558 north of Khe Sanh in January 1968, we continued to do platoon size patrols even though all the signs pointed to a large NVA presence. On January 26, 1968, a two platoon patrol from Fox Company made heavy contact on a small hill 4000 meters out—a hill suspected of having NVA mortar positions. The next day the whole company was to envelop the hill.

Getting into position was not easy or uneventful. The air prep was late and we were too close to it. At one point our flank squad disappeared behind a wall of napalm (never trust Air Force close air support). Miraculously, the bomb landed between two squads without harm except to the platoon leader's nerves.

We gave up on stealth (always a mistake in Viet Nam) for command and control. That is, the NVA were dug in around thick brush for cover and we tried to maintain control by voice and radio because you couldn't move more than a few meters without losing sight of each other. As we reached a bomb cleared area, the NVA opened fire. Up front of the on line assault was Private First Class Heifner. He had seen them

but too late to save himself himself, however in time for the rest of us to get down and return fire.

The collage of memories from combat are not always rational or logical. I remember the branches being cut down all around my head ... "Get Heifner" ... "Get Heifner" ... a squad leader starts forward after giving me the "why me" ... a bullet through the arm for his effort ... a corpsman started forward without being asked ... "First squad cover us now" ... everyone seems to be on full automatic ... "Help the corpsman," but the Marine decides to get up and run forward instead of crawling ... he fell beside me, boot and lower part of his leg resting on his back pocket ... the captain is running up from somewhere behind us

screaming, "Cease fire, cease fire!" ... "No, no keep it up. We've got to get our wounded" ... "Guns up" (M-60 machine guns) ... first burst from our M-60 gets quite a response from the NVA with a bullet through the stock and the gunner's arm.

In ever so slow motion now ... an NVA stands up and starts running, carrying an M-60, as if we can't believe we're dealing with a real person—no one seems to shoot, then someone does .. "Pull back" ... the wounded ... Doc's tagging Private First Class Heifner ... "Why can't you help him?" ... "Look, it went through the heart." In anger I pick him up and start walking only to have the reality, the heat, and the weight take over. **Kent Wonders, Tucson, AZ**

8" Howitzer

Following Orders - A Marine Tradition

I did, and my Marines did what we were sent there to do; followed orders ... and considering the restraints under which we were operating, I think we did damn well.

We carry on our shoulders a proud tradition that has been molded by hundreds of thousands who have gone before us, and I'll be damned if we are going to let them down. There have been several times when the odds were phenomenal and how did Marines come through it? Semper Fi.

Marines don't say they can do the job; they do it.

That's why Americans believe in us. I swear to God.

That is one thing I always remembered; if I got shot, a Marine would never leave me on the battlefield. And I didn't leave anybody on the battlefield. Marines got killed going to the aid of other Marines. That is something we know and we are proud of it.

And lead. If you can't stand up and lead, then get the hell out of the way. Someone else will do it. Don't tell your people with your hands on your hips to run around the grinder. You be up front running and let them follow you.

Since I've been in Washington, people come to visit and I take them to the Wall, and it affects me differently every time.

The first time I went to the Wall was during the inauguration two years ago. Wheew! That was a tough day, really tough. That visit shot the rest of the day. I couldn't go out. The gal I was with couldn't believe the effect it had on me.

First of all, I say this ... the location of it overpowers the design—the Lincoln Memorial on one hand and the Washington Monument on the other. What a place of honor. There is not another war memorial in Washington with that prominence. I'm proud of it and what it represents.

Back in November, I was there with my mom and dad, and I got a little misty eyed. Sometimes, emotionally, I just go to pieces, privately.

A couple of weeks ago I was there, and I walked away with a little bounce in my step. I really felt good. **Harvey "Barney" Barnum received the Medal of Honor for his action with the 3rd Marine Division in Vietnam.**

Ouch (Loudly)!

While on recon patrol southwest of Da Nang in late 1966, we traversed "Leech Valley" and all of us picked up quite a few of the pestiferous bloodsuckers all over us. We took turns getting them off each other when we finally were able to set up a hilltop perimeter and drop our gear.

Our squad leader, Corporal Lumpkin, asked the corpsman for some mercurochrome, but was given Iodine to combat a leech attached to a very personal part of his anatomy.

Lumpy, as he was called, applied the iodine. When it hit the affected area, we heard his agonized screams and saw him with his pants down around the ankles, standing on a rock in full view of the valley below, holding the affected area with both hands.

Although the lieutenant was concerned with the effect of the noise on security, we felt that no gook in his right mind would attack up a steep hill with a lone visible Marine standing bare-assed on a rock and screaming.

The rest of the patrol was uneventful after that. **Lance Corporal G. Edward Merrihew, C-3rd Recon, Santa Rosa, CA; the corporal affected was Albert R. Lumpkin Jr., Tyrone, PA**

Attack On Hill 861A

The 2nd and 3rd Battalions of the 26th Marines were heavily involved in defending the high ground dominating the Khe Sanh combat base in January and February 1968. Some observations of participants, excerpted from a February 1989 Marine Corps Gazette article by Eric Hammel:

"The NVA couldn't take the Khe Sanh Combat Base without first taking Hill 861, and once we got there, Hill 861A. If the NVA had owned those two hills, they would have been looking down on the combat base and on 2/26, which was on Hill 558. It was very important to hold 861 and 861A." - Captain Earle Breeding, E-2-26.

"One night shortly after we got to Hill 861A, the troops on the west side of the company perimeter called the command post to tell me they had movement in the saddle between us and Hill 861. When I went out there, they told me they had heard movement and shouted a challenge. When the response came back in English, they assumed that Kilo Company was running a patrol down there.

"I knew that Kilo Company wouldn't run a patrol down there without telling us, but I called Kilo anyway. Sure enough, they didn't have any patrols out. As soon as I heard that, I got every weapon that could bear firing into the draw ... They sure as hell weren't running a patrol of English speakers for the hell of it." - Captain Breeding.

"Approximately January 18, a team from MACV Headquarters in Saigon visited Khe Sanh and offered the use of some electronic devices, which would give indications of enemy presence ... later called sensors. Within 48 hours we began receiving reports that the devices were being implanted in likely avenues of approach to the combat base ... We also began receiving reports that the devices were indicating enemy activity (Everything in the area was considered enemy). These reports increased in volume to over 100 a day." - Major Jerry Hudson, 26th Marines S-2.

"During the nights of 3-4 and 4-5 February sensors reported numerous heavy movements from the northwest of Hill 881S ... The total count of enemy troops, reported by the sensors, added to a possible 1,500 to 2,000 men in the course of those two nights ... We concluded that an attack in the thick mist was imminent." - Captain Mirza Baig, 26th Marines Target Info Officer.

"(The enemy was getting ready to assault Hill 881S or Hill 861A, or both together.) A decision had to be made which one to interdict. The choice was in favor of 881S as the artillery there could be employed in support of 861A as required." - Major Hudson.

"It never occurred to me that ... the enemy's intent was and always had been to attack Hills 881S and 861 simultaneously. I had forgotten the NVA battle plan. There were no sensors near Hills 861 and 861A ... the target information office and alleged expert on NVA doctrine was caught flatfooted."—Captain Baig.

"I got the word from all three of my platoon commanders. It sounded to me like they were coming through the wire all around the hill. I thought I was getting hit from all sides at once. I had the troops throw gas grenades, and I called for ... artillery support.

"The gas probably did us more harm than it did the NVA. It started filtering into the low-lying areas—our fighting holes and trenches."—Captain Breeding.

"The whole skyline to my left ... was starting to light up. There was yellow and orange color coming off the horizon. There were things going off—the 60mm mortar over there, the Chicom grenades, machine guns—and I could hear people talking and yelling ... All kinds of stuff was hitting the fan."—Private First Class Mike DeLaney, E-2-26.

"The way the teargas didn't affect the NVA at all leads me to believe they were hopped up on drugs ... they should have been bothered. We were bothered. During the first lull, I found one NVA soldier with his AK-47 slung over his shoulder. He and others were going through our living hootches, more interested in reading *Playboy* magazines than in fighting the war. That's when we counterattacked.

"... We were drawing fire support from five separate locations."—Captain Breeding.

"Finally it slowed up and they backed off, I looked around ... Stuck in a sandbag only a foot or so from where I was handing rounds into the gun pit was an NVA rifle grenade that never went off.

"There were bodies everywhere. Their bodies were in full uniform. That scared me. Until then, I thought we were fighting Viet Cong guerillas ... I didn't know they had NVA soldiers up there until I saw them dead on the ground inside our perimeter ... These people were well armed."

"... We found drugs—syringes and chemicals."—Private First Class DeLaney.—Association member Earle G. Breeding, Washington, DC, and member Jerry Hudson, Charlotte, NC.

A Marine washing his clothes in a mud hole because water is rationed at Con Thien. (Defense Dept. Photo - Marine Corps)

Adventures Of A Phuy Bai B-N-G

I had been in the field a few days. It is raining, the December monsoon. Around Phu Bai there were rice paddies and small villages (90 percent friendly).

A BNG (brand new guy), I was still trying to figure out how a grunt carries all that gear, keeps up, and stays alert. At least I had jungle boots. There was a shortage. When you saw a guy swandive off the paddy into water, there was a rush to keep him from drowning. All you could see were stateside boots, spit shine, and doing the choking chicken kick.

There it is. Rain, slip and slide, all day ... running ambush at night. Not the good dry ambushes but damp, drizzly ones. Flares glowing here and there, trying to show someone something. Always slight foul-ups where we made no kills/contact. BNG's errors.

Movement down the trail, BNG ... setting next to machine gun ... slides back the .45 ... SHU CLUNK! Gooks off and running. (No thought of shooting the machine gun, not allowed to touch, being a cherry you know.)

Another BNG ... "Halt, who goes there?" Gooks off and running. Typical boot camp training and the lack of Vietnam brains when we hit the bush.

On A Dark Night

Off we go on a big ambush, on a dark night, raining hard and steady ... black, we link up holding on to the man's pack in front. Blindman's bluff in the extreme. I am wondering, "Does he have a limber stick out trying to feel for trip wires while we stumble along?" The radio is finished, too wet to work. What will happen if we get hit? Booby trap? Lost?

A 10-foot wide stream appears. Fast moving, and still a dark night. Two grunts, braced, help all cross the stream. Me, holding M-16, pack up over the head, plus trying not to drown or lose all the ammo, flares, letters and spare (dry?) clothes. The old salt had pack all the way up, but M-16 at his side under four feet of water. "hell with the gun, save the letters." No one says a thing.

Off we go slogging along, still hanging on to each other. Lightning now ... I can barely see, glasses streaked wet with rain.

There is a small hut. We wheel over and check it out quietly. There may be gooks looking for shelter too.

We're ready. Crash in the front door, at the ready ... No shooting, we are in a friendly area. No one wants to go to jail for murder or be reported for abusing civilians.

Someone is there but will not come out, so we set up. One half in the front room and one half in the other room.

I am immediately assigned to an outside wall. Standing up so you do not go to sleep. Peering out, watching the lightning, looking for gooks, and remembering the warning, "If you see anything, come get me. Do not shoot on your own! Come and get me."

Two hours later, I am relieved and told where to slip off to crash. "Show no light and try not to make any noise. Always know where your weapon is. Don't take off your boots."

I feel my way in, talking quietly to another new guy in the room. I'm wet, can't seem to dry out, cold too.

I put a heat tab on the ground and light it off. Squat over with your poncho on and you'll get warm. Do not put your head under the poncho or you will pass out and die from the fumes. No way to get dry and no way to get comfortable. I give up trying to get warm/dry and lay down, damp and miserable.

It Happens

... Movement on my left, just enough to hear.

The other Marine has gone on watch. Crunch, step, straw being moved very quietly. Gooks coming to slit my throat? Pit viper sliding over to bite me? Snake to jump on me to see if I make a good midnight snack?

I feel around and find my M-16. Make sure it is off safety. I sit what seems hours waiting to be attacked. I cannot hack it anymore. I'm ready ... Death before dishonor. I light a match. ("No bayonets this week, but you'll get one soon," he had said.)

The match light is ruining my night vision. It moved again. One hand with match, one ready with a "John Wayne" hold on the M-16.

A pig, the biggest pig I have ever seen, comes out of the glare. I had been rolling close to the pig ... and it was scared.

Ha, ha and off to sleep I went. Killer Marine survived. **Bob Crass.**

The Rat Race

Our recon platoon left California for jump school at Okinawa in July 1966, but transferred there to an LST headed for Da Nang to replace a badly decimated 3rd Recon unit. At Camp Reasoner, we were designated 2nd Platoon of C Company.

Recon patrols were normally five days in the bush and five at camp. Garrison duty incorporated things hated by Marines—digging trenches, filling sandbags, mess duty, burning out the 55-gallon outhouse drums, and boredom.

In free time, most of us were too hot or tired for football or baseball, so we invented our own less strenuous entertainment—knife-throwing contests into plywood backstops, weight lifting, and other diversions.

Since we were not allowed out of the compound to visit the PX, we sat by the concertina wire barricade waiting for passing grunts with dried apricots one of us had been sent from home. Hung on a string, we said they were "Charlie's ears."

We would even entrust an "honest looking" grunt with money to bring us back ice cream from the PX, but usually never saw him again. Only one that I can recall ever returned with the cones (the ice cream had not survived the 115-degree temperature).

Finally, one Recon member suggested rat races. We captured a number of rats in cages, then lined them up and marked a finish line down the road. We arranged our bets, and on signal, the cages were opened.

The uncooperative rats scurried in every direction but toward the finish line, making it mass confusion as we tried to keep track of "our rat" while trying to get out of the way of the frantic rodents.

There were no winners. None of the rats made it to the finish line in that rat race. **Lance Corporal G. Edward Merrihew, C-3rd Recon, Santa Rosa, CA**

A Letter To Parents, Wives, And Relatives

In February 1968, Lieutenant Colonel William Weise, commanding the 2nd Battalion of the 4th Marines, wrote a letter to parents, wives, and relatives of members of the unit, whose emblem called them "The Magnificent Bastards."

This letter was republished as recently as 1989 in a national veterans' publication. Here is the letter of Lieutenant Colonel Weise (Brigadier General William Weise, Alexandria, Virginia), slightly condensed, which was quite helpful in informing folks on the home front when criticism was rife:

"As always, our thoughts and prayers are with you. Since my last letter, we have become Battalion Landing Team 2/4 and have participated in seven major operations (Osceola II, Ballistic Armor, Neosho II, Fortress Attack, Kentucky, Lancaster II, and Napoleon/Saline. In addition, we spent ten days at Subic Bay in the Philippine Islands to retrain, rehabilitate our equipment, and relax (although not too much).

"... Our BLT is a lot larger and more powerful than the standard Marine infantry battalion. (It) includes tanks, ONTOS, amphibian tractors, artillery, 4.2 mortars, trucks, reconnaissance, shore party, communications, medical, and headquarters elements in addition to our infantry battalion. 'The Magnificent Bastards' have almost doubled in size and firepower. The main punch of Special Landing Force Alpha, we are organized for special amphibious operations. We can land across the beaches by both boat and tractor, or over the beach by helicopters (Courtesy Medium Helicopter Squadron 363).

"Our home base is aboard amphibious ships of the powerful 7th Fleet, although lately we have ben doing most of our living in foxholes ashore and showers have been few and far between. I am writing this letter sitting beside my hole in northern Quang Tri province near a village called Thai Thuong. I would like to thank the commanding officer, officers, and crew of the USS *Iwo Jima* whose assistance and facilities have made the reproduction and distribution of this letter possible.

"As you have undoubtedly heard from the various news media, the enemy has been making a determined effort to drive the Marines out of the I Corps Zone. He is taking terrible casualties in this attempt, and we are here to see that he does not succeed. BLT 2/4 had done very well so far, and I am proud of your Marines and Sailors. Never in her history has America fielded a finer group of dedicated and courageous young men. In many ways, their sacrifices are nobler than those of their forefathers at Concord and Valley Forge. For your men are not fighting for their own freedom, but for the freedom of people thousands of miles from their own homes.

"We continue to receive excellent spiritual care. Our chaplain is always out in the field with us and makes frequent visits to the men in the front lines. Because we cannot gather in large numbers, he holds many services for groups as small as three or four men. During one week alone, he held 30 services. Needless to say, he also does a lot of private counselling ... The Navy doctors and corpsmen are right here with us, and I can't say enough for the tremendous job they are doing. We are getting the best medical service any armed forces has ever received.

"Over the past two months, especially at Christmas, we have received thousands of packages, letters, cards, and many other tokens of support from families, friends, churches, and organizations. They have come from all over the United States and from all walks of life.

"The donors and writers have ranged from grade school children to the president of a large corporation. They all say the same thing, although none more eloquently than a nine-year-old girl from Paterson, New Jersey, who said, 'Thank you for fighting for us, Marine. I pray to the Lord to keep you safe.' I know, and your men know, that this is the real spirit of America and for that we thank God.

"In this letter, I cannot possibly give you all the news about your men. We are doing our utmost to provide them with the best facilities we can under present conditions. Because of our frequent moves, mail—incoming and outgoing—is often delayed. But please keep writing.

(At this point in the letter, Lieutenant Colonel Weise explained how to write him in case they haven't heard from their men, and what to do in various family crisis situations.)

"Above all else, keep your faith in us and what we are doing. I am certain that history will prove us right. Never before has a nation done so much for another with so little expected in return."

Khe Sanh: An Overview

(Reprinted From The February 1989 Issue Of Marine Corps Gazette)

Concerned that its strategy of attrition would ruin it before American will collapsed, North Vietnam decided to attempt a dramatic politico-military victory in 1968 by hurling its own regulars and all the VC (Viet Cong) it could muster against South Vietnam's major cities, military installations, and symbolic targets like the American Embassy in Saigon. Although Hanoi's leadership knew that 1968 was an American presidential election year, its main target was not American public opinion but the morale of the GVN (Government of Vietnam) and ARVN (Army of the Republic of Vietnam) and their urban supporters.

As the monsoon rains pelted the Marine combat base at Khe Sanh, American intelligence in late 1967 identified two NVA (North Vietnamese Army) divisions closing upon the exposed and undermanned base. The 26th Marines headquarters had already outposted the critical hills north of the base, but the Khe Sanh perimeter—which encircled an airfield, supply dumps, and artillery positions—stretched beyond the capacity of its ground defenders ... By January

1968 the 26th Marines had added an ARVN ranger battalion, a tank platoon, and five artillery batteries, in all about 6,000 men. The rain and the enemy had closed Route 9 to the base, but the Marine generals were not very worried about resupply; the base defense force would probably require 185 tons a day, which could be air-delivered even under fire. Helicopter resupply of the hill outposts which absorbed half of Khe Sanh's infantry, might be somewhat more difficult. As for fire support, Khe Sanh had ample help from Air Force and Marine air and Army long-range artillery positioned at Ca Lu.

From III MAF's perspective the basic question was not whether Khe Sanh could be held but whether it should be. Any attack down Route 9 could be stopped more efficiently at Ca Lu, and (General Rathvon McC. Tompkins) suspected that a Khe Sanh attack might be only a diversion preceding another drive at Con Thien and Gio Linh. (General Robert E.) Cushman shared Tompkins' concern but, at MACV's insistence, promised to hold Khe Sanh because (General William) Westmoreland had concluded that he had a superb chance to destroy two NVA divisions and win his own psychological victory. Both he and the Marine generals were astounded to learn that (President) Johnson so feared a defeat that he required the JCS (Joint Chiefs of Staff) to pledge personally that Khe Sanh would not fall. Such acts and a media thirst for

dramatic news gave the battle undeserved importance.

The siege of Khe Sanh (January 21-March 30, 1968) ended in an overwhelming American victory, brought about by the sturdy defense of the base and outposts by Marine infantry and massive air and artillery support. Although the NVA pounded the base with as many as a thousand shells and rockets a day, its infantry attack on the hill posts and the perimeter itself did no lasting damage to the position. The most worrisome aspect of the siege became the coordination of American fire support. Using target information gathered from both ground and airborne sensors and photographs, American air pulverized NVA positions with 100,000 tons of bombs. Artillery and mortar expenditures may have gone as high as 200,000 rounds. On the ground, Marines and ARVN rangers beat back all assaults on the perimeter and outposts despite some sharp night fighting. In fact, the only small American defeat occurred when tank-supported NVA overran the Lang Vei Special Forces camp, which should have been abandoned earlier. Although the NVA attack was real enough, Khe Sanh came no closer to being a Dien Bien Phu than Iwo Jima was to a Wake Island. The relative casualties spoke volumes: 205 defenders dead and about 800 seriously wounded as against probably 10,000 NVA killed in action. **Allan R. Millett, from *Semper Fidelis*, The Free Press, 1980**

Destroying A Myth

A popular myth at Khe Sanh was that the NVA were digging tunnels under the wire and coming up within the base to overrun us from inside. Photos had been published of Navy corpsmen with stethoscopes "listening" to the ground at different areas around the base.

Here's the real story.

After the morning of January 21 (start of the 77-day Khe Sanh siege during the 1968 Tet offensive), all of the hootches in the 3rd Recon Battalion area were destroyed as we moved into underground bunkers prepared prior to the siege. We hadn't planned on living underground and had therefore built very small bunkers. There was little room in them, and most of our personal gear remained in what was left of the hootches.

Some stereos and personal gear came up missing. We tried without success to come up with an idea that would enable us to search other units' bunkers for the missing gear. Finally, somebody came up with the idea of posing as corpsmen and wandering around the base listening to the ground. If we wanted to check someone's bunker, we pretended to hear something and asked to enter the bunker to check it out.

We were never turned down because who wants a tunnel beneath their bunker? A few photographers saw us and took pictures that were in Stars and Stripes and stateside newspapers and publications. This convinced the outside world that the NVA tunnel threat was real! **Private First Class Dave Doehrman, B Company, 3rd Recon Battalion, Fort Wayne, IN**

A Surprise At Christmas Year 1968

Vietnam wasn't funny, But Marines served and fought there, so there was bound to be a lot of funny things happening—even in wartime, and even in wartime during Christmas, 1968. This piece was written by Brian Sweeney, now residing in Nogales, Arizona.

Probably the most vivid memory I have of the 3rd Marine Division and Vietnam is the Christmas I spent there in 1968. Besides the obvious reasons for remembering Christmas there was one incident that best summed up my feelings about Vietnam, the war, and especially the pain of spending Christmas away from family and friends.

My unit was A Battery, 1st Battalion, 12th Marines, and I was assigned to A Company, 1st Battalion, 3rd Marines as the FO. Just prior to Christmas the Battalion was sent on Operation Taylor Common in an area known as base area 1-12 which was west of Da Nang. The area of operation was in an inaccessible jungle region that supposedly no Allied troops had entered since the French. We were chosen for the operation because of our experience in fighting and building LZ's in the jungle.

And build we did. When Christmas eve arrived, the company I was assigned to had just completed two new LZ's. The building of the LZ was done mostly by the Weapons Platoon since the line platoons were either patrolling or preparing their own defenses. The work was hard. It required a lot of manual cutting of trees, digging and moving of stumps and leveling of the ground. When the weapons platoon finished the LZ's, all they wanted to do was rest and enjoy the Christmas truce.

The Company XO had different ideas. He was an Annapolis graduate and had different ideas than the rest of us reserve Marines. Somehow he had obtained some white paint from the rear and he wanted the Weapons Platoon to paint some logs that had been removed from the LZ area. Then he wanted them to be imbedded into the mud floor of the LZ to spell out Merry Xmas for the helicopter pilots that would bring us our Christmas meal the next day. He gave the white paint and his instructions to the weapons platoon lieutenant.

The weapons platoon leader, Lieutenant Robinson, was as different from the XO as you could possible get. He was a prep school graduate and a modern Marine who wanted to be told not only what to do but why. He was extremely displeased at the prospect of additionally needless work for his very tired men. But they started their assignment and soon after dark finished the job.

The next morning there a real ruckus at the LZ with everyone in the company running down to see what was happening. I was attracted by the laughter and walked down. The LZ had its giant message imbedded into the mud floor with the white painted logs. But instead of a Merry Christmas, there was a giant F_ _ _ U.

Marines defending Khe Sanh take a break to be lead in prayer by Chaplain Ray Stubbe of Milwaukee, WI. (Defense Dept. Photo - Marine Corps)

Vietnam Vignettes

Rounding up some other items of interest from the 3rd Marine Division experience in Vietnam:

In Eric Hammel's Khe Sanh, *Siege of the Clouds*, one of the vivid combat experience accounts of that operation is that of Private First Class Lawrence Seavy-Cioffi.

Seavy-Cioffi, a 3rd Marine Division Association member, was a forward artillery scout observer from D-2-12 assigned to A-1-9, and his detailed recollection of the events of 8-9 February 1968 comprises Chapter 16 of the book.

Hammel adapted Seavy-Cioffi's unpublished manuscript, *Our Victory for Alpha One*, interspersed with command comments by Lieutenant Colonel John Mitchell, 1-9 CO.

Danny Lee Perkins, Norman, Oklahoma chronicled "moments" of his 3rd Marine Division experience in Vietnam through a collection of free style poetry which he has assembled.

Among a number of highly regarded books on the Vietnam combat experience was *The Proud Bastards*, by E. Michael Helms, Youngstown, Florida. Helms, a 3rd Marine Division Association member, served with E/2/4.

A paperback version of his book was issued by Zebra Books in the spring of 1990.

Unknown Hero - Robert Waltrich

The character and courage shown by this young man is an example that has always been true of American servicemen whenever called by their country to do their duty. This man is truly an unknown hero.

At age 20 I was considered a combat veteran, not so much for what I had done, but because I had survived eight months in the field. My unit, Hotel Company, 2nd Battalion, 4th Marine Regiment had fought the North Vietnamese for several months along the DMZ. We had been moved to an area outside of Da Nang in order to receive new men and give us a little less contact with the enemy. As usual, this did not work out as you will see.

On March 27, 1967, my platoon received several new men. One man, Robert J. Waltrich was assigned to my squad. Robert was a real handsome fellow with blond hair and broad shoulders. He stood about six feet tall and was the type guy that you see on the Marine Corp recruiting posters. I was glad to have him in my squad as he had a great attitude about being in Vietnam.

Robert had been drafted into the Army, but joined the Marine Corps, as his father had been in the Corps during World War II. Robert was a "new guy" and not sure of himself yet, so I told him to stay close to me and I would take care of him. I remembered the hell I went through when I was new and wanted to help the new men as much as possible.

On March 30, we moved off our hill to a small village named DaLoc to provide security for the people to harvest their rice. As we moved to the outside of the village, I noticed that all the huts were empty and I knew something was up. We had just moved through a hedgerow when word came from the point squad to hold up. We were now in the open between two hedgerows in a dried rice paddy. I went down on my knee to wait and at this moment Private First Class Robert

Waltrich saw the North Vietnamese soldiers waiting in ambush for us. Robert, without hesitation, yelled for me to watch out—"Look out Corporal Erwin" words that I shall never forget. As he spoke those words, he stepped in front of me and was cut to pieces by the enemy machine gun.

The few precious moments he gave me was enough to allow me to crawl to a bunker near by. My entire squad was killed that morning along with many other Marines and North Vietnamese soldiers. The fighting was fierce and violent, and then it was over. I had survived because of the greatest act of courage a Marine could muster was displayed that morning by a young, inexperienced, drafted, American boy. Robert J. Waltrich had laid down his life for me—without hesitation he had made the supreme sacrifice.

Robert has never been given any medal (of honor) or recognition for his bravery. He probably never will be recognized. I suspect that there are many thousands of American soldiers in their graves that exhibited the same type of courage that Robert Waltrich did, never to be seen by others.

They did their duty ... There is no more profound tribute to those fallen in battle than the words spoken by General J.A. Garfield during the first National Memorial Day observances in 1868.

"We do not know one promise these men made, one pledge they gave, one word they spoke; but we do know they summed up and perfected, by one supreme act, the highest virtues of men and citizens. For love of country they accepted death, and thus resolved all doubts, and made immortal their patriotism and virtue." This is why we have Memorial Day!

Robert J. Waltrich, of Chicago, worked for the Illinois Central Railroad before enlistment ... Son of Carl and Frances Waltrich (his father was a World War II Corsair pilot).

Frank G. Erwin, Nashville, TN

There's Nobody I Want In My Marine Corps

It was Military Day at Cheshire High School, Cheshire, Connecticut, and the military service representatives were attempting to recruit students into their respective branches. The junior and senior boys were assembled in the school auditorium, with faculty members observing from the rear of the room as each recruiter got up to give his pitch.

The Air Force recruiter got up to explain the advantages of joining the United States Air Force. He was greeted with catcalls and whistles from the young high schoolers.

The Army recruiter received the same treatment as did the Navy recruiter.

Then the Marine recruiter, a seasoned gunnery sergeant, rose and glared.

"There is no one here worthy of being a United States Marine," he growled. "I deplore that the faculty in the back of the room would let the students carry on like this. There isn't anybody here I want in my Marine Corps."

When he sat back down, several eager students swarmed around his table.

One of those hovering around the gunny's table was Cheshire High School Senior Class President Harvey (Barney) Barnum Jr. He did the paper work to enlist as a senior in high school and joined the Platoon Leadership Class when he went to St. Anselm College in Manchester, New Hampshire.

He joined the Marine Corps—raised his hand—November 12, 1958.

Chu Lai - Chinese For Krulak

When the 3rd Marine Expeditionary Brigade landed 57 miles south of Da Nang in early 1965, there was no designation on the map for the location.

Resourceful Marine Lieutenant General Victor H. (Brute) Krulak quickly found a name for the spot which caught on—Chu Lai.

Chu Lai is a Mandarin Chinese translation of the general's last name!

Not Nice To "Point"

I had been in Vietnam a month or so and had gained enough experience to be point man. This was not an honor but the least desired duty since the one who went first encountered more dangers first. On our next patrol, which was company sized, 2nd Platoon was the lead group, the squad I was in was lead and I was point.

The word came to move out, through a fairly large clearing that once had been a rice paddy. The dikes were still there, being used as footpaths, and I started walking on one in the general direction we had been informed to go. No one had said, "On or off."

The Viet Cong knew that we used the dikes to move on and would mine them. After learning this one way or another, we would then walk on the sides of them. Then the Viet Cong would mine the sides of them. And we'd then go back on top of them.

We moved out with a 15-20 meter interval between each man in our movement so I was able to get more than 50 meters along the top of the dike before my squad leader passed on the word to get off the dikes. So we did.

I then moved on no more than 10 meters when I heard an explosion behind me. I hit the deck. Another one. They had mined the sides of the dike and I had walked by them when I walked on top. Immediately came the order to move to the tree-line and form a perimeter. And more explosions. I could not see or hear anymore chaos behind me as we had picked up sniper fire and were now shooting indiscriminately towards various spots to our front in an attempt to discourage, if not kill him. This effort met with little success as even the gunships accompanying the medevacs only managed to slow him down during the long time it took to evacuate the causalities.

After the gunships left an A4 was called in and made three strafing runs. The sniper fired at the plane after the first two passes. As we formed up to move out I discovered that my squad had six men, our platoon had less than 20 and F Company, strength was less than 50 percent with the causalities we just took, and not one by bullets.

This was going to be a tough war against a formidable opponent. **David Torrel, Eveleth, MN**

Political Problem Solving

On the occasion of one of our monthly meetings with representative rank and file members of the battalion, the troops of my maintenance organization were really upset. They obviously wanted the CO to know they were very angry. A young Hispanic Marine truck driver stood up and said he didn't understand why we were trying to help these Vietnamese people when they were stealing our gear.

Almost every truck going into the Da Nang Airbase was being ripped off. Not only the cargo, but also the drivers' personal radios were being taken. The troops all joined in impressing me that this was really a big problem. The worst place was at a traffic light where a one-legged Vietnamese teenager could be identified as the worst offender.

The Marines were frustrated because their officers had stressed the "no incidents" policy. It was to the point, they said, someone might get killed.

I went to see the chief of the hamlet who I had many other dealings with. He was an interesting man about 40 who was always smiling with lots of gold showing. He never actually worked that I could see, but he seemed to control everything.

As usual, I found him gambling at some kind of traditional game with tiles. He heard me out, and knew I was upset and wanted action. He smiled as we parted.

The one-legged youngster was never seen again. **Colonel W.F. Sheehan, USMC (Ret.) later Division Supply Officer, Fayette, MO**

Marines of G Company, 4th Marines are in action near Hills 459 and 500, 4,000 meters southwest of Phu Bi, Vietnam in 1967. (Defense Dept. Photo - Marine Corps)

We Didn't Lose In Vietnam

An unknown 3rd Marine Division combat veteran as a private first class, sick and tired of hearing that America was defeated in Vietnam, submitted this historical summary (which has been edited slightly) for this book:

1. The Vietnam veteran's duty was police action—arrest the Communist takeover of South Vietnam. The veteran serving in Vietnam accomplished his duty. Moreover, the Vietnam veteran won every battle he fought, including Khe Sanh. After the United States military withdrew in 1972, a cease-fire was agreed to in Paris. When it was signed, more population and land were under Saigon control than at any previous time. We left winning and with military victory in 1973.

2. Six months after our troops left, the International Control Commission reported daily North Vietnamese violations inside South Vietnam. In 1973, a commission helicopter was shot down by the NVA. The commission withdrew in July 1973. Full scale war returned. The ceasefire failure was North Vietnam's fault. Then, the United States Congress ended military hardware aid for South Vietnam.

3. In early 1975, the world press reported a massive multi-division tank and troop invasion from North Vietnam into the South. Without continued United States aid, South Vietnam surrendered, April 30, 1975.

4. Unsuccessful foreign policy is different than battlefield defeat. The United States was not invaded, occupied, or signatory to a surrender. Surrender is defeat for a nation in war, such as Germany and Japan in World War II. The dilemma fallacy that either the United States won or lost the war was illogical in this case because it didn't allow for the existing alternative—successful participation until the 1973 ceasefire and the United States withdrawal before the third war in Vietnam in 1974-1975.

5. The Vietnam veteran was the United States in Vietnam. How may a nation be defeated in war if its military is not participating when the defeat occurs? South Vietnam lost the war, along with unsuccessful foreign politics—not the United States military.

A retired United States Army colonel, David L. Jones of Annandale, Virginia, summarized it this way in a letter published in the *Washington Post*:

"The United States was not defeated on the battlefields of Vietnam. Its armed forces were never even remotely threatened with defeat there; they were capable of doing whatever their civilian leadership asked of them. In the end, they were withdrawn by that leadership. The United States was defeated at home—in the halls of Congress, in the streets, and, yes, in the press—but not on the battlefield."

From an article by George Kirts in the August 19, 1991, issue of *Stars and Stripes - The National Tribune*, these quotes are excerpted:

"In reality, the Vietnam vet had no control over the course of the war. The United States forces won every major battle they were engaged in during Vietnam."

"The failure in Vietnam was more a failure of the United Nations peacekeeping concept than of United States military policy ... the failure was largely a failure of Vietnamization ... Vietnamization never really worked and probably never would have."

"The South Vietnamese people lacked the resolve to remain free. The only thing that kept Saigon propped up until 1975 was the United States military presence."

"It is a wise man that knows when to walk away from a battle he cannot win."

"In the end, the real truth is that we tried our best to give the people of South Vietnam freedom. They just didn't want it bad enough."

Going Up!

A Marine Sea Night (CH46) helicopter hoists Marines from dense vegetation near DaNang, Vietnam. (Defense Dept. Photo, Marine Corps)

Twelve Fatal Decisions

General Raymond G. Davis had the opportunity to view the Vietnam War from a number of crucial vantage points and the opportunity to discuss it with many in authority. In 1989, he summarized this experience and subsequent findings in an analysis published in the Marine Corps Gazette.

As Marine Corps Manpower Coordinator, he worked initially in the buildup and deployment of forces. As a commanding general of the 3rd Marine Division in Vietnam, he experienced the effects of policies, goals, and efforts. Lastly, as commander of the Marine Corps Development and Education Command, he studied and analyzed decisions and results.

In substance, those responsible for political decisions must also comprehend the effect of their decisions on the military, he observed. In brief, the decisions:

1. Role of Military forces. Early on, it was obvious that a half million troops would be required. Prompt application of a force that size would have enabled early success, but the buildup took until 1968.

2. Limit on funds. Deployments had to be slowed down to remain within expenditure limits.

3. Withholding the Ready Reserve. The draft buildup took three years, when organized Ready Reserve units could have greatly accelerated the process.

4. Pacification in reverse. With inadequate forces, large segments of South Vietnamese had to be moved to camps which disrupted the population and its faith in the process. Until adequate forces became available in 1968, no real progress was made in pacification.

5. Sanctuary for enemy forces. Enemy forces had sanctuary in Cambodia, Laos, and North Vietnam, enabling freedom of maneuver and inability on our part to destroy heavy artillery and rocket bases there.

6. A $6 million blunder. A 26-mile northern front was built to control infiltration, but the enemy had a free run everywhere else along the 1400-mile border. Forces inefficiently tied down to man this costly line would have been much more effective mobile, as shown in 1968 when the rules were changed.

7. Air defense buildup. The substantial enemy air defense buildup was greatly aided by the decision to withhold military attacks on that system.

8. Cease-fires. Cease-fires permitted the Viet Cong to "visit" home areas while in actuality they were used for resupply, reorganization, and recruiting. The massive early 1968 TET offensive was a graphic example of this deception.

9. Artillery/rocket/bombing halt. The enemy "wanted to talk" in 1968, so we agreed to stop shooting across the DMZ if they ceased firing rockets into large cities. But we had captured his rocket supply and he was hurting. Since he couldn't shoot rockets at us, he outmaneuvered us, and also used bombing halts to interrupt war progress when our aircraft were particularly clobbering him.

10. ARVN regional concept. The ARVN became a fixed division regional application, not a mobile national army.

11. Withholding of support in Laos. After success in Cambodia, preparations to destroy NVA forces in Laos were undercut by premature publicity which caused decisions to withdraw key planned support operations.

12. Premature United States withdrawal. Since the NVA was only partially destroyed when we withdrew, the results were predictable.

Can War Be Boring???

In May 1966, I was a private first class and supplyman for Bravo Company, 3rd Tank Battalion at Marble Mountain.

It had been awfully quiet at night for several months, but one night around 2300 two of our machine gun bunkers opened up. I dove for the foxhole outside our supply tent.

The next morning I found that because of the boring nights of watch at machine gun bunkers, two Marines decided to create a little excitement by firing their M-60's at the sand which stretched for some distance in front of the bunkers. The one Marine said he would start firing first, then the second Marine was to join in.

When they were asked the reason for the firing, they vouched for each other and said they had seen Viet Cong in front of them. **Alan L Bachman**

There's Now In Georgia A "General Ray Davis Highway"

General Ray Davis, who commanded the 3rd Division in Vietnam and received a Medal of Honor for relief and rescue of a besieged company at the Chosin Reservoir (Korea), now has a section of highway in Georgia named for him.

Georgia Governor Zell Miller fixed the General's name to State Route 42. "That section." commented one Third vet, "just has to be the safest road in the world." General and Mrs. Davis live in Georgia.

Coming Home To The Fourth

My most memorable tour of duty in Vietnam was with the 4th Marines during late 1968. It was of special significance because I had previously served with that regiment as a platoon leader and company commander in World War II. In 1944, the Commandant had directed reactivation of the Raider battalions, which became the main infantry element of the reactivated 4th Regiment. The "old Fourth" was captured at Corregidor.

My tour of duty with the then new Fourth took me through the Guam and Okinawa campaigns. At Guam, the 4th was part of the 1st Provisional Marine Brigade, which landed with the 3rd Marine Division.

I was privileged to command the 4th Marines in Vietnam. It was like coming home. Again, the 4th Marines did a terrific job. **Colonel M.J. (Stormy) Sexton, USMC (Ret.) 4th Marines, Carlsbad, CA**

CARE

One of my most vivid memories of Nam happened on March 6, 1966. Private First Class Gregory James Moore of Chicago hit a booby trap and was blown away. I was a dozen or more steps behind him as usual and escaped injury. My sergeant had a cloth bag from a CARE package and we put Monroe's gear in it.

I was carrying this and looked down at the bag as his blood seeped through. These words were stamped on the bag: "Gift of the People of the United States of America." Monroe died en route to the hospital. **Lance Corporal Don M. Stephenson, USMC (Ret.), Hamilton, TX**

A Dramatic Good-Bye
The "Fighting Ninth's" Vietnam Farewell

On the morning of July 25, 1969 tribute was paid to the 9th Marines in ceremonies at the Quang Tri Combat Base with colors of every unit of the 3rd Marine Division and its supporting organizations assembled and passing in review.

The Marines of these units, still engaged in combat duties, remained in the field or at their combat bases, and were represented in the ceremony by their color guard and troop commander.

The "Fighting Ninth" was the first of the division's regiments to leave Vietnam as part of the United States troop drawdown. It ended more than four years of combat which began on March 8, 1965, when the 3rd Battalion, under Lieutenant Colonel Charles McPartlin, spearheaded the landing at Da Nang. The battalion provided security at the airbase and its immediate area until relieved by 1/9 on June 17 when the battalion returned to Okinawa.

On July 4 the 2nd Battalion, under Lieutenant Colonel George Scharnberg, and the 9th Marines Regimental Headquarters arrived from Okinawa. Colonel Frank E. Garretson was the first of twelve regimental commanders who would lead the "Fighting Ninth" in combat in Vietnam. By mid-August the 3rd Battalion was back in country and the regiment was again completely formed.

A Moving Ceremony

The July farewell ceremony was a moving and solemn one. As the 3rd Marine Division band played, the spectator stands were filled with Marine Corps, Army, Navy, and Vietnamese officials who realized that the departure of the 9th Marines signalled a dramatic change in America's role in Vietnam.

> *... Placing his helmet on the butt of the weapon, and with hands covering the helmet, he lowered his head in silent prayer. This ceremony was repeated 37 times.*

Colonel Edward F. Danowitz, commanding officer, 9th Marines, and Colonel Robert H. Barrow review the farewell parade at Quang Tri Combat Base on July 25, 1969 prior to withdrawal of RLT-9 from RVN to Okinawa. At right are Sergeant Major L.T. Porter and the 9th Marines color guard.

On July 13 Secretary of Defense Melvin R. Laird had announced that the 9th Marines, in addition to Army and Navy units, would be withdrawn beginning in mid-July. On July 14 Battalion Landing Team 1/9, under command of Lieutenant Colonel Tom Culkin, sailed from Da Nang for Okinawa on board amphibious ships of the 7th Fleet initiating Phase I of President Nixon's troop withdrawal plan.

37 Operations

The 2nd Battalion and sections of the regimental headquarters represented the "Fighting Ninth" at Quang Tri since the 3rd Battalion, commanded by Lieutenant Colonel Don Wood, was still in the field and would not leave RVN until August 14.

Once the parade was formed, troops from Lieutenant Colonel Robert L. Modjeski's 2nd Battalion paid special tribute to comrades who had fallen in the 37 operations in which the 9th Marines had participated. As the name of each operation was called, a Marine from the battalion stepped forward smartly and drove the bayonet of his rifle into the ground beneath the pierced steel planking of the air strip. Placing his helmet on the butt of the weapon, and with hands covering the helmet, he lowered his head in silent prayer. This ceremony was repeated 37 times.

The emotional scene of 37 Marines standing silent with bared, bowed heads was well captioned in the prayer of the 3rd Marine Division's Chaplain, Captain J.E. Zoller, United States Navy, as he intoned the final words ... "It is unto Thee that we commend the souls of all who gave their last full measure of devotion."

Prior to ordering the parade to "Pass in Review," the commanding officer of the 9th Marines, who had been designated the reviewing officer for the parade, left his post and proceeded to the spectator stands to invite Colonel Bob Barrow, the regiment's previous commander, to join in the review of the division's honoring units. Colonel Barrow had commanded the 9th Marines from July 1968 to April 1969 and had led the regiment in "Operation Dewey Canyon," one of the most successful United States operations in Vietnam. At the conclusion of the review his well deserved selection to brigadier general was made known. He was later to become the Marine Corps 27th Commandant.

Honored guests at the ceremonies included General Creighton Abrams, Commander United States Military Assistance Command Vietnam, Lieutenant General Herman Nickerson Jr., Commanding General III MAF; Lieutenant General Melvin Zais, USA, Commanding General of the XXIV Corps; and the 3rd Marine Division's Commanding General, General William K. Jones, USMC. The Vietnamese representatives were headed by Lieutenant General Hoang Xuan Lam, Commanding General of I Corps.

Joint Recognition

General Jones, in his farewell address to the 9th Marines, stressed the importance of all elements in the Regimental Landing Team which supported each Marine in combat. "No one knows better than the Ninth Marines that infantry alone cannot prevail. Air, artillery, engineers, shore party, tanks, motor transport, and all of the other elements that contribute to the combat effectiveness, health and welfare deserve this recognition of RLT-9." The General concluded by bidding the regiment "Godspeed, fairwinds, and following seas."

> ... Captain Kelly and his company were escorted into the village ... the district chief expressed gratitude of the people for making the area secure from the enemy threat.

In addition to this official ceremony the Vietnamese people, when apprised of the departure of Marines from their areas, tried as best they could to express their appreciation and farewells. The citizens of than Quat Xa hamlet near Chu Lai showed their sentiment to C Company when they held a going away party for the Marines. It provided the opportunity for them to meet the people and officials for whom they had provided security during their stay in Vietnam. Captain John H. Kelly and his company were escorted by a large group of children into the village to meet with the district chief who expressed the gratitude of the people for making the area secure from the enemy threat. Following refreshments and entertainment—provided by a culture drama team, musicians and singers—the afternoon concluded with an exchange of gifts. The Marines were offered fruit and local handicrafts while in return they gave the villagers candy, gum, and much needed household items.

The emotion of Marines and the Vietnamese people was evident as Charlie Company climbed aboard their trucks to head for their embarkation port. Such gatherings took place in many other hamlets and villages. In addition, each sailing from the deep water port at Da Nang was preceded by a memorable and impressive final farewell as Vietnamese military and civilian officials, accompanied by schoolchildren, gathered to present mementoes and tokens of appreciation to Marines of each battalion preparing to climb aboard ship.

"Keystone Eagle"

The 7th Fleet had designated the deployment of RLT-9 from Vietnam as "Operation Keystone Eagle." The move in which more than 4,000 troops and over 2,300 short tons of equipment were lifted to Okinawa was carried out smoothly due in great part to the planning and coordination with the 3rd Marine Division staff.

The USS Paul Revere (LPA 248) moved the three battalions of the 9th Marines in three lifts during the period July 13 through August 17. The ship provided a dramatic change from the life the Marines had known in the field. New white skivvy shirts were provided by the ship which became the uniform of the day. Showers were open on a 24 hour basis and the galley poured out steak, ice cream and good navy chow on almost a continuous basis. Most of all, it was the warm and cheerful hospitality of the ship's crew—from captain to seaman—which made the Marines feel appreciated. It was a trip without the chipping, scraping and painting to which many Marines had been detailed while aboard ship.

The Paul Revere's commanding officer, Captain M.J. Karlowica, United States Navy, noted his impressions of Operation Keystone Eagle in his letter to the commanding general, 3rd Marine

> ... Skivvy shirts became the uniform of the day ... showers were open on a 24 hour basis and the galley poured out steak, ice cream, and good Navy chow on almost a continuous basis.

Division, on September 8 stating: "In the truest sense, the degree of discipline, cooperation and leadership that was exhibited by the officers, NCO's and men of the 9th Marines was truly outstanding and as shipmates they were second to none. Please extend to the 9th Marines the sincere gratitude of a grateful ship for a superb job well done. The officers and men of the 9th Regiment are always most welcomed on board the Paul Revere. Wherever she may be, she will be honored to be their floating home away from home."

A Warm Welcome

The "Fighting Ninth" was warmly welcomed during its move into Camp Schwab as the first of the division's infantry units to return to their home base. Lieutenant Colonel Calhoun J. Killeen, commanding officer of the 2nd Battalion, 12th Marines, the 9th Marines direct support artillery unit in Vietnam, noted the same hospitality as he moved his artillerymen onto the Okinawa base. Major General Killeen would return to Okinawa as the commanding general of the 3rd Marine Division from July 1978 to July 11, 1979.

In noting the arrival of the Marines, the press on August 24, 1969 said: "Allied Forces throughout Vietnam and all over the world can be justly proud of the accomplishments of the Ninth RLT." It was a fitting and final tribute to the thousands of enlisted men and officers who had served in the Republic of Vietnam with the "Fighting Ninth." It was a tribute which could have also been properly extended to all Marines and naval personnel who served in each unit of the 3rd Marine Division in the war in Vietnam.

This story was provided by Colonel Edward F. Danowitz, who served as commanding officer of the 9th Marines from April 8 to September 8, 1969.

(They Called It ... Thor)

The distant guns
Sounded their forthcoming fury.
My "Iron Coffin"
Was my safe home.
My brothers in the crew
Knew well the sound.
Soon we would be encompassed
In a terrible rain of shrapnel
Served by the distant guns.
Thank you, Charlie,
For keeping us alert
During a tranquil moment.
Tomorrow we will find you.

On a hill called Con Thien, the sun's heat began its daily punishment as my crew prepared for the pursuit of Charlie. This operation was going to be one hell of a search and destroy mission; the platoon leader had emphasized DE-STROY as we were psyched up for the best, or worst; it's all in how you view killing.

As the sun rose, we mounted up and got positioned with our assigned grunt units. This time we were with the 9th Marines. These guys hit the shit wherever they go. *(I even heard once that the whole outfit pulled an in-country R&R and caught incoming mortars before they could buy one beer!)*

Beer, the magic word, I haven't had one since my tank hit a mine and we had to go back to Cam Lo to get it repaired. *(I was ready for a nice long stay in the rear, but that damn gung-ho mainte-nance officer made us work seventeen hours straight and we were out of there before we could even see a USO show. Heat and hot beer can really screw up your head in a hurry though; better put off the drinking for a while.)*

The sun hits the red-scarred moun-tains west of our hill as we move out, and it looks like an early morning scene in Colorado. What I'd give to be home with my wife and kid. *(Only three months to go, and I'll get to see that namesake of mine. Julie says that he's a real terror. I can't figure out how a six month old baby can terrorize a full-grown woman, but I guess anything is possible.)*

The day gets hotter as we push on. It's about 114 degrees to our best estima-tion; but in our tank it has to be 125 and the humidity is at least 98 percent. The sun beats down on us; we sweat and gulp salt tablets to keep from getting sun stroke.

Carl, my gunner, is really a poor bastard. He has to stay inside the tank, prepared for a sudden rain of bullets and rocket propelled grenades coming from any direction, at any time. He sweats and bitches twice as much as my driver Steffee, or I. It's frightening and boring down there, resting beside 350 gallons of napalm. He can't see what's going on outside; he can only rely on me for directions. He has to trust me with his life. But he's prepared; I've seen to that. *(After hours and hours of practice on the proper disassembly and assembly of the machine gun, fire practice, search and traverse of the turret, and rapid response drills ... I know he's ready; he knows his stuff, but he still sits in a stupor caused by the heat, humidity and boredom.)*

My driver, my six-foot, six inch gi-ant, is another story. Steffe is cocky, always ready for a fight, and has almost no fear for his life. He's an extension of my own hands and feet. *(I say, "Right brake," over the intercom and the tank moves to the right almost as soon as soon as I open my mouth. I say stop, and he will stop our 53 ton "Iron Coffin" on a dime, even if we are going the top speed of 30 miles per hour.)*

These two men, my suffering, loyal gunner and my cocky, obedient driver are my crew ... they are me. In a fire fight we function as one. My driver can take us anywhere the "Slopes" go and my gunner can find a "Slope" in the densest jungle, I only have to say, "Burn 'em," and there are no more troubles in front of our gun. We function like a perfect gear system, meshing with precision, even in this goddamned Vietnamese heat.

The radios hum with squelched static. A grunt radio operator comes over the air, "Yankee Six, we've made con-tact." The sound of firing through the receiver sound like fire crackers on the Fourth of July. *(My first fire crackers, what a kick in the ass! Throwing them at all the dogs, cats and little old ladies in town!)*

More sounds over the receivers. A war breeze picks up. *(Calm water, gentle breeze, my wife and I are at the lake. The sounds in the distance of young boys terrorizing the dogs, cats, and little old ladies. We laugh and make love on a blanket on the beach.)*

"Fox Trot Three-One, pull up on line. When I give you the word, burn the

hedgerow directly in front of you. Charlie is giving us hell, so get your ass in gear!"

I come back with a baffled, "Roger," and we move up and burn the hedgerow and the men in their bunkers below. The fire looks like the homecoming bonfires we have at school. *(The heat feels great on a frosty autumn night. My wife can keep me just as warm. Cold winter night, two bodies snuggle close together for warmth. Waking in the morning and making love for breakfast.)*

"Fox Trot Three-One and Alfa One-Five, come to your left and give us some support." We move and burn some more. We make runs to the landing zone with wounded grunts and heat casualties. The sun is "cooling" more men than the gooks ever hoped to. The grunts hang from the side of the turret, and I don't mind missing some of the action going on up ahead to carry the wounded men who need our help, but my tank really gets rusty from all the blood the wounded lose on the way to the L.Z. *(One day I remember, a grunt got in the way of Joe (Jesus) Bonilla's gun just as he let off a 'beehive' round, and the sorry son-of-a-bitch got half his neck and face blown away. He was still alive when we pulled up. His buddy asked us to carry him to the L.Z., so we pulled the poor guy on the tank and his buddy jumped on to hold him down over the bumps and bomb craters. While my back was turned, the wounded man's buddy grabbed my blanket out of the gypsy rack and used it as a pillow for his messed up friend! He got my only blanket soaked in blood! I mean, why me? The guy died before we could move a hundred yards to the L.Z. I really got pissed off and kicked the dead man's buddy off the tank; this pissed him off and he threatened to shoot me; my driver cooled him before he could carry out the threat. War can be hell, espe-cially if you've only got one crusty blan-ket and no chance for a replacement before the bone chilling monsoons be-gin.)*

We push on; our advance is really a huge circle around Con Thien. The op-eration is called "Thor" and we are em-ploying the tactic of "the hammer and anvil." There is a blocking force in the area somewhere, and they are supposed to stop the retreating gooks that we don't get. The fighting has been fairly light, but the sun has taken its toll. My gunner is about to go out of his tree. Just as I begin to calm him down for the fifteenth time today, an R.P.G. slams into the side of the turret. I duck behind my .50 caliber machine gun and spray into the area where I think the gook is that fired at us. I yell at Carl to do the same with the .30 caliber, but there's no response. After all the goddamn schooling I've given him, has he clutched?

I stick my head inside the turret and start to kick him in the head, the stupid son-of-a-bi_ _ _. Then I see blood everywhere, the white interior of our tank is red." Carl is sitting there, leaning on the napalm bottle making a gurgling, moaning sound. Steffe knows there's something wrong and asks me for directions.

"Kick it in the ass, Steffo. One of those son-of-a-bitchin', slant-eyed, slope-head, fender-bellied, rice-powered pieces of crap just blew away Carl! Get this garbage burner to the L.Z. before he dies!"

On the way to the L.Z. I give Carl a shot of morphine and try to stop some of the bleeding. I take my brand new, clean blanket out of the gypsy rack and wrap him up as best I can. "You poor son-of-a-bitch, live! You can't screw up our team. We've got it all over everybody else. We're tight as a frog's ass and we gotta stay close together!" Carl's eyes sort of roll back inside his head and he starts to turn white. "Move it Steffo, move it!" We can't find the L.Z. The grunts have moved ahead and their rear guard is where the L.Z. used to be. Then a medical evacuation helicopter lifts off from the next rice paddy, and we get over there and pull Carl out, wrapped in my blanket, to get tagged by the corpsman for a medical evacuation.

"You take good care of him, Doc, he's a goddamned good man," I yell down to the corpsman.

The corpsman smiles, looks down at Carl and then shakes his head, "This dude is dead," and he covers Carl's white face with my bloody blanket.

Dead!" Good God Almighty! Three minutes ago he was inside our "Iron Coffin" bitchin' about the heat! Ten months we've been together. Ten months! What a waste! He only had a few weeks to go and he gets blown away now! He was going to get married when he got home! What'll his parents do? What'll his chick do? What'll I do?

"Shit!" "We can't help him now, Sherm. Maybe our next gunner can be as good as Flash," Steffe comes over the intercom.

He's right, waste or no waste, we have to look out for our own asses. The Slopes show no mercy on us; why should we show any, to anyone? A good drunk will help when and if we get back to the rear.

"Let's go get a few for Carl, Steffo." And we move our bloody "Iron Coffin' back on line with the grunts, the tanks and the gooks.

The heat of the action and the heat of the sun makes my eyes water, or does it?

I haven't cried since I was in the eighth grade!

John F. Wear II was a Marine Sergeant when he had this personal experience while assigned to Alpha Company, 3rd Tank Battalion, 3rd Marine Division in Northern I Corps of the Republic of Vietnam in the summer of 1968. He lives in Penn Valley, PA

Reflections For Veteran's Day

I served in Vietnam with the 2nd Platoon, F Company, 9th Marines in 1966-1967. During that time I participated in more than 300 combat patrols ranging in size from two man LPs to battalion sweeps. I still have some gook metal in my knee from one of them.

Long after that war, things are going on that I wonder about: Some Vietnam veterans are trying to sue the government over the use of Agent Orange. Some of the war's veterans are complaining about the lack of homecoming parades. The women who were there want their own special monument, even though the names of the nine women who died there are on the existing Vietnam monument along with the 58,100 names of the men who died there.

As I consider these happenings and remember the past, I have formed my opinion:

On two-thirds (estimated) of the combat patrols I was on, nothing happened. No contact. Our patrol went out, came back and no one was injured. I have now reached the conclusion that the use of Agent Orange to defoliate areas of the jungle restricted the movements of the VC/NVA. This lessened their opportunities to ambush us which gave me a greater chance for survival. Even if I have the stuff in me, I have lived for 22 plus years since. I am grateful for that.

After leaving Nam, I was met at the Minneapolis Airport by three brothers and a cousin. Their handshakes, and those of others were enough. I didn't need a parade. It was good enough just to be back.

Like veterans of other wars, I would not talk about some things and would try to forget the bad things. But there are some things that I try to remember.

I try to remember the men who I served with and to remember those who died in combat, both in this war and in the past wars. I try to remember my time of service for my country, remembering that our intentions were honorable. Many of the results we achieved were commendable, but the end was deplorable. **David Torrel, Eveleth, MN**

Christmas Was Tough

Christmas in Vietnam was tough. We pretended as best we could. The loudspeaker in the battalion area broadcast Christmas carols for the ten days before the 25th, but with 86 degree temperature, Jingle Bells and White Christmas just didn't do much for the spirit of most Marines.

We had religious services and made a big push to get everybody involved. We had the traditional special meal in the mess hall—turkey with all the trimmings. You couldn't have improved on the feast the cooks put together. We stood down and had the day off.

Marines who wanted to go were trucked to the orphanage at China Beach. A gift for each child in the orphanage had been arranged. The Marine presented the gift and then took the child through a special food line we set up on the beach. It certainly brightened the day for the kids and the Marines, and brought out the best in people. It was also the hardest of times. **Colonel W.F. Sheehan, USMC (Ret.), Division Supply Officer, Fayette, MO**

Scholarships For Education

... A True And Lasting Memorial

The November 1967 death of then division commander Major General Bruno A. Hochmuth in a helicopter crash in Vietnam action sparked the development of the 3rd Marine Division Association College Scholarship Fund as a memorial.

This highly successful association program has provided more than 100 Marine children, most of them survivors of division Marines KIA in Vietnam, with up to $2,500 a year for as many as four years of college study.

Over the more than 20-year history of this widely lauded activity, division Marines' children have been helped in their efforts to become lawyers, engineers, accountants, and a wide variety of professional and technical specialists.

The program until 1989 provided for children of division Marines who gave their lives in Vietnam, then was expanded to care for all qualifying children of association members, including those disabled in Vietnamese action.

The idea came from awareness of the general's long standing interest in education, which he was planning to follow in his impending retirement that never materialized.

Original contributions from 3rd Marine Division members in Vietnam of $12,065 just after the general's death provided the financial impetus. Many memorial contributions have followed through the years, some of quite generous nature, along with fund-raising and contributions of the association and its chapters.

Administration of the program is handled by a board of trustees which includes:

Chairman: Bert R. Barton of Hermitage, Pennsylvania (H&S-21 in World War II).

President Emeritus: CWO (Ret.) Thomas O. Kelly of LaPlata, Missouri (H&S-21 and Division Headquarters Battalion in World War II; also Korean War).

Secretary: Lieutenant Colonel (Ret.) Richard E. Jones of Inverness, Florida (1-9 in Vietnam War; also World War II and Korean War).

Member: Colonel (Ret.) Edward F. Danowitz of Altamonte Springs, Florida (Division Headquarters in Vietnam War; also World War II and Korean War).

Member: Joseph J. Kundrat Jr., of Reisterstown, Maryland (K-3-9 in Vietnam War).

Member: Robert B. Van Atta of Greensburg, Pennsylvania (H&S-12 in World War II; also WWII and Korean War).

Barton, Kelly, and Van Atta have been members of the board of trustees since it was formed in 1968. To CWO Kelly goes much of the credit for organizing the activity, and providing much of the work that made the program successful.

He spent long hours at Marine Corps headquarters searching out 3rd Marine Division casualties in Vietnam, identifying those with children, and corresponding with widows to acquaint them with the program, among other things. Without that effort, many of the scholarship awardees would not have been aware of the benefits possible through the association.

Further information on the 3rd Marine Division Memorial Scholarship program is available from the secretary, Lieutenant Colonel Richard E. Jones, USMC (Ret.), P.O. Box 634, Inverness, Florida 32651. From memorial story below.

Memorial To A Fallen General

When Major General Bruno A. Hochmuth, then commanding the Third Marine Division in Vietnam, died November 14, 1967, in the explosion and crash of a flaming helicopter in a flooded rice paddy, the tragedy triggered the highly successful association college scholarship program as a lasting memorial.

The commanding general was returning from a meeting at Hue with South Vietnamese military officials to the Marine CP at Dong Ha. Four others, including a member of the general's staff, were killed.

General Hochmuth had been the division commander since March 20, 1967, when he succeeded Major General Wood B. Kyle.

The general's long standing interest in education prompted the scholarship program as a memorial. Initial funds came from members of the division who raised $12,065 as its financial start.

The program has been in operation since 1969, and has furnished needed funds to more than 80 college students, a high percentage of whom were children of Third Marine Division members killed in Vietnam, for up to four years.

General Hochmuth was born May 10, 1911, at Houston, Texas, and graduated from Texas A&M College in June 1935. He was commissioned a second lieutenant in the Marine Corps the following month.

He participated in the Saipan, Tinian, and Okinawa operations in WWII, after which he served in Japan. He was given regimental command with the Second Marines in 1951, and became a general officer in 1959.

A Standup Operation

I stood up a lot during Operation Double Eagle. As a fire team leader and heli-team leader with H-2-3 aboard the USS *Valley Forge* getting ready to jump off on Operation Double Eagle, as I recall early in 1966, we spent some time mainly sitting on metal decks. I developed a very serious case of hemorrhoids.

I could hardly walk. To sneeze, I feared, would mean instant disintegration or slow death. They were going to have to throw me into one of those bouncy landing craft—there was no way I could go down those nets.

Turning into sick bay, a surgical cut and a kotex later and with the doc's "Go get 'em Marine," off I went. However, during what turned out to be a highly successful operation, I stood up a lot.

He participated in the Saipan, Tinian, and Okinawa operations in World War II, after which he served in Japan. He was given regimental command with the 2nd Marines in 1951, and became a general officer in 1959. **(Former S/Sgt. Ed Nicholls, H/2/3 (P.O. Box 473, Kendalia, TX 78027); Note:** Nicholls has done extensive work on computerizing a record and the participants in Operation Starlite, and seeks further information.

An ARVN With Problems

On bridge security north of Da Nang with some South Vietnamese Army men, we were able to communicate with them through one of their men and one of my squad being able to speak French. In that conversation, we learned that one of the ARVN had fought for the Viet Minh in the battle at Dien Bien Phu. He described the battle and told us he fought against the French because they were bad.

Then, he said he was fighting against the Communists now because they were bad. He was doing this despite the fact that his wife, mother, and children were living in Hanoi! **David Torrel, F-9th Marines, Eveleth, MN**

Communications men laying telephone wire in a Jap sniper area. Pvt. First Class Eric Erickson, Pvt. First Class Edgar Rupkey, Pvt. First Class James Gutieviez, Jr., and Pvt. First Class Joseph Beelart. (Defense Dept. Photo - Marine Corps)

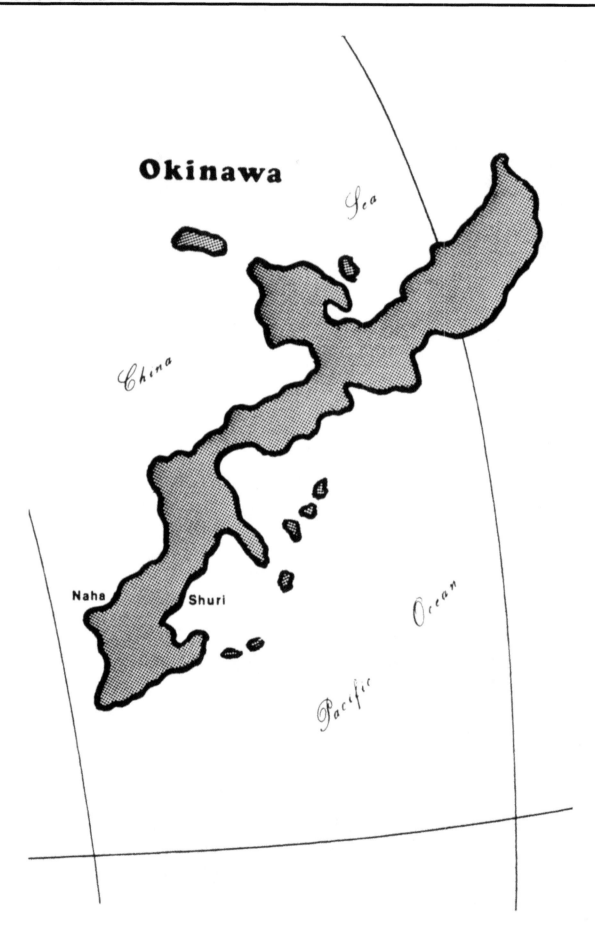

The Flavor Of Okinawa

Carroll Garnett of Chester, Virginia provides a comprehensive overview of 3rd Marine Division highlights on Okinawa with stories excerpted from the "Okinawa Marine." For consistency's sake, the tense used in the original story, is generally used here as well.

These stories will be nostalgic for Marines who where stationed there ... and informative to 3rd Division veterans of other days to show how Marines trained and lived on Okinawa ... to remain a force in readiness. The expression "Nothing is constant except change" is most applicable in this section—as the Marine Corps has stayed in tune with a changing world.

A Division In Motion
Happy Birthday— 3rd Marine Division
Sept. 16, 1942 - Sept. 16, 1972

Some people would like to count backwards once they reach their 30th birthday, but the 3rd Marine Division continues to count forward, celebrating three decades of meritorious service to God, country and corps.

Formed in San Diego on September 16, 1942 under the command of Major General Charles D. Barrett the division received its baptism of fire on Bougainville, then proceeded through the campaigns of Guam and Iwo Jima.

8 Streamers

With the ending of World War II, the division was deactivated in December of 1945. For its superb actions during those first three years of its existence, the division was decorated with eight streamers for participation in the Asiatic-Pacific campaigns.

The deactivation was short-lived. The sudden outbreak of the Korean Conflict, along with General MacArthur's request for Marines, resulted in the reformation of the Division in January 1952 at Camp Pendleton.

In August of 1953, after 19 months of training at Camp Pendleton, the bulk of the division arrived in Japan to augment its sister organization, the 1st Marine Division.

A Force In Readiness

Men and women of the 3rd Marine Division are proud of their Division's long time role on Okinawa as a "Force in Readiness in the Western Pacific" as well as a "Division in Motion," with units at sea, on land and in the air.

These same Marines are very conscious and respectful of the immense cost of capturing Okinawa during the closing months of World War II.

Except to the trained eye of a military expert, Okinawa was considered a rather poor chunk of land comprising some 700 square miles, but its strategic value was immense. It was only 350 miles from Japan proper, and it had good harbor facilities and plenty of room to stage troops. It would provide an ideal jumping-off place for any invasion of Japan. This island was the stake of the war's last and biggest amphibious assault.

D-Day ... 1945

The invasion of Okinawa began on the morning of April 1, 1945, with units of Lieutenant General Simon Buckner's 10th Army and Lieutenant General Roy S. Geiger's 1st Marine Division. By the time the battle ended, 82 days later, over 300,000 troops were on the island.

At the closing, 110,071 Japanese were dead, nine for every American; a similar number of civilians perished, most of them needlessly.

The original landing was unopposed since the Japanese commander, Lieutenant General Mitsuru Ushijima, chose to concentrate his forces on the southern part of the island, believing a defense in depth would attract the main assault forces which then would allow him to cut them off. For the Americans, five days elapsed with no opposition. But then on April 6-7, nearly 700 enemy aircraft, including some 350 kamikazes, pounded the American beachheads and task forces assembled offshore.

The first five hours took a toll of six American ships and 135 kamikaze pilots. From then on, the conquest of Okinawa proceeded in a bloody series of pushes. The 1st Marine Division suffered heavy casualties, 7,665 killed, wounded and missing.

Buckner was among the Americans killed. Ushijima, and his principal subordinate Lieutenant General Cho, knelt in full dress uniform before their headquarters cave and cut out their entrails. Geiger took Buckner's place as top commander and thus became the first Marine in the nation's history to command a field army.

The Japanese lost several thousand aircraft, the Americans 763. The Kamikazes cost the Americans 36 ships sunk and 368 damaged. The Americans had the base they needed for their planned invasion of Japan.

With the end of World War II, the 3rd Marine Division was deactivated in December of 1945. However, with the outbreak of the Korean Conflict, the division was reactivated in January 1952 at Camp Pendleton. The first elements of the 3rd Marine Division were moved to Okinawa in July 1955; headquarters followed in March, and the remainder of the division moved there during 1957. Okinawa has been the division's home since.

More Honors

Vietnam was the addition of several more pages to the history of the 3rd Marine Division ... plus the Presidential Unit Citation and the Vietnamese Service Medal with one Silver and three Bronze Stars.

In its role as a "force in readiness in the Western Pacific," forward elements of the Division landed on the beaches of DaNang March 8, 1965, marking the arrival of the first large-scale ground combat units of the United States Armed Forces in the Republic of Vietnam.

During its four years, eight months spent on Vietnamese soil, the 3rd Marine Division participated in 120 major operations.

On November 7, 1969, the 3rd Marine Division returned home to Okinawa. Upon its homecoming, the division assumed its traditional role as America's "force in readiness," continuously deploying a battalion of Marines afloat with the United States Seventh Fleet patrolling the Western Pacific waters.

Colonel Accepts A Change With Built-in Emotion
28 Years Of Memories

On July 21, 1944 the 3rd Marine Division stormed the beaches of Guam, in the Marianas, which the Japanese had captured in December, 1941.

A 15-year old Guamanian boy watched eagerly as Marine troops "hit the beach" for this was a day his family and other inhabitants of Guam had long awaited. Among the invading 3rd Marine Division Regiments was, of course, the 9th, one of the corps' most famous. Little did this 15-year old realize he would some day be named the commander of this elite unit.

On Saturday, March 4, 1972 a ceremony—with all its pomp and circumstance—was held at Camp Schwab at which Colonel Vincente T. Blaz, acknowledged receipt of command of the 9th Regiment and its 3,500 troops. Colonel Blaz is the highest ranking officer in the United States military of Guamanian descent. He succeeded Colonel R.T. Lawrence who would be assuming new duties as 3rd Marine Division assistant chief of staff, G-3.

Colonel Blaz graduated from the University of Notre Dame and received commission of 2nd lieutenant in the Marine Corps in 1951. From that day back in 1944, Colonel Blaz has now realized his cherished goals: association with the Marine Corps and the 9th Marine Regiment.

Marine English Teacher

Sergeant Robert J. Bodisch of 9th Marines Regimental S-3 office, Camp Schwab, after his regular daily chores are over, has a second job of teaching English to Okinawan children six nights a week in 30-minute sessions.

He noted, "These kids were afraid of Marines and it took me a long time to win their trust and friendship. They're really interested and eager to learn, even though the class is work and no play."

A native of Gulfport, Mississippi, Bodisch finds teaching children is also a learning experience for him as he writes Japanese on the blackboard from a pocket-sized dictionary. His students help him when he makes a mistake. His wife the former Yoko Uehara, a PX employee, also assists him with this language.

Bob's youngest student is seven and the oldest is 16. However, he plans to hold a class for older persons at a different time so they won't feel embarrassed. Bodisch has been on Okinawa since March 1971, and has extended his stay once. **Okinawa Marine, 1/22/72**

Boys Orphanage Aided by BLT 2/4

The Ologopo Boy's Town Orphanage, Subic Bay, Philippine Islands was given a donation of nearly $1,000 by the men of Basic Landing Team 2/4. The team, based on Okinawa, had been on training exercises in the Subic Bay area.

They returned to Okinawa the week of January 20 (1972). Father John Moran of the orphanage expressed appreciation for the gift and for the assistance the Marines had given the agency since it was built in 1967. He noted without the help of the Marines, it could not have survived.

Earlier, several of the children of the orphanage were treated to a Patton tank demonstration by members of the 2nd Platoon, Company A, 3rd Tank Battalion.

"Hot Lips" Cox Of The Amtrackers

When Private First Class John H. Cox Jr. reported to his 1st Amphibian Tractor Battalion at Camp Schwab, he made Sergeant Major M.P. Carcirieri a happy man ... for the unit had a bugle, but no one to play it.

Cox, as it developed, had 10 years experience with the trumpet; so, with slight adjustment, he became the battalion bugler. He had a little trouble because his lip muscles were out of practice, but gradually this was overcome.

He made calls for reveille, chow, liberty and taps from a centrally located area outside the battalion living quarters, and played the same tune twice into a four-foot long megaphone which was on a moveable pivot. That way, all the troops heard, whether they wanted to or not.

Cox enjoyed his extra assignment, but did take a lot of kidding plus mock screams and good-natured threats. On one occasion, some listeners even threatened to put him inside the megaphone.

3rd Recon Saves Two Japanese From Typhoon

About 20 members of the 3rd Reconnaissance Battalion saved the lives of two Japanese students who were swept away from shore August 15 (1973) by waves kicked up by Typhoon Iris. The students were wading when a large wave knocked them down and carried them out about 100 yards off shore.

The Recon Marines, on the United States controlled portion of the 15-mile long beach of Numazu, Japan, heard their calls for help and attempted to launch a rubber raft. At the same time, 2nd Lieutenant Robert M. Biddle Jr., and Sergeant Charles A. Abbey, donned life jackets and swam to their aid.

The two Marines helped the Japanese stay afloat for nearly 30 minutes until other Marines reached them in the rubber boat, which overturned five times before reaching them. The two Japanese boys, both college students, were uninjured and left the beach unassisted.

Lieutenant Biddle, Sergeant Abbey and the other Marines involved in the rescue are members of the 2nd Platoon, B Company, 3rd Recon Battalion, 3rd Marine Division. The platoon is presently attached to Battalion Landing Team 2/9 for training at Camp Fuji. **From Okinawa Marine, August 24, 1973**

Fuji Fire Victims Receive Medals

Two Marines who helped rescue some of their comrades during the October 18 Camp Fuji flash fire have been given the highest medal awarded during peacetime.

Corporal Patrick T. Schaefer, 22, and Lance Corporal Steven D. Haishuk, 21, who are listed in satisfactory condition at Brooke Army Medical Center, Ft. Sam Houston, Texas, were awarded the Navy-Marine Corps Medal.

The fire resulted in what has been termed the worst peacetime disaster ever suffered by Marines. and started when Typhoon Tip split open a gasoline storage tank and sent a river of fire through the Marine encampment. As a result 12 Marines are dead.

The latest to die was Lance Corporal Willie Davis Jr., 22, of Detroit, who died November 19. He was one of 37 Marines flown from Japan to the specialized burn unit. Of the 30 still hospitalized, 12 remain in critical condition and 11 are listed as satisfactory.

Meanwhile, back on Okinawa more than $2,450 in donations have been collected by 3rd Marine Division chaplains to be sent to the burn victims in Texas. **Okinawa Marine, November 1979**

Okinawa Reverts To Japan

A Yen For Money

On May 15, 1972, Okinawa reverted to Japanese government control following agreement between President Richard Nixon and Japanese Prime Minister Eisaku Sato.

The President said, "There is a natural interdependence between Japan and the United States. We are both nations of the Pacific, we are nations that have responsibility for peace in the Pacific."

In reverting to Japan, Okinawa and the other Ryuku Islands became Japan's 46th Prefecture. This included the land where the United States bases stood on Okinawa; and the United States was given the right to maintain installations and conduct various activities there as prescribed by treaty.

Up to that time the United States had permitted a large degree of self government of the island by the Government of the Ryuku Islands (GRI) so long as it did not interfere with the security needs of the United States. United States servicemen could be tried only by military courts-martial.

With the reversion, the laws of Japan applied everywhere on the island, including military bases.

A military legal specialist explained, "We on Okinawa have maintained a special set of laws and regulations which are comparable to martial law or military rule for the masses.

But, reversion to Japan is no different than transferring to Continental United States. While in the United States, we are bound to the laws and regulations of the state in which we serve. In Japan if you break a law you are subject to trial by Japan.

"Keep in mind that except for the exchange of currencies, life in Japan (Okinawa) will not be too different from life in the states."

The Japanese Money Transfer

The Japanese Ministry of Finance announced that in late April 1972, the largest trans-oceanic shipment of money in history would be transported from Tokyo to Naha, Okinawa to initiate the massive currency turn-over.

Two LSTs (landing ship, tank) carried the 60 tons of yen, the equivalent of approximately $150 million dollars, in denominations ranging from 1-yen aluminum coins to 10,000 yen banknotes.

At Naha the money was handled at the newly-built Bank of Japan office there, and distributed for exchange among 190 selected places—including 140 banks, 37 municipal offices, and 13 post offices.

Later, the rate of currency exchange was announced at 305 yen per one dollar.

Marines Delve Into Guam Past

A number of Camp Schwab based Marines of "India" County, 3rd Battalion, 9th Marines, had the opportunity to take a close look at Guam's history—the traditions and culture of the island—when they took a break from their busy training schedule earlier this month (August 1973) and participated in a USO sponsored tour of the island where "America's day begins."

During the tour Marines saw remnants of Guam's Chamorran history, her Spanish era and evidence of World War II—all found among the jungle and on the numerous beaches and open bays.

An especially interesting area visited was "Two Lovers' Leap" and the story behind it. Legend has it that two Guamanian lovers, attempting to escape separation and the marriage of the female to the captain of a Spanish galleon, took refuge atop the cliff. A search by the Spaniards ensued. When the couple's capture appeared inevitable, they tied their hair together and leaped some 400 feet to their deaths.

The "India" Company of Marines were intrigued by the evidence remaining on the island from World War II. It was in July 1944 that thousands of United Stated Marines, including "India" Company's parent unit, the 9th Marines, 3rd Marine Division, stormed the island to help liberate it from Japanese forces. All that remained for those on the USO tour were glimpses of old caves and bunkers, a few weathered and corroded guns, the rubble of gun emplacements, a tank graveyard and a number of historical markers. **From *Okinawa Marine*, August 31, 1973**

General Cushman To Visit Okinawa

The commandant of the Marine Corps, General Robert E. Cushman Jr., will visit Leatherneck units and facilities on Okinawa next week (July 1974). This will be General Cushman's third tour of the island since becoming commandant in January 1972. His last visit was in June 1973.

The 39-year Marine veteran is slated to arrive on Okinawa July 23 to begin a two-day tour of camp facilities and training operations.

General Cushman, who earned the Navy Cross, the nation's second highest combat award during World War II, when he commanded the 2nd Battalion, 9th Marines, later commanded the 3rd Marine Division on Okinawa in 1961-1962. *Okinawa Marine*, **July 19, 1974**

Meanwhile ... Back In The States

Return To Boot Camp

The 18th annual reunion of the 3rd Marine Division Association was held for five days in July 1972 at the picturesque Savannah Inn and Country Club.

However, the highlight of the entire affair was held at the Marine Corps Recruit Depot, Parris Island, South Carolina as the 33-man volunteer "retread" platoon proudly marched past the reviewing stand before an appreciative audience.

In preparation, the group had spent a rigorous day of boot retraining conducted by three drill instructors. Their shouted orders as well as insults, rang a familiar bell for these men: "Hurry up, you meatheads, you dummies, on the double, get moving!"

The retreads, clothed in utilities, helmets and even World War II leggings, soon showed they had not forgotten close and extended order drills. A few went all the way in this re-enactment by acquiring "skinhead" haircuts. Even practice with the M-16 rifle and 45-pistol was held.

The entire unit received expert marksmanship badges and later graduation diplomas from Major General Carl W. Hoffman, MCRD commander. The platoon received standing ovations at the parade and reunion banquet as well as praise from Marine Commandant General Robert E. Cushman Jr. (himself a veteran of the 9th Marine Regiment) who also voiced admiration for the division's battle feats on Bougainville, Guam, Iwo Jima and Vietnam.

9th Marines Add Two More To Colorful Battle Standard

Two completed chapters of combat history were bound together in September 1973 as the Presidential Unit Citation (Army) and Meritorious Unit Commendation were fastened to the already colorful battle standard of the 9th Marine Regiment at Camp Schwab, Okinawa.

Making the presentation of the streamers during the afternoon pageantry was Major General Fred Haynes, new commander of the 3rd Marine Division.

The citation accompanying the Presidential Unit Citation stated (in part): "The assigned and attached units of the 9th Marine Regiment distinguished themselves by extraordinary heroism, professionalism, and achievement in military action against the North Vietnamese Army in the Da Krong and Northern A Shau Valleys, Quang Tri Province, Republic of Vietnam, during the period of January 22 to March 18, 1969."

The second award recognized the regiment for (in part): "Meritorious achievement in action against the enemy in the Republic of Vietnam during Operations Dawson River/Aftor and Dawson River West from October 24, 1968 to January 19, 1969 ... in the extreme western and southern regions of Quang Tri Province."

The regiment had then earned a total of four unit citations during its 45 year history. **From *Okinawa Marine*, September 14, 1973**

Major General Ryan Commanding MAF, 3rd Division

During a colorful, mid-afternoon ceremony, Major General Michael P. Ryan took the reins of the III Marine Amphibious Force (III MAF) and 3rd Marine Division in a dual changeover at Camp Courtney, January 8, 1973. Major General Ryan succeeded Lieutenant General Louis Metzger as III MAF commander and Major General Joseph C. Fegan Jr., as 3rd Marine Division commander. General Metzger plans to retire while General Fegan will assume command of the Marine Corps Recruit Depot, San Diego, California.

General Robert E. Cushman Jr., commandant of the Marine Corps, offered his personal congratulations to both out-going commanders for a job well done. **From *Okinawa Marine*, January 12, 1973**

Who Else???
Merchant Ship And Crew Rescued By The Marines

Third Marine Division Marines played a key role last week (May 1975) in the rescue of the United States merchant ship *Mayaguez* and its crew after the ship had been seized by a Cambodian gunboat on the high seas.

While Marines from the 2nd Battalion, 9th Marines, were landing on Koh Tang Island, a re-enforced platoon from Delta Company, 1st Battalion, 4th Marines, was boarding the pirated United States Ship *Mayaguez*.

Led by Major Ray Porter and Captain Walt Wood, the platoon boarded the *Mayaguez* and within 10 minutes had completed the initial search for the crewmen. "We found some food on the bridge," said Captain Wood, "but there was no sign of the crewmen."

Major Porter and Captain Wood raised the American Flag and later in the day, the original Mayaguez crewmen returned to the ship and sailed to Singapore.

Lieutenant Colonel Randall W. Austin, commander of the ground forces which invaded Koh Tang Island, said the attack force consisted of 179 men, the size of one reinforced rifle company and from the 2nd Battalion, 9th Regiment. They were flown to the target area by eight Air Force helicopters.

The mission, according to Colonel Austin, "Was to seize and occupy the island and to locate and recover any American seamen that might be there."

Of the eight helicopters that made the initial trip, one was shot down and burned with no one from it making the shore. The second helicopter was hit and went down in almost the same landing zone as the first but all the Marines and crew members from that made it safely to the shore.

Another of the helicopters who had already discharged its Marines, had taken such heavy fire it crashed in the sea on the way back. Another loaded helicopter took such heavy hits it had to return to base without disembarking troops. However, these troops were able to board another helicopter and eventually returned to action later in the day.

Colonel Austin summarized the action on the island which ended at about 2100 that evening: "It certainly was heavy combat; it was a very classic heliborne assault I think, in which, unfortunately, we encountered heavy fire from the ground. Once on the ground we started to use some classic fire maneuver efforts to link up our people—very conventional fighting with conventional normal infantry weapons. It was indeed, very close combat. I have several reports from my officers and men on the ground that there were many instances where the enemy had, in fact, thrown hand grenades at our people and we picked them up and threw them back. This will give you an idea of how close the combat was."

He estimated 150 enemy troops in the area and deemed the mission a success, adding: "I'd like to believe that this in some way contributed to the release of the crewmen. It was certainly part of the overall plan and I think we did, in fact, accomplish the part that we set out to do. *Okinawa Marine*, **May 23, 1975**

Bring Christmas To Lepers

Representatives of the 9th Marine Regiment of Camp Schwab made a Christmas visit, carrying food and gifts, to the 600 patients of the Airaku-En (Garden of Joy for Love) Leprosarium, Nago, Okinawa.

The Leprosarium was founded in 1927 by an Episcopalian missionary and continued under his direction for 10 years until the Japanese Government made it a national hospital. World War II, with its death and destruction, demolished the city; but shortly after the surrender, the United States Army Medical Corps began rebuilding the hospital. In 1952 the Ryukyuan government took charge and has continued until the present.

The gifts included $500 in cash and money orders, two color television sets and a solid state stereo system. Following presentation of gifts, Marines distributed cookies, ice cream and Kool-Aid. The hospital senior doctor said: "These gifts, and your visit, have made these people very happy." Marines and the patients then sang a rendition of "Silent Night" in their respective languages. **From *Okinawa Marine*, December 22, 1972**

Tankers Celebrate

At Camp Hansen, Okinawa, the 3rd Tank Battalion celebrated its 33rd anniversary with meritorious promotion and a cake cutting ceremony.

During the ceremony, Major General Herbert L. Wilkerson, commanding general, 3rd Marine Division, handed the first piece of cake to the oldest and youngest members of the battalion, Master Gunnery Sergeant James C. Smith, 44, Headquarters and Service Company, and Private First Class George J. Garnett, 18, Alpha Company.

Then he meritoriously promoted seven Marines from the battalion.

The 3rd Battalion was activated at Camp Elliot, San Diego, California on September 16, 1942. It was moved to Camp Pendleton, and later to New Zealand for jungle and amphibious training. It was deployed to historic Bougainville, Guam and Iwo Jima campaigns in World War II, and for their actions at Iwo Jima, the 3rd Tank Battalion was awarded the Presidential Unit Citation Streamer with one Bronze Star for heroic dedication to duty, prior to returning to Camp Pendleton in December 1945 and deactivation in January 1946.

On March 5, 1952, the battalion was reactivated at Camp Pendleton, and deployed to Camp Fuji where a year later it supported the United Nations Forces in Korea. The Korean Service Streamer was awarded to the battalion for their efforts during that conflict.

Redeployed to Camp McGill, Japan in 1956, the battalion moved to Camp Hansen in 1957 where it was detached from the 3rd Marine Division and assigned to Force Troops, Pacific. In December 1963, the battalion was reassigned to the 3rd Marine Division.

The battalion left Okinawa for Vietnam in March 1965, where between July 1965 and October 1969, it was deployed, as needed, to many operations.

For their outstanding actions in that war, the unit received two Meritorious Unit Commendations, the National Defense Service Streamer with one Bronze Star, Vietnam Service Streamer with two Silver Stars and the Vietnam Cross of Gallantry with Palm.

In October 1969, the 3rd Tank Battalion, 3rd Marine Division returned to Camp Hansen. *Okinawa Marine*, **October 3, 1975.**

Colonel Gray Assumes Command 4th Marines

Colonel Alfred M. Gray Jr. will become the commander of the 4th Marine Regiment, 3rd Marine Division at Camp Hansen, Okinawa, on July 29, 1974.

Colonel Gary Wilder, who commanded the unit since August 1973, will report to the Marine Corps Recruit Depot, Parris Island, South Carolina.

Colonel Gray enlisted in the Marine Corps in 1950 and was commissioned in 1952. He saw combat duty with the 7th and 11th Marine Regiments in Korea. In addition, he served three tours in the Republic of Vietnam. Among his personal decorations are the Silver Star Medal, Bronze Star Medal with Combat V, and Legion of Merit with Combat V. (He later became Marine Corps Commandant.) *Okinawa Marine*, **August 9, 1974**

Major General K.J. Houghton Commands 3rd Marine Division

Major General Kenneth J. Houghton, winner of the Navy Cross, two Silver Stars, two Bronze Stars and three Legions of Merit, assumed command of the 3rd Marine Division at Camp Courtney, Okinawa, on August 23, 1974.

Major General Fred Haynes who commanded the division since August 1973 will report to Camp Lejeune, North Carolina, as the commanding general, Marine Corps Base.

The new commander comes to the division from Camp Pendleton, California, where he served as commanding general, 1st Marine Division, since May 1973.

Major General Houghton enlisted in the Corps in April 1942 and was commissioned the following September. The general saw combat action at Tarawa and Saipan during WWII and saw further combat in the Korean conflict and Vietnam. *Okinawa Marine*, **August 23, 1974**

Assist in Refugee Evacuation

Marines of the 3rd Marine Amphibious Force were involved in an evacuation in the Republic of Vietnam during April 1975. Four companies from the 4th Marines were acting as security and control forces aboard both United States Naval ships and commercial ships evacuating refugees down the coast of Vietnam.

Working hand-in-hand with Navymen, Marines from 1/4 helped load the first group of refugees aboard the amphibious cargo ship USS *Durham* on April 3, 1975. Eyewitness accounts told of many acts of heroism by Navymen and Marines.

During subsequent days, some 35,000 refugees were off-loaded from five American ships which had carried them to Phu Quoc Island off the coast of Vietnam.

The Marines and Navymen worked around the clock to alleviate the suffering; 53 tons of food, water and medical supplies were delivered to the merchant ships.

Rear Admiral Donald B. Whitmire, commander of the task group, said Marines and Navymen undoubtedly saved many lives. He added, "I can't say enough about the tireless and selfless efforts that our Navymen and Marines are expending to relieve the suffering of the people who have been forced to flee their homes." *Okinawa Marine*, **April 25, 1975.**

Rescue Heroes ...
24 Receive Medal For Mayaguez

Twenty-three Marines and a Navy Corpsman were awarded Navy Achievement Medals at Camp Schwab on March 19 for their participation in the rescue of the SS *Mayaguez* last year.

Major General Herbert I. Wilkerson, commanding general, 3rd Marine Amphibious Force/3rd Marine Division, presented the awards to the members of 2nd Battalion, Ninth Marines (February 9) who stormed ashore on Koh Tang Island during the Mayaguez rescue operation.

The citations commended the men for their individual actions during the actual assault and for their rescue efforts in aiding wounded comrades. Lieutenant Colonel Roger E. Simmons, 2/9 commanding officer, assisted in presenting the awards. *Okinawa Marine*, **March 26, 1976**

Oki Marines Evacuate Americans In Final Effort

Early Wednesday morning a CH-46 helicopter landed on the roof of the United States Embassy in Saigon and plucked off the last Marines of a security force involved in evacuating Americans from the besieged city. With that act, the United States Marine Corps ended almost 15 years involvement in the Republic of Vietnam.

Okinawa-based Marines operating from ships of the United States Seventh Fleet conducted the final evacuation of some 6,000 Americans and South Vietnamese which began Tuesday afternoon and ended Wednesday morning.

Secretary of Defense James R. Schlesinger announced in Washington that the final withdrawal of Americans from Vietnam was ordered by President Ford on advice of the United States Ambassador. *Okinawa Marine*, **May 2, 1975**

Copter Training ...
"Plain Old-Fashioned Fun"

Marines attending the 3rd Marine Division Noncommissioned Officers Leadership School at Camp Hansen look forward toward one particular phase of their training as "plain old-fashioned fun."

The Marines, representing units from Guam, receive orientation on helicopter operations, safety features of the CH-46 Sea Knight and its other roles in support of ground troops.

The primary importance of this training is to familiarize troops with the helicopter and its capabilities, "But," according to NCO School Tactics Instructor Staff Sergeant Bill Stott, "the young troopers enjoy this phase of training most and look at it not as training, but rather as plain old fashioned fun."

During the helo orientation phase, the Marines receive a detailed briefing on safety features and equipment of the aircraft. They learn how to enter and leave the aircraft quickly and safely. Broken down into squads, one squad at a time enters the chopper for briefing and, upon leaving the helicopter, simulates an actual set down in a landing zone and setting up a perimeter.

The "fun" comes at the end of the training mission as each squad is airlifted around a small area for the practical application of flying in a helicopter. *Okinawa Marine*, **February 13, 1976**

Toys, Toys, Toys,

More than 850 Christmas toys were delivered to the Ishimine Children's Home in Naha December 22 by Marines from Supply Company, Supply Battalion, 3rd FSR. The Marines, who collected the toys in an island wide drive beginning last month, delivered a Christmas tree to the children's home along with the gifts from Santa Claus.

The "Toys For Children" drive was one of several projects sponsored by Marines for Okinawa children during the Christmas season. *Okinawa Marine*, **December 31, 1975.**

A Warm Welcome

The "Fighting Ninth" was warmly welcomed during its move into Camp Schwab as the first of the division's infrantry units to return to their home base. Lieutenant Colonel Calhoun J. Killeen, commanding officer of the 2nd Battalion, 12th Marines, the 9th Marines direct support artillery unit in Vietnam, noted the same hospitality as he moved his artillerymen onto the Okinawa base. Major General Killeen would return to Okinawa as the commanding general of the 3rd Marine Division from July 1978 to July 11, 1979.

In noting the arrival of the Marines, the press on August 24, 1969 said, "Allied Forces throughout Vietnam and all over the world can be justly proud of the accomplishments of the Ninth RLT." It was a fitting and final tribute to the thousands of enlisted men and officers who have also been properly extended to all Marines and Naval personnel who served in each unit of the 3rd Marine Division in the war in Vietnam.

This story was provided by Colonel Edward F. Danowitz, who served as commanding officer of the 9th Marines from April 8 to September 8, 1969.

"Bee" Company Is Home With 9th Engineers

"Bee" Company's flower followers recently found a comfortable home for their queen, thanks to Master Sergeant Merlin A. Miller, construction foreman for "Alpha" Company, 9th Engineer Support Battalion.

"Bees started swarming around our supplies about three weeks ago," Miller explained. "They presented a work hazard to the men, but everytime the bees were chased away, they returned and swarmed around an embarkation box.

"I thought how good it would be to have natural honey," Miller said. "So I decided to keep the bees around."

Private First Class Kenneth E. Sutherland and Lance Corporal Halbert L. Cooper gingerly corralled the swarm, dumped the bees in the box, nailed the box shut, and carried it to a desolate corner of their working area.

Miller is looking forward to collecting fresh natural honey. "I haven't done this since I was a kid in Ponchatoula, Louisiana," he said. "I was surprised to find honey bees on Okinawa." Miller said the provost Marshall hasn't found any restrictions on keeping bees and many people have offered to take care of his sweet new venture. However, the soon to come natural honey persuades him to hold on the beehive.

Miller said honey can usually be collected twice a year and he expects his 1,000 plus bees to yield their first harvest in about a month, with more to come next fall. *Okinawa Marine,* **May 1976**

Charlie Company
"Gone To The Dogs"

Charlie Company, 4th Marines, has "gone to the dogs" since the addition of four new troopers. The new "troopers" are puppies found a couple of weeks ago in the shrubs behind the company barracks and taken in by 1st Sergeant George Olivar and his men.

"Many times we heard the pups making noises in the bushes," 1st Sergeant Olivar explained. "We started feeding Charlie (the largest one) and after that she just kept hanging around."

Later, Privates First Class James H. Taylor and Donald J. Sampson assisted Olivar in recruiting a few more pups from their foliated home. The new recruits were named Privates Blacky, Duke, and Bear. Charlie had three more days time-in grade, so she made lance corporal.

The men are very enthusiastic over their company's new mascots, but nobody matches 1st Sergeant Olivar's enjoyment with them. "The first sergeant kind of adopted them," said Taylor.

The puppies are being brought up in true Marine Corps tradition. Olivar said, "Our new fire team just won't follow orders, but Charlie should straighten the rest out. She's their duty noncommissioned officer."

The puppies haven't lost their family ties since their relocation. "Their mother comes out of the bushes and visits my room," 1st Sergeant Olivar said, "because she knows I take care of the pups." *Okinawa Marine,* **March 1976**

Okinawa Marines Clean up Guam After Typhoon Pamela

Combat ready units of the 31st Marine Amphibious Unit (MAU) from Okinawa spent two weeks giving disaster recovery assistance to the military and civilian communities on Guam early this month.

The units from the 4th and 9th Marine Regiments, 3rd Marine Division, steamed from Subic Bay in the Philippines to Guam as soon as the request for assistance was received to help clean up the damage and debris left in the wake of Typhoon Pamela.

"As soon as we arrived, we set up a recovery operations command post aboard the USS Tripoli, explained Colonel Simon J. Kittler, commanding officer, 31st MAU. "We determined the areas needed to be cleaned up immediately, giving them the highest priority. The next morning working parties from both battalion landing teams (BLT) were concentrating on alleviating the problems in those areas."

By the time the Marines were ready to leave, they had cleared debris and removed damaged buildings at Anderson Air Force Base, Naval Regional Medical Center, antenna communications sites, Bishop of Guam's House and Chancellary, water treatment and sewage plants, the recreational parks and beaches of Guam, Guam Memorial Hospital, schools, the downtown area and some residential sections.

Corporal Alvin Gilbert, amphibious tractors, BLT 1/9, summed it up well:

"Most people have the impression the Marines are just combat-ready to fight in armed conflicts. That's just not true. We're ready for any emergency, whether humanitarian or military. I think the people of Guam understand that the Marines are ready, willing and able to serve our country whenever and wherever we are needed—no matter what capacity we serve in." *Okinawa Marine,* **June 1976**

9th Marines Meet Filipino Counterparts

The 1st Battalion, 9th Marines Battalion Landing Team recently got a first-hand look at their Filipino counterparts in a joint amphibious assault at the Zambales Training area near Subic Bay.

Philippine Marine Commander, Lieutenant Commander Grederico C. Banaban, was "impressed' with the United States Marines' cooperation with his 1st Marine Battalion Landing Team. "There are no major differences between the Philippine and United States Marines," Lieutenant Commander Banaban said. "We follow United States Marine doctrines in training and even send some of our Marines to schools in the states."

Philippine Marine officers have attended Officer's Candidate School at Quantico, Virginia, and other officers and enlisted personnel have gone to the Landing Force Planning Course and Electronics Communication School at Coronado, California. "We have only 6,000 active Philippine Marines," Lieutenant Commander Banaban stated, "thus we lack the strengths in size that the United States Marines enjoy."

Otherwise, their system almost copies the United States Marine Corps operation. Recruiting standards are much more stringent. "Approximately 20 to 30 out of every 1,000 applicants are accepted," Banaban said. Philippine Marine recruits must be high school graduates, a criteria the United States Marine Corps is now augmenting.

Philippine Marines also have quarterly physical fitness tests, comprised of pull ups, push ups, squat jumps, sit ups, and a three mile run.

They replaced the United States Marines' motto of "Semper Fidelis" (Always faithful) with Karangalan, Kabayanihan, Katungkulan" meaning "honor, integrity, duty."

"We look forward to training together with United States Marines, since we do try to follow their doctrines." Banaban said. "We try to hold joint operations annually to exchange new techniques." *Okinawa Marine*, March 1976

Marines Share Christmas Spirit

When a battalion from the 12th Marines goes to Camp Saint Barbara, Korea, located 11 miles south of the DMZ, for training exercises, they always seem to find time to visit the Po-Wha Orphanage.

The orphanage, located 25 miles north of the camp, is visited on an average day by 15 to 40 Marines. The home, has about 110 children ranging in ages form two to 18 years old. "The children don't speak any English at all," said Navy Chaplain Lieutenant (junior grade) James W. Fahey, assigned to the 3/12. Communications are made entirely through a lot of love and a whole bunch of hugs and kisses.

"Once a child spends some time with any one particular Marine," continued the chaplain, "that Marine becomes at-tached to the orphan. Everytime the Marine visits after that, the child will automatically come to him and him only."

In the past year, 12th Marines donated one-half ton of childrens clothing and blankets collected from surrounding chapels on MCB, Camp Butler, $2,719 in cash and a rice storage house built by the Marines in August. Desks, chairs, tables, and sewing machines were purchased with the donated funds.

At the present time the orphanage is having a new heating system installed with the money that was donated to them. "The children loved it immensely and the troops even more so!" said Chaplain Fahey, adding he hopes to continue visiting the orphanage many more times in the future. *Okinawa Marine*, December 13, 1977

New Sergeant Major Began Career With 3rd Marine Division

There is a Marine in the 3rd Marine Division who claims he has just about done and seen everything the Marine Corps has to offer.

During his more than 30 years of active duty, he has fought in three wars, been both a drill instructor and recruiter, served at a Marine Barracks in Europe and has been stationed in Thailand and Puerto Rico, plus almost every major Marine installation in the States.

This Marine is Sergeant Major Peter J. Marovich, 45th sergeant major of the division, who began his Marine Corps career with the same division in which he now holds the top enlisted rank. "I started with this division in World War II," said Marovich. "I was with it when it formed, went overseas with it, and made the landings at Bougainville, Guam, and Iwo Jima.

As for the division today, Marovich likes what he has seen since he assumed his duties on January 28. "I've been to all the division units here on the island and I'm very much impressed with the appearance, courtesy and physical fitness of the Marines, explained the 53-year young Marovich, who looks like he is in better condition than most Marines half his age.

Holder of a Bronze Star with Combat V, Marovich is proud of what he considers a few personal achievements throughout his career, adding: "I'll retire after this tour and I think the highlight of my career is to become the division sergeant major of the same division I fought with as a snuffy," *Okinawa Marine*, May 20, 1977

Marines launch their boats, preparing for a midnight assault on an enemy beachhead. (Official USMC photo by LCpl. David A. Hiersekorn)

Maflex 1978 ...
Largest Exercise Since Vietnam

In the largest post-Vietnam training exercise conducted in the Pacific, more than 6,000 Marines demonstrated their combat readiness in the Republic of the Philippines during Marine Amphibious Force Landing Exercise (MAFLEX) 1978.

Marines from Okinawa and Iwakuni, Japan and Hawaii formed the 3rd Marine Amphibious Force (MAF) which combined with more than 11,000 sailors aboard 31 naval vessels of the United States Seventh Fleet to make up the amphibious task force.

Civil Action Too

Code named Bayanihan IV/Fortress Lightning, the exercise was designed to provide training in amphibious landing techniques and operations ashore. In addition to the MAFLEX, a civic action project was accomplished. The project was a join operation of the United States Marine Corps and Navy and the Philippine Armed Forces in the province of Occidental Mindoro.

Designed to help promote harmonious relations between the United States Armed Forces and the Republic of the Philippines, the civic action project benefited the people of Mindoro by enhancing their living conditions, providing health-care items, and supplying educational facilities and material.

Under the command of Major General Adolph G. Schwenk, 3rd Marine Aircraft Wing, 3rd Force Service Support Group (FSSG) and 1st Marine Brigade.

The exercise began October 9 with the task force sailing from Okinawa to the South China Sea coast of the dense jungle island of Mindoro. On D-Day, October 16, a massive pre-dawn amphibious and vertical assault was launched.

Two of the three battalions involved, 2nd Battalion, 3rd Marines, 1st Marine Brigade and 3rd Battalion, 4th Marines, 3rd Marine Division, was the aggressor.

For many of these infantrymen MAFLEX 1978 was the first opportunity to apply what they had learned during small unit tactics. Hawaii Marines, 2nd Battalion, 3rd Marines, came ashore in amphibious tractors (AAVP) and Navy landing craft. They forced their opposition from the beaches, reviewed their plan of attack and moved inland.

Their companions in the assault, 3rd Battalion, 4th Marines, Okinawa, were flown 15 miles inland by CH-46 Sea Knight and CH-53 Sea Stallion helicopters. The choppers were launched in waves from the flight deck of the USS *New Orleans*, amphibious platform helicopter ship.

Aggressors Ready

The aggressors, 3rd Battalion, 9th Marines, based on Okinawa, were caught between the two forces. But they were ready. Having been on Mindoro for more than two weeks, they knew the terrain, had picked their defensive positions and were accustomed to the 90 degree heat. It was the sweltering heat and high humidity that proved to be the worst enemy for the invading force during the first day. From the beach, 2nd Battalion, 3rd Marines marched 12 miles inland and assaulted a heavily defended hill forcing the aggressors to retreat.

By the final day of the five-day field problem, the opposing infantrymen had been pushed back to their last position, trapped between the two battalions. One more assault sandwiched in the aggressors bringing the exercise to a close.

Though the combat troops had finished their work, others were not so fortunate. In came the helicopters. Lifting troops from hilltop landing zones, they flew them to positions in the rear to prepare for debarking. The reloading aboard the ships was completed October 27 and the vessels sailed for Okinawa.
Okinawa Marine, **November 4, 1977**

Team Spirit '78 ...
Marines Test Readiness

Aboard the USS *Tripoli* synthetic daylight filled the troop berthing areas. Marines from the 31st MAU began to stir and roll out of their racks and wipe the sleep from their eyes. D-Day had arrived and the leathernecks were about to begin one of the largest, most publicized training exercises in recent history.

Similar scenes were taking place aboard the USS *Fort Fisher* (LSD-40) and the USS *Mount Vernon* (LSD-39) as other units of the 31st MAU began to awaken and prepare for the coming assault—on the southeastern coast of the Republic of Korea.

As implied by its name, the exercise was designed to test and sharpen the skills and team spirit of United States forces (Army, Air Force, Navy and Marine Corps) combined with South Korean forces.

Just before daybreak, tactical jets from the USS *Midway* broke the stillness of the bay at Toksok-ri, dropping parachute flares to illuminate the beach. Moments later, the first wave of leathernecks assaulted the beach.

Once ashore the attacking Marines met sporadic resistance. A fireteam from Company G, 2nd Battalion, 9th Marines, fired upon them, slowing their progress and creating some confusion within the assaulting force. In a treeline 100 yards from the shoreline, the forward assault elements met the main aggressor forces and the fighting intensified. The Marines, however, had gained a foothold and the aggressors were forced to retreat.

As the sun began lighting the morning sky, the Marines pushed the battle further inland. Overhead the first of the helicopter borne assault Marines were on their way to areas still further inland. The secured beach began to be formed into a logistical support center from which the Marines ashore would receive everything necessary for the assault to continue.

During the ensuing six-day exercise, the aggressors fought a delaying action giving up as little ground as possible to the superior forces. Defending the coastal area east of Highway 7, which runs parallel to the coast, one to three miles inland from the East China Sea, was Company G, 2/9. West of the highway, Company L, 3rd Battalion, 4th Marines, resisted against 31st MAU helo assault units.

The 31st MAU was encompassed into the 9th MAB as the exercise progressed. Other units making up 9th MAB included MAG-30, LSG-9 and RLT-9.

The brigade continued to push south across the mountainous Korean terrain toward the city of Pohang, located 165 air miles southeast of Seoul.

Concurrently, elements of the Republic of Korea (ROK) 1st Marine Division made an amphibious assault and then advanced through enemy lines north along Highway 7.

Aggressor units conducted night attacks against the far superior landing forces to "avoid the losses that would result in a daytime attack against similar odds."

Two weeks of pre-exercise training helped the aggressors become familiar with the terrain and adapt to the cold Korean environ. These factors proved invaluable assets to the aggressors as they counterattacked the assaulting forces, often making their way up steep and stony mountainsides in near darkness and freezing temperatures.

Though the high-spirited enemy forces had advantages, the assault troops were not without some of their own. Fire superiority, reliable in the form of close air support, artillery, and tanks, proved a key factor in the eventual victory over the aggressors.

With its amphibious and helicopter assault capabilities, the Marine Corps has the flexibility to project ground forces at points where there is little or no enemy resistance. Though it would be ideal to land against no enemy fire, United States Marines have, throughout its 202 year history, proven that they can successfully attack a defended beach, push inland and take their objectives.

The Marines of the 9th MAB, in order to be well-trained and ready for any contingency, chose to assault an enemy that was prepared and waiting. In the end, the leathernecks of 9th MAB showed once again they remain "A Force in Readiness." *Okinawa Marine,* **April 7, 1978**

Is your weapon ready, Marine?

Marine Sniper: A Coveted Title

The Marine Corps Sniper Manual defines a sniper as "A Marine that has been carefully screened, selected and has undergone comprehensive training in advanced infantry and markmanship techniques. The sniper's training, combined with the inherent accuracy of his rifle, firmly establishes him as a valuable addition to the weapons available to the infantry commander."

More than 30 Marines are undergoing the rigorous training schedule demanded of the men that wish to become snipers. They are members of the first class enrolled in Sniper School at Camp Hansen. The training involves classes in camouflage and concealment, infantry tactics, marksmanship techniques, along with the care and cleaning of weapons.

Learning everything to become a sniper in itself is a difficult task. Making it even more difficult, is the fact that the men are learning 30 days worth of material in just 14 days. It is not uncommon for the students to put in more than 60 hours per week to obtain the coveted title of a Marine Sniper.

Surprisingly, morale at the school is extremely high. However, morale isn't the only thing that is high in the school. Because they are an elite group, the standards expected of them are quite high also. According to the Marine Corps Snipers Manual, students should be expert marksmen, right handed shooters, non-smokers and not wear glasses. In addition the students should not be emotional. *Okinawa Marine*, **June 2, 1978**

Hospital Corps Marks 80 Years

It takes guts to go into battle. Even professional soldiers, sailors and Marines will feel the cold horror of death's proximity. A man must be brave to leap out of a foxhole with bullets whizzing overhead or to clamber out of a landing craft onto a beach already strewn with bodies and blood. Yet this has been done bravely by fighting men throughout history, because they know they must, and because they have weapons with which to defend themselves and attack the enemy.

The fighting man's mind is usually so occupied with the thoughts of defense and offense and his hands so busy with the weapons that his brain seldom has the opportunity to worry about anything other than the immediate objective.

The hospital corpsmen of the Navy share the anxieties of the fighting man, yet they are not defenders, but savers of lives. They have no screaming attack and battle on which to release their tensions but only the knowledge that their healing hands and soothing words of encouragement have saved the lives of many men.

It was during the Spanish-American War that the Navy's Hospital Corps was finally organized as a unit under the provisions of an Act of Congress, approved June 17, 1898. Two years after the founding of the Hospital Corps, one of their own, was awarded the nation's highest award: the Medal of Honor. It was during the Boxer Rebellion in China that Hospital Apprentice Robert Henry Stanley "distinguished himself with meritorious conduct" in action with relief units of Allied Forces in China.

In August 1902, the Hospital Corp's first school was established at Portsmouth Naval Hospital in Virginia. Hospital Corpsmen have served with the Marine Corps since their inception eighty years ago, always present and sharing in their triumphs and tragedies.

Even when "Uncommon Valor was a common virtue" on the island of Iwo Jima; corpsmen were there. In Associated Press photographer Joe Rosenthal's picture immortalizing the flag raising on Mount Suribachi a corpsman was there. Pharmacist Mate Second Class, John H. Bradley helped raise the flag during one of the bloodiest campaigns in Marine Corps history.

June 17 marked the 80th anniversary of the Navy's Hospital Corps and with this anniversary comes the realization that their jobs get tougher, more dangerous, and more demanding. More technical skill is required now, more than ever before in history. *Okinawa Marine*, **article by Corporal Wayne Hollis, USMC, Camp S.D. Butler, Okinawa, June 1978**

Solomons Get Independence

On July 6, 1978, the Solomon Islands became the world's newest nation ending 85 years of British colonial rule. During festivities marking the birth of the new nation, a plaque was presented by Major General J.K. Davis, Commanding General, 1st MAW, to Solomon Islands Prime Minister, Peter Kenilorea, during a reception on the USS *Holt* at Guadalcanal.

Commenting on the plaque which read, "From the officers and men of the United States Marine Corps. To the people of the Solomon Islands on occasion of their independence," Major General Davis said, "The plaque was presented as a token of friendship from the Marine Corps to the islanders. As you all know these islands witnessed some of the bloodiest fighting of World War II."

The plaque, weighing approximately 115 pounds was designed by Staff Sergeant S.E. Ellison, FMFPac, Audio Visual Department, Camp Smith, Hawaii. *Okinawa Marine*, **July 28, 1978**

Major General Killeen New 3rd MAF/ 3rd Marine Division

Major General Calhoun J. Killeen assumed command of the 3rd MAF and the 3rd Marine Division during a change of command ceremony at Camp Courtney July 10, 1978. The 3rd MAF/3rd Marine Division was formerly commanded by Major General Adolph G. Schwenk, who was reassigned as deputy general, FMFPac in Hawaii.

Major General Killeen graduated from the United States Naval Academy in June 1949, earning a BS degree in engineering, and was commissioned a Marine second lieutenant. Prior to his present assignment, he served as inspector general of the Marine Corps, Headquarters Marine Corps, Washington, DC. *Okinawa Marine*, **July, 1978**

How It Was in 1978
3rd Marine Division 36 Years Readiness

The Western Pacific force in readiness, the 3rd Marine Division will celebrate its 36th birthday tomorrow September 16, 1978. The division has been serving in this part of the world since 1942.

Trained and organized for combat during World War II, under the command of Major General Charles Barrett, the division received its baptism of fire on November 1, 1943. On that date, the division landed on Bougainville, the last Japanese stronghold in the Solomon Islands.

The newly formed division also saw action on Guam and Iwo Jima.

Due to the location of Iwo Jima, which is only 670 miles from Japan, the battle was of great importance to both sides. This was the division's last major operation of the war, and battle weary Marines helped take control of the island after 26 days of fierce fighting.

Shortly after the battle on Iwo Jima, World War II came to an end and the division was officially deactivated in December 1945. Seven years later, in January 1952, the division was reactivated due to escalating events in Korea. The division arrived in Japan in 1952 to support the 1st Marine Division in the defense of the Far East.

The 3rd Marine Division first became involved with Okinawa in June 1955 when elements of the division began moving from Japan. Within two years, the move to Okinawa was completed by the division.

Adding another chapter to its growing history, the division made an amphibious landing at Da Nang, Vietnam in March 1965. They later expanded their operations from Hue to the Demilitarized Zone (DMZ).

Vietnam gave the division an opportunity to gain valuable combat experience during more than four and a half years there. In that time, its Marines conducted more than 120 operations in fulfilling the division's mission.

In September 1969, the division began stand-down operations. On November 7, 1969, the division's colors were returned to Okinawa, signifying its return to home base. The division headquarters was set up at Camp Courtney where it presently resides.

Since the Vietnam Conflict, the division has shown its capabilities by reacting on a moment's notice to participate in Operations Eagle Pull and Frequent Wind, evacuating refugees from Cambodia and Vietnam in April 1975. A month later, the division's Marines were involved in the recapture of the merchant ship *Mayaguez*.

The division is presently under the command of Major General C.J. Killeen. Even though it is a peacetime environment, the division stands ready to act as a force in readiness anytime and anyplace the President and the Congress so directs. *Okinawa Marine*, **September 1978**

Pushing "Bullies" Around; All In A Day's Work

There's a lance corporal over at Headquarters Battery, 12th Marines who's pushing some bullies around. Those bullies are bulldozers, fork lifts and other heavy equipment and that lance corporal is Cynthia Boucher.

Coming into the Corps on an open contract, the Westbrook, Maine, native found herself facing the Heavy Equipment Operator's School after boot camp. She didn't have too much confidence in being able to handle her new assignment but was assured the nine week course at Fort Leonard Wood Army Post, Missouri, would train her in proper handling of all the heavy equipment the Marine Corps uses. "And they did," according to Boucher.

"Each week we learned about a different type tractor or machine, spending time on dozers. Now, since coming here, I find I am using all that information and am actually driving and handling every piece of heavy equipment 'only the male Marines used to use."

"I like working in this field," she said. "The male Marines seem to be motivated by my being here. They see a woman doing the same job as they are doing and they immediately work harder."

"Operating this heavy equipment has little to do with actual strength as most of it is know how and common sense," she said.

"I feel well-trained and capable to drive any heavy equipment 'bully' that the male Marine can." And with that, she hopped up on a huge 8230 full sized dozer and guided it around for a demonstration. *Okinawa Marine*, **October 1978**

Commandant Visits Unit Where He Was Medal Of Honor Winner

General Louis H. Wilson will wind up his three day sayonara tour of Okinawa tomorrow morning.

During his visit he attended briefs and toured the 3rd Marine Amphibious Force, 3rd Marine Division, 1st Marine Aircraft Wing, 3rd Force Service Support Group and Marine Corps Base, Camp Butler. He also held meetings with officers and staff noncommissioned officers before leaving Okinawa for Guam.

Wilson, a Medal of Honor recipient of World War II, assumed office as Commandant of the Marine Corps during the Marines' bicentennial year,

July 1, 1975. Upon becoming commandant, he received his fourth star.

During his 38 years of Marine Corps service, he saw action in three wars and rose to become the 26th commandant. He served with the 3rd Marine Division in World War II and the 1st Marine Division in Korea and Vietnam.

General Wilson received the nation's highest medal, the Medal of Honor, while serving as commanding officer of F Company, 2nd Battalion, 9th Marines, 3rd Marine Division. He received the medal for risking his own life while repelling enemy forces at Fonte Hill, Guam in 1944. *Okinawa Marine*, **March 23, 1979.**

Marine Corps Restructuring Infantry Battalions

In a move to improve its combat readiness, the Marine Corps is restructuring infantry units.

According to Headquarters, Marine Corps officials, the restructuring will provide increased firepower, centralized fire support controls, enhance the sustainability on the battle field and improve the uniform readiness throughout the four Marine divisions.

The major restructuring of the infantry battalion will be elimination of one rifle company and replacing it with a weapons company. In the new improved battalion, there will be three rifle companies, a weapons company and a smaller headquarters and service company.

The rifle company will have increased firepower with two additional M-60 machine guns and one 60mm mortar. This will give the new rifle companies eight machine guns and four mortars. Weapons company will consist of the current 81mm mortar and dragon anti-tank platoons, plus an assault section which will contain three multi-shot flame thrower weapons. There will also be a battalion fire support coordinator with the weapons company.

The headquarters and service company will be smaller because the Dragon and 81mm mortar platoons will be transferred to the weapons company.

The new infantry battalion, according to Headquarters Marine Corps, will be implemented throughout the four divisions by August 1, 1979. *Okinawa Marine*, April 20, 1979

Typhoon "Judy" Delays Exercise "Fortress Gale"

A combined armada of Marines and sailors demonstrated America's sea power in the Pacific during a two-week military exercise dubbed "Fortress Gale." Involving 25 ships, 280 combat aircraft and 40,000 men, "Fortress Gale" was powerful enough to only be delayed 24 hours by Typhoon Judy and large enough to not only encompass Okinawa itself, but the seas around the island.

While the highlight of the exercise took place last Saturday and Sunday, when 7,000 Marines of the 3rd Marine Division stormed ashore and spilled over the island, the exercise also gave the Navy a prime opportunity to test its amassed fleet capacity in a multi-threat environment.

Fortress Gale was conducted to provide forces with training in amphibious landing techniques and operations ashore.

According to landing force commander, Major General Calhoun J. Killeen, "This exercise is complex to plan and has so many sides to the training benefits. The main importance is that we have the shipping available, and that is vital to an exercise of this scale."

Clearly, however, the most visible and focal point of the exercise was the landing of Marines on various parts of Okinawa. The landing force was comprised of Okinawa based Marines and units from Hawaii and California. *Okinawa Marine*, August 31, 1979.

CH 53 Super Stallion

We Remember ...
Marine Corps Code Talkers And Other American Indians

These Marine code talkers practice archery with bows and arrows, not using them as basic weapons. Bougainville Campaign was the first time Navajos worked with 3rd Marine Division.

Editor's Note: Today is American Indian Day. The following article recognizes American Indians achievements in the Marine Corps. (Article from Okinawa Marine, September 28, 1984).

Throughout the fighting in the Solomon Islands, Guam, Peleliu and Iwo Jima, during World War II, the Japanese were totally confounded by a strange language picked up on their tactical radios.

It was Navajo, a language that has been described as American double-talk mixed with the sound of a hot water bottle being emptied. It was the language of the Marine Corps' famous code talkers.

Most codes are based on the user's native language. If the language is widely known, the code can be broken—regardless of how clever it is.

However, Navajo is considered one of the world's "hidden languages." That is, its original alphabet and symbols have been lost to history. When World War II broke out, there were only about 55,000 Navajos in the United States—mainly in Arizona, New Mexico, and Utah.

They tended to stay to themselves, which kept their language from spreading. Estimates are that only 28 non-Navajos, mainly white American scientists and missionaries, could speak this unique tongue where four inflections of the same word would give the word four different meanings.

The Navajos were the only Indian tribe in the United States that had not been intensively studied by art students and anthropologists from Germany—Japan's ally—as other native Americans had been during the previous two decades prior to the outbreak of the war.

When the first 30 Navajo recruits finished training at FMF Training Center, Camp Elliott, California, they were assigned to the 1st and 2nd Marine Divisions. Others went to the 3rd Marine Division and to raider battalions.

These Navajo "code talkers," as they ultimately became known, increased to 420 before the end of the war. And the original contingent made its own unique contribution to military history when it took the basic Navajo language and modified it into an effective wartime code. Samples of that effort: a fighter aircraft became a chicken hawk; the month of March became squeaky voice; a battleship a whale; and rabbit trail was the equivalent of a route. Words were devised for officers, communications, organizations, and so on.

No wonder the Japanese couldn't break the "code," which proved indispensable for rapidly transmitting classified dispatches during the invasions of Guam and Peleliu. Because Navajos were on the sending and receiving ends, there was no time lost in enciphering and deciphering messages—a situation that is absolutely critical when moving troops about swiftly and with precision.

Major Howard M. Conner, a Marine Corps signal officer, credited the code talkers with the success in the battle for Iwo Jima. The whole invasion was directed in Navajo code. Orders went from the Corps command posts on the beach. During the first 48 hours, while troops were landing and securing, the Navajos sent and received more than 800 messages.

Several weeks later, when the flag was raised on Mount Suribachi, the victory message was sent—appropriately enough—in Navajo code. The "hidden language" was no longer lost in history.

Rare Okinawan Woodpecker Given Haven

The Noguchigera, a rare woodpecker who nests high up on Yonaha Dake Mountain in northeast Okinawa, can claim he "faced down the Marine Corps ... and won."

In 1970 a Fire Support Base was constructed in the northern training area, but when confronted with the possibility of destroying the habitat of this almost extinct Okinawan woodpecker, the Corps decided on December 31, 1970 not to continue its efforts to fire live ammunition at the site.

In the beginning the Marine Corps sought to share this area with the woodpecker, but received strong opposition from local villagers and ornithologists who argued that firing weapons there would jeopardize the bird's survival. This bird's conduct, his nesting habits, and generally, his way of life, have remained much of a mystery to man.

The importance of the bird may be seen by the fact that during mid-February (1972), Dr. Lester Short, an ornithologist for City University of New York and the American Museum of Natural History, visited Okinawa to study this rare species, along with others.

Tackle Cliffs And Learn New Skills

"In every clime and place ..." says the Marine's hymn. That includes mountains and cold weather.

During the final five days of their training, L Company, 3rd Battalion, 4th Marines, combined these two aspects of training at the Mountain Warfare School north of Pohang, Korea. The school had been instructing Korean and American Marines in mountaineering skills such as reppelling, rope bridges and rock climbing since 1962.

The final day of training, the entire company challenged the last two training objectives, the helicopter reppel and slide-for-life. The simulated helicopter reppel was done from a platform suspended 120 feet above a dried out river bed. Each Marine was required to do one seat reppel and one Australian (head first) reppel.

Perhaps more frightening than the reppel itself was the walk up the mountain and across the bridge to the platform. Once they got to the platform, a Korean instructor hooked them up to the reppel line. It's against every natural instinct to jump from 120-feet high onto a bed of rocks. But that is what they did, backward and head first.

A single rope more than 750 feet high wrapped up the week's training. It was the slide-for-life and it was stretched from one mountain top down to another, across a road and dried-up river bed and rocks. While hooked up to a pulley, the Marines sailed down the rope on a trip which lasted an average of 15-18 seconds. It was a fitting end to a week of tough, but exciting training. **Okinawa Marine, February 11, 1981**

Major Calhoun Killeen To The Realm Of The Noguchigera

Major General Calhoun J. "Cal" Killeen has reason to highlight as exclusive one unusual award he received on Okinawa. It was there that he was named as "Master of the Realm of Noguchigera."

Tullis J. Woodham Jr., in August ceremonies and as Grand Registrar Supreme of the Realm of the Noguchigera, named the American to the high office.

First off, understand, the Noguchigera is a woodpecker—distinctive, rare, noble, and honored in the East. Savants repeat its attributes ... "a poise and dignity and dominance earning it unrivaled a permanent high perch in the ornithic social ladder and in fact the pinnacle of all pecking orders."

The recognition to General Killeen was because "he stands alone and apart from the profusion of accolades and honors which routinely have shouldered the yoke of responsibility.

"It—and seriously—is awarded only to those with uppermost responsibility who have served to enhance the social, moral and physical and economic character of the Island of Okinawa."

The award, says residents and members of the realm, credit General Killeen for contributions in the area of international relations and community services and "because he never forgot the little people."

Native of Pittsburgh, graduate (1949) of the United States Naval Academy, Calhoun J. "Cal" Killeen, served in Korea, in pre-Vietnam, Okinawa, then in Vietnam with 12th Marines. After the war he served as director, Operations Divisions, Operations and Training Department at Headquarters, USMC. In July 1979 he was assigned commanding general, 3rd Marine Division, Okinawa.

Among other awards he holds are: Bronze Star received in Korea, Legion of Merit, Vietnam Cross of Gallantry, and Navy Commendation Medal for other action. He lives in McLean, Virginia.

Engineer's Job ... Not Always Explosive

"Be very careful, I'm in no hurry to go home tonight," said the sergeant as he scanned a group of Marines five feet away fumble with one pound of TNT each, "I just want to go home," he finished, this time moving to help one of the Marines.

That's why the sergeant, Tim Voorhees, from the 3rd Combat Engineer Battalion (CmbtEngBn) at Camp Hansen, was present that day. He was helping Marines from an assault battalion learn how to position and detonate various explosive charges. "It's part of my responsibility as an engineer." he explained.

But engineers don't just blow things up; they build them, too. Whether it's a bridge, road, building or even an airfield, engineers have the skill to put it up. "Demolition is just one part of an engineer's job," said Corporal David Polacios, an engineer, "and a fun part, too. But there's a lot more we have to know. We are all familiar with many different engineering areas, like erecting bridges, building roads or temporary buildings and devising booby traps. We have to know it all."

The 3rd CmbtEngBn builders and destroyers for the 3rd Marine Division provide combat units with engineering capabilities—ranging from minor road repairs to laying or clearing minefields, with a myriad of tasks in between. Their specific mission is to increase combat effectiveness of the landing force by providing close combat engineer support directly to assault units. **Okinawa Marine, February 11, 1982**

Officers, Staff NCO's On Mess Duty For A Day

Saying "thank you" can be said in many ways, but when the officers and staff noncommissioned officers of the 3rd Recon Battalion at Onna Point say "thanks," they mean it.

In order to show their appreciation for a job well done, the staff of the battalion decided to give those individuals on mess duty and the cooks a day off as a group.

Lieutenant Colonel Peter J. Rowe, commanding officer of the battalion, suggested that the "tables be turned" for an upcoming Sunday meal, and that took place October 29.

Twenty officers and staff NCOs, including the battalion commanding officer and sergeant major, literally "took over" the mess hall for the day doing all of the tasks that the regular workers would do.

Lieutenant Colonel Rowe had the privilege of being the baker during the morning, at which time he baked and frosted two cakes for the evening meal. Later that day, he was the chief cook.

"Frequently an individual or group will go unrecognized for a job taken for granted. The cooks in the battalion dining facility put in many long hours every day. This was just our way of saying thanks," said Lieutenant Colonel Rowe, as he stood with cake frosting on his hands.

Sergeant Major Willie J. Trawick, chief messman for the day, added: "This was fun, each meal, breakfast and evening chow, turned out really well." **Okinawa Marine, November 9, 1978**

Area of Diversity ...
Okinawa Is More Than A 'Rock'

You can't say you've seen all of Okinawa until you've been there. The island of Okinawa (the big "rock" you're sitting on), is only one of 106 islands in Okinawa Prefecture.

Some of these islands are big, and boast their own language, in addition to Japanese. Other islands range in size from "pebbles" that poke their heads above water at high tide, to islands big enough to rate their own television station and barber shop.

Off To Kumejima

Just 25 minutes air time west of Naha, is the island of Kumejima. That's not too far, but for the lack of American tourists, it may as well be another planet. People are quite friendly there. The island is so small everyone knows each other.

The bus drivers are courteous, frequently making unscheduled stops, if asked—a convenience for roving tourists as well as the islanders. Rice from here is rated highly in Japan—a country of rice connoisseurs. A by-product of rice is sake, and a good place to sample Kumejima sake is in the town of Nakadomari.

Tourist And Night Life

Then about 45 jet minutes south of Naha is the lively paced island of Miyako. Its big city, Hirara, can rival any Okinawan city's night life. Hirara during the day offers many attractive souvenir shops and good restaurants.

A beautiful spot is the Higashi Hennazaki Lighthouse, located on a long, narrow peninsula jutting into the sea. Steep cliffs along the ocean offer a spectacular view of reefs, seen through the clear blue water. The islanders have their own language too. Unlike Japanese, high and low pitch is stressed in words.

From Hirara, you can catch a 25-minute flight south to Ishigaki. The world's largest tidal wave—278 feet high and travelling at 490 mph, hit the island on April 24, 1871, tossing an 850 ton block of coral 1.3 miles. The island also claims the highest hill, 526 meter-high Mount Omoto.

Long-Lifers

There's a boat from Ishigaki City that goes to the little island of Taketomi—about 20 minutes away. Okinawans claim the longest life span in Japan, but Taketomi claims it in Okinawa. There are 320 people on this island, 90 who are more than 70 years old. There are bicycles for rent and it takes only 20 minutes to peddle from one end of the island to the other. There's one beach on the island where star sand is found. The sand is actually shaped like a five point star and supposedly comes from fossils.

Also from Ishigaki City, a two hour boat ride can get you to Iriomote, known as the mystery island as it is largely unexplored. Dark, jungle covered mountains dominate Iriomote. Tourists who left the main road have been "swallowed-up" in the jungle, never to be seen again.

Okinawa offers an island for everyone's taste. Among the legitimate excuses for going on an island tour might be shell or souvenir collecting, cultural osmosis or for the lack of any other reason, conducting a professional sampling of each island's unique sake. *Okinawa Marine*, **February 1982**

'Tractor-Rats' Take LTVP-7 Out Surfing

Unique to the Marine Corps and other allied amphibious assault organizations, the amphibious tracked vehicle (Landing Tracked Vehicle Personnel-7) not only floats and maneuvers in the water, but its tracked power train makes it practically unstoppable on land.

Armor-plated and armed with a .50 caliber machine gun, the LTVP-7 provides Marines with effective land and water transportation, while protecting them from small arms fire. Anyone who has been buttoned up in an amphibious tractor for a hot and cramped ride from ship to shore will swear there has to be a better way.

However, tractor-rats sneer at such remarks.

"We are the amphibious operation," said Staff Sergeant Vincent Cunningham of 3rd Platoon, B Company, 1st Tracked Vehicle Battalion. "We assault the beachhead targets, let the grunts out and hold it. The first wave pushes the enemy off the objective, then the second hits so as not to allow them to set-up again. It's this combination of tracks and grunts that keeps the enemy off-balance, giving us Marines the upper hand.

When other waves hit, they can fill in the gaps," Cunningham said.

"Once the beach is secured, we become part of an armored column with tanks. We don't have as much firepower or armor as a tank, but then, let's see a tank swim!" he said. "We have it over a helicopter, too. We can fill up and go 300 miles with a full load, in almost any weather and no matter what the temperature. We can carry 25 combat loaded Marines or 10,000 pounds of cargo and, support troops under small arms fire in many places where helos or jets can't."

In addition to assault, Amtrac companies also have recovery and communication vehicles that provide mobile repair and command post capabilities.

Chopping through the surf or rolling overland is what tractor-rats like best. However, it takes a lot of work to keep their vehicles running. But, the end result makes the hours of maintenance worthwhile. "Like airplanes or any other machine," Cunningham said, "the more you run them, the better they go. In the water, the rougher it is, the better we like it." *Okinawa Marine*, **August 1982**

Check fire! It's the 12th Marines! This regiment of cannon cockers—who've loaded canister with practically no fuse and sighted down water—cooled machine guns while smoking cigars in the China sun—are as Asiatic as their infantry counterparts in the 4th Marines.

They've lowered the muzzles on their artillery, fired point blank, then slugged it out toe-to-toe against frenzied attacks on Pacific islands.

They rained thunder on "Charlie" in "Nam" and hauled their regimental colors up and down the Far East so much that they're more at home on Okinawa than most Americans have a right to be.

Take a look at those scarlet colors and you'll see eight streamers with countless battle stars all sounding more oriental than Chinese gongs. There's China, Bougainville, the Solomon Islands, the Marianas, Iwo Jima, Japan, Thailand, Indonesia, Da Nang, Chu Lai, Cam Lo and Khe Sanh. Those are only a few places in a long series that ring out valor louder than a 105 round on impact.

Impact. Artillerymen call it splash now-a-days, but those rounds echo back to October 1927, when old Asia hand, Smedley D. Butler, was ordered to scrounge a brigade for China, "In order to protect American lives and property." Pulling men from units throughout the Pacific, the two-time Medal of Honor winner put machine guns, howitzers and riflemen together in Tientsin, China under a lieutenant colonel and formed the 12th Marines. In the early 1930s, they hauled down their colors, leaving China and the rest of the Far East to the 4th Marines.

The regiment was then deactivated but World War II changed that. With the formation of the 3rd Marine Division, the 12th was reactivated. The 3rd Battalion sailed to Pago Pago, American Samoa; the 4th Battalion landed in New Zealand where the regiment was eventually reunited.

In November 1943 the 12th received its baptism of fire on Bougainville. After this the next campaign was Guam, and still later Iwo Jima. After World War II, the regiment found itself back in China as part of the 3rd Marine Brigade which was standing between uneasy Nationalist and Communist Chinese forces. Tensions were still unresolved in 1946, when the regiment sailed for home.

Back In Far East

By 1953 they were back in the Far East working out of Camp McNair on the slopes of Mt. Fuji.

In 1956 the regiment arrived on Okinawa and set up housekeeping at old Camp Hague and later, Camp Hansen. The 2nd Battalion was sent to Thailand in 1962 at that government's request. Then in 1965 on to Da Nang and other vicious engagements.

By mid-1969 Vietnam was coming to an end for the United States and the 12th Marines. Eventually, they would return to Okinawa and to Camp Zukeran, (Foster).

They have a long lineage, those cannon cockers of the 12th Marines. They may often be seen hiking 10 or more miles along the coast. They can also be heard running in formation past the headquarters units of Camp Butler, their cadence echoing across their Far East base like it did years ago in China. It echoes of firepower that stretches the vast Pacific. *Okinawa Marine,* **September 16, 1982**

Scout Directs Fire

He's dug in deep on the side of a mountain, and is the closest man to an impact area. He is an artillery battery's scout observer.

For Fox Battery, 2nd Battalion, 12th Marines, such a man is Lance Corporal Tom Davis, who's the eyes for the battery's 105mm Howitzers.

Armed with binoculars, map, compass, radio and an M-16 rifle, the Kansas City, Kansas native expertly plies the tools of his trade. His responsibility begins with him calling target coordinates from a hidden forward position to the Fire Direction Control Center, who controls the big guns, and it ends when the artillery rounds have made a direct hit.

Besides artillery, Davis is also trained to score direct hits by calling in naval gunfire and close air support. These skills were acquired by the observer during the two-and-one-half month basic artillery course at Fort Sill, OK.

There, Davis learned the importance of quick and accurate map reading and land navigation, which allows him to locate the enemy he sees and accurately plot their position on a map. Once targeted, he radios the position back to his FDC. "Supporting arms can only be as accurate as the grid coordinates they are given," Davis noted.

Freddie Fender ...
Country Star Relives Corps Days

Freddie Fender, the 46-year old country western music star, of *Wasted Days* and *Before the Next Tear Drop* fame claims he's had many thrills in his life, but, according to Fender, one of his biggest thrills ever, came last week, December 3, when he relived his youth as a Marine and drove a 3rd Marine Division tank through the hills of Central Training Area.

"The last time I rode a Marine Corps tank was as an 18 year old private first class," said Fender as he donned a tanker's helmet for the M-60 tank ride. "I was a gunner on an M-47 ... I was good, too," he added with pride. "I could hit an ant on the side of a mountain with the tank's 90mm gun." That was in 1955, when Fender reported to Okinawa from Camp Fuji with 4th Tank Battalion, and then, newly arriving 3rd Marine Division.

"They shipped us here from South Camp Fuji on Landing Ship Tanks," recalls Fender. "We arrived at Naha Port and drove our tanks to Easley Range (now Camp Hansen). There was no camp then. We slept on the ground while the engineers built Quonset huts."

One of the first huts Fender remembers was the mess hall, where officers and enlisted dined and partied together. Learning to play the guitar while growing up in Texas, Fender was known to often pick up a guitar and perform. He also performed at less opportune times.

"I remember one night I was standing guard duty when a couple of Army friends from Camp Zukeran (now Foster) talked me into a little party. Well, I put down my rifle, picked up my guitar and started to play. That performance landed me in the brig."

Although Fender admits his Marine Corps career was less than illustrious, he still holds a deep affection for the Corps. "When I was in, the Corps was the finest fighting force on earth. From what I've seen on this tour, it still is."

After his discharge in August 1956, Baldemar Huerta, (Fender's real name) played beer joints, dances and " ... anywhere else I could make a buck." Those performances eventually landed him his first recording contract with Falcon Records, a Mexican-American company in McAllen, Texas. In 1958 he teamed up with Wayne Duncan, his first manager, and recorded his two million seller, *Wasted Days and Wasted Nights* on the Duncan label. After this hit, he stopped recording until 1971. Then, in late 1974, after recording *Before the Next Tear Drop Falls,* another two million seller, he began touring and has been on the road ever since.

On Okinawa, fans came not only to hear him play, but also to get a glimpse, an autograph and possibly have their picture taken with him. Though Fender played before full houses during his visit on Okinawa, the crowds were much smaller than the thousands he's accustomed to. But, Fender said, "When you play for small audiences, it's easier to develop an intimacy and rapport with them."

Baldemar Huerta, alias Freddie Fender, returned to Okinawa after a 27-year absence. This time he was paid to sing and play the guitar. And he didn't even have to stand guard duty. ***Okinawa Marine,* December 10, 1982.**

Marines Give Aid In Sea Rescue

Marines of Marine Medium Helicopter Squadron-161 at MCAS, Futenma, were quick to respond in the rescue of two pilots from Marine Fighter Attack Squadron-212, MCAS Iwakuni, when their aircraft crashed in the waters off Okinawa Friday.

"We got the call that an aircraft was down approximately 65 miles east-southeast of Okinawa, at 9:55 a.m.," said Captain James Todd, pilot of one of two helicopters involved in the rescue operations. "We refueled and were underway in about 10 minutes. By 10:30 a.m., we were on the scene," said Captain Todd, who was responsible for picking up the second survivor. "We were two to three minutes behind Captain Hill's aircraft. We made just one pass to make visual contact with the survivor and he seemed to understand us well, so we lowered the hoist." Sergeant John Carrell, crew chief on Todd's aircraft, added, "We lifted the man in. He seemed in good spirits and he was obviously happy to be aboard."

The survivors from the F-4 Phantom were rescued just 65 minutes after the initial call for help. After the rescue, the two were taken to the Naval Regional Medical Center for routine examinations.

Motivation, Adventure, Training

If mud and muscle are essential elements in the making of a United States Marine, then the Northern Training Area serves as one tough refresher course for deployed units.

Pulling off the highway here on the northeastern part of Okinawa, a narrow, winding road of dirt and gravel leads to the Quonset huts, rustic barracks and outhouses, weight room, modest dining facility and pull up bars of the base camp. Upon entering the camp, which is nestled in one of the few clearings, a sense of isolation overwhelms the body and soul.

The camp is peaceful, quiet, and seemingly deserted—the perfect atmosphere for serious Marine Corps training.

It is here in this heavily, wooded, red clay, subtropical environment—amid the muck, mire and isolation—that elements of the 3rd Marine Division come to hone combat skills in all kinds of weather. ***Okinawa Marine,* February 1983**

The Rougher
The Better Tractor Rats Like It

Unique to the Marine Corps and other allied amphibious assault organizations, the amphibious tracked vehicle (Landing Tracked Vehicle Personnel-7) not only floats and maneuvers in the water, but its tracked power train makes it practically unstoppable on land. Armor plated and armed with an 50-caliber machine gun, the LTVP-7 provides Marines with effective land and water transportation, while protecting them from small arms fire.

Anyone who has been buttoned up in an amphibious tractor for a hot and cramped ride from ship to shore will swear there has to be a better way. However, tractor-rats sneer at such remarks. "We are the amphibious operation," said Staff Sergeant Vincent Cunningham of 3rd Platoon, B Company, 1st Tracked Vehicle Battalion. "We assault the beachhead targets, let the grunts out and hold the beachhead.

The first wave pushes the enemy off the objective, then the second hits so as not to allow them to set-up again. It's this combination of tracks and grunts that keeps the enemy off balance, giving us Marines the upper hand. When other waves hit, they can fill in the gaps." Cunningham said.

"Once the beach is secured, we become part of an armored column with tanks. We don't have as much firepower or armor as a tank. but then, let's see a tank swim!" he said. "We have it over a helicopter, too; we can fill up and go 300 miles with a full load, in almost any weather and no matter what the temperature. We can carry 25 combat loaded Marines or 10,000 pounds of cargo and, support troops under small arms fire in many places where helos or jets can't."

In addition to assault, Amtrac companies also have recovery and communication vehicles that proved mobile repair and command post capabilities.

Chopping through the surf or rolling overland is what tractor-rats like best. However, it takes a lot of work to keep their vehicles running. But, the end result makes the hours of maintenance worthwhile. "Like airplanes or any other machine," Cunningham said, "the more you run them, the better they go. In water, the rougher it is, the better we like it." **Okinawa Marine, August, 1982.**

Challenges Of A Recon Team

They're the "eyes and ears" for 3rd Marine Division. They're the Marines that comprise the 3rd Reconnaissance Battalion. They remain in constant training to keep their skills razor sharp, whether they be in an inflatable boat off the coast of Korea, in the jungles of the Philippines or a host of other countries.

"Each country we train in offers different challenges. You get to know the people and learn how to overcome difficulties presented by these challenges," said Corporal Allen Segal.

Recon Marines used diverse methods to accomplish the job. If their mission requires them to arrive by air there's several options; parachute, repel, or SPIE (Special Patrol Insertion and Extraction) rig. If by sea, they're capable of departing any floating vessel in an inflatable boat, or scuba dive to the shore.

"When we make a reconnaissance prior to the amphibious assault, we record such obstacles as height of the tides, the patterns of the waves, and the length of the beachhead," said Corporal Peter Koe, team leader. "A lot of our work in done at night because our goal is to get in, do the job, and get out without being detected," Koe said.

There are two types of recon missions: keyhole and stingray. The keyhole requires recon teams to observe the enemy position to gather information on the size, location while advancing and avoiding contact.

During the stingray mission, the team locates enemy positions and fortifications ... then calls in supporting fire. Supporting fire can consist of mortars, artillery, air support and naval gunfire. "Important decisions are made from the information a recon unit obtains," said Staff Sergeant Anthony Agars, platoon sergeant.

A recon team consists of a team leader, assistant leader, radioman and corpsman. The team members are cross trained and each is capable of continuing the mission if one is injured or killed. **Okinawa Marine, July 1, 1983**

Pride ... Tradition ...
He's A Grunt

The Marine Corps arsenal has constantly changed to keep up with the times. However, there remains one weapon that has changed little during the past two centuries. Known by many titles, the most common 20th century nickname for a United States Marine infantryman is "Grunt."

There is noting particularly pleasant about the title. But, then neither is his job, known to be one sought by more than a few adventurous youths. Yet, on the grunt's pack worn shoulders lies the awesome responsibility of fighting his nation's wars. The grunt is the one who engages the enemy at close range in an age of push button warfare and computer soldiers, the grunt is trained to fight in primitive environments and equally primitive terrains.

The grunt is the kid from next door. Only a few months ago he was transformed from the disheveled ranks of the civilian world to the world of finely aligned formations that responded instantly to a sergeant's terse commands. In an age of liberal ideas, emphasis on human rights, the nation's taxpayers are served by a corps of men who run contrary to type and whose ironclad discipline will not find them wanting when the nation needs Marines to preserve those rights.

He is the tool of American diplomacy with enough naivete to be pure at heart and believe men are still gallant.

He is the pride of a nation which at times scorns him, but which nonetheless requires and demands he answer its every call. **Okinawa Marine, April 29, 1983**

Shooters Praise M16A2 Rifle

Until recently, infantry Marines rotating to Camp Hansen from the 2nd Marine Division at Camp Lejeune, North Carolina, were the only ones on Okinawa who used the Corps new M16A2 rifles on a regular basis.

But, at the Far East Division Matches at Camp Hansen, Marines from all over the Pacific enjoyed the advantages of firing the new rifle; some of them for the first time. The consensus among the shooters has been one of great praise for the weapon. The M16A2 is being used for only the second time in match competition on Okinawa. It was introduced last year.

"The first time I squeezed a round off I knew I would like the weapon," said Lance Corporal Raymond Chavarria, 12th Marine Regiment. "It didn't take me long to get my dope down, because the accuracy of the rifle allowed me to shoot a bull's eye immediately."

Featured improvements on the M16A2 are a wider and more square, front sight post with only four elevation notches as opposed to the M16A1's five. The rear sight has the biggest change. The old windage drum has been replaced with an elevation knob located on the right side of the weapon. This allows a shooter to maintain his front sights, without alteration at different firing lines, once he obtains battlesight zero.

"I can set my front sight post at the 200 yard line and never have to worry about it again," said Sergeant Timothy Nesbitt of 3rd Battalion, 9th Marines (3/9). "With this new sight system, I can devote the time I used to spend on sight changes, to better my concentration and shoot for higher scores." The elevation knob is a welcome change by all shooters because it allows adjustments to be made by a simple turning of the knob. "It's like turning on a radio," said Nesbitt. "No more nails from now on."

Additional features are a burst of fire, allowing three rounds to be fired with one squeeze of the trigger, heavier plastic and barrel, a longer stock, interchangeable hand guards, a modified pistol grip to prevent slippage, a built-in deflector for left-hand shooters and a new flash suppressor to decrease the kick of the new rifle.

An overall improvement, the M16A2 is gradually being procured throughout FMF units, yet competitions such as the Far East Division Matches, allow Marines of all commands and MOSs to fire and enjoy the vastly improved weapon. "The opportunity is great for me because I only fire during annual re-qualification," said Corporal Anthony Fair, 9th Marine Regiment. "Yet I can still see and feel a big difference in the weapon. It's the best I've ever fired." *Okinawa Marine*, **March 1, 1985**

Women Marines Find Guard Duty Challenge

Sexual equality in the military community on Okinawa took a unique step forward with the recent opening of the women's cells at the Joint Forces Correctional Facility here (Camp McTureous).

There are six women Marines who even see the situation as a chance to improve themselves, through their new duties as correctional guards. "I'm glad to see they finally got a facility for women here," said Corporal Vanessa Russ, senior member of the women guards. "Equality is what it's all about; women should, if they break the law, receive the same punishment as guys do."

According to Master Sergeant Robert Stone, correctional supervisor for the facility, the women guards have held every job there is inside the correctional facility, including supervision of male prisoners.

"There were surprisingly few problems with the male prisoners adjusting to our presence," said Russ. *Okinawa Marine*, **November 5, 1981**

"Take A bite" ...Change From C Rats To MRE's

The new Meals-Ready-To-Eat (MRE) were introduced in late 1982; however, Marines will continue to use C-rats until supplies are depleted before being issued the new combat meals.

MREs are lighter than their counterparts, which should be good and welcome news for Marines carrying their chow to the field. The older meals weighed 26 pounds per case, versus 17 pounds for a case of MREs. MREs are packaged in flexible "retort" pouches, similar but lighter than those pouches found in C-rats.

The food in the new combat meals is heat sterilized, much like the canning process. Unlike the tin can, the flat shape of the pouch permits even and rapid heating of the contents, thus providing superior quality. Another big advantage is that they can be warmed by body heat, if necessary, or can be eaten without water. Also, there are no tin cans to worry about.

Development of MREs was initiated in 1959, after a requirement was established for a family of processed foods which would satisfy combat feeding for individuals and large groups.

With the introduction of MREs, another Marine tradition falls by the wayside. Since the packages can be opened simply by tearing off the pouch seal, "John Wayne" or P-38 can openers will not be needed or issued. Each meal contains meat, crackers, cheese, jelly or peanut butter spread, a dessert and instant coffee. In addition, each of three varieties contain beans with tomato sauce, a freeze dried potato patty or cocoa beverage powder. The units also include sugar, cream and toilet paper.

Dessert portions are either a high calorie cake, brownies or cookies. Seven of the 12 MRE menus have one of four different freeze-dried fruits. Calorie content ranges from 1,125 to 1,336 per MRE complete meal.

Desert Marines should take note that MREs require a slight increase in water requirements. In hot deserts, an average active person requires a minimum of three gallons of water per day; otherwise, dehydration sets in. Natural thirst causes the average person to want only two-thirds the water needed, and for this reason, MREs help get the required water.

Cold weather training presents a slight problem, with a need for more fuel bars. However, heating the pouches in a canteen cup is considered the best method currently available, or by warming the contents from body heat by carrying them inside the Marine's clothing. So, forget the good old days, when you fought over ham and eggs, or beef and rocks (potatoes), and throw away or keep as a collector's item, your John Wayne P-38 can opener. It's out with the old and in with the new. *Okinawa Marine*, **September 9, 1983**

Hello, This Is The President

"Lance Corporal Bartlett, the President wants to speak to you." Smiling, the command duty officer handed the phone to Lance Corporal Edwin Bartlett, duty watch stander at the 3rd Marine Amphibious Force/3rd Marine Division.

"Of course, I thought he was kidding," said Bartlett, who had expected the Christmas Eve watch to be quiet and uneventful. But the female voice at the other end of the line said, "This is the White House operator. Please hold for the President of the United States." As he waited, Bartlett's feeling alternated between amazement and amusement. He suspected he was the victim of an elaborate Yuletide hoax. Then his jaw dropped to his knee. He heard the voice of Ronald Reagan, President of the United States and Commander in Chief of the United States Armed Forces, speak his name.

"Lance Corporal Bartlett, are you there?" as if in a dream, the 20 year old Guilford, Connecticut, native heard himself reply, "Yes, Mr. President, I am here."

Bartlett said it took about 10 seconds for his grip on reality to return. But then his face went pale, his knees became weak. He sat down quickly, otherwise he'd fall. "He asked me how long I'd been over here and what was my military occupational specialty. Those were the words he used," said Bartlett. "He said he wanted to express that both himself and Mrs. Reagan really appreciate the job we are doing over here and wish us the best for the holidays and to be careful."

He also asked if I would like him to call my family. I told him, "I would like you to, but I understand that you have a very busy schedule." President Reagan said, "he was never too busy to make a call for something like that."

By the time the conversation ended, Bartlett's head was in the clouds. When his watch ended, he went home to the barracks so thrilled with himself that his feet barely seemed to touch the ground.

He could hardly wait to tell the guys in the barracks. But he might as well have said Santa parked his sleigh in the commanding general's reserved spot. "Yeah, right. The President called you. Sure, Bartlett."

They sang a different tune, however, when the December 26 Stars and Stripes carried Bartlett's name on the front page as one of the five United States servicemen around the world to receive Christmas calls from the President.

Then, like Monday morning quarterbacks, his peers wanted to know why he didn't ask the President about military pay and benefits, the defense budget, and other political issues. "He was doing all the asking," said Bartlett. "I was on the spot right there and then. I was still in shock and I didn't think of asking him any question. I did wish him the best and a happy holiday and all. I thought it was really nice he took the time out to give us a call. It was only five of us but then he can't call everyone." Bartlett continues to take good natured ribbing but whenever the subject is raised, he's all smiles. "You can ask anyone, I was really happy. I can say, 'Yes, I've talked to the President.'"

Christmas is supposed to be a time for magic and miracles, caring and giving. All who know Bartlett are happy for him. The thoughtful gesture of the President made Bartlett's first Christmas overseas an occasion he'll always remember. ***Okinawa Marine,*** **January 4, 1985.**

Last General Officer With World War II Combat Experience

Major General James L. Day took over the reins of MCB, Camp Butler on July 27, 1984.

With a colorful career spanning nearly 42 years, including seven tours on the "Pearl of the Pacific," he brings a wealth of experience ... and a genuine love for the Okinawan people. That affection stems, according to the General, not only from his numerous tours here, but from what he calls his "most vivid memories of combat."

"Okinawa was probably the battle that I remember more than any other," said the veteran of Korea and Vietnam. "I suppose that's true only because of the amount of casualties suffered by both sides." The three month battle caused widespread destruction and the deaths of over 100,000 non-combatant Okinawans, in addition to military casualties of 60,000 Americans and 110,000 Japanese.

"I think my affection for the Okinawan people emanates from the fact that I saw the suffering they went through during the war. They are a tremendously courageous people and the fact that they lost at least one third of their population—nearly everyone lost somebody in their family—speaks highly of that courage," said Major General Day.

Spending nine years as an enlisted Marine and rising to the rank of gunnery sergeant, Major General Day's career was off and running. But it wasn't until the Korean War that he traded his stripes for the bars of a second lieutenant.

He holds the distinction of being the last Marine Corps active duty general officer with combat service during World War II. ***Okinawa Marine,*** **December 21, 1984.**

The 3rd Marine Division Shootist ... Double Distinguished

As a Marine "double distinguished" rifle and pistol marksman, Major Charles H. Thornton Jr. is highly adept at handling the Corps small arms inventory.

The recipient of numerous awards in military small arms competition was first introduced to competitive marksmanship in 1965, while attending college at Hampton Institute in his native Virginia. There, his efforts as team coach and as a competitor on the firing line contributed to three conference rifle championships.

A veteran of several division, Corps, inter-service and national matches, the major earned his Distinguished Rifle Medal in 1974. He was designated "Double Distinguished" when he received the Distinguished Pistol Medal four years later.

The 3rd MAF/3rd Marine Division deputy comptroller and budget officer, Camp Courtney recently captained the Far East Team at the 1984 Marine Corps Matches at Camp Lejeune. While there, the 38 year old marksman was also a shooting member of the division team which won the prestigious FMF Combat Infantry Trophy for the 3rd Marine Division during the rifle team segment of the matches. ***Okinawa Marine,*** **June 15, 1984**

Revolutionary Landing Craft Is Air-Cushioned and Fast

Commandant General Paul X. Kelley, officially received delivery May 3 of the Landing Craft, Air Cushion (LCAC) from the New Orleans operations plant of the Bell Aerospace Division of Textron.

"This improvement in landing craft technology is the most significant advance in warfare since the introduction of the helicopter," General Kelley testified before Congress.

The LCAC is capable of speeds up to 50 knots on calm seas and can carry 150 troops or the main battle tank and several other vehicles through surf onto and across beaches inland. It gives the Marine Corps amphibious doctrine new life with its ability to assault 70 percent of the world's beaches as opposed to the current 17 percent for today's amphibious assault vehicles.

The craft is capable of doing 40 knots with a 60 ton payload and is compatible with existing ships currently used by the Navy to deliver landing forces. As described in Jane's Fighting Ships: "LCACs are supported above the land or water surface by the continuously generated cushion of air held by flexible 'skirts' that surround the base of the vehicle."

It is powered by four gas turbine engines that turn two four bladed, twelve foot propellers. It can clear a four-foot obstacle and is approximately five times as fast as any of its predecessors. According to an article in the July 1983 edition of "Amphibious Warfare Review" by Colonel Jack Scharfen, USMC (Retired), "The LCAC may be the only vehicle that can safely traverse mud flats, ditches and marsh-lands. Beaches that formerly would be denied to the landing force because of a lack of egress routes now become accessible ..."

The vehicle operates on Marine diesel fuel or aviation fuel (JP-5) and will measure 87'11" long by 47' wide by 23'3" high when "on cushion." It's "a revolutionary dimension to amphibious warfare," according to Secretary of the Navy John F. Lehman Jr. *Okinawa Marine*, **May 18, 1984**

Marines "Turn To ..."

Help Farmer Harvest Sugar Cane

The Marines weren't carrying rifles when they boarded the trucks and headed north. In fact, they weren't even wearing the usual camouflage uniform. Instead of going to the field to train for combat, the Marines of 4th Battalion, 12th Marines were going to attack a field of sugar cane.

They volunteered part of their weekend, January 14, to help cut sugar cane for an Okinawan family.

Kazuo Nohara, a farmer needed help. With only one day to cut his sugar cane crop, and his wife in the hospital, Nohara faced the task alone. Some neighbors offered to help, but they had their own fields to tend.

Word of Nohara's plight passed through the community churches and eventually reached Pastor Mark Duarte, who passed it on to his congregation. Chaplain Ruben A. Ortiz of 4/12 was there and offered to round up some Marines to help. Other members of the church said they would like to help also.

"I was on the verge of tears when I heard all of these people were coming out to help me," said Nohara. The weather had cleared by 9 a.m. when the trucks arrived. The Marines picked up tools, received instructions, and swarmed the fields.

First they cut the tops off the cane, then cut the stalks down. The roots were hacked off and the leaves were stripped. Stalks were then put in small piles and tied together to await the trucks to take them away.

It was hard work, but Lance Corporal Kevin E. Howard of "Mike" Battery didn't mind. "I'm here to prove a point. Marines aren't here just to fight when there's conflict," he said. "We're here to help those who are in need. It's the Marine Corps way." *Okinawa Marine*, **January 19, 1990**

Recognized For Heroics In Iowa Tragedy

Camp Hansen—In April 1989, Captain Jeffrey Bolander was the commanding officer of the USS *Iowa's* Marine Detachment, conducting routine "call-for-fire" training of the ship's main battery.

"We (Marines) were standing on the 0-4 and 0-5 level of the bridge simulating that we were ashore in an actual combat environment," commented the infantry officer. "We called for fire to turret one, and in the process of sending a message to turret two it happened ..."

Turret one exploded from within, killing 47 crew members. "I didn't know what to do at first. I was in complete shock and for some reason I donned my gas mask," he recalled. "I didn't know what happened."

General quarters sounded over the loud speaker.

"No way—as soon as I saw the turret on fire, I ran and tried to get the escape hatch open. It was almost upside down and lodged shut because of the explosion," he said. "I remember screaming, 'get me a hammer!' because all I could think of was ... there were people in there.

"I had a crowbar and just kept hitting this 500-pound hatch like it would really help—but I refused to give up. I kept hitting and hitting and hitting. All of a sudden, I whacked it one more time with this little crowbar and miraculously the hatch fell open.

The fire and rescue parties were ready to go, and they quickly entered the hatch but immediately came back out. "For eight or nine levels down, it was blocked with rubble. It was just awful, but we had to go in hoping for survivors."

There were no Marine casualties. "It took a while for the fires to be contained," Bolander recalled. "So much water had to be used that it was starting to flood the ship. Everyone worked throughout the night to do what they could.

"I just wish there was more we could have done."

Experts agree that if the blast that shook the Iowa had spread to the ship's magazine, the resulting explosions could easily have destroyed the ship, and cost more of its crew, if not for their quick response.

For his heroic actions that day, Bolander earned the Navy/Marine Corps Medal. *Okinawa Marine*, **July 6, 1990**

Convoy Training

"Rough Riders" Rough It In The Field

Central Training Area, Okinawa—Marines from Truck Company, Headquarters Battalion, 3rd Marine Division, tested their skills as they rumbled throughout the hills here July 28 through 31 (1990).

More than 60 "Rough Riders" practiced basic battle skills as well as combatant motor transport techniques including night time defense, patrolling, convoy operations and field expedient repairs on tactical vehicles.

"This is training they can't get in garrison," said 1st Sergeant Michael J. Senecal, Truck Company 1st sergeant. "It's something every Marine should get often, regardless of MOS.

"The most important training was the convoy operations," Senecal explained."We taught our Marines what to do during enemy attacks in a tactical convoy—protect the vehicles by setting up flank security."

The exercise tested their ability to support 9th Marine Expeditionary Brigade. The scenario called for a convoy to pick up personnel and cargo from Landing Zone Dodo and carry them to a drop-off point with ambushes set up along the way.

"We want to give the NCOs a chance to learn the responsibilities of a convoy commander," Lieutenant William D. Collins said. "They need to know what to do in any situation and be able to teach their Marines as well—from calling in air strikes to setting up a defensive perimeter when attacked."

Truck Company 5-tons rumble through the Central Training Area in a convoy during the company's field exercise. The exercise was to familiarize noncommissioned officers with the responsibilities of a convoy commander.

Marines Clean Up Leprosarium Yard

A dozen 9th Engineer Support Battalion, 3rd Force Service Support Group Marines volunteered to police the grounds of an Okinawan leprosarium on O-Shima.

Leprosy is a progressive, infectious disease caused by a bacterium that affects the skin, flesh and nerves and is characterized by ulcers, white scaly scabs, deformities and the eventual loss of sensation. It is apparently only communicated through long, close contact.

Armed with rakes, weed eaters and machetes, the Marines slashed at tall weeds and brush on the small island, eight kilometers north of Nago.

For at least 70 or 80 years, lepers have lived in the area," said Lieutenant Commander John Samb, 9th ESB's chaplain. "Because of their disease, they were forced to live on small off-shore islands and would comb Okinawan villages at night for food. Then in the 1930s, the Government of Japan built a barracks style leprosarium on O-Shima Island.

During World War II, many of the lepers lost their lives during the fighting. "Once the fighting stopped, the present leprosarium compound was built," the chaplain continued. "Before the war, more than 1,100 people lived there. About 600 people live there now. Sixty-three is the average age of the residents."

The Marines also took the time to give a nearby beach a face lift, picking up and cutting grass. "I was really surprised at what I saw," said Sergeant Joel Dover, H&S Company, 9th ESB. "I thought we would see a series of dilapidated ones. What I saw were some very nice ones. I'm glad we're able to do this for these people."

Although other Marines echoed this opinion, they couldn't share this with the leprosarium residents who shied away during the cleanup. The 9th ESB Marines intend to return several times this year, according to the chaplain, and perhaps then the Marines will meet the shy patients of Airaku-en. ***Okinawa Marine*, February 9, 1990**

Revisions Bring Women Marines A Step Closer

Women Marines will no longer be excluded from certain sections of the annual essential subjects test, according to a revised order on the training policy for women Marines.

Marine Corps Order 1500, 24D follows-up the Women Marine Review conducted in 1984. The review stated, "Since women Marines serve in many different units and MOSs, their exposure to danger in a hostile environment cannot be precluded. They must be trained in defensive techniques and operations in the event of unforeseen hostile activity."

The 3rd FSSG has the bulk of Fleet Marine Force, Pacific women Marine on Okinawa. In fact, five percent of the Group consists of women Marines.

"I like the new changes to an extent," said Sergeant Teresa A. Balizan, 23, of Las Cruces, New Mexico, the assistant group career planner. "We now have the opportunity to prove ourselves. I feel we can shoot, run and do almost everything male Marines can do. I hope this will help to desegregate the women Marines and make our mission more compatible to the role of male Marines."

"It used to be women recruits at Parris Island never fired weapons, then they were allowed to FAM (Familiarization) fire them and now we are going to actually qualify with the rifle," said Captain Sheila M. Quadrini, 28, of Hilton, New York, manpower management officer for the group. "It seems we've come full circuit." *Okinawa Marine*, **August 30, 1985**

Splashing With The Trackers

Almost anyone who has been around the Navy/Marine Corps team for any amount of time will easily admit that the most difficult of all naval operations is the amphibious assault.

While no member of such an operation can justifiably be called the most important, there are some warriors whose role seems to be somewhat more vital than others.

One such unit is the Marines who actually transport the infantry from ship to shore. They are called trackers, short for amphibious tractor, the vehicle they operate. While most trackers make their job look pretty easy, it's not because they have an easy job. It's because they train, and train, and train some more.

One such day of training was conducted at Camp Schwab's Ourawan

A Return To Okinawa

Twenty years after his command, Major General Calhoun J. "Cal" Killeen, USMC (Ret.) addresses a new breed of 3rd Division Marines following a live fire demonstration at their camp on Okinawa.

The pride, the quality was as firm as ever ... "No doubt about it, they're good," he enthused. He was CG 3rd Marine Amphibious Force and 3rd Marine Division on Okinawa 1978-1979. He lives in McLean, Virginia.

Beach by A Company, 1st Tracked Vehicle Battalion. "It's valuable for us to get out here and splash our tracks," explained Sergeant John Glauner, a crew chief with A Company. "The vehicles have to be run to maintain our readiness. They must be checked under pressure." Along with sea rescues, A Company practiced driving in water formations, with each crew member taking a turn at the wheel. Formations included circling, right and left flanking movements, the combat V and moving to on line. "All in all, today's water ballet was another good training session for us," noted 2nd Lieutenant Jeffrey Andrews, 1st Platoon commander. *Okinawa Marine*, **June 27, 1986**

Artillery Goes Aerobic

The cartoon depicts a dress blue-clad Marine officer standing erect and proud beside a smoking howitzer. Around him are a group of Marines involved in heated hand-to-hand combat. Inked across the scene are the words, "artillery lends dignity to what would otherwise be a bloody brawl."

But once a week, Marines from Fox Battery, 2/12 put dignity on a back burner and humble themselves before a petite brunette whose commands are barked like a Marine drill instructor.

While other Marines are struggling through an early morning formation run, Fox Battery Marines can be found on their hands and knees bending, twisting and stretching their bodies into positions that rival a circus contortionist, while upbeat music sets their tempo.

An occasional groan can be heard as they struggle through the early morning routine. But even with the pain and sometimes obvious lack of coordination, they approach their workout with good humor. These guys are marching to a different drum. It's called aerobics, yet when practiced by a group of grown men whose entire lives center around an image of masculinity it takes on a form all its own. Sometimes painful, other times comical, it has become a weekly ritual for the "cannon cockers."

Aerobics have been a part of the weekly routine for more than a year according to Captain John Aten, Fox Battery Company. "It's a break from the standard routine for us," he said.

Aten said the idea came from one of the lieutenants in the battery, who attended a class with his wife and emerged with a new respect for aerobic exercise. "He suggested it to me and we gave it a try," said Aten. "We've been doing it ever since."

"We work every muscle group in the body, plus condition the cardiovascular system. That's something our normal PT doesn't do. We look forward to it every week." *Okinawa Marine*, **May 31, 1985**

Steering A New Course

Lieutenant General Charles G. Cooper, commanding general FMFPac/MCBs, Pacific, arrived at Camp Foster to visit various Marine Corps installations on Okinawa during a seven day visit. As senior Marine in the Pacific, Lieutenant General Cooper commands some 80,000 Marines and assigned sailors.

Question: The Marine Corps appears to be in the midst of an almost unprecedented modernization program spanning the spectrum of weapons and equipment. Why are we seeing so much new gear in hand or on the horizon when for years it seemed like Marines almost used "hand-me-downs?"

Lieutenant General Cooper said: "We're bringing in new weapons starting with a new pistol. The old .45 is being replaced with the 9mm Baretta, an Italian pistol. It's a fine pistol, about the same weight as the .45, but has a more compatible grip, twice the range and accuracy and carries twice as many bullets. It won't also be prone to malfunctions.

"With the M16 rifle, we've made some improvements, such as the three-round burst control so the troops can't spray ammunition everywhere. We've also done a few things to it that should have been done 20 years ago. It's a super weapon and will be our rifle for the foreseeable future.

"We've finally replaced the Browning Automatic Rifle (BAR) with the squad automatic weapon, a light machine gun. We got a bunker buster called a shoulder launched, multi-purpose assault weapon (SMAW). We're going to use the .50 caliber machine gun more that we used to.

"We also have a new artillery piece, the M198. It's a long, heavy, accurate and clumsy compared to the old 105, but it can do a lot more.

"On the air side, the FA/18 will replace the F-4. The A-4 1 light attack will be replaced by the advanced Harrier-B. That's going to be a beautiful, sweet bird.

"I could go on and on, but these are the highlights of our new weaponry." *Okinawa Marine*, **March 1, 1985.**

The New "Doc" Makes The Grade

The quote goes somewhat like this—"The Navy corpsman ... a long haired, undisciplined smart aleck ... who'll run through the gates of hell to help a wounded Marine."

"The quote's true, but I don't think about it much," explained Seaman Terry Burdett, a hospital corpsman and "Doc" to the Marines in 1st Platoon, "Bravo" Company, 1st Battalion, 6th Marines (1/6). "I'm here to do my job, which is to provide medical support to the Marines in my platoon."

The Texan reported aboard last August and was immediately accepted by the leathernecks as their new "Doc." "Everyone was very friendly and showed me around," he remembers. "The Marines also told me the ups and downs of the company and what was expected of me."

Since it was his first overseas tour, Burdett was a little worried that he might not be able to handle the mental stress and physical demands of being the corpsman for a platoon of "Grunts." "When we're in the field, I do everything the Marines do, trekking through the countryside with a heavy Alice pack and staying awake all night. Plus, I have to be alert and notice any medical problems and know how to treat them quickly," he explained.

Burdett doesn't mind being the lone Navy guy in a platoon of Marines. "My wife's father is a retired Marine gunny, so I was kind of prepared to be with the Marines. I really enjoy being in a company of leathernecks," he remarked proudly. The Marines under his watchful care feel the same way about "Doc," too. *Okinawa Marine*, **May 31, 1985.**

Godfrey Becomes CG 3rd MAF/3rd Marine Division

Major General E.J. Godfrey assumed command of the nearly 22,000 Marines and sailors of the 3rd Marine Amphibious Force/3rd Marine Division from Major General Harold G. Glasgow during ceremonies June 4.

Major General Godfrey is returning to Okinawa for his fifth tour of duty with the Division. Prior to taking command of 3rd MAF, he was the director, Operations Division and the assistant deputy chief of staff for Plans, Policies and Operations, HQMC. *Okinawa Marine*, **June 6, 1986**

Plunge Into "Peters Pit" Great Training

More than 130 Marines of L Company, 3rd Battalion, 3rd Marines, swam their way to confidence as they negotiated the Northern Training Area's infamous "Peter's Pit" recently. Training along the jungle slopes of the NTA, Marines had mixed feelings before plunging into the pit, which is part of the 2.5 mile long Combat Skills Endurance Course.

This aspect of training concluded the company's 11 day training package and included land navigation obstacles prior to the pit.

Exhausted but eager to experience what they've heard so much about, Lima Marines low-crawled toward the final obstacle. Ahead, they saw their platoon commanders standing in four feet of murky cold water waiting to assist them.

As the commanding officers motioned for their Marines to jump into the pit, instructions were also bellowed out to ensure safe and successful procedures. After each Marine entered the pit, he took a deep breath, submerged and firmly grabbed the guide rope under the muddy waters. He continued to stay underwater, passing the concertina wire above him until his awaiting platoon commander touched him at the other end. The touch signaled him to pop up, climb over the pit's wall and attack the enemy.

"The most difficult part of the pit is a Marine's inability to see where he's going," said 2nd Lieutenant Mike Donovan, Weapons Platoon commander. "He may be somewhat disoriented at first, but it's realistic training."

For others, the realistic training seemed to be testy yet motivating. "The water wasn't too inviting but that's expected in combat," said Lance Corporal William Luke, a fire team leader. "The pit is a confidence builder," said Staff Sergeant Richard Moss, Weapons Company platoon sergeant. "After today, swimming under those conditions will be second nature. They know they can do it now," he said.

After taking on the Peter's Pit experience, Lima Marines concluded the course by squeezing through a narrow, muddy infiltration course and conducting a medevac at their landing zone. *Okinawa Marine*, **November 10, 1988**

Truck Company Marines conceal their 5-ton trucks with cammie netting. (USMC Photo by LCpl. Chago Zapata)

Bridge Over Troubled Waters And Other Tank Obstacles

A platoon of tanks on their way up a steep hill only to reach a 40 foot wide, 10 foot deep river.

Within five minutes the Armored Vehicle Launched Bridge unfolds its 15 ton span across the obstacle—the assault continues. Tankers from Charlie Company, 1st Armored Assault Battalion, 3rd Marine Division, demonstrated the AVLBs capability at the Central Training Area's tank obstacle course.

Similar to an M60 tank, the AVLB has an M-60 A-2 chassis, but there's no turret; instead, an aluminum cast bridge sits on top of the tank. The 60-foot bridge is unfolded using hydraulics and placed across obstacles such as a ditch or river. It takes two marines to operate the 59 ton armored vehicle—a bridge operator and bridge commander.

The commander is responsible for selecting the final launch site, and is in constant radio contact with the battalion. The AVLB commander looks for at least three feet of support on each side of an obstacle the bridge is to span. A tank can cross an eight-foot gap, but anything wider can delay the tank's mission. The AVLB provides a quick crossing within three to five minutes, adding to tank maneuverability, according to Lieutenant Colonel James Hawkins, commanding officer, 1st AAB. **Okinawa Marine, July 7, 1989**

Retired Marine Runs So The Weak May Walk

Camp Butler—"It is better to give, than to receive. The more you give the more you get back," says retired Master Gunnery Sergeant Tom Knoll—and he gives his most for crippled children.

For the tenth time in 11 years, the 57 year old retired Marine is running 240 miles one last time to raise funds to transfer patients to the Shriner's Crippled Childrens Hospital in Honolulu. He will finish his circuit of Okinawa this afternoon at Camp Foster's Kitimae Gate.

At last count, Knoll had earned about $210,000 for various children's organizations and hopes to earn a million for them, before quitting his runs.

With the money he earns from a book he published, titled *Why Not a Million?* and this last run on Okinawa, he hopes to get closer to the million dollar mark. He plans to continue running for charity, but not at the level he ran in the past. Knoll says pledges are received, not just from Okinawa, but also from civilian and military personnel from mainland Japan, Korea, Hawaii, California and even Germany.

Knoll began his annual runs here in 1978 during his last tour in Okinawa. Having run several marathons, he said he wanted one last big run around the island. "Sergeant Major Larry Henry, a good friend, suggested that I run it for donations to help crippled kids, so I did. And we raised about $10,000 that year," said Knoll, a Milwaukee, Wisconsin native. "It was supposed to be a one-time thing, but when a little girl gave me a kiss and told me she loved me at the finish, I was hooked," said Knoll.

Knoll retired from the Marine Corps at Munich, West Germany in 1983, where he was stationed for three years. He lived there for three more years before moving to Camp Zama, Japan, where he is now the liaison officer for the 500th Military Intelligence Brigade.

Knoll ran an average of 100 miles a week for the past few months, in preparation for this last run around Okinawa and considers himself in better shape than when he was 21. "It's for the kids. It's the time of year when we should think of someone else, this is one of the messages of Christmas," said Knoll. **Okinawa Marine, December 22, 1989**

Everyone Loves The Plumbers

"You give people a hot shower after a few days in the field and they treat you good," said Private First Class Mark Wernli. "We like to let them see us when they come out."

Probably the most popular Marines during Beach Crest 1989 (A Marine Air Ground Task Force exercise) were six field plumbers from Marine Wing Support Squadron-172, utility operations. They provided drinking water and hot showers for all the Marines deployed here.

Wernli is part of a team that sets up and operates the Reverse Osmosis Water Purification Unit. The $500,000 machine is capable of turning eight gallons per minute of salt water or 18 gallons per minute of fresh water from any puddle, pond or pool into pure drinking water. The machine is encased in a metal framework which makes it portable by truck, helicopter or plane, and can be set up by a six man crew and working within an hour.

Wernli said he often proves his point to interested visitors by purifying the nastiest water he can find, and after drinking some himself, he offers it to the visitor. "Most people think it tastes different, but that's only because it's purer than the water out of the tap." **Okinawa Marine, December 23, 1988**

Helicopter Crashes Off USS *Denver*

A Marine Corps helicopter with 22 Marines on board crashed during take-off from the USS *Denver* Tuesday during night operations in support of Exercise Valiant Mark 90-4.

Exercise Valiant Mark is an amphibious exercise, comprised of at-sea operations and small scale landings at designated training areas on Okinawa.

The crash occurred approximately 20 miles south of Okinawa and involved a single CH-46E helicopter from Marine Medium Helicopter Squadron 265, based at Kaneohe Bay, Hawaii. On board were four crewman and an additional 18 Marines from 2nd Battalion, 3rd Marines, also based in Hawaii. Both units are on Okinawa as part of the six-month unit deployment program.

Thus far, eight Marines have been rescued and 14 remain missing. **Okinawa Marine, June 2, 1989**

Volcano

When Mt. Pinatubo Erupted Marines Provided Disaster Relief

The evacuation of more than 20,000 members of American forces with their dependents from Clark Air Base and Subic Bay Naval Base, Republic of the Philippines, threatened by a series of volcanic eruptions from Mt. Pinatubo, ended in June (1991), with the help from the Marines.

Both U.S. facilities on the Philippines' largest island of Luzon were buried under six to 12 inches of volcanic ash from the mountain which started spewing on June 10.

After a week of volcanic eruption became life threatening, military officials ordered the evacuation of Americans. Pinatubo's threat was intensified by earthquake activity and Typhoon Yunya's winds and rains.

Clark Air Base was evacuated first, when Pinatubo oozed lava and spewed ash over the countryside, ending more than six centuries of dormancy. "When it blew, it blasted smoke and ash up to about 35,000-40,000 feet," said Lieutenant Colonel John M. Gautreaux, Marine Fighter Attack Squadron 122's commanding officer.

"I was on the flight line when it blew. It got dark because it basically covered the whole sky."

Fortunately, the squadron had already taken the precaution of sending most of its FA-18 Hornet jets to Kadena Air Base, Okinawa. Other squadrons departed Cubi Point along with United States Navy aircraft and planes from Clark Air Base. The Marines flew to Kadena and Marine Corps Air Station, Iwakuni, Japan, while the others went to other Pacific bases or aircraft carriers. On station over the Philippine and East China Seas to provide transiting aircraft with fuel for their journey, were KC-130 refuelers of Marine Aerial Refueler Squadron 152 out of Marine Corps Air Facility, Futenma, Okinawa.

Reports stated nearly every tin roofed aircraft hangar, building and storage shed at the two facilities had collapsed under the weight of the volcanic ash. At Clark Air Base, damage to offices, the commissary exchange, hospital and numerous homes was compounded by mud slides and flood waters that came in the wake of Typhoon Yunya's gales and rain.

The Navy and Marine Corps team responded quickly, along with the Air force and other United States agencies in the Philippines. Under Operation Fiery Vigil, 17 ships of the United States 7th Fleet transported evacuees to the nearby island of Cebu for further air transportation out of the endangered areas. A contingent of 275 Marines and sailors from Okinawa bases units assisted the 7th Fleet in providing logistical, medical and evacuation support for the operation.

All told, about 6,000 Marines and sailors were involved in providing disaster relief and assisting with security. More than 200 flights of medical evacuees from Subic Bay arrived at Kadena Air Base, Okinawa where United States forces and families were more than prepared for the influx of dependents. More than 700 families from all services offered to help by volunteering to give of their time or to take families from the Philippines into their homes. ***Leatherneck*, September 1991**

"Peleliu" Arrives During Evacuation

In June 1991, United States Marines, as well as the Navy and Air Force, went into action in rendering assistance and evacuating more than 20,000 members of American forces with their dependents from the Philippines due to volcanic eruptions of Mt. Pinatubo.

Medical evacuees began arriving at Kadena Air Base, Okinawa. More than 200 were from Subic Bay and one of the flights radioed ahead for special assistance. United States Air Force teams of 313th Medical Group were on hand to help. They brought down Cecilia Randall on a stretcher. She was holding her newborn son, Peleliu, named after the ship where he was born only hours before. The Air Force medics went to work ensuring the baby and mother were in good health.

Unaware of the events unfolding around him, Peleliu opened his eyes and yawned. ***Leatherneck*, September 1991**

Scuba Practice Takes Recon Down Under

Submerged 15 feet underwater, a Marine navigates his reconnaissance team close to shore, disregarding the fish swimming just a few feet away. The leader sends two scout swimmers to land as the rest await the signal to come ashore.

Leathernecks from Delta Company, 3rd Reconnaissance Battalion, 3rd Marine Division practiced tactical scuba insertions in six man teams in the Pacific Ocean.

"We can do ascents from a submarine where we send out six man teams to land on a beach to check the area for danger before sending more people," explained 1st Lieutenant Daniel Hodges, platoon commander.

The Marines simulated ascending from a sub by taking Zodiac rubber rafts a few hundred yards out to sea and then entering the water. Divers then shot azimuths while submerged and kept a 15 foot depth as they headed for shore. Teams swam in wedge formations with two scout swimmers in the lead. Once ashore, lead scout swimmers patrolled, ensuring the area was secure. One swimmer provided security while another went back to the beach and signaled the team to come ashore.

Signaling is important, and a swimmer from each team popped his head out of the water for two seconds every two minutes to spot the signaling scout swimmer. "It's just enough time to see if the beach is secure," said Sergeant Bill Mies, team leader.

The rest of the team exited the water in single file, then camouflaged their diving gear in the bushes before setting out to accomplish the mission.

Recon Marines train in many aspects of patrolling, parachuting and scuba diving—and all these require expertise and practice. ***Okinawa Marine*, December 14, 1990**

Scamp Monitors Troop Movements From Afar

Central Training Area, Okinawa—Marines from Third Sensor Control and Management Platoon dropped what they were doing—from the air—during a training exercise here July 25, 1990.

The Air Delivered Seismic Intrusion Device is designed to detect earth vibrations caused by vehicles or troops in movement.

Third SCAMP, 3rd Surveillance, Reconnaissance, Intelligence Group, 3rd Marine Division is trained to place sensors and monitor them for enemy movement within the sensor's detection radius.

"For the sensors to work, we must drop them from at least 500 feet," said Corporal Noel Navanjo, a SCAMP Marine. "Sensors are either hand-dropped from the back of a helicopter or mechanically dropped from a UH-1N Huey, as if it were dropping a bomb."

For this particular exercise, SCAMP Marines hand-dropped sensors from the tailgate of a CH-53 Sea Stallion.

This method requires a team of four Marines called a stick. Each stick is composed of a dropper, feeder, spotter and a plotter. It takes the combined effort of the entire stick to place sensors in the desired location.

"To get an accurate reading from the sensor, we string a series of sensors, normally three to five, along the side of the road to be monitored," said Gunnery Sergeant Allen Szczepek, platoon commander.

A SCAMP monitoring team determines the size of the unit or approximate amount of vehicles, what direction they're moving and their rate of speed, once the sensors are in place. SCAMP can monitor sensors from up to 50 miles away, depending upon the type of obstructions between the sensors and monitors.

"However," Szczpek added, "by dropping and using a series of monitors, we can watch the sensors seismographic data from hundreds of miles away."

When sensors detect movement, a series of lines appear on a roll sheet for SCAMP monitors to translate. Each line represents a different sensor. From this data, SCAMP determines a myriad of information about the unit crossing through the monitored area.

"These sensors have a lot of advantages over placing Marines in the field doing the same job," Szczepek said. "Sensors don't get hungry and don't need to sleep."

An Air Delivered Seismic Intrusion Device lays on the deck. The ADSID sensor detects earth movements made by troops or vehicles within the sensor's detection area. (Official USMC Photo Cpl. Bob Hall)

Snap Shooting Adds New Twist To Hitting Target

Marines take pride in their skill with a rifle, no matter how many "marksman" badges they have earned, and the ability to calmly deliver aimed fire from 500 meters is very important. A new twist in marksmanship training is snap shooting, or accurately hitting targets at very close range in a very short time.

When a Marine wades through streams in the Central Training Area, he would probably have to shoot from less than 25 yards. That's the battlefield skill that India Company, 3rd Battalion, 3rd Marines, sharpened recently at Range 4.

"We're using a 'quick-kill' course we set up," said Captain Jeff Patterson, company commander. "We start shooting at seven yards, then move back to 10, 20 and 25 yards." Accuracy was judged by points, with 20 rounds worth five points each.

"You don't have time to use the sights," said 1st Lieutenant Bill Foy, Weapons Platoon commander. Fire had to be delivered quickly by instinctively aligning the top of the carrying handle and front sight assembly. At the 25-yard mark, they had 10 seconds to fire two rounds, then fire two more in the prone position. For a perfect score of 100, every second counted, but only two made perfect scores. Spotting your enemy at 500 meters gives you time to shoot; however, when he's at close range, you have to be fast on the draw. ***Okinawa Marine*, December 22, 1988**

Lance Corporal Seth Calahan sits in the back of a CH-53 and watches for the signal to drop the Air Delivered Seismic Intrusion Device and pass it onto Lance Corporal Hector Cuadrado (left). The ADSID detects earth movements made by troops or vehicles within the sensor's detection area.

Marines Go "Mud-Boggin" Along

Marines from 3rd Combat Engineer Battalion's motor transport section ventured into Okinawa's vast jungle recently to hone their driving skills so they would be just as combat ready as any infantry unit.

More than 20 Marines from CEB's motor T section spent three days here driving along winding dirt roads typical of Okinawa's countryside.

"The purpose of the training was to improve the skills of our motor transport Marines," said 2nd Lieutenant Edward H. Romasko, MT's maintenance officer.

Mock ambushes were set up and motor transport maintenance Marines barricaded roads that a MT convoy would drive down.

As the convoy slowly rolled along the hazardous trails, a team of MT Marines scouted ahead and removed obstacles that would detain vehicles, all the while looking for possible enemy ambushes.

"Mud boggin" tests drivers' skills during 3d Combat Engineer Battalion's truck operation training. (USMC Photo by Cpl. Bob Hall)

"Our Marines must learn to overcome obstacles in the road and know how to react if their convoy is attacked," Romasko said.

Along with overcoming obstacles and ambush operations, MT Marines conquered traverse terrain.

However, 'mud boggin' wasn't just fun and games. Motor transport Marines must know how to drive through deep mud without getting their vehicles stuck. If they do get their vehicles stuck, they have to know how to get it out.

Marines Blast Weapons During Fam-Fire

Zimbales Training Area, Republic of Philippines—The deep roar of M-2 .50 caliber and the bark of M-60 machine guns broke the morning silence. Leathernecks of Engineer Support and Headquarters and Service Companies, 3rd Combat Engineer Battalion, 3rd Marine Division, conducted weapons familiarization firing and demolition training here.

During the week of firing, they fired more than 25,000 ball and 1,600 tracer rounds from their M-26s; 2,600 rounds from their squad automatic weapons; 1,400 9mm pistol rounds; 11,000 rounds from the M-60, and 10,000 rounds through the .50 caliber heavy machine guns.

They also threw more than 300 practice and 125 live fragmentation hand grenades. In the demolition phase, they blew up TNT, C-4, dynamite, bangalore torpedoes, shape and cratering charges and the Zunni rocket line charge system.

"The purpose of this training is to familiarize us with all types of weapons," Gunnery Sergeant Mike E. Kopanski, Engineer Superintendent Company said. "Some of the Marines have never thrown a grenade or fired a machine gun."

Before firing, Marines attended classes on both machine guns including general data, assembly, disassembly and the specific role for each.

After the classes, they quickly broke-down the ammunition and prepared to fire. Soon the smell of cordite filled the air as they sent thousands of high explosive rounds downrange toward makeshift targets.

During the training they practiced laying in guns on the firing line and how to provide final protective fire. According to Chief Warrant Officer-2 Robert S. Mizner, utilities platoon commander, EngrSupt Company, they must know how to handle the weapons because machine guns are their primary weapons.

"We don't have a weapons platoon or company attached to us," Mizner said. "We must know how to do everything with machine guns. They're our sole offense and defense."

The Marines invited members of the Philippine Security Forces to also fire each gun and throw grenades.

Cpl. Dale J. Siebert (left) of Heavy Equipment Platoon throws a practice grenade while 1st Lt. Martin D. LaPierre of Motor Transport Platoon watches his technique. (USMC Photo by Sgt. John C. DiDomenico)

MEF Success Due To Teamwork

Most Marines and sailors on Okinawa are attached to or support the 3rd MEF, but a vast majority don't even realize what its overall mission includes.

This concerns Major General H.C. Stackpole III, 3rd MEF commanding general, who feels the MEF is the most expeditionary force in the world today.

The 3rd MEF, as the United States' only forward deployed expeditionary force, has an area of responsibility 10 times the size of the United States.

Because the 3rd MEF works closely with the Navy's 7th Fleet and is capable of responding to low and mid intensity conflicts, Major General Stackpole said, "we are the epitome of the Navy-Marine Corps team. Being based on Okinawa, we can deploy—anywhere within the 7th Fleet. Unlike stateside units, we are not dependent upon a base or overflight rights." With the 3rd MEF role in combating terrorism and narcotics, the general says there is an increasing need for small unit training.

The charter for 3rd MEF is the Marine Air-Ground Task Force. It is a unique combination of an aviation combat element (1st MAW), ground combat element (3rd Marine Division), combat service support element (3rd FSSG) and a command element that ties them all together. *Okinawa Marine*, **April 13, 1990**

Girl Scout Cookies Even On Okinawa

The Girl Scouts of America will supply their counterparts here with their famous cookies for the first time, beginning this month.

In the past, Government of Japan trade regulations prohibited importation of the cookies. The West Pacific Girl Scouts will sell six varieties of cookies for $2 a box. They began neighborhood door-to-door and booth sales.

Girl Scouts, who range in age from six to 17, have been selling cookies to support their program for more than 50 years.

Proceeds will help both individual troops and the local Lone Troop Committee which provides services and facilities to the girls. *Okinawa Marine*, **February 9, 1990.**

Private First Class Hugh Archer, a Squad Automatic Weapons gunner from 3d platoon, Charlie Company, 1st Battalion, 24th Marines, guards the LZ perimeter. Marines from 1st Battalion, 24th Marines enhanced their combat readiness with helicopter indoctrination training at Camp Hansen. (USMC Photo by Cpl. Thomas J. LaPointe)

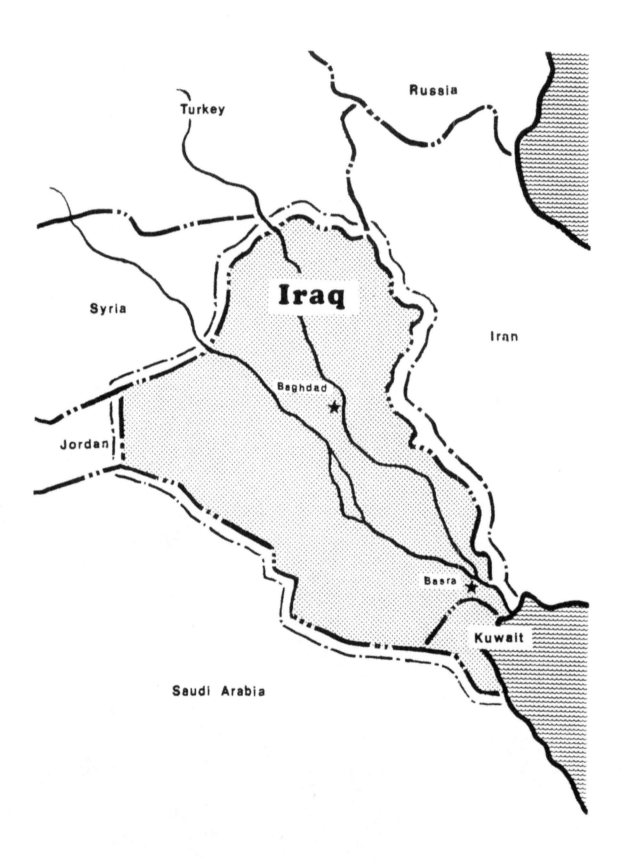

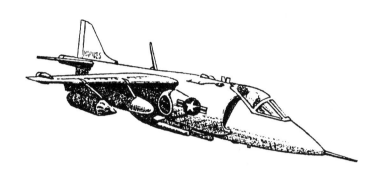

Desert Shield ...
Okinawa Based Marines
Head For Saudi Arabia

The recent call for additional reinforcements supporting Operation Desert Shield has resulted in the deployment of a number of Okinawa and mainland Japan-based III MEF units to Southwest Asia.

The deployments followed a November 30 announcement identifying 2,200 Marine reservists to deploy in support of III MEF.

Units identified for deployment during December and January are:

*1st Battalion, 1st Marines based at Camp Pendleton, California.

*3rd Battalion, 7th Marines, UDP to Okinawa from the Marine Corps Air Ground Combat Center, Twenty Nine Palms, California.

*2nd Battalion, 3rd Marines, UDP to Okinawa from MCAS Kaneohe Bay, Hawaii.

*Marine Attack Squadron 231, UDP to MCAS Iwakuni from MCAS Cherry Point, North Carolina.

*Marine Fighter Attack Squadron 212, based at MCAS Kaneohe Bay.

*Marine Fighter Attack Squadron 232, based at MCAS Kaneohe Bay.

*1st Battalion, 11th Marines, based at Camp Pendleton.

*2nd Battalion, 12th Marines, based on Okinawa with the 3rd Marine Division.

*3rd Battalion, 12th Marines, based on Okinawa with the 3rd Marine Division.

For Operation Desert Shield, Marine forces will be combined into 1 MEF, commanded by Lieutenant General Walter E. Boomer. I MEF's ground combat element will consist of two maneuver elements: the West Coast-based 1st Marine Division, commanded by Brigadier General James A. Myatt (currently in Southwest Asia) and the East Coast-based 2nd Marine Division, commanded by Major General William M. Keys.

1 MEF's aviation combat element will consist of a Marine Aircraft Wing, consolidating assets from the West coast-based 3rd MAW and the East Coast-based 2nd MAW. It will be commanded by Major General Royal N. Moore Jr.

1 MEF's combat service support element will be provided by the West Coast-based 1st FSSG, augmented by the East Coast-based 2nd FSSG. The reinforced 1st FSSG will be consolidated into a single combat service support organization, with 1st FSSG handling general support responsibilities, while 2nd FSSG manages direct support tasks. Brigadier General James A. Brabham Jr., commands 1st FSSG, with Brigadier General Charles C. Krulak acting as his deputy commander.

In addition to Marine forces assigned to 1 MEF, the 4th MEB has been assigned to Operation Desert Shield as part of the United States Naval Force Central Command, commanded by Vice Admiral Stanley Arthur.

Upon arrival in the area of operations, the Camp Pendleton-based 5th MEB, commanded by Brigadier General Peter J. Rowe, will consolidate with 4th MEB, under the command of Major General Harry W. Jenkins Jr. The 5th MEB set sail from West Coast ports on its way to Operation Desert Shield November 30. **Okinawa Marine, December 21, 1990.**

The Third Marine Division In The Gulf War

By Nicholas Kominus

This review of the Persian Gulf activity, and 3rd Marine Division participation was prepared by Nicholas Kominus, a 1st lieutenant who served in the Marine Corps Reserve from 1950 until 1962. His last active duty was as the intelligence officer of the 9th Marines when they were stationed in Japan in the mid-fifties.

A former Marine Corps Combat Correspondent, he is now president of U.S. Cane Sugar Refiners' Association, Washington, DC.

Things were looking up for the United States as we entered the 1990s. The Cold War was coming to an end, and world Communism was collapsing. Good was triumphing over evil. At long last, world peace appeared to be near at hand.

But, as they used to say in the Old Corps, and probably still say today, "every time things seem to be going right some SOB comes along and messes it up."

Saddam Hussein Appears

This time it was a despot named Saddam Hussein, a lifelong terrorist who seized control of Iraq in the late 1970s.

He had just emerged from a costly eight-year "holy" war with Iran in which he gained nothing, and became saddled with a huge debt. As a result, conditions were chaotic at home, and he resorted to the time-honored ploy of all tyrants by focusing the attention of his people abroad.

Although Kuwait had helped finance his war against Iran, he used old boundary disputes and grudges over OPEC oil pricing policies to justify his aggression against his small neighbor to the south.

Kuwait Is Overrun

At 0200 on August 2, 1990, he unleashed his surprise attack. By the end of the day, his army, the fourth largest in the world, and armed to the teeth by the Soviets, overran most of Kuwait.

(Continued on page 276)

60 mm Mortar

Island Units Mount Out; Next Stop Saudi Desert

Camp Butler—Focused attention on the Persian Gulf entered its second month last week, and no breakthrough appears imminent.

The military build up continues as more American ground forces, including Marine units home based here, arrived in the Middle East.

One of the highlights of the week involved the release from Iraq of more than 300 Westerners, including 47 American women and children.

Local Marine Corps units participating in the Persian Gulf crisis, dubbed Operation Desert Shield, comprise Regimental Landing Team 4 (reinforced).

A United Nations embargo on Iraq that started August 6 appears to be having some impact. Baghdad officials balked at allowing more airlifts of foreign hostages and told nations with citizens trapped there to urgently send food. It also suspended payment of its foreign debt September 3, in response to the embargo.

"There's not going to be any war unless the Iraqis attack," said Army General H. Norman Schwarzkopf, commander of United States Forces in the Middle East. Joint Chiefs of Staff Chairman General Colin Powell said, "Do not interpret our interest in peace as weakness. Do not think for one moment that we will be cowed or coerced by your actions or threats. We're made of tougher stuff." **Okinawa Marine, September 7, 1990**

It was too easy. His troops were now poised on the border of Saudi Arabia, ready to pounce on another neighbor that helped him against Iran, but one that he also despised. The hope for a world at peace suddenly evaporated. The civilized world had a problem on its hands. If Saddam got hold of Saudi Arabia, he'd control half of the world's known oil reserves.

President Bush quickly seized the initiative and effectively organized support in the United Nations for action against Saddam. Before it was over, 30 nations became a part of the United Nations coalition designed to defend Saudi Arabia, and throw Saddam out of Kuwait.

A Plan Of Action

The coalition buildup, Operation Desert Shield, was based upon a contingency plan developed earlier in anticipation of such a crisis, and was implemented quickly.

The Marines arrived early, by airlift and sea-lift, and along with United States airborne forces, snuffed any plans Saddam might have had for pushing on to Saudi Arabia, a la Adolph Hitler.

At the time Saddam struck, the 3rd Marine Division was scattered. The 3rd Marines were in Hawaii, and the 4th, the 9th, and 12th Marines were in Okinawa. The 3rd Marine Division was not deployed in the Gulf as a Division. However, all of its battalions, except 2/9, were deployed. The 3rd and the 4th Marines were attached to the 1st Marine Division.

Climate ... Horrible

They were sent to Saudi Arabia by air and sea. And once they got there, they discovered that the climate and terrain could hardly be described as hospitable. In addition to the Iraqis, they had to fight the heat and the sand. In August, when the troops first started to arrive, the daytime temperatures were well over 100 degrees with high humidity. At night, things cooled down to 95 degrees. As a result, water discipline took on new meaning.

Heat or no heat, the training began immediately, and in earnest. In addition to the normal training that the old timers will remember, gas or chemical warfare training had to be taken seriously. Saddam had reportedly used gas on the Iranians, and his own people, the Kurds. Iraq's chemical capability caused con-

(Continued on page 277)

M-60E
Machine gun

In the last analysis, what the Marine Corps becomes is what we make of it during our respective watches. And that watch of each Marine is not confined to the time he spends on active duty. It lasts as long as he is 'Proud to bear the title of United States Marine.'

Gen. Louis H. Wilson
26th Commandent of the Marine Corps
(Reprinted from Marine Corps Gazette)

siderable anxiety among the coalition. Once the campaign started, matters weren't helped by the fact that when an artillery shell hit the sand the explosion was muffled, and made a sound similar to the impact of a gas round. But, for some unknown reason, the Iraqis never used gas.

"Cross Training"

Because thirty nations were united against Iraq during the Gulf War, there was a unique and tremendous need for coordination, and so-called "cross training" with the Arab coalition.

The 3rd Marines, which became known as Task Force Taro, started cross training with the Saudi Marines located near their positions. This training went so well that the 3rd Marines became the focal point for cross training with the Arab Coalition Forces. In addition to the Saudis, this training involved numerous Arab forces, including those from nearby Qatar and Afghan "Freedom Fighters," the Mujahedin.

Colonel John H. Admire, the commanding officer of the 3rd Marines, said "a unique camaraderie developed as a natural result of the challenges and sacrifices of desert life. American Marines were invited into Saudi Bedouin tents for meals and began to experience Arab culture and hospitality. Marines hosted Arab Coalition Forces during our traditional Marine Corps Birthday Ceremony and acquainted Arabs with the heritage of our Corps. The success of the cross training was no small accomplishment and contributed materially to later victories.

Desert Storm Begins

In January 1991, the 3rd Marines deployed in an area just 20 kilometers from the border. On January 17, 1991, after all attempts failed to get Saddam out of Kuwait peacefully, Desert Shield ended, and Desert Storm began with intensive coalition air attacks. This softening up continued until February 24 when the ground offensive officially began.

In the interim, there were, however, a number of border clashes. On January 18, the 3rd Marines became the first ground unit to come under heavy fire from Iraqi artillery, rocket, and missile fire. The 3rd responded with artillery and, by so doing, began the first coalition ground action of the war. The Iraqis initiated a number of probes and Ma-

(Continued on page 278)

Wives Share Deployment Concerns

Camp Butler—Deployment is part of military life—and separation is not new for thousands of families left behind. But for women here whose husbands have left for Operation Desert Shield, the current departure has been different. They don't know when their husbands will return, or even if they will.

The wives left here share the same circumstances as military wives across the United States, they must endure an extended separation from loved ones who are in a potentially hostile environment.

Marines who've deployed to places unknown for this operation should feel confident about the help their families get, according to officials here. That help includes wives support groups, the Family Service Center and family assistance programs. "None of the wives we've dealt with are newly arrived. All have been here awhile and have been associated with the regiment for a while," said Lieutenant Colonel Joel L. Goza, 3rd Battalion, 12th Marines, commander. "We've been doing this for a few weeks and haven't had any major crises. I think just being there helps. We understand this is different from a regular deployment."

Like other deployed units with wives on Okinawa, 3/12 conducts deployment briefs and has a spouse support team that's limited to the commanding and executive officers, sergeant major and battalion chaplain.

The Deployed Family Assistant Program also assists deployed Marines' wives. "I try to work as a liaison between dependents and all supporting agencies on the island," said Captain Frank A. Cassiano Jr., program coordinator.

Workshops on surviving military separations as well as courses in stress management, active individual parenting and individual and family counseling are also offered.

"We've discussed what we thought our roles would be," said Gerry Warner, Family Service Center's information referral service director. "We know there's no overlap, that we're not re-creating the wheel. We're tapping wives clubs for resources as volunteers to train. The FSC has ten lines set up to handle calls from Desert Shield. Right now, we have four volunteers, but we're looking for 20."

"Often, when people are back in the states, wives go back to their home communities when their husbands deploy," Warner said. "We realize here it'll cost money to ship families back. That may affect the mission of the operation. Our FSCs can provide a lot of help here and other wives understand the problems of deployments. That won't be the case back in Wisconsin or wherever. Here, we're the only game in town," Warner said. *Okinawa Marine*, October 26, 1990

rines suffered their first casualties, some of which were inflicted by friendly fire.

A week later, the 3rd Marines became involved in what was probably the most notable ground battle of the war, the battle of Khafji. The Iraqis attacked and seized the border town on the coast, which had been abandoned by the Saudis earlier. Two six-man reconnaissance teams from the 3rd were temporarily trapped in an abandoned hotel in Khafji. During that time, they managed to send intelligence reports back to headquarters.

Initially, the 3rd Marines were slated to recapture Khafji, but, the Arab Coalition Forces were eager to prove their mettle, and United States leaders deferred to them. Troops from Saudi Arabia and Qatar recaptured the town as the 3rd Marines backed them up. The 12th Marines fired in support of the Saudi and Qatar forces.

Now Infiltrators

As a result of the Arab Coalition's outstanding performance, plans for the final campaign were altered, and the 3rd Marines moved west to rejoin the 1st Division for the first time in nearly two months. Upon rejoining the 1st, the Regiment received a new mission as an infiltration force. It would now be deployed to penetrate the Iraqi defenses, and protect the main assault force from Iraqi armor counterattacks on the right flank. The 4th Marines, called Task Force Grizzly, were given a similar mission on the left flank of the 1st Division.

Meanwhile, other 3rd Marine Division units were involved in the amphibious exercises off the coast of Kuwait, which were designed to keep Saddam's generals off balance. They did. The Iraqis had to retain a number of divisions on the coast in anticipation of amphibious landings that never materialized. Colonel Admire said the 3rd Marines "accepted our supporting attack role with the understanding that we would have no armor, no assault amphibian vehicles, and no major mechanical or explosive breaching assets. We would simply infiltrate at night, on foot, with bayonets and rifles as our principal weapons."

Colonel Admire reported that "the evening of February 22 we crossed the border into Kuwait on foot to attack positions south of the Iraqi defensive barrier. Throughout the daylight hours of February 23 we remained undetected in harbor sites and prepared for the infiltration. Then, the evening of February

(Continued on page 279)

Nicholas Kominus

Nicholas Kominus is the President of the U.S. Cane Sugar Refiners' Association, which is headquartered in Washington, D.C. Prior to joining the Association 31 years ago, he was an editor with the U.S. Department of Agriculture. Earlier, he worked for a number of newspapers.

He enlisted in the Marine Corps Reserve in 1950, and was released in 1962 as a First Lieutenant. His last active duty assignment in 1955 when he was the Intelligence Officer of the 9th Marines when they were stationed in Japan.

23, crawling on hands and knees, Task Force Taro infiltration forces penetrated the substantial Iraqi minefields, barbed-wire obstacles, tank traps, and earthen berms. By sunrise the lead elements had penetrated the barrier and initiated the clearing and proofing of three vehicle lanes for follow-on-forces."

Changes In Weather

United States commanders anticipated that the Iraqis would defend the first barrier lightly, and the second barrier strongly. After the first barrier was breached the weather was bad. Winds and smoke from the oil well fires started by the Iraqis reduced ground visibility to around 100 meters, and made crucial Marine air support nonexistent.

But, shortly before a decision had to be made as to whether to delay the assault on the second barrier, the winds shifted, the clouds disappeared, and the skies cleared. Colonel Admire later said that this change in the weather "was as if by divine intervention." The 3rd Marines attacked on schedule, and the second barrier was secured. These successful thrusts enhanced the main assault. There were brief, but fierce, encounters. All sorts of enemy equipment was destroyed, numerous prisoners were taken.

Helicopter Assault

The 3rd Marines also had the distinction of conducting the only Marine helicopter assault of the War. On February 25 the 3rd secured the 1st Division's right flank by flying into the burning Burgan oil fields. The following day and into the night, the unit helped secure the Kuwait National Airport, which was the Corps' final objective of the war. A cease-fire was proclaimed on February 28.

The Gulf War was over sooner than the prophets of doom had predicted. Saddam's prophecy about the mother of all battles fizzled. Thankfully, casualties were light.

Third Division Marines contributed materially to the success of the campaign. The troops and their leaders performed admirably.

A burning Iraqi tank

279

3rd Marine Division
Medal Of Honor Winners

World War II

Private First Class Henry Gurke, USMC, Neche, ND
Second Lieutenant John H. Leims, USMCR, Chicago, IL
Private First Class Leonard F. Mason, USMC, Middleborough, KY
Sergeant Robert A. Owens, USMC, Spartanburg, SC
Private First Class Luther Skaggs, USMCR, Henderson, KY
Sergeant Herbert J. Thomas, USMCR, Columbus, OH
Private Wilson D. Watson, USMCR, Tuscumbia, AL
Corporal Hershel W. Williams, USMCR, Quiet Dell, WV
Captain Louis H. Wilson Jr., USMC, Brandon, MS
Private First Class Frank P. Witek, USMCR, Derby, CT

Vignettes Of Heroism ...

Private First Class Henry Gurke, 3rd Raider Battalion, November 9, 1943 at Empress Augusta Bay, Bougainville. He smothered an enemy grenade with his body, saving lives of his comrades, and allowed successful defense of a position which was about to be overrun.

2nd Lieutenant John H. Leims, Company B, 1st Battalion, 9th Marine Regiment, March 5, 1945, Iwo Jima. He led an attack against heavily defended and fortified enemy positions, withdrew his forces when they were about to be surrounded, but returned repeatedly to the besieged position to rescue wounded from certain death.

Private First Class Leonard F. Mason, Company E, 2nd Battalion, 3rd Marine Regiment, July 22, 1944 on Guam. He single-handedly destroyed two Japanese machine gun positions allowing his platoon to accomplish its mission. Mason died of wounds suffered in the encounter.

Sergeant Robert F. Owens, 3rd Marine Regiment, November 1, 1944, Empress Augusta Bay, Bougainville. Sergeant Owens charged into the fire port of a 75mm gun bunker, silencing the gun and its crew and sparing the invasion beach innumerable casualties. He was mortally wounded.

Private First Class Luther Skaggs, 3rd Battalion, 3rd Marine Regiment, July 21-22 on Asan-Adelup Beachhead, Guam. He assumed command of a mortar unit on loss of its section leader, advanced to better counter the enemy, then persisted to defend the critical mortar position in the face of strong enemy counterattacks ... although severely wounded.

Sergeant Herbert J. Thomas, 3rd Marine Regiment, November 7, 1943 at Koromokina River, Bougainville. After leading his squad to abolish two machine gun positions, Sergeant Thomas threw himself on his own hand grenade which had rebounded from jungle vines as he prepared to attack a third position. His sacrifice saved the lives of his men who destroyed the enemy machinegun.

Private Wilson D. Watson, 2nd Battalion, 9th Marine Regiment, February 26-27, 1945, Iwo Jima. Destroying an enemy pillbox, Sergeant Watson then stormed up a steep incline to the top of a fortified hill where, amidst the enemy, he killed 65 before he ran out of ammunition for his Browning Automatic Rifle and was joined by the rest of his squad.

Corporal Hershel W. Williams, 1st Battalion, 21st Marine Regiment, February 23, 1945, Iwo Jima. Volunteering for a critical four-hour mission, Corporal Williams—assaulted a series of unyielding emplacements with his flamethrower. He eradicated some of the most fanatically-held strong points faced by his regiment.

Captain Louis H. Wilson Jr., F Company, 2nd Battalion, 9th Marine Regiment, July 1944, Guam. He initiated repeated attacks with daring combat tactics to capture and hold against savage counterattacks the strategic high ground on Fonte Ridge. Captain Wilson contributed immeasurably to the success of the regimental mission and to breaking the back of enemy resistance above the beachhead.

Private First Class Frank P. Witek, 1st Battalion, 9th Marine Regiment, August 3, 1944, Guam. When his platoon was met with surprise fire from a concealed enemy position, Private First Class Witek was able to protect most of them, and a wounded comrade, by the incessant response of his Browning Automatic Rifle. Later he was to move ahead of his own tanks and infantry to destroy a blocking machine gun position and lose his life in the action.

Medal Of Honor Winner Thomas Was Not Forgotten Back Home

When Sergeant Herb Thomas was killed in action on Bougainville and awarded the Medal of Honor, the people back home in Charleston, WV and at his school—Virginia Polytechnic Institute—did not forget him.

- A new dormitory at VPI was named in his honor.

- A hospital under construction in S. Charleston, WV was named the Sergeant Herbert J. Thomas Memorial Hospital.

- The Charleston Detachment of the Marine Corps League was named the Herbert J. Thomas Detachment.

- The South Charleston Veterans of Foreign Wars Post changed its name from Lane Anderson (a hero of World War I) to the Anderson-Thomas Post.

- The destroyer *Herbert J. Thomas* was launched March 25, 1944 at the Bath Iron Works Corp. yard. The destroyer was christened by his sister, Miss Audrey Irene Thomas, of South Charleston, WV.

Herbert J. Thomas had been an outstanding football player at Virginia Polytechnic Institute in Blacksburg, VA. In the fall of 1940 (his senior year), he was leading high scorer in Virginia and second leading scorer in the Southern Conference. He received honorable mention All-American, and All-Southern and All-State honors.

Herb left college in April 1941, two months before he would have received his BS degree in business administration, to join the Air Corps. From there he went to the Marine Corps and eventually joined the 3rd Marine Regiment of the 3rd Marine Division.

It was at the the Torokina River, Bougainville on November 7, 1943 when Sergeant Thomas led his squad, under heavy fire, through jungle growth-covered terrain. The squad had destroyed two Japanese machine gun nests, and had discovered a third strongly-emplaced position.

Reconnoitering an avenue of approach, Thomas ordered his men to charge the gun after he had thrown a grenade into the emplacement.

When ready to attack he threw a grenade which struck against vine growth overhead and fell directly among the closely-grouped members of his squad. Without hesitation he leaped and threw himself upon the grenade. The explosion killed him instantly.

Thomas was posthumously awarded the Navy Cross "for conspicuous gallantry" on that day, and later awarded the Congressional Medal of Honor. The

This photo of Herbert J. Thomas came from his yearbook at Virginia Polytechnic Institute.

presentation was made to his father, Herbert J. Thomas Sr. by Secretary of the Navy James Forrestal.

A July 1 editorial in his college newspaper noted "Those of us who sit safely at home, fretting about petty inconveniences, and spending money on extravagances when we could be putting it into War Bonds, ought to give a thought to Herb Thomas and the example he has set for us all ... We on the home front must prove ourselves worthy of his sacrifice."

Medal of Honor Winner and Man He Saved

James Ganopulos (right) owes his life to Capt. John H. Leims. Ganopulos was one of the wounded Marines saved by Leims on 7 March 1945 during the battle for Iwo Jima and for which Leims was awarded the Medal of Honor. Ganopulos attended the presentation ceremony and the luncheon following it when Leims was presented the Medal of Honor by President Harry S. Truman in 1946. Ganopulos resides in Homestead, PA. Leims died in the late 80's.

The Medal Of Honor In Vietnam

Fifty-seven Marines were awarded the Medal of Honor for exceptional heroism in Vietnam, 31 of which were serving with the 3rd Marine Division. Nine of those 31 survived the actions for which their awards were made. The 31 awardees included six privates first class, six lance corporals, four lieutenants, five captains, and a Navy medic.

Division Medal Of Honor Awards - Vietnam

Medal Awardee (Birthplace)	Unit	Date and Location
*ANDERSON James Jr., PFC (Los Angeles, CA)	F-2-3	28 Feb 67 near Cam Lo
*ANDERSON, Richard A., LCpl (Washington, DC)	3rd Recon	24 Aug 69 Quang Tri
BALLARD, Donald E., H2C USN (Kansas City, MO)	M-3-4	16 May 68 Quang Tri
*BARKER, Jedh C., LCpl (Franklin, NH)	F-2-4	21 Sep 67 near Con Thien
BARNUM, Harvey C., 1st Lt (Cheshire, CT)	H-2-9	18 Dec 65 Ky Phu
*BOBO, John P., 2nd Lt (Niagara Falls, NY)	3-9	30 Mar 67 Quang Tri
*CARTER, Bruce W., PFC (Schenectady, NY)	H-2-3	7 Aug 69 Quang Tri
*COKER, Ronald L., PFC (Alliance, NE)	M-3-3	24 Mar 69 Quang Tri
*CONNOR, Peter S., SSgt (Orange, NJ)	F-2-3	25 Feb 67 Quang Tri
*CREEK, Thomas E., LCpl (Joplin, MO)	I-3-9	13 Feb 69 near Cam Lo
*DICKEY, Douglas E., PFC (Greenville, OH)	C-1-4	26 Mar 67
*FOSTER, Paul H., Sgt (San Mateo, CA)	2-4	14 Oct 67 near Con Thien
FOX, Wesley L., Capt (Herndon, VA)	A-1-9	22 Feb 69 Quang Tri
*GRAVES, Terrence C., 2nd Lt (Corpus Christi, TX)	3rd Recon	16 Feb 68 Quang Tri
*JENKINS, Robert H. Jr. PFC (Interlachen, FL)	3rd Recon	5 Mar 69 near DMZ
LEE, Howard V., Maj (New York, NY)	E-2-4	8-9 Aug 66 near Cam Lo
LIVINGSTON, James E., Capt. (Towns, GA)	E-2-4	2 May 68 Dai Do
*MAXAM, Larry L., Cpl (Glendale, CA)	D-1-4	2 Feb 68 Cam Lo
McGINTY, John J., III, SSgt (Boston, MA)	K-3-4	18 July 66
MODRZEJEWSKI, Robert J., Capt (Milwaukee, WI)	K-3-4	15-18 Jul 66
*MORGAN, William D., Cpl (Pittsburgh, PA)	H-2-9	25 Feb 69 Quang Tri
*NOONAN, Thomas P. Jr., LCpl (Brooklyn, NY)	G-2-9	5 Feb 69 A Shau

Division Medal Of Honor Awards - Vietnam (continued)

Medal Awardee (Birthplace)	Unit	Date and Location
O'MALLEY, Robert E., Cpl (New York, NY)	I-3-3	18 Aug 65 near An Cuong
*PAUL, Joe C., LCpl (Williamsburg, KY)	H-2-4	18 Aug 65 near Chu Lai
*PERKINS, William T. Jr., Cpl (Rochester, NY)	C-1-1	12 Oct 67 Quang Tri
*PROM, William R., LCpl (Pittsburgh, PA)	I-3-3	9 Feb 69 near An Hoa
*REASONER, Frank S., 1st Lt (Spokane, WA)	3rd Recon	12 Jul 65 near Da Nang
*SINGLETON, Walter K., Sgt (Memphis, TN)	A-1-9	24 Mar 67 Quang Tri
*TAYLOR, Karl G. Sr., SS (Laurel, MD)	I-3-26	8 Dec 68
VARGAS, M. Sando Jr., Capt (Winslow, AZ)	G-2-4	30 Apr-2 May 68 Dai Do
*WILSON, Alfred M., PFC (Olney, IL)	M-3-9	3 Mar 69 Quang Tri

* Posthumous

NOTE: Corporal Perkins is shown as having been in service with the 1st Marine Division when he won the Medal of Honor. The compilation in the 3rd Marine Division 1992 Directory of Membership shows him as a member of Service Co., Headquarters Battalion, 3rd Marine Division. Likewise, Staff Sergeant Connor is shown as a member of the 1st Division when he won the award, and in the Directory of Membership as having been with F-2-3.

The source of their units and information was *The Congressional Medal of Honor: The Names ... The Deeds*, published in 1988 at Chico, CA.

Vignettes Of Heroism In Vietnam

Division Medal of Honor winners whose stories are not featured elsewhere in the Vietnam War section, with brief excerpts from their citations:

Private First Class James Anderson Jr. (Los Angeles, CA) F-2-3: "... Company F was advancing in dense jungle northeast of Cam Lo ... to extract a heavily besieged reconnaissance patrol ... Suddenly, an enemy grenade landed amidst the Marines and rolled alongside Private First Class Anderson's head. Unhesitatingly, and with complete disregard for his personal safety, he reached out, grasped the grenade, pulled it to his chest, and curled around it as it went off."

Lance Corporal Richard A. Anderson (Washington, DC), 3rd Recon: "... While conducting a patrol ... his reconnaissance team came under a heavy volume of fire ... Although painfully wounded in both legs, Lance Corporal Anderson continued to deliver intense fire ... wounded a second time, he continued to pour an intense stream of fire ... Observing an enemy grenade land between himself and another Marine (who was treating him), Lance Corporal Anderson immediately rolled over and covered the lethal weapon with his body, absorbing the full effects of the detonation."

HM2C Donald E. Ballard (Kansas City, MO) M-3-4: "... Ballard was returning to his platoon from the evacuation landing zone when the company was ambushed by an NVA unit employing automatic weapons and mortars ... After rendering medical assistance, he directed four Marines to carry the casualty to a position of relative safety ... an enemy hand grenade landed near the casualty ... Ballard fearlessly threw himself upon the lethal explosive device to protect his comrades. When the grenade failed to detonate, he calmly arose ... and resolutely continued his determined efforts in treating other Marine casualties."

Lance Corporal Jedh C. Barker (Franklin, NH), F-2-4: "... While serving as a machine gunner ... during a reconnaissance operation, Lance Corporal Barker's squad was suddenly hit by sniper fire ... Although wounded ... Lance Corporal Barker, in the open delivered a devastating volume of accurate fire ... again wounded ... an enemy grenade landed in the midst of the few surviving Marines. Unhesitatingly and with complete disregard for his personal safety, Lance Corporal Barker threw himself upon the deadly grenade."

2nd Lieutenant John P. Bobo (Nigagara Falls, NY) I-3-9: "Establishing night ambush sites ... the command group was attacked. He organized a hasty defense ... when an exploding enemy mortar round severed Lieutenant Bobo's right leg below the knee, he refused to be evacuated ... with his leg jammed into the dirt to contain the bleeding, he ... delivered devastating fire into the ranks of the enemy attempting to overrun the Marines ... mortally wounded ... his valiant spirit inspired his men ... who repulsed the enemy onslaught."

Private First Class Bruce W. Carter (Schenectady, NY) H-2-3: "... Private First Class Carter's unit came under a heavy volume of fire from a numerically superior enemy force ... Private First Class Carter and his fellow Marines were pinned down by vicious crossfire when, with complete disregard for his safety, he stood in full view ... to deliver a devastating volume of fire ... leading them (Marines) from the path of a rapidly approaching brush fire ... a hostile grenade landed between him and his companions ... fully determined to protect the men following him, he unhesitatingly threw himself on the grenade."

Private First Class Ronald L. Coker (Alliance, NE) M-3-3: "... As Private First Class Coker's squad neared a cave, it came under intense hostile fire, seriously wounding one ... Private First Class Coker moved across the fire-swept terrain toward his companion. Although wounded ... he began to drag his injured comrade toward safety when a grenade landed on the wounded Marine ... Private First Class Coker grasped it ... and turned away ... it exploded. Severely wounded, but undaunted ... as he moved toward friendly lines (with his comrade) ... two more enemy grenades inflicted still further injuries. Private First Class Coker with supreme effort continued to crawl and pull the wounded Marine with him."

Staff Sergeant Peter S. Connor (Orange, NJ) F-2-3: "... Staff Sergeant Connor spotted an enemy spider hole emplacement approximately 15 meters to his front. He pulled the pin from a fragmentation grenade, intending to charge the hole boldly ... he realized that the firing mechanism was faulty ... he held the safety device firmly, but the fuse charge was already activated ... Manifesting extraordinary gallantry and with utter disregard for his personal safety, he chose to hold the hand grenade against his body to absorb the explosion and spare his comrades."

Lance Corporal Thomas E. Creek (Joplin, MO) I-3-9: "... Providing security to a convoy ... the Marines came under a heavy volume of fire ... Observing a position from which he could more effectively deliver fire ... he fearlessly dashed across the fire-swept terrain and was seriously wounded ... an enemy grenade was thrown into the gully where he had fallen ... Lance Corporal Creek rolled on the grenade and absorbed the full force of the explosion ... saving the lives of five fellow Marines."

Private First Class E. Dickey (Greenville, OH) C-1-4: "... Private First Class Dickey had come forward to replace a radio operator wounded in intense action ... Suddenly an enemy grenade landed in the midst of a group of Marines, including the wounded radio operator who was immobilized ... Private First Class Dickey, in a final valiant act, quickly and unhesitatingly threw himself on the deadly grenade."

Sergeant Paul H. Foster (San Mateo, CA) 2-4: "...As an artillery liaison operations chief ... his position in the fire support coordination center was dangerously exposed and he was wounded when an enemy hand grenade exploded ... Sergeant Foster resolutely continued to direct mortar and artillery fire ... an enemy hand grenade landed in the midst of Sergeant Foster and five companions ... he shouted a warning, threw his armored vest over the grenade, then unhesitatingly placed his body over the armored vest."

Captain Wesley L. Fox (Herndon, VA), A-1-9: "... Under intense fire from a well-concealed enemy force ... Captain Fox was wounded along with all of the other members of the command group (except the executive officer) ... Advancing through heavy enemy fire, he personally neutralized one enemy position and calmly ordered an assault ... coordinating aircraft support with the activities of his men. When his executive officer was mortally wounded, Captain Fox reorganized the company and directed the fire ... Wounded again, Captain Fox refused medical attention, established a defensive posture, and supervised the preparation of casualties for medical evacuation ... He inspired his Marines to such aggressive action that they overcame all enemy resistance and destroyed a large bunker complex."

Second Lieutenant Terrance C. Graves (Corpus Christi, TX) 3rd Recon: "... He and two patrol members ... suddenly came under a heavy volume of hostile small arms and automatic weapons fire ... After attending the wounded, 2nd Lieutenant Graves ... launched a determined assault, eliminating the remaining enemy troops ... The unit again came under heavy fire which wounded two more Marines and 2nd Lieutenant Graves. Refusing medical attention, he once more adjusted air strikes and artillery fire ... He led his men to a new landing site into which he skillfully guided the incoming aircraft and boarded his men ... exposed to the hostile fire ... Realizing that one of the wounded had not embarked, he directed the aircraft to depart ... and moved to the side of the casualty ... directed fire until a second helicopter arrived ... enemy fire intensified, hitting the helicopter ... all aboard were killed."

Private First Class Robert H. Jenkins Jr. (Interlachen, FL) 3rd Recon: "... Private First Class Jenkins' 12-man reconnaissance team ... were assaulted by an NVA Army platoon employing mortars, automatic weapons, and hand grenades ... Private First Class Jenkins and another Marine ... delivered accurate machine gun fire ... a North Vietnamese soldier threw a hand grenade into the emplacement ... Private First Class Jenkins quickly seized his comrade and ... leaped on top of the Marine to shield him from the explosion, absorbing the full effect of the detonation."

Captain James E. Livingston (Towns, GA) E-2-4: "... He fearlessly led his men in a savage assault against enemy emplacements ... Although twice painfully wounded by grenade fragments ... he led his men in the destruction of over 100 mutually supporting bunkers ... halted by a furious counterattack of an enemy battalion ... (under) a heavy volume of enemy fire, Captain Livingston boldly maneuvered the remaining effectives and halted the enemy counterattack. Wounded a third time and unable to walk, he steadfastly remained ... deploying ... supervising."

Corporal Leonard L. Maxam (Glendale, CA) D-1-4: "... At the weakened section of the perimeter, completely exposed to concentrated enemy fire, he sustained multiple fragmentation wounds ... as he ran to an abandoned machine gun position ... commenced to deliver effective fire ... received a direct hit from a rocket-propelled grenade ... inflicting severe wounds to his face and right eye ... courageously resumed his firing position and ... was struck again. With resolute determination, he gallantly continued to deliver intense fire ... additional wounds ... too weak to reload his machine gun ... he continued to deliver effective rifle fire ... repeatedly hit, he succumbed ... having defended nearly half of the perimeter singlehandedly."

Corporal William D. Morgan (Pittsburgh, PA) H-2-9: "... Observing that two wounded Marines had fallen in a dangerously exposed position ... all attempts to evacuate them halted ... Corporal Morgan shouted encouragement to them as he initiated an aggressive assault against the hostile bunker ... clearly visible ... hostile soldiers turned their fire in his direction and mortally wounded him ... diversionary tactic enabled the remainder of his squad to retrieve their casualties and overrun the NVA position."

Lance Corporal Thomas P. Noonan (Brooklyn, NY) G-2-9: "... Four men were wounded, and repeated attempts to recover them failed because of intense hostile fire ... Lance Corporal Noonan ... shouting words of encouragement ... dashed across the hazardous terrain and commenced dragging the most seriously wounded man ... toward minimal security ... before he was mortally wounded ... his heroic actions inspired his fellow Marines to a (successful) spirited assault."

Corporal Robert E. O'Malley (New York, NY) I-3-3: "... Raced across an open rice paddy ... jumping into a trench ... singly killed eight of the enemy ... then led squad to assistance of an adjacent unit suffering heavy casualties ... regrouping remnants ... gathered besieged and badly wounded under fire to helicopter for withdrawal ... three times wounded, steadfastly refused evacuation, while covering his squad's boarding while from an exposed position he delivered fire against the enemy."

Lance Corporal Joe C. Paul (Williamsburg, KY) H-2-4: "... Lance Corporal Paul ... chose to disregard his safety and boldly dashed across the fire-swept rice paddies, placing himself between his wounded comrades and the enemy, and delivered effective suppressive fire from his automatic weapon to divert the attack long enough to allow the casualties to be evacuated ... Although critically wounded, he resolutely remained in his exposed position and continued to fire ... until he collapsed."

Corporal William T. Perkins Jr. (Rochester, NY) C-1-1: "... While serving as a combat photographer attached to Co. C ... in the course of a strong hostile attack, an enemy grenade landed ... he shouted the warning ... and hurled himself on the grenade, absorbing the impact of the explosion with his body, thereby saving the lives of his comrades at the cost of his."

Lance Corporal William R. Prom (Pittsburgh, PA) I-3-3: "... Realizing that the enemy would have to be destroyed before the injured Marines could be evacuated, Lance Corporal Prom ... was

instrumental in routing the enemy, permitting his men to regroup ... The platoon again came under heavy fire ... Reacting instantly, Lance Corporal Prom moved forward to protect an injured comrade. Unable to continue his fire because of severe wounds, he continued to advance ... in full view of the enemy, he accurately directed the fire of his support elements until mortally wounded."

First Lieutenant Frank S. Reasoner (Spokane, WA) A-3rd Recon: "... Under fire from numerous concealed positions, and virtually isolated from the main patrol body, he organized a base of fire for an assault ... Repeatedly exposing himself ... he skillfully provided covering fire ... silencing an automatic weapons position in a valiant attempt to effect evacuation of a wounded man ... His radio operator was wounded, and 1st Lieutenant Reasoner moved to his side and tended his wounds ... when the radio operator was hit a second time attempting to reach a covered position, 1st Lieutenant Reasoner, courageously running to his aid, fell mortally wounded ... provided the inspiration to enable the patrol to complete the mission."

Sergeant Walter K. Singleton (Memphis, TN) A-1-9: "... Sergeant Singleton quickly moved from his relatively safe position ... and made numerous trips through the enemy killing zone to move injured men out of the danger area ... he seized a machine gun and assaulted the key enemy location ... forced his way into the enemy strong point. Although mortally wounded, his fearless attack killed eight of the enemy and drove the remainder ... saved the lives of many of his comrades."

Staff Sergeant Karl G. Taylor (Laurel, MD) I-3-26: "... Staff Sergeant Taylor crawled forward to the beleaguered unit through a hail of hostile fire ... repeat-

edly maneuvered across an open area ... to rescue seriously wounded ... In another area, in proximity to an enemy machine gun position ... his rescue group was halted by devastating fire ... he took his grenade launcher and in full view of the enemy charged across the open rice paddy ... although wounded several times, he succeeded in reaching the machine gun bunker and silencing the fire from that section, moments before he was mortally wounded."

Captain M. Sando Vargas Jr. (Winslow, AZ) G-2-4: "... Though suffering from wounds ... Captain Vargas led ... an attack on the fortified village of Dai Do ... Pinned down by intense enemy fire ... again wounded by grenade fragments, he reorganized his unit ... into a strong defensive perimeter ... launched a renewed assault ... enemy retaliated with a massive counterattack ... in the open, encouraging and rendering assistance ... hit for the third time ... Observing his battalion commander sustain a serious wound, he disregarded his excruciating pain ... carried his commander to a covered position ... resumed supervising his men while simultaneously assisting in organizing battalion perimeter defense."

Private First Class Alfred M. Wilson (Oleny, IL) M-3-9: "... Acting as squad leader ... skillfully maneuvered his men to form a base of fire and act as a blocking force ... machine gunner and assistant were seriously wounded ... Realizing the urgent need to bring the weapon into operation ... fearlessly dashed across the fire-swept terrain to recover the weapon ... as they reached the machine gun ... observing a grenade fall between himself and the other Marine ... Private First Class Wilson ... unhesitatingly threw himself on the grenade, absorbing the full force of the explosion with his own body."

Medals Of Honor Earned At "Rockpile"

The "Rockpile," a torturous 700-foot mountain in the middle of an open area about 16 miles west of Dong Ha near the DMZ, was a keystone of an area in which the most savage battle of the Vietnam War to that point took place in the summer of 1965.

Operation Hastings began in earnest the morning of July 15, with Task Force Delta. Two of its three battalions (2/4 and 3/4) and its artillery support (3/12) were from the 3rd Marine Division.

A disaster-plagued heliborne troop assault followed an intense aerial assault, and additional Marine troops ultimately joined the heated combat with fresh and fanatically aggressive NVA units fighting to save their homeland.

Enemy forces battling Captain Robert J. Modrzejewski's K-3-4 attacked in mass formations and died in the hundreds. Staff Sergeant John J. McGinty described the action:

"We started getting mortar fire, followed by automatic weapons from all sides. The NVA were blowing bugles, and we could see them waving flags. They moved in waves with small arms (fire) right behind their mortars, and we estimated we were being attacked by 1,000 men.

"We just couldn't kill them fast enough. My squads were cut off from each other, and together, we were cut off from the rest of the company. I had some of my men in the high grass where the machine gunners had to get up on their knees to shoot, which exposed them. The enemy never overran us, but he got one or two of his squads between us."

In an after action report, the captain recalled that an NVA company attempted to overrun his Kilo Co. one night.

"It was so dark, we couldn't see our hands in front of our faces, so we threw our trip flares and called for a flare plane overhead. We could hear, smell, and occasionally see the NVA. In the morning, we found 25 bodies, some of them only five yards away, stacked on top of each.

"We could hear bodies being dragged through the jungle for four hours after the shooting stopped. A thorough search at first light revealed 79 enemy dead by body count."

Captain Modrzejewski's Co. K, especially McGinty's platoon, was hit hard. "Our company was down from 130 (Marines) to 80, and I had kids who were hit in five or six places."

Both Captain Modrzejewski and Staff Sergeant McGinty were later awarded Medals of Honor.

The operation ended August 3, with over 824 enemy dead. The Marines lost 126 killed and 448 wounded. The NVA withdrew, and a small Marine task force remained. Then on August 6, a recon team saw NVA troop movements in the area, sightings which increased the next day. The enemy assaulted the remaining Marines.

The small force called for artillery and close air support. At Dong Ha, Captain Howard V. Lee (E-2-4) asked permission to lead a relief force to the beleaguered Marines, then gathered seven volunteers which were coptered to the Marine perimeter.

Captain Lee immediately took command, reorganized the defenses and saw to the re-distribution of ammo which the choppers unloaded into the defenses. The small unit continued repelling NVA ground probes. That night, Captain Lee radioed that he had only 16 men left still able to fight. He was twice wounded, with grenade fragments in his right eye and the right side of his body.

The next morning, Captain Lee relinquished command to Major Vincil Hazelbaker of VMO-2, whose copter had been shot down in resupply efforts.

Meanwhile artillery and napalm strikes on enemy positions kept the NVA at bay, and E and F companies with a command group arrived. They spread out, but the enemy had disappeared once again.

Captain Lee was awarded the Medal of Honor.

****

Excerpts from the Medal of Honor citations of Captain Modrzejewski, Staff Sergeant McGinty and Captain Lee, all of whom survived their harrowing brushes with death while performing so heroically:

"... Major (then Captain) Modrzejewski led his men in seizure of the enemy redoubt, which contained large quantities of ammunition and supplies. That evening, a numerically superior enemy force counterattacked ... The enemy assaulted repeatedly in overwhelming numbers, but each time was repulsed ... The second night, the enemy struck in battalion strength, and Major Modrzejewski was wounded in this intensive action ... fought at close quarters.

"Although exposed to enemy fire, and despite his painful wounds, he crawled 200 meters to provide critically needed ammunition to an exposed element ... and was constantly present wherever fighting was heaviest, despite numerous casualties, a dwindling supply of ammunition, and the knowledge they were surrounded. He skillfully directed artillery fire ... and courageously inspired the efforts of his company.

"(the next day) Co. K was attacked by a regimental-size enemy force. Although ... vastly outnumbered and weakened ... Major Modrzejewski reorganized his men ... and directed their efforts to heroic limits."

Air and artillery strikes that he had called in, together with Co. K efforts, finally repulsed the fanatical attacks.

****

"... Second Lieutenant (then Staff Sergeant) McGinty's platoon ... came under heavy arms, automatic weapons, and mortar fire from an estimated enemy regiment. With each successive human wave which assaulted his 32-man platoon during the four hour battle, Second Lieutenant McGinty rallied his men ... two of his squads became separated from the remainder of the platoon ... Second Lieutenant McGinty charged through intense automatic weapons and mortar fire to their position. Finding 20 men wounded and the medical corpsman killed, he quickly reloaded ammunition magazines and weapons for the wounded men and directed their fire.

"Although painfully wounded ... he continued to shout encouragement and to direct their fire so effectively that the attacking hordes were beaten off. When the enemy tried to outflank his position, he killed five of them at point blank range with his pistol ... he skilfully adjusted artillery and air strikes within 50 yards of his position. This destructive firepower routed the enemy, who left an estimated 500 bodies ..."

****

"... A platoon of Major (then Captain) Lee's company ... was attacked and surrounded by a large Vietnamese force ... Major Lee took seven men and proceeded by helicopter to reinforce the beleaguered platoon. Major Lee disembarked from the helicopter ... and braving withering enemy fire ... fearlessly moved from position to position directing and encouraging the overtaxed troops.

"The enemy then launched a massive attack ... Although painfully wounded ... in several areas of his body including his eye, Major Lee continued undauntedly throughout the night to direct the valiant defense ... The next morning he collapsed from his wounds.

"... However, the small band of Marines had held their position and repeatedly fought off many vicious enemy attacks for a grueling six hours until their evacuation ... Major Lee's actions saved his men from capture, minimized the loss of lives, and dealt the enemy a severe defeat."

*****

(Adapted in part from *In the Shadow of the Rockpile,* by Tom Bartlett, in the September 1991 issue of *Leatherneck* magazine.)

Outstanding Leadership
3rd Division Commanding Generals

MajGen Charles D. Barrett	16 September 1942-14 September 1943
MajGen Allen H. Turnage	15 September 1943-14 September 1944
BGen Alfred H. Noble	15 September 1944-13 October 1944
MajGen Graves B. Erskine	14 October 1944-20 October 1945
BGen William E. Riley	21 October 1945-28 December 1945
BGen Merrill B. Twining	7 January 1952-14 February 1952
MajGen Robert H. Pepper	15 February 1952-9 May 1954
MajGen James P. Risely	10 May 1954-30 June 1955
MajGen Thomas A. Wornham	1 July 1955-26 July 1956
BGen Victor H. Krulak	27 July 1956-6 September 1956
MajGen Alan Shapley	7 September 1956-1 July 1957
MajGen Francis M. McAlister	2 July 1957-28 March 1958
MajGen David M. Shoup	29 March 1958-1 April 1959
Col Rathvon McC. Tompkins (Acting)	2 April 1959-8 May 1959
BGen Lewis C. Hudson	9 May 1959-19 June 1959
MajGen Robert B. Luckey	20 June 1959-31 August 1960
MajGen Donald M. Weller	1 September 1960-1 September 1961
MajGen Robert E. Cushman Jr.	2 September 1961-3 June 1962
MajGen Henry W. Buse Jr.	4 June 1962-9 May 1963
MajGen James M. Masters Sr.	10 May 1963-16 June 1964
MajGen William R. Collins	17 June 1964-4 June 1965
MajGen Lewis W. Walt	5 June 1965-17 March 1966
MajGen Wood B. Kyle	18 March 1966-17 March 1967
MajGen Bruno A. Hochmuth	18 March 1967-14 November 1967
BGen Louis Metzger (Acting)	15 November 1967-27 November 1967
MajGen Rathvon McC. Tompkins	28 November 1967-20 May 1968
MajGen Raymond G. Davis	21 May 1968-14 April 1969
MajGen William K. Jones	15 April 1969-30 March 1970
MajGen Louis H. Wilson	31 March 1970-22 March 1971
MajGen Louis Metzger	23 March 1971-7 January 1972
MajGen Joseph C. Fegan	8 January 1972-7 January 1973
MajGen Michael P. Ryan	8 January 1973-31 August 1973
MajGen Fred E. Haynes Jr	1 September 1973-22 August 1974
MajGen Kenneth J. Houghton	23 August 1974-13 August 1975
MajGen Herbert L. Wilkerson	14 August 1975-19 July 1976
MajGen George W. Smith	20 July 1976-16 July 1977
MajGen Adolph G. Schwenk	17 July 1977-10 July 1978
MajGen Calhoun J. Killeen	11 July 1978-11 July 1979
MajGen Kenneth L. Robinson Jr.	12 July 1979-24 July 1980
MajGen Stephen G. Olmstead	25 July 1980-21 June 1982
MajGen Robert E. Haebel	June 1982-June 1984
MajGen Harold G. Glasgow	June 1984-June 1986
LtGen Edwin G. Godfrey	June 1986-September 1987
LtGen Norman H. Smith	September 1987-September 1989
LtGen Henry C. Stackpole III	September 1989-June 1991
BGen Michael J. Byron	June 1991-Present

....and **In Grateful Acknowledgement: Two Score and Ten Contributors**

H. W. Adams	Hurley, MS
James E. Adams	Butler , PA
Daniel C. Adkins	Washington, D.C.
Amy Adler	Washington, D.C.
John M. Admire	Washington, D.C.
Karl C. Appel	Baywood Park, CA
Robert W. Attridge	Beverly, MA
George Aucoin	Evergreen, CO
Ernest Baals	Erial, NJ
Joseph R. Baricko	Suitland, MD
Joseph J. Barishek	Lewes, DE
H. C. Barnum, Jr.	Reston, VA
Tom Bartlett	Woodbridge, VA
Bert Barton	Hermitage, PA
Frank Benis	Washington, D.C.
Albert J. Beveridge	Copake, NY
E. N. Bieri	Alliance, OH
Lawrence Binyon	
Harry Bioletti	Warkworth, N.Z.
Steven C. Bishop	Kansas City, MO
Cipriano Blas	Yuma, AZ
John O. Blas	Agana, GU
Ben Blaz	Tamuning, GU
Frank Brandemihl	Farmington Hills, MI
George Briede	La Belle, FL
E. K. Brinkley	Rogersville, Al
Clarence Brookes	Cortland, NY
Howard J. Brooks	Pembroke, NC
Jerome W. Brown	Chantilly, VA
Benjamin J. Byrer	Valencia, PA
William D. Carothers	Weston, WY
George W. Carr	Durham, NC
James J. Carvino	Brooklyn, NY
Art Cassaretto	Delhi, CA
Paul F. Colaizzi	Pittsburgh, PA
Steve J. Cibik	Virginia Beach, VA
Michael R. Conroy	Hominy, OK
Charles Cork	Birmingham, AL
Cliff Cormier	Gainesville, FL
Richard M. Coulter	Daingerfield, TX
Edward E. Craig	El Cajon, CA
Richard B. Crerand	York, PA
Chris J. Crowley	Etiwanda, CA
J. Robert Cudworth	Camillus, NY
Elvis J. Curtis	Los Alamitos, CA
Bart Daly	Williston Park, NY
Edward F. Danowitz	Altamonte Springs, FL
Frederick K. Dashiell	Alexandria, VA
Raymond G. Davis	Stockbridge, GA
Canio J. DiGerardo	Brooklyn, NY
Emiddio J. DiMartino	Las Vegas, NV
Alfred G. Don	Pensacola, FL
Gene Doughty	Bronx, NY
Chris Doumis	Hollywood Beach, FL
Robert F. Downes	Montvale, NJ
Edward J. Duch	Chicago, IL
Wendell H. Duplantis	Battle Creek, MI
Woodrow W. Easterling	Moselle, MS
John J. Eddy	Yigo, GU
Frank G. Erwin	Nashville, TN
John W. Evans	St. Albans, WV
Leo A. Farrow,	Yonkers, NY
Charles Felix	Abilene, TX
Kay Sterling Felton	Long Beach, CA
Howard Fix	Harrisonville, PA
August Fopiano	Brooklyn, NY
Charles R. Ford	Ocala, FL
Conrad M. Fowler	Lanett, AL
James F. Freeman	Sacramento, CA
Harry A. Gailey	San Jose, CA
James M. Galbraith	Lorain, OH
James J. Gallegos	Sante Fe, NM
James Ganopulos	Homestead, PA
Frank Gardner	Lake Ridge, VA
Carroll Garnett	Chester, VA
Joseph Garza	Covina, CA
Austin P. Gattis	Washington, D.C.
Roy Gerken	Hollywood, FL
Patrick F. Gilbo	Rockville, MD
Myron A. Gildersleeve	Zearing, IA
Harry L. Gilliam	Sunnyvale, CA
Robert A.Grant	Cholame, CA

ACKNOWLEDGEMENTS

ACKNOWLEDGEMENTS

George J. Green	Webster Groves, MO
Colleen Greer	Arlington, VA
John D. Guilfoyle	New Hampton, NY
C. L. Guthrie Jr.	Spartanburg, SC
John N. Habay	Boca Raton, FL
Brent Hancock	Hershey, PA
James R. Harper	Atlanta, GA
James V. Haney	Rochert, MN
Roger H. Hanson	Bullhead City, AZ
Cecil M. Harvey	Ardmore, OK
Newport E. Hayden	Piedmont, CA
Robert M. Hayter	Port Ludlow, WA
Robert J. Healy	Placentia, CA
Howell T. Heflin	Washington, D.C.
Michael Helms	Youngstown, FL
Leo E. Hetzman	Pensacola, FL
William T. Hewitt	Binghamton, NY
Clifford B. Hicks	Brevard, NC
Francis J. Hoban	Santa Rosa, CA
Carl Horton	Pioneer, CA
Everad F. Horton	Johnson City, TN
Louis J. Hum	Santee, CA
Albert L. Jenson	San Diego, CA
Steve A. Johnson	Jamestown, NY
Clifford L. Jones	McGehee, AR
William K. Jones	Alexandria, VA
Alvin M. Josephy	Greenwich, CT
T. O. Kelly	La Plata, MO
Charles Kennedy	Dumfries, VA
John J. Kerins	Terre Haute, IN
Calhoun J. Killeen	McLean, VA
Henry J. Klimp	St. Simon's Island, GA
Nicholas Kominus	Springfield, VA
Henry J. Krueger	McPherson, KS
James G. Kyser III	Dumfries, VA
A. Jay Langford	Washington, D.C.
John H. Leims	Chicago, IL
John S. Letcher	Glasgow, VA
Eleanore C. Little	N. Freedom, WI
Howard E. Long	North Port, FL
Robert Lowry	Hillsboro, CA
James P. Lynch	Palm Coast, FL
Douglas Lyvere	Hollister, CA
Louis R. Machala	Dallas, TX
Shibat Makio	Washington, D.C.
Dennis Mansour	Iselin, NJ
Noelle Mason	Killara, NSW, AUS
Eugene E. May	Indianapolis, IN
Fred H. McCrory	Vestavia Hills, AL
Otho McDaniel	Paragould, AR
Frank V. McKinless	
John E. McKinnon	Oakland, MS
Gerald F. Merna	Alexandria, VA
Rochus Meyer	Turner, OR
Kenneth E. Mills	Wellsville, OH
Thomas F. Murphy	Andover, MA
George R. Nadolny	Tucson, AZ
Edgar Neer	Columbus, OH
John B. Nelson	Marion, IN
Clayton R. Newell	Washington, D.C.
Frank W. Nunan	Lansdowne, PA
W. A. O'Bannon	Mariposa, CA
Cyril J. O'Brien	Silver Spring, MD
William Parrie	Benecia, CA
E. T. Passons	Sulphur Springs, TX
John W. Pelletier	Warner, NH
William I. Pierce	Merrillville, IN
William W. Putney	Woodland Hills, CA
Roger Radabaugh	Willmar, MN
Antonio Ramos	Cabo Rojo, PR
Carey A. Randall	Jackson, MS
Robert G. Ream	Dundee, OR
Vincent J. Robinson	Norfolk, VA
Michael Rodriguez	Kansas City, MO
Eugene Rosplock	Elmira, NY
Frank E. Ross	Framingham, MA
Eric S. Ruark	Baltimore, MD
Michael B. Ryan	Chesterfield, MO
Andrew M. Sabol	Shavertown, PA
Dominick Savino	Oceanside, NY
Edward S. Schick	Fallbrook, CA
Harold H. Schwerr	N. Mankato, MN
L. J. Seavy-Cioffi	New York, NY
Ira M. Shelton	Brawley, CA

290

Edwin H. Simmons   Morgan Hill, CA   Joseph A. Tyrpa   Depew, NY
Gaylon Souvignier   Canton, SD   Steve Vajda   Westwood, NJ
James R. Sprungle   Annapolis, MD   Robert B. Van Atta   Greensburg, PA
Henry C. Stackpole   Honolulu, HI   James Vance   Washington, D.C.
Don M. Stephenson   Hamilton, TX   Richard A. Walker   Monroe, LA
Francis P. Storm   New Milford, CT   Gerald W. Wallack   Meriden, CT
Jay J. Strode   Jerome, ID   Joseph E. Walters   Center Ossipee, NH
Kenneth F. Sullivan   Lakeview, AZ   John F. Wear   Penn Valley, PA
Brian Sweeney   Nogales, AZ   Dale W. Whaley   Loomis, CA
Herbert J. Sweet   Alexandria, VA   Maury T. Williams   Dayton, OH
Joe Thein   Wauwatosa, WI   Guy D. Wirick   Crown Point, IN
Eugene B. Thompson   Chillicothe, MO   Walter E. Wittman   Bradenton, FL
Herbert H. Thompson   Mt. Vernon, IL   John J. Wlach   Commack, NY
David Torrel   Eveleth, MN   William K. Wonders   Tucson, AZ
Anthony J. Troiano   Valley Stream, NY   John W. Yager   Perrysburg, OH
Norman N. Tyndall   Coeur d' Alene, ID

ACKNOWLEDGEMENTS

OMC mechanics test the smoke-screening device of an M60A1 tank. OMC also provides support to the Marine Barracks in the Republic of the Philippines, mainland Japan and Guam. (USMC photo by Cpl. Bob Hall)

Two Score and Ten Index

A

B

Jensen, George   101
Jenson, Albert L.   49, 50, 52, 75, 77, 112, 290
Johnson   116
Johnson, Chandler W.   129
Johnson, Gilbert "Hashmark"   164
Johnson, Leslie   61
Johnson, Lyndon   197
Johnson, Ralph   164
Johnson, Steve   206, 290
Jones, Amos   157
Jones, Clifford L.   63, 290
Jones, Davey   33
Jones, David L.   234
Jones, Franklin P.   201
Jones, Richard E.   240
Jones, William K.   211, 237, 288, 290
Jordan, Clay W.   125
Josephy, Alvin   53, 100, 131, 290
Joy, C. Turner   85
Joy, Quentin   77

K

Kadena Air Base   270
Kaipara Flats   22
Kaltenborn, H.V.   37
Kaneohe Bay   183
Karig, Walter   178
Karlowica, M. J.   237
Katsumi, Edward   182
Kavieng   59
Kavieng Island   121
Kelley, Paul X.   264
Kelly, John H.   237
Kelly, T. O.   8, 23, 34, 38, 62, 81, 159, 240, 290
Kelso   112
Kemp, Frank   32
Kenilorea, Peter   253
Kennedy, "Boondocker" Charlie   63
Kennedy, Charles   290
Kennedy, John F.   65, 69
Kenyon, Howard   133
Kerins, Jack   35, 36, 48, 87, 115, 116, 125, 290
Keys, William M.   275
Khafji   278
Khe Sanh   16, 195, 204, 205, 206, 207, 224, 225, 227, 230, 231
Kieta   49
Killeen, C. J.   237, 254, 255, 257, 266, 288, 290

Kim, Ki Hwa   192
Kim, Nam Ho   192
Kimber, Maureen   30
Kin Airfield   193
King   21, 23, 53, 70
King Edward VII   37
King, Ernest J.   21
King Neptunis Rex   33
King, Ralph M.   43
Kinniard   117
Kinston   22
Kinugawa Maru   21
Kirk, Bernard   87, 135
Kirts, George   234
Klimp, H.J.   91, 148, 290
Knoderer, Frank   155
Knoll, Tom   269
Knox, Frank   166
Koe, Peter   261
Koh Tang Island   246, 248
Koiari   67
Kokumbona   59
Kolombangara   75
Kominus, Nicholas   5, 275, 279, 290
Kopanski, Mike E.   272
Koromoka River   47
Koromokina   21
Koromokina River   63
Kriendler, Robert   23, 34, 37, 38
Krueger, Henry J.   118, 290
Krulak   200
Krulak, Charles C.   275
Krulak, Victor H.   43, 188, 216, 288
Kuhn, John   224
Kumejima   258
Kundrat, Joseph J. Jr.   240
Kuribayashi   127, 129, 137, 145
Kuribayashi, Tadamichi   125, 148
Kuwait   10, 11
Kwajalein   13, 92, 120, 121
Ky Phu   283
Kyle, Wood B.   201, 205, 240, 288
Kyser, James G. III   5, 7, 290

L

LaBohn, Louis E.   106
Laht, Frank F.   104
Laird, Melvin R.   236
Lake Yamanaka   252
Lam, Hoang Xuan   237
Lambton Quay   31
Lamour, Dorothy   80
Lancaster   205
Langford, A. Jay   290
Laos   195, 203, 235

LaPierre, Martin D.   272
LaPointe, Thomas J.   273
Larson, August F.   193
Laruma   51
Laruma River   45, 47, 65, 162
Las Plugas   22
Las Pulgas Canyon   36
Lawrence, R. T.   244
Layton, Howard   159
Lazama, Jesus Dydasco   116
Lee, Eddie   164
Lee, Howard V.   283, 287
Leeds, Peter   13
Lehman, John F. Jr.   264
Leims, John   100, 135, 136, 137, 145, 155, 281, 282, 290
Lejeune, John A.   19
Leniart, Joseph   159
Letcher, John S.   54, 56, 290
Lewis, Alfred E.   64
Liapes, George   152
Liberty Ship   20
Lincoln, G.A.   91
Lindblad   138, 139
Linden, Richard A.   79
Linger, J.W. (Bill)   28, 29
Little, Eleanore   166, 167, 290
Little, Jack   80
Livesay, Slim   63, 83
Livingston, James E.   283, 285
Lonergan, Vincent J.   152
Long, Howard E.   42, 290
Loring, Gloria   13
Lowery, Louis R.   129, 130, 144
Lowry, Robert   73, 290
LST-396   75
Lucas, Ulysses   164
Luckey, Robert B.   288
Luke, William   267
Lumpkin, Albert R. Jr.   226
Lundquist, John A.   196
Lunga Point   37, 59, 72, 76
Lynch, James P.   290
Lyon, George   88, 109, 115, 116, 154
Lyvere, Douglas   43, 47, 61, 290
LZ Margo   223

M

MacArthur, Douglas   21, 23, 25, 41, 65, 83, 156, 174
Machala, Louis R.   157, 290
MacLean, Fred Jr.   189
Macon   203

© 1986 VOLK

USS Thetis Bay   193
USS Tripoli   252
USS Utah   143
USS Valley Forge   240
USS Windsor   81
USS Zelin   84
Utah Mesa   211
Utley, William   118

V

Vajda, Steve   151, 291
Van Atta, Robert B.   5, 7, 175, 240, 291,
Van Orden, G.O.   22
Van Tuong Peninsula   201
Vanairsdale, James B.   193
Vance, James   291
Vance, Robert T.   53
Vandegrift, Alexander   19, 32, 111
Vandegrift, Archer   65
Vargas, M. Sando Jr.   284, 286
Vella Gulf   75
Vella Lavella   65, 75, 120
Villa, Albert M.   136
Voight, George R.   89
Voorhees, Tim   257
Vouza, Jacob   65

W

Wacky-Racky (Waiaraka) Park   21
Wade, Orrin R.   48
Waikaraka   38
Waikowhai   34
Wake Island   13
Walden, George   21, 99
Walker, Richard   69, 291
Wallack, Gerald W.   291
Walt, Lewis W.   12, 201, 205, 216, 288
Walters, George   90, 104
Walters, Joseph E.   291
Walters, Richard J.   189
Waltman, Robert   189, 232
Warkworth   29, 30, 36, 37, 38
Warner, Gerry   278
Warner, Gordon   42, 62

Warren, Virgil   100
Washing Machine Charlie   21, 158, 162
"Water Buffalo"   218
Watson, Jack R.   68
Watson, Wilson D.   281
Watsonville, California   19
Wayne, John   199
Wear, John F.   239, 291
Weber, Lee P.   224
Weeks   196
Weise, William   229
Weiss, George   55
Welch, Raquel   13
Weller, Donald M.   288
Wellington   27, 31, 32
Wernli, Mark   269
West, Frazer   74
West, Joshua C. III   73
West, Jack   73
West View Danceland   19
Westmoreland   224
Westmoreland, William   230
Westwood, Fred   148
Whaley, Dale W.   123, 291
Whangarei   21, 27
Wharton   90
Wheeler, Richard   127
White, B.   109
White Beach   186
White, John A.   188
Whitlock, James M.   165
Whitmire, Donald B.   247
Whittemore, E.L.   91
Widener   145
Wiedhahn, Warren H.   215
Wilder, Gary   247
Wilkerson, Herbert   247, 248, 288
Wilkes, Paula   114
Wilkes, Warren Y.   114
Wilkinson, Theodore S.   43
Williams, Gregon   177
Williams, Hershel   149, 281
Williams, Lytle G. (Bud)   73
Williams, Marlowe   131, 142, 159
Williams, Maury T.   23, 83, 128, 147, 291
Williams, Robert H.   55
Williford, Carroll J.   107
Wilson   101
Wilson, Alfred M.   284, 286
Wilson, Earl   13
Wilson, Goober   111
Wilson, John B.   35, 47, 54, 181
Wilson, Louis   63, 83, 91, 94, 113, 114, 254, 255, 277, 281, 288
Wirick, Guy   153, 291

Witek, Frank P.   281
Withers, Harnold J.   131
Wittman, Walt   48, 155, 291
Wlach, John J.   22, 149, 291
Wolf, Horace L. Sr.   36, 37
Wonders, Kent   221, 223, 225
Wonders, William K.   291
Wonsan   12
Wood, Belleau   23
Wood, Don   236
Wood, Samuel   164
Wood, Walt   246
Woodham, Tullis J. Jr.   257
Woods, Harry   157
Woodworth   20
Wornham, Thomas A.   288
Worrell   102
Wright, Homer Cornelius   95

X

X-Ray Hill   98

Y

Yager, John W.   51, 57, 291
Yamamoto, Isoroku   93, 125
Yamanaka Orphanage   182, 189
Yawata   93
Yellow Beach #3   46
Ylig Bay   115
Yoders, Stanley A.   136
Yona   114
Yonabaru   187
Yonaha Dake Mountain   256
Yongil Bay   178
Young, Joe   98

Z

Zais, Melvin   237
Zapata, Chago   268
Zeitlin, Dave   94
Zimbales Training Area   250, 272
Zimmer, William   137, 145
Ziska   186
Zoller, J.E.   236